Core Mathematics

Dennis Bila
Washtenaw Community College

Ralph Bottorff
Washtenaw Community College

Paul Merritt
Highland Park Community College

Donald Ross
Washtenaw Community College

WORTH PUBLISHERS, INC.

Core Mathematics

Copyright © 1975 by Worth Publishers, Inc.

Printed in the United States of America

Library of Congress Catalog Card No. 74-82696
ISBN: 0-87901-035-5

Fourth printing, August, 1978

WORTH PUBLISHERS, INC.
444 Park Avenue South
New York, New York 10016

To our students

Preface

Core Mathematics is designed for students with a wide diversity of mathematical backgrounds and abilities. It presumes little prior knowledge of mathematics and is presented in a semiprogrammed worktext format that allows students to proceed at their own pace. Students who have retained some knowledge from previous study in math can go rapidly through the material that is familiar to them, thereby gaining the benefit of a useful review of their skills. Students without this advantage will want to proceed at a slower pace.

Core Mathematics can be used in a course for the general or liberal arts students who will not continue on to more advanced mathematics. For students who will take further courses in mathematics, but who do not have the necessary background in mathematics to enter the introductory algebra course, this text provides all the skills and knowledge necessary to prepare them for that course. And, although *Core Mathematics* is intended for use in a laboratory situation with an instructor, its format is also well suited for use by those who need to review the fundamental concepts of arithmetic, algebra, and geometry in a more traditional classroom situation prior to continuing their mathematical education.

For many years we have struggled with the problem of providing well-written programmed materials for those students who wish to learn mathematics. Many of these students have had poor experiences in their earlier mathematics courses and are reluctant to take further courses. Our attempts to help them have been hindered by the lack of adequate text material. As a result, we began writing semiprogrammed booklets on single-concept ideas to supplement the material we were using. The positive reactions of students to our materials were encouraging. As a result, we decided to develop a series of semiprogrammed texts written specifically for those who need this level of instruction. *Core Mathematics* is the first in a series which also includes *Introductory Algebra* and *Intermediate Algebra*. We sincerely hope that your experiences with these texts will be as rewarding as ours have been.

Chapters 1–4 introduce and review the essential basic operations of whole numbers, fractions, decimals, and percents.

Chapters 5–8 provide the essential concepts and the skills in numerical manipulation that are necessary for an understanding of algebra.

Chapters 9–11 introduce and review geometry and measurement. Of particular importance are Chapters 10 and 11 on the metric system. Because the metric system is the principal system of measurement in use throughout the world, everyone should be well acquainted with it.

Chapter 12 presents the essentials of ratios and proportions. Chapter 13 provides the knowledge and skills necessary to interpret circle, bar, and line graphs. Chapter 14 introduces the elementary statistical concepts which are particularly relevant to an understanding of many articles in contemporary newspapers and journals.

The symbols and notation used in the chapters on the metric system are those of the International System of Units (SI). The National Bureau of Standards has adopted the

SI in all its research work and publications, except where use of these units would impair communications or reduce the usefulness of the material.

Chapters 10 and 11 were reviewed by the National Bureau of Standards. What appears in this text are the SI units with few exceptions. For example, the commonly used symbol "cc" has been used for cubic centimeter parenthetically following the correct "cm^3" because we feel that both forms will be used for a time in actual practice, particularly in the health fields. The term "weight" has been used instead of "mass" because we feel that this will be more widely understood and used. Following the recommendation of the National Council of Teachers of Mathematics, we have used the English spelling of "meter" and "liter."

Features of *Core Mathematics*

Format

Each unit of material (a single concept or a few closely related concepts) is presented in a section. Within a section, short, boxed, numbered frames contain all instructional material, including sample problems and sample solutions. Frames are followed by practice problems, with workspace provided. Answers immediately follow, with numerous solutions and supplementary comments.

Exercises

Exercises at the end of each section are quite traditional in nature. However, they are generally shorter than those found in most texts because the student has already done numerous problems and answered many questions during the study of the section. Hence, the exercises serve as a review of the content of the section and as a guide that will help students to recognize whether they have mastered the material. Answers to the problems are provided immediately, permitting the student to advance through the text without having to turn to the back to check answers. Word problems are used throughout.

Sample Chapter Tests

At the end of each chapter we have provided a Sample Chapter Test. It is keyed to each section with the answers provided immediately for student convenience. It may also serve as a pretest for each chapter. However, if an instructor wishes to pretest without the availability of the answers (to the student), one form of the post-test for the chapter (provided in the *Test Manual*) could be used for this purpose.

Upon finishing the chapter, the student should complete the Chapter Sample Test, and the results should be shown to the instructor and discussed with the student. If the instructor and student are confident about mastery of the content of the chapter, a post-test can then be administered. Used in this manner, outstanding results can be anticipated on the first attempt for each post-test.

Objectives

The Sample Chapter Tests serve as objectives for both student and instructor. The instructor can readily ascertain the objectives of any chapter by examining the sample test at its end, and the student is in a good position to see what is expected of him by examining problems and questions that are going to be asked at the completion of the chapter. We believe that objectives stated verbally are of less benefit to the student than the statement of a problem that must be solved.

Glossary

The glossary at the back of the text provides the student with the pronunciation and the definition of all mathematical words and phrases used in the text. This is particularly important wherever the text is used in a laboratory situation where the student will not hear the words used in class discussions.

Acknowledgments

We would like to thank those whose reviews were particularly helpful in the writing of this text: Eldon Baldwin, Prince George's Community College; Ted Carlson, Bunker Hill Community College; David E. Conroy, Northern Virginia Community College; Cecilia Cooper, Kendall College; William J. Gordon, Roxbury Community College; Anne E. Martin, Delta College; Janet M. Rich, Miami-Dade Community College; Richard Spangler, Tacoma Community College; Jay Welch, San Jacinto College; and L. E. Barbrow, National Bureau of Standards.

We would like to express our thanks to our colleagues at Washtenaw Community College and Highland Park Community College for their contributions, with special thanks to Janet Hastings and Percy Mealing for their assistance in developing the supplemental testing program. Also to student reviewers John Dowding, Alan F. Turner, Donald R. Bolen, and Diana Bakich. Our thanks also to our typists, Phyllis Bostwick, Vonda Bottorff, Marilyn Myers, Kathy Stripp, Lillian Thurston, and Carolyn Williams.

The class testers of the manuscript and preliminary edition were Gordon Bremenkampf, Dearborn High School; Emile Conti, Edsel Ford High School; Elwyn D. Cutler, Ferris State College; Anne E. Martin, Delta College; and Eldon Baldwin, Prince George's Community College. The staff members of Washtenaw Community College who assisted in class testing were Tas Dowding, Janet Hastings, Bill Lewis, Larry Prichard, Percy Mealing and Robert Mealing.

Dennis Bila
Ralph Bottorff
Paul Merritt
Donald Ross

January, 1975

To the Student

You are about to brush up on some skills you may already have learned in a previous course; but they may be a bit rusty. You will learn some new techniques, too. This book is designed to help pinpoint those areas for which you need to sharpen your skills and to permit you to move quickly through those topics you know well.

If you follow the suggestions below, you will make best use of your time, proceeding through the course as quickly as possible, and you will have mastered all the material in this book.

Chapter Sample Tests

The Chapter Sample Test at the end of each chapter will help you to determine whether you can skip certain sections of the chapter. It is a cumulative review of the entire chapter, and questions are keyed to the individual sections of the chapter. To use this test to best advantage, we suggest this procedure:

1. Work the solutions to all problems neatly on your own paper to the best of your ability. You want to determine what you know about the material without help from someone else. If you receive help in completing this self-test, the result will show another person's knowledge rather than your own.
2. Once you have done as much as you can, check the test with the answers provided. On the basis of errors made, determine the most appropriate course of study. You may need to study all sections in the chapter or you may be able to skip one or two sections.
3. When you have completed the chapter, rework all problems originally missed on the Chapter Sample Test and review the entire test in preparation for the post-test that will cover the entire chapter.

Instructional Sections

To use each section most effectively:

1. Study the boxed frames carefully.
2. Following each frame are questions based on the information presented in the frame. When completing these questions, do the work in the spaces provided.
3. Use a blank piece of paper as a mask to cover the answers below the stop rule. For example:

Q1   How many eggs can 21 chickens lay in 1 hour, if 1 chicken can lay 7 eggs in $\frac{1}{2}$ hour?

(Work space)

STOP • **STOP** • **STOP** • **STOP** • **STOP** • **STOP** • **STOP** • **STOP** • **STOP**

Paper Mask

4. After you have written your calculations and response in the workspace, slide the paper mask down, uncovering the correct solution, and check your response with the correct answer given. In this way you can check your progress as you go without accidentally seeing the response before you have completed the necessary thinking or work.

Q1   How many eggs can 21 chickens lay in 1 hour, if 1 chicken can lay 7 eggs in $\frac{1}{2}$ hour?

$$7 \times 2 = 14 \qquad \begin{array}{r} 14 \\ \times 21 \\ \hline 14 \\ 28 \\ \hline 294 \end{array}$$

STOP • **STOP** • **STOP** • **STOP** • **STOP** • **STOP** • **STOP** • **STOP** • **STOP**

A1   294: because 1 chicken can lay 14 in 1 hour, 21 chickens can lay 21(14) in 1 hour.

Paper Mask

Notice that the answer is often followed by a colon (:). The information following the colon is one of the following:
a. The complete solution,
b. A partial solution, possibly a key step frequently missed by many students,
c. A remark to remind you of an important point, or
d. A comment about the solution.
If your answer is correct, advance immediately to the next step of instruction.

5. Space is provided for you to work in the text. However, you may prefer to use the paper mask to work out your solutions. If you do, be sure to show complete solutions to all problems, clearly numbered, on the paper mask. When both sides of the paper are full, file it in a notebook for future reference.

6. Make all necessary corrections before continuing to the next frame or problem.

7. If something is not clear, talk to your instructor immediately.

8. If difficulty arises when you are studying outside of class, be sure to note the difficulty so that you can ask about it at your earliest convenience.

9. When preparing for a chapter post-test, the frames of each section serve as an excellent review of the chapter.

Section Exercises

The exercises at the end of each section are provided for additional practice on the content of the section. For your convenience, answers are given immediately following the exercises. However, detailed solutions to these problems are not shown.

You should:

1. Work all problems neatly on your own paper.
2. Check your responses against the answers given.
3. Rework any problem with which you disagree. If you cannot verify the response given, discuss the result with your instructor.
4. In the process of completing the exercise, use the instructional material in the section for review when necessary.
5. Problems marked with an asterisk (*) are considered more difficult than examples in the instructional section. They are intended to challenge the interested student. Problems of this difficulty will not appear on the Chapter Sample Test or the post-test.

Contents

Core Mathematics

Chapter 1

Operations with Whole Numbers

1.1   Place Value

---

1   Many systems are used throughout the world to represent numbers. For example, seven is written, using roman numerals, as VII. Primitive man might have represented seven using the tally system, ⊬⊦ ||. The numeral we use is 7. The symbols we use were first developed in India and expanded by the Arabians. For this reason they are called the *Hindu–Arabic numerals*. The numerals are: 0, 1, 2, 3, 4, 5, 6, 7, 8, and 9. The number system is called a *decimal* ("deci" is a Latin prefix meaning "tenth") *system* because any number can be written as a numeral that uses only these 10 symbols, which are called *digits*. The value indicated by any digit in a numeral depends on the digit itself and its place in the numeral. For example, in the numeral 1,235:

```
           ┌──────────thousands
           │  ┌────────hundreds
           │  │  ┌──────tens
           ↓  ↓  ↓
          1,235 ←──────ones
```

The place value of the digit 1 is thousands. The place value of the digit 2 is hundreds. The place value of the digit 3 is tens. The place value of the digit 5 is ones.

Q1   What is the place value of the 5 in the following numerals?

    a. 285 _____   **b.** 503 _____   **c.** 2,358 _____

STOP • STOP • STOP • STOP • STOP • STOP • STOP • STOP • STOP

A1   **a.** ones   **b.** hundreds   **c.** tens

Q2   What is the place value of the 2 in the following numerals?

    a. 2 _____   **b.** 20 _____   **c.** 200 _____   **d.** 2,000 _____

STOP • STOP • STOP • STOP • STOP • STOP • STOP • STOP • STOP

A2   **a.** ones   **b.** tens   **c.** hundreds   **d.** thousands

---

2   Notice in Q2 that zero is used as a *placeholder*. That is, in the numeral 200 the 2 indicates two hundreds and the zeros indicate no tens and no ones, respectively. Higher place values can be found by studying the following chart. The numeral 4,501,273,968 can be represented as follows:

Billions			Millions			Thousands			Ones		
Hund.	Tens	Ones	Hund.	Tens	Ones	Hund.	Tens	Ones	Hund.	Tens	Ones
		4	5	0	1	2	7	3	9	6	8

> The place value of the 4 is one billions.
>
> The place value of the 5 is hundred millions.
>
> The place value of the 0 is ten millions.
>
> The place value of the 1 is one millions.
>
> The place value of the 2 is hundred thousands.
>
> The place value of the 7 is ten thousands.
>
> The place value of the 3 is one thousands.
>
> The place value of the 9 is hundreds.
>
> The place value of the 6 is tens.
>
> The place value of the 8 is ones.

Q3   What is the place value of the 3 in the following numerals?

    a. 413,208 _____   **b.** 234,008,176 _____

    c. 387,005 _____

STOP • STOP • STOP • STOP • STOP • STOP • STOP • STOP • STOP

A3   **a.** thousands   **b.** ten millions
    c. hundred thousands

Q4   Determine the place value of each digit in the numeral 710,234,895.

    7_____   1_____   0_____

    2_____   3_____   4_____

    8_____   9_____   5_____

STOP • STOP • STOP • STOP • STOP • STOP • STOP • STOP • STOP

A4   7 (hundred millions), 1 (ten millions), 0 (one millions), 2 (hundred thousands), 3 (ten thousands), 4 (thousands), 8 (hundreds), 9 (tens), 5 (ones)

Q5   For the numeral 308,006,507:

    a. 8 is the _____ digit.
                    place value

    b. 6 is the _____ digit.

STOP • STOP • STOP • STOP • STOP • STOP • STOP • STOP • STOP

A5   **a.** millions   **b.** thousands

Q6   For the numeral 457,306,218:

    a. 1 is the _____ digit.   **b.** 0 is the _____ digit.

    c. 5 is the _____ digit.

STOP • STOP • STOP • STOP • STOP • STOP • STOP • STOP • STOP

A6   **a.** tens   **b.** ten thousands   **c.** ten millions

---

Q7   What digit is in each of the following positions?

    a. thousands position in 317,483   _____

b. tens position in 914,321 _____

c. hundred thousands position in 6,781,235 _____

d. hundreds position in 367,941 _____

e. ten millions position in 120,500,300 _____

STOP • STOP • STOP • STOP • STOP • STOP • STOP • STOP • STOP

A7   **a.** 7   **b.** 2   **c.** 7   **d.** 9   **e.** 2

3

After identifying place value it is possible to write the name for a number given in decimal form. For example, 420,465,982 is read "four hundred twenty *million,* four hundred sixty-five *thousand,* nine hundred eighty-two." The commas in the decimal form break the nine digits into three groups:

Millions	Thousands	Ones
4 2 0,	4 6 5,	9 8 2

To read the number, you read each group of three digits and follow that name with the proper grouping name. The number 420 is read "four hundred twenty" and is followed by the word million because 420 occupies the three positions used to describe millions. The 465 is read "four hundred sixty-five" and is followed by the word "thousand" because the 465 occupies the positions used to describe thousands. The 982 is read "nine hundred eighty-two." It is not followed by any special word to indicate positions.

Q8   Write the name for each of the numbers that appear below:

a. 3,124 _____

b. 50,424 _____

c. 106,500 _____

d. 317,150,000 _____

_____

e. 24,037 _____

STOP • STOP • STOP • STOP • STOP • STOP • STOP • STOP • STOP

A8   **a.** three thousand, one hundred twenty-four
b. fifty thousand, four hundred twenty-four
c. one hundred six thousand, five hundred
d. three hundred seventeen million, one hundred fifty thousand (*Note:* When three zeros occur in one grouping, the entire grouping is omitted in the name.)
e. twenty-four thousand, thirty-seven

4

It will also be necessary to write the decimal form of a number given in word form. For example, two hundred seventy million, sixty-five thousand, four hundred twelve is written 270,065,412 in decimal form. You can use the grouping procedure suggested above to help write this decimal form of the number. Three groups will be written because the name given contained the word "million." If the name contains only the word "thousand," only two groups will be used.

Millions	Thousands	Ones
_ _ _	_ _ _	_ _ _

Each of the positions must be filled by a digit (0, 1, 2, . . . , 9). The words "two hundred seventy" precede the word "million." So 270 is placed in the three positions beneath the word "Millions." Sixty-five precedes the word "thousand," so 065 is placed in the three positions beneath the word "Thousands." Notice that a 0 is used so that all three positions are occupied. The last words, "four hundred twelve," indicate that the 412 must be placed in the positions below the word "Ones."

Millions	Thousands	Ones
2 7 0	0 6 5	4 1 2

Q9 Write the decimal form of the number seventy-one million, four hundred three thousand, six hundred seventy-five. Three groupings will be used because the word "million" appears.

Millions	Thousands	Ones
— — —	— — —	— — —

STOP • STOP • STOP • STOP • STOP • STOP • STOP • STOP • STOP

A9 71,403,675, because below the word "Millions" will be written 071. (Notice in the final answer that the 0 is omitted.) Below the word "Thousands," 403 will be written, and the final three places will contain 675.

Q10 Write each of the following numbers in decimal form:

  a. five hundred three thousand, eight hundred seventy _____

  b. two million, forty-three thousand, one hundred _____

  c. two thousand, seven hundred six _____

  d. twenty-nine million, thirty-one thousand, four hundred _____

  e. thirty-one thousand, seven hundred nine _____

STOP • STOP • STOP • STOP • STOP • STOP • STOP • STOP • STOP

A10 **a.** 503,870 **b.** 2,043,100 **c.** 2,706
  d. 29,031,400 **e.** 31,709

Q11 Write each of the following numbers in decimal notation:

  a. two hundred forty-one _____

  b. five thousand, thirty-six _____

  c. twenty thousand, three hundred sixty-eight _____

  d. five million, sixty-two thousand, fifteen _____

STOP • STOP • STOP • STOP • STOP • STOP • STOP • STOP • STOP

A11 **a.** 241 **b.** 5,036 **c.** 20,368 **d.** 5,062,015

Q12 Write each number in word form (remember to write each group as a separate number and then attach the name of the group):

  a. 936 _____

  b. 4,025 _____

  c. 620,519 _____

  d. 93,004,815 _____

STOP • STOP • STOP • STOP • STOP • STOP • STOP • STOP • STOP

A12 **a.** nine hundred thirty-six
  b. four thousand, twenty-five

c. six hundred twenty thousand, five hundred nineteen

d. ninty-three million, four thousand, eight hundred fifteen

(*Note:* The word "and" is *not* used in writing these numerals.)

Q13   Write each number in decimal notation:

    a. seven hundred ninety-eight   _____

    b. seven hundred three thousand, five   _____

    c. ten thousand, ten   _____

    d. eight million, six thousand, twenty-four   _____

STOP • **STOP** • **STOP** • **STOP** • **STOP** • **STOP** • **STOP** • **STOP** • **STOP**

A13   **a.** 798   **b.** 703,005   **c.** 10,010   **d.** 8,006,024

Q14   Write each number in word form:

    a. 805   _____

    b. 803,012   _____

    c. 70,976   _____

    d. 7,002,118,906   _____

                     _____

STOP • **STOP** • **STOP** • **STOP** • **STOP** • **STOP** • **STOP** • **STOP** • **STOP**

A14   **a.** eight hundred five

    b. eight hundred three thousand, twelve

    c. seventy thousand, nine hundred seventy-six

    d. seven billion, two million, one hundred eighteen thousand, nine hundred six

---

This completes the instruction for this section.

1.1   Exercises

1. **a.** What is the name of our number system?
   b. Why is it so called?
2. Change each of the following numbers from tally notation to decimal notation:
       a. ||||| || **b.** ||||| ||||| ||| **c.** ||||| ||||| |||||
       d. ||||| ||||| ||||
3. Express each of the following numbers in tally notation:
       a. 5   **b.** 8   **c.** 16   **d.** 10
4. Write each of the following in words:
       a. 543   **b.** 42,107   **c.** 300,007
       d. 20,030,560   **e.** 77,707,007   **f.** 1,002,000,300
5. Write each of the following numbers in decimal notation:
       a. three thousand, five
       b. fifty thousand, sixty
       c. eight hundred thousand, two hundred
       d. five million, two hundred sixty-three thousand
       e. eighty million, forty thousand, forty
       f. seventeen million, five hundred thirty thousand, six hundred fifteen

1.1 Exercise Answers

1. **a.** Hindu–Arabic, or decimal, system
   b. because there are 10 digits
2. **a.** 7   **b.** 13   **c.** 15   **d.** 14
3. **a.** ⌕⌗⌗   **b.** ⌗⌗⌗ ||| **c.** ⌗⌗⌗ ⌗⌗⌗ ⌗⌗⌗ |
   d. ⌗⌗⌗ ⌗⌗⌗
4. **a.** five hundred forty-three
   b. forty-two thousand, one hundred seven
   c. three hundred thousand, seven
   d. twenty million, thirty thousand, five hundred sixty
   e. seventy-seven million, seven hundred seven thousand, seven
   f. one billion, two million, three hundred
5. **a.** 3,005   **b.** 50,060   **c.** 800,200   **d.** 5,263,000
   e. 80,040,040   **f.** 17,530,615

1.2 Addition and Subtraction

1

One of the basic problems in the *addition* of whole numbers* is the lack of knowledge of the basic addition facts. The following table can be used to find 100 *addition facts*.

	0	1	2	3	4	5	6	7	8	9
0	0	1	2	3	4	5	6	7	8	9
1	1	2	3	4	5	6	7	8	9	10
2	2	3	4	5	6	7	8	9	10	11
3	3	4	5	6	7	8	9	10	11	12
4	4	5	6	7	8	9	10	11	12	13
5	5	6	7	8	9	10	11	12	13	14
6	6	7	8	9	10	11	12	13	14	15
7	7	8	9	10	11	12	13	14	15	16
8	8	9	10	11	12	13	14	15	16	17
9	9	10	11	12	13	14	15	16	17	18

To find the *sum* of two numbers in the table, locate one of the numbers in the left-hand column and the other in the top row. The light shading indicates the sum of 4 and 5:

$$4 + 5 = 9$$

The heavy shading indicates the sum of 5 and 4:

$$5 + 4 = 9$$

The use of the property that $5 + 4 = 4 + 5$ makes it necessary to memorize only one half of the table. The table also shows an important property of zero. That is, the sum of zero and any number is that number. For example,

*The whole numbers are 0, 1, 2, 3, 4, and so on. The next whole number is formed by adding 1 to the previous whole number.

$$0 + 1 = 1 \quad \text{and} \quad 1 + 0 = 1$$
$$0 + 2 = 2 \quad \text{and} \quad 2 + 0 = 2$$
$$0 + 9 = 9 \quad \text{and} \quad 9 + 0 = 9$$

Q1 Give the sum of each of the following pairs of numbers:

    a. $2 + 3 =$ _____   **b.** $6 + 9 =$ _____   **c.** $7 + 7 =$ _____

    d. $9 + 8 =$ _____   **e.** $8 + 9 =$ _____   **f.** $9 + 6 =$ _____

    g. $4 + 8 =$ _____   **h.** $9 + 5 =$ _____   **i.** $0 + 5 =$ _____

    j. $7 + 8 =$ _____   **k.** $6 + 7 =$ _____   **l.** $6 + 0 =$ _____

STOP • STOP • STOP • STOP • STOP • STOP • STOP • STOP • STOP

A1
    a. 5   **b.** 15   **c.** 14

    d. 17   **e.** 17   **f.** 15

    g. 12   **h.** 14   **i.** 5

    j. 15   **k.** 13   **l.** 6

Q2 What number must be placed in the box to make the statement true?

    a. $8 + \square = 15$ **b.** $\square + 3 = 10$ **c.** $\square + 5 = 13$

    d. $6 + \square = 6$ **e.** $8 + 9 = \square$ **f.** $4 + \square = 5$

STOP • STOP • STOP • STOP • STOP • STOP • STOP • STOP • STOP

A2
    a. 7   **b.** 7   **c.** 8

    d. 0   **e.** 17   **f.** 1

Q3 Give each of the following sums:

    a. $4 + 7 + 8 =$ _____   **b.** $9 + 6 + 8 =$ _____

    c. $4 + 5 + 7 + 9 + 1 + 8 =$ _____   **d.** $3 + 6 + 4 + 7 + 9 + 8 =$ _____

    e. $9 + 2 + 8 + 6 + 5 + 9 =$ _____

STOP • STOP • STOP • STOP • STOP • STOP • STOP • STOP • STOP

A3   **a.** 19   **b.** 23   **c.** 34   **d.** 37   **e.** 39

Q4 Give each of the following sums:

a. 3	**b.** 8	**c.** 9
7	7	4
9	2	8
6	9	8
8	6	7
5	5	9
	7	6

STOP • STOP • STOP • STOP • STOP • STOP • STOP • STOP • STOP

A4   **a.** 38   **b.** 44   **c.** 51

2 Addition problems are usually arranged in columns rather than in a horizontal line. For example, $243 + 25$ is written $\begin{array}{r} 243 \\ \underline{25} \end{array}$. Since numbers are added only when they represent the same place value, it is necessary to align all ones in the ones column, the tens in the tens column, and so on. Often the sum of two one-digit numbers will be a two-digit number. In many addition problems it will be necessary to regroup this two-digit sum. For example, in the problem $8 + 4 = 12$, the sum 12 can be rewritten as 1 ten plus 2 ones. This regrouping could be used as follows:

$$
\begin{array}{r}
368 \\
274 \\
\hline
12 \\
13 \\
5 \\
\hline
642
\end{array}
$$

Since 8 ones + 4 ones = 1 ten and 2 ones, write 1 in the tens column and 2 in the ones column.

Since 6 tens + 7 tens = 1 hundred and 3 tens, write 1 in the hundreds column and 3 in the tens column.

Since 3 hundreds + 2 hundreds = 5 hundreds, write 5 in the hundreds column.

The digits in each column are then added to obtain the sum 642.

A more convenient method of addition is to carry the excess units to the top of the next column to the left. For example,

$$
\begin{array}{r}
1\ 1 \\
3\ 6\ 8 \\
2\ 7\ 4 \\
\hline
6\ 4\ 2
\end{array}
$$

8 ones + 4 ones = 1 ten and 2 ones; write 2 in the ones column and carry the 1 to the top of the tens column.

1 ten + 6 tens + 7 tens = 1 hundred and 4 tens; write 4 in the tens column and carry 1 to the hundreds column.

1 hundred + 3 hundreds + 2 hundreds = 6 hundreds; we write 6 in the hundreds column.

This method is called *carrying*.

Q5   Find the sums using the carrying method:

a. 36	**b.** 67	**c.** 39	**d.** 74
22	28	45	59

STOP • STOP • STOP • STOP • STOP • STOP • STOP • STOP • STOP

A5   **a.** 58   **b.** 95:
$$
\begin{array}{r}
1 \\
6\ 7 \\
2\ 8 \\
\hline
9\ 5
\end{array}
$$
   c. 84   **d.** 133:
$$
\begin{array}{r}
1 \\
7\ 4 \\
5\ 9 \\
\hline
1\ 3\ 3
\end{array}
$$

Q6   Find the sum:

a. 47	**b.** 88	**c.** 327	**d.** 406
56	45	462	391

STOP • STOP • STOP • STOP • STOP • STOP • STOP • STOP • STOP

A6   **a.** 103   **b.** 133   **c.** 789   **d.** 797

Q7   Find the sum:

a. 508	**b.** 2006	**c.** 352	**d.** 769
349	7895	680	101
		438	912
		905	214

STOP • STOP • STOP • STOP • STOP • STOP • STOP • STOP • STOP

A7　　**a.** 857　　　　　**b.** 9,901:
$$
\begin{array}{r}
1\ 1\ \ \ \ \\
2\ 0\ 0\ 6\\
7\ 8\ 9\ 5\\
\hline
9\ 9\ 0\ 1
\end{array}
$$
c. 2,375　　　　**d.** 1,996

Q8　　Find the sum (be sure to align the numbers correctly):

a. 273 + 2,350 + 32　　　　　　**b.** 37,051 + 308 + 3,905 + 10,023

STOP • STOP • STOP • STOP • STOP • STOP • STOP • STOP • STOP

A8　　**a.** 2,655:
$$
\begin{array}{r}
273\\
2350\\
32\\
\hline
2655
\end{array}
$$
b. 51,287:
$$
\begin{array}{r}
37051\\
308\\
3905\\
10023\\
\hline
51287
\end{array}
$$

Q9　　Find the sum:

a. 205 + 37,296 + 8,105 + 14,716　　　　**b.** 37,285 + 4,206 + 621 + 97,486

STOP • STOP • STOP • STOP • STOP • STOP • STOP • STOP • STOP

A9　　**a.** 60,322　　　**b.** 139,598

Q10　　The areas of the five national parks in California listed in order of size are: Yosemite, 756,441 acres; King's Canyon, 452,825 acres; Sequoia, 385,100 acres; Lassen, 103,429 acres; and Redwood, 58,000 acres. What is the total acreage of national parks in California?

STOP • STOP • STOP • STOP • STOP • STOP • STOP • STOP • STOP

A10　　1,755,795 acres

3　　*Subtraction* of whole numbers is the opposite of addition. That is,

8 − 5 = 3 because 3 + 5 = 8
9 − 2 = 7 because 7 + 2 = 9
5 − 0 = 5 because 5 + 0 = 5

> One method of finding the *difference* between two whole numbers is to state the difference as a sum. For example, to find $8 - 3 = \square$, think $\square + 3 = 8$. Since $5 + 3 = 8$, then $8 - 3 = 5$.

Q11   Determine the following differences:

a. $7 - 5 = $ _____   b. $9 - 2 = $ _____   c. $6 - 3 = $ _____

d. $4 - 4 = $ _____   e. $9 - 7 = $ _____   f. $6 - 0 = $ _____

STOP • STOP • STOP • STOP • STOP • STOP • STOP • STOP • STOP

A11   a. 2   b. 7   c. 3   d. 0   e. 2   f. 6

Q12   Determine the following differences:

a. $18 - 3 = $ _____   b. $12 - 6 = $ _____   c. $17 - 7 = $ _____

d. $18 - 9 = $ _____   e. $12 - 5 = $ _____   f. $19 - 6 = $ _____

STOP • STOP • STOP • STOP • STOP • STOP • STOP • STOP • STOP

A12   a. 15   b. 6   c. 10   d. 9   e. 7   f. 13

---

4   The difference between two whole numbers that each have more than one digit is determined by writing the numbers in columns, the larger number on the top, and finding the difference between the individual digits. For example, the problem $853 - 621$ is

$$\begin{array}{r} 853 \\ -621 \\ \hline 232 \end{array}$$

written. To find the difference, you think

3 ones − 1 one = 2 ones
5 tens − 2 tens = 3 tens
8 hundreds − 6 hundreds = 2 hundreds

Q13   Find the difference:

a. $\begin{array}{r} 79 \\ -63 \\ \hline \end{array}$ b. $\begin{array}{r} 82 \\ -31 \\ \hline \end{array}$ c. $\begin{array}{r} 63 \\ -43 \\ \hline \end{array}$ d. $\begin{array}{r} 9648 \\ -9347 \\ \hline \end{array}$

STOP • STOP • STOP • STOP • STOP • STOP • STOP • STOP • STOP

A13   a. 16   b. 51   c. 20   d. 301

---

5   When subtracting whole numbers, a larger number cannot be subtracted from a smaller one. That is,

$5 - 8 = \square$ because $\square + 8 = 5$

no possible whole number

To find the difference of $\begin{array}{r} 734 \\ -528 \\ \hline \end{array}$ it will be necessary to *borrow* 1 ten from the tens column and add this amount to the 4 ones. For example,

$$\begin{array}{r} {\scriptstyle 2\ 10} \\ 7\ \cancel{3}\ 4 \\ -5\ 2\ 8 \\ \hline \end{array} \quad \text{or} \quad \begin{array}{r} {\scriptstyle 2\ 14} \\ 7\ \cancel{3}\ \cancel{4} \\ -5\ 2\ 8 \\ \hline 2\ 0\ 6 \end{array}$$

This leaves 2 tens in the tens column and 14 ones in the ones column. Then think: 14 ones − 8 ones = 6 ones. Finally, continue the subtraction to obtain the difference, 206.

Q14   Complete the following subtraction problem:

$$\begin{array}{r} \overset{3\ 13}{7\ \not4\ \not3} \\ -5\ 1\ 7 \\ \hline \end{array}$$

STOP • STOP • STOP • STOP • STOP • STOP • STOP • STOP • STOP

A14   226

Q15   Determine the difference:

    a.   85   **b.**   78   **c.**   90   **d.**   74

       −49   −69   −78   −58

STOP • STOP • STOP • STOP • STOP • STOP • STOP • STOP • STOP

A15   **a.** 36   **b.** 9: $\begin{array}{r}\overset{6\ 18}{\not7\ \not8}\\-6\ 9\\\hline 9\end{array}$ **c.** 12   **d.** 16: $\begin{array}{r}\overset{6\ 14}{\not7\ \not4}\\-5\ 8\\\hline 1\ 6\end{array}$

6   It is often necessary to borrow from more than one column when subtracting. For example:

$$\begin{array}{r} \overset{1\ 13}{7\ \not2\ \not3} \\ -4\ 8\ 5 \\ \hline 8 \end{array}$$
First borrow 1 ten from the tens column and add to the ones column. 13 ones − 5 ones = 8 ones.

$$\begin{array}{r} \overset{6\ 11\ 13}{\not7\ \not2\ \not3} \\ -4\ 8\ 5 \\ \hline 2\ 3\ 8 \end{array}$$
Since 1 ten − 8 tens is impossible, borrow 1 hundred from the hundreds column and add it to the tens column. Since 1 hundred = 10 tens, add 10 + 1 in the tens column, leaving 11 tens. Then continue the subtraction.

Q16   Complete the subtraction problem:

$$\begin{array}{r} \overset{3\ 18}{6\ \not4\ \not8} \\ -4\ 5\ 9 \\ \hline 9 \end{array}$$

STOP • STOP • STOP • STOP • STOP • STOP • STOP • STOP • STOP

A16
$$\begin{array}{r} \overset{5\ 13\ 18}{\not6\ \not4\ \not8} \\ -4\ 5\ 9 \\ \hline 1\ 8\ 9 \end{array}$$

Q17   Determine the differences:

  a.   6234   **b.**   7463   **c.**   3872
           −5847   −1597   −1998

STOP   •   STOP   •   STOP   •   STOP   •   STOP   •   STOP   •   STOP   •   STOP   •   STOP

A17   **a.** 387:   5 11 12 14   **b.** 5,866   **c.** 1,874:   2 17 16 12
                            6̸ 2̸ 3̸ 4̸   3̸ 8̸ 7̸ 2̸
                         −5   8   4   7   −1   9   9   8
                         ─────────────   ─────────────
                            3   8   7   1   8   7   4

Q18   Determine the difference:

  a.   451   **b.**   744   **c.**   27343   **d.**   43312
           −362   −398   −18251   −37543

STOP   •   STOP   •   STOP   •   STOP   •   STOP   •   STOP   •   STOP   •   STOP   •   STOP

A18   **a.** 89   **b.** 346   **c.** 9,092   **d.** 5,769

---

7   Zeros present additional difficulties when borrowing. For example, in the problem $\begin{matrix}406\\-238\end{matrix}$
it is not possible to borrow 1 ten from the tens column. Therefore, borrow 1 hundred from the hundreds column and write it as 10 tens in the tens column.

        3 10
        4̸ 0̸ 6
     −2 3 8
     ─────────   1 ten can now be borrowed from the tens column, leaving 9 tens and 16
              9   ones.
        3 1̸0̸ 16
        4̸ 0̸ 6̸
     −2 3 8

This process can be shortened by borrowing 1 hundred and writing it as 9 tens (tens column) and 10 ones (ones column). Some examples are:

    3 9 16   5 9 12   6 9 15   2 9 9 15
    4̸ 0̸ 6̸   6̸ 0̸ 2̸   7̸ 0 5̸   3̸ 0 0 5̸
  −2 3 8   −3 0 8   −4 1 6   −2 1 6 9
  ─────────   ─────────   ─────────   ────────────
    1 6 8   2 9 4   2 8 9   8 3 6

Q19   Determine the difference:

  a.   406   **b.**   503   **c.**   108
           −218   −246   −59

STOP   •   STOP   •   STOP   •   STOP   •   STOP   •   STOP   •   STOP   •   STOP   •   STOP

A19   **a.** 188:
```
        3 9 16
        A̶ 0̶ 6̶
       − 2 1 8
       ─────────
         1 8 8
```
b. 257:
```
        4 9 13
        5̶ 0̶ 3̶
       − 2 4 6
       ─────────
         2 5 7
```
c. 49

Q20   Determine the difference:

a.   3052
       −2468

b.   7023
       −3141

c.   5048
       −3071

STOP • STOP • STOP • STOP • STOP • STOP • STOP • STOP • STOP

A20   **a.** 584   **b.** 3,882   **c.** 1,977

Q21   Determine the difference:

a.   8003
       −3823

b.   2004
       −1514

STOP • STOP • STOP • STOP • STOP • STOP • STOP • STOP • STOP

A21   **a.** 4,180:
```
        7 9 10
        8̶ 0̶ 0̶ 3
       − 3 8 2 3
       ──────────
         4 1 8 0
```
b. 490:
```
        1 9 10
        2̶ 0̶ 0̶ 4
       − 1 5 1 4
       ──────────
           4 9 0
```

Q22   Determine the difference:

a.   7030
       −1347

b.   70010
       −28273

STOP • STOP • STOP • STOP • STOP • STOP • STOP • STOP • STOP

A22   **a.** 5,683:
```
        6 9 12 10
        7̶ 0̶ 3̶ 0̶
       − 1 3 4 7
       ──────────
         5 6 8 3
```
b. 41,737:
```
        6 9 9 10 10
        7̶ 0̶ 0̶ 1̶ 0̶
       − 2 8 2 7 3
       ───────────
         4 1 7 3 7
```

Q23   Determine the difference:

a.   6005
       −2146

b.   50040
       −26872

c.   30030
       −17293

STOP • STOP • STOP • STOP • STOP • STOP • STOP • STOP • STOP

A23   **a.** 3,859   **b.** 23,168   **c.** 12,737

Q24　　A college has a total enrollment of 8,093 students. If 3,926 of these students are men, how many are women?

STOP　•　STOP　•　STOP　•　STOP　•　STOP　•　STOP　•　STOP　•　STOP　•　STOP

A24　　4,167

8　　Mistakes in addition and subtraction problems are often caused by one of the following errors:

1. Lack of knowledge of the basic addition and subtraction facts.
2. Copying the problem incorrectly.
3. Not writing the numerals clearly.
4. Placing the digits in the wrong columns.
5. Forgetting to carry or borrow.
6. Errors in carrying or borrowing.
7. Lack of concentration.

If you are aware of the possible difficulties, you can watch for them and avoid errors.

This completes the instruction for this section.

1.2　　Exercises

1. Determine the following sums:

   a. 863　　　　　　**b.** 459　　　　　　**c.** 728
   　　749　　　　　　　　863　　　　　　　　137

   d. 234　　　　　　**e.** 867　　　　　　**f.** 611
   　　198　　　　　　　　693　　　　　　　　208
   　　765　　　　　　　　345　　　　　　　　843

   g. 680　　　　　　**h.** 1928　　　　　　**i.** 236 + 87 + 1,008
   　　 34　　　　　　　　6024
   　　287　　　　　　　　4376
   　　　　　　　　　　　 5928

   j. 45 + 8 + 687　　　**k.** 7 + 80,006 + 53

2. Determine the difference:

   a. 　372　　　　　　**b.** 　503　　　　　　**c.** 　653
   　　−186　　　　　　　 −217　　　　　　　 −278

   d. 24266　　　　　　**e.** 　7008　　　　　　**f.** 　706
   　−7318　　　　　　　 −1443　　　　　　　 −168

g. 60113
−51864

h. 2731
−1816

i. 50000
−48697

j. 683 − 29

k. 48,000 − 47,999

1.2 Exercise Answers

1. **a.** 1,612 **b.** 1,322 **c.** 865 **d.** 1,197 **e.** 1,905 **f.** 1,662 **g.** 1,001
   h. 18,256 **i.** 1,331 **j.** 740 **k.** 80,066
2. **a.** 186 **b.** 286 **c.** 375 **d.** 16,948 **e.** 5,565 **f.** 538 **g.** 8,249
   h. 915 **i.** 1,303 **j.** 654 **k.** 1

1.3 Multiplication and Division

---

1 *Multiplication* is actually a short form of addition. The sum $3 + 3 + 3 + 3 + 3 = 15$ could be written as a multiplication problem as follows:

$5 \times 3 = 15$ (read "5 times 3 equals 15")

factor   product

The 5 and 3 are called *factors* and 15 is the *product*. The first factor states how many times the second factor is added to form the product. Other examples are:

$3 \times 5 = 15$ because 3×5 means $5 + 5 + 5$
$4 \times 2 = 8$ because 4×2 means $2 + 2 + 2 + 2$
$3 \times 9 = 27$ because 3×9 means $9 + 9 + 9$

Q1 In the problem $5 \times 9 = 45$, list the factors and product.

factors_____ product_____

STOP • STOP • STOP • STOP • STOP • STOP • STOP • STOP • STOP

A1 factors 5, 9; product 45

Q2 In the problem $63 = 7 \times 9$, list the factors and product.

factors_____ product_____

STOP • STOP • STOP • STOP • STOP • STOP • STOP • STOP • STOP

A2 factors 7, 9; product 63

Q3 Write the following multiplication problems as repeated addition problems (do not compute the sums):

a. 3×8_____

b. 4×0_____

c. 3×1_____

d. 1×3_____

STOP • STOP • STOP • STOP • STOP • STOP • STOP • STOP • STOP

A3 **a.** $8 + 8 + 8$ **b.** $0 + 0 + 0 + 0$ **c.** $1 + 1 + 1$ **d.** 3

Q4   Write the following repeated addition problems as multiplication problems (do not compute the product):

a. $4 + 4 + 4 + 4 + 4$ _____

b. $5 + 5 + 5 + 5$ _____

c. 8 _____

d. $0 + 0 + 0 + 0 + 0 + 0$ _____

STOP • STOP • STOP • STOP • STOP • STOP • STOP • STOP • STOP

A4   **a.** 5×4 **b.** 4×5 **c.** 1×8 **d.** 6×0

---

2   Always finding the product of two numbers by repeated addition would be too time-consuming. It is much quicker and easier to use the *multiplication* facts provided in the following table. They should be learned.

	0	1	2	3	4	5	6	7	8	9
0	0	0	0	0	0	0	0	0	0	0
1	0	1	2	3	4	5	6	7	8	9
2	0	2	4	6	8	10	12	14	16	18
3	0	3	6	9	12	15	18	21	24	27
4	0	4	8	12	16	20	24	28	32	36
5	0	5	10	15	20	25	30	35	40	45
6	0	6	12	18	24	30	36	42	48	54
7	0	7	14	21	28	35	42	49	56	63
8	0	8	16	24	32	40	48	56	64	72
9	0	9	18	27	36	45	54	63	72	81

To use the table, locate the first factor in the left-hand column and the second factor in the top row. The light shading shows the product 5×7:

$5 \times 7 = 35$

The heavy shading shows the product 7×5:

$7 \times 5 = 35$

The property $5 \times 7 = 7 \times 5$ aids in learning the facts, because you only need to memorize one half of them. The table also shows important properties of 0 and 1. That is,

(any number) $\times 0 = 0$

Examples are:

$5 \times 0 = 0$ $0 \times 8 = 0$ $0 \times 0 = 0$ $9 \times 0 = 0$

Also,

(any number) $\times 1 =$ that number

Examples are:

$4 \times 1 = 4$ $1 \times 7 = 7$ $0 \times 1 = 0$ $9 \times 1 = 9$

Q5   Determine the product (use the table as little as possible):

a. $6 \times 2 =$ _____

b. $9 \times 6 =$ _____

c. $6 \times 0 =$ _____

d. $5 \times 7 =$ _____   **e.** $8 \times 1 =$ _____   **f.** $7 \times 4 =$ _____

g. $7 \times 8 =$ _____   **h.** $0 \times 4 =$ _____   **i.** $8 \times 6 =$ _____

j. $2 \times 3 =$ _____   **k.** $4 \times 1 =$ _____   **l.** $5 \times 0 =$ _____

m. $7 \times 2 =$ _____   **n.** $5 \times 4 =$ _____   **o.** $0 \times 8 =$ _____

p. $6 \times 8 =$ _____

STOP • STOP • STOP • STOP • STOP • STOP • STOP • STOP • STOP

A5   **a.** 12   **b.** 54   **c.** 0   **d.** 35   **e.** 8   **f.** 28   **g.** 56

      h. 0   **i.** 48   **j.** 6   **k.** 4   **l.** 0   **m.** 14   **n.** 20

      o. 0   **p.** 48

---

3   The most frequently missed facts are:

$$6 \times 6 = 36 \qquad 8 \times 7 = 7 \times 8 = 56$$
$$7 \times 6 = 6 \times 7 = 42 \qquad 9 \times 7 = 7 \times 9 = 63$$
$$8 \times 6 = 6 \times 8 = 48 \qquad 8 \times 8 = 64$$
$$9 \times 6 = 6 \times 9 = 54 \qquad 9 \times 8 = 8 \times 9 = 72$$
$$7 \times 7 = 49 \qquad 9 \times 9 = 81$$

Continued practice will help you learn these facts. Remember that the order of the factors is not important.

Q6   Determine the product:

a. $6 \times 9 =$ _____   **b.** $6 \times 6 =$ _____   **c.** $9 \times 8 =$ _____

d. $7 \times 7 =$ _____   **e.** $8 \times 7 =$ _____   **f.** $7 \times 8 =$ _____

g. $9 \times 6 =$ _____   **h.** $0 \times 9 =$ _____   **i.** $1 \times 9 =$ _____

STOP • STOP • STOP • STOP • STOP • STOP • STOP • STOP • STOP

A6   **a.** 54   **b.** 36   **c.** 72

      d. 49   **e.** 56   **f.** 56

      g. 54   **h.** 0   **i.** 9

---

4   The product of two numbers, where one of the numbers has more than one digit, can be determined as follows: The product 34×2 is written:

$$
\begin{array}{r}
34 \\
\times 2 \\
\hline
68
\end{array}
\quad
\begin{array}{l}
\text{larger factor} \\
\text{smaller factor} \\
\text{product}
\end{array}
$$

Multiplication process: 2×4 ones $= 8$ ones. Write the 8 in the ones column. 2×3 tens $= 6$ tens. Write the 6 in the tens column.

Q7   Determine the product:

     a. $\begin{array}{r} 12 \\ \times 4 \\ \hline \end{array}$ **b.** $\begin{array}{r} 33 \\ \times 3 \\ \hline \end{array}$ **c.** $\begin{array}{r} 23 \\ \times 2 \\ \hline \end{array}$ **d.** $\begin{array}{r} 42 \\ \times 4 \\ \hline \end{array}$

STOP • STOP • STOP • STOP • STOP • STOP • STOP • STOP • STOP

A7   **a.** 48   **b.** 99   **c.** 46   **d.** 168

5   The product of two numbers will often involve carrying. One method of performing this type of multiplication is shown below. The product 5 × 17 is written:

3   carrying row
1 7

×5
―――
8 5   product

Multiplication process: 5 × 7 = 35 (3 tens and 5 ones). Write the 5 in the ones column and carry the 3 to the top of the tens column. 5 × 1 tens = 5 tens; now add the number carried: 5 tens + 3 tens = 8 tens. Write the 8 in the tens column. Remember that the number carried is added after the next digit is multiplied.

Q8   Complete the multiplication problem:

4
2 7
×7
―――
9

STOP • STOP • STOP • STOP • STOP • STOP • STOP • STOP • STOP

A8
4
2 7
×7
―――
1 8 9

―――――――――――――――――――――――――――――――

Q9   Determine the product:

a.   56   b.   39   c.   83   d.   78
     ×8   ×9   ×5   ×7

STOP • STOP • STOP • STOP • STOP • STOP • STOP • STOP • STOP

A9   a. 448:
4
5 6   b. 351   c. 415   d. 546:
5
7 8
×8   ×7
―――   ―――
4 4 8   5 4 6

―――――――――――――――――――――――――――――――

Q10   Determine the product:

a. 317   b. 486   c. 732   d. 691
    ×6   ×8   ×3   ×7

STOP • STOP • STOP • STOP • STOP • STOP • STOP • STOP • STOP

A10 a. 1,902:
1 4
3 1 7   b. 3,888:
6 4
4 8 6   c. 2,196   d. 4,837
×6   ×8
―――――   ―――――
1 9 0 2   3 8 8 8

6   Multiplying two numbers, each with two or more digits, is completed as follows:

```
        3     second carrying row
        4     first carrying row
      5 8
     ×4 5
     ─────
      2 9 0
    2 3 2
    ───────
    2 6 1 0
```

Multiplication process: $5 \times 58 = 290$ (multiplying as usual). 4 tens × 8 ones = 32 tens (3 hundreds and 2 tens). Write the 2 in the tens column and carry the 3 to the second carrying row.

4 tens × 5 tens = 20 hundreds. Add 20 hundreds to the 3 hundreds carried. 20 + 3 = 23 hundreds (2 thousands and 3 hundreds). Write the 3 in the hundreds column and the 2 in the thousands column. Add the digits in each column to obtain the product.

Q11   Complete the multiplication problem:
```
        2
      4 3
     ×6 7
     ─────
      3 0 1
```

STOP • STOP • STOP • STOP • STOP • STOP • STOP • STOP • STOP

A11
```
        1
        2
      4 3
     ×6 7
     ─────
      3 0 1
    2 5 8
    ───────
    2 8 8 1
```

Q12   Determine the product:
  a.   83   **b.**   85
         ×29   ×37

STOP • STOP • STOP • STOP • STOP • STOP • STOP • STOP • STOP

A12   **a.** 2,407:
```
        2
      8 3
     ×2 9
     ─────
      7 4 7
    1 6 6
    ───────
    2 4 0 7
```

b. 3,145:
```
        1
        8̸
      8 5
     ×3 7
     ─────
      5 9 5
    2 5 5
    ───────
    3 1 4 5
```
It is common practice to cross out the first carrying row so as not to confuse these numbers with the numbers in the second carrying row.

Q13   Determine the product:

a.   715
     ×54

b.   684
     ×34

c.   196
     ×92

STOP   •   STOP   •   STOP   •   STOP   •   STOP   •   STOP   •   STOP   •   STOP   •   STOP

A13   **a.** 38,610:

```
          2
          2
        7 1 5
      × 5 4
      ───────
      2 8 6 0
    3 5 7 5
    ─────────
    3 8 6 1 0
```

b. 23,256:

```
        2 1
        8 1
        6 8 4
      × 3 4
      ───────
      2 7 3 6
    2 0 5 2
    ─────────
    2 3 2 5 6
```

c. 18,032:

```
        8 5
        1 1
        1 9 6
      × 9 2
      ───────
        3 9 2
    1 7 6 4
    ─────────
    1 8 0 3 2
```

Q14   Determine the product:

a.   389
     ×88

b.   673
     ×23

c.   875
     ×335

d.   335
     ×62

STOP   •   STOP   •   STOP   •   STOP   •   STOP   •   STOP   •   STOP   •   STOP   •   STOP

A14   **a.** 34,232   **b.** 15,479

c. 293,125:

```
        2 1
        2 1
        3 2
        8 7 5
      × 3 3 5
      ───────
      4 3 7 5
    2 6 2 5
  2 6 2 5
  ───────────
  2 9 3 1 2 5
```

d. 20,770

---

7   Zeros sometimes present difficulties when multiplying:

1.
```
      7
    7 0 9
   ×    8
   ──────
    5 6 7 2
```

Multiplication process: $8 \times 9 = 72$ (7 tens and 2 ones). Write the 2 in the ones column and carry the 7 to the tens column. $8 \times 0 = 0$. Now add the carried number. $0 + 7 = 7$ (tens). Write the seven in the tens column. Continue the multiplication in the usual way.

2.
$$
\begin{array}{r}
\overset{\overset{3}{\cancel{6}}}{6\ 0\ 9} \\
\times 4\ 7 \\
\hline
4\ 2\ 6\ 3 \\
2\ 4\ 3\ 6 \\
\hline
2\ 8\ 6\ 2\ 3
\end{array}
$$

Q15   Complete the multiplication problem:

$$
\begin{array}{r}
\overset{2}{4\ 0\ 3} \\
\times 5\ 8 \\
\hline
3\ 2\ 2\ 4
\end{array}
$$

STOP • **STOP** • **STOP** • **STOP** • **STOP** • **STOP** • **STOP** • **STOP** • **STOP**

A15
$$
\begin{array}{r}
\overset{\overset{1}{\cancel{2}}}{4\ 0\ 3} \\
\times 5\ 8 \\
\hline
3\ 2\ 2\ 4 \\
2\ 0\ 1\ 5 \\
\hline
2\ 3\ 3\ 7\ 4
\end{array}
$$

Q16   Determine the product: 709
$$
\begin{array}{r}
\times 9
\end{array}
$$

STOP • **STOP** • **STOP** • **STOP** • **STOP** • **STOP** • **STOP** • **STOP** • **STOP**

A16   6,381:   709 ← It is not necessary to always
$$
\begin{array}{r}
\times 9 \\
\hline
6381
\end{array}
$$
write the carrying number.

Q17   Determine the product:
   a.   807   **b.**   2301   **c.**   908
$$\times 49 \qquad\qquad \times 13 \qquad\qquad \times 7$$

STOP • **STOP** • **STOP** • **STOP** • **STOP** • **STOP** • **STOP** • **STOP** • **STOP**

A17   **a.** 39,543   **b.** 29,913   **c.** 6,356

8   Other difficulties with zero arise when the zero is in the bottom factor. Consider the following example.

$$\begin{array}{r} 512 \\ \times 405 \\ \hline 2560 \\ 000 \\ 2048 \\ \hline 207360 \end{array}$$

Multiplication process: $5 \times 512 = 2,560$ (in the usual way). Now multiply 0×512. Recall that $0 \times$ (any number) $= 0$. That is, $0 \times 2 = 0$, $0 \times 1 = 0$, $0 \times 5 = 0$, and so on. Now complete 4×512 in the usual way.

Q18   Determine the product: $\begin{array}{r} 9418 \\ \times 305 \\ \hline \end{array}$

STOP • STOP • STOP • STOP • STOP • STOP • STOP • STOP • STOP

A18   2,872,490: $\begin{array}{r} 9418 \\ \times 305 \\ \hline 47090 \\ 0000 \\ 28254 \\ \hline 2872490 \end{array}$

Q19   Determine the product:

a. $\begin{array}{r} 7927 \\ \times 509 \\ \hline \end{array}$ b. $\begin{array}{r} 3879 \\ \times 403 \\ \hline \end{array}$

STOP • STOP • STOP • STOP • STOP • STOP • STOP • STOP • STOP

A19   a.   4,034,843: $\begin{array}{r} 7927 \\ \times 509 \\ \hline 71343 \\ 0000 \\ 39635 \\ \hline 4034843 \end{array}$ b.   1,563,237

9   If the bottom factor ends with one or more zeros, the zeros may be brought down as shown:

1. $\begin{array}{r} 52 \\ \times 30 \\ \hline 00 \\ 156 \\ \hline 1560 \end{array}$ or $\begin{array}{r} 52 \\ \times 30 \\ \hline 1560 \end{array}$ 2. $\begin{array}{r} 57 \\ \times 20 \\ \hline 00 \\ 114 \\ \hline 1140 \end{array}$ or $\begin{array}{r} 57 \\ \times 20 \\ \hline 1140 \end{array}$ 3. $\begin{array}{r} 84 \\ \times 400 \\ \hline 00 \\ 00 \\ 336 \\ \hline 33600 \end{array}$ or $\begin{array}{r} 84 \\ \times 400 \\ \hline 33600 \end{array}$

Q20 Complete the multiplication problem:
$$\begin{array}{r} 76 \\ \times\,60 \\ \hline 0 \end{array}$$

STOP • **STOP** • **STOP** • **STOP** • **STOP** • **STOP** • **STOP** • **STOP** • **STOP**

A20
$$\begin{array}{r} 76 \\ \times\,60 \\ \hline 4560 \end{array}$$

Q21 Determine the product:

a. $\begin{array}{r} 32 \\ \times\,10 \\ \hline \end{array}$ b. $\begin{array}{r} 36 \\ \times\,20 \\ \hline \end{array}$ c. $\begin{array}{r} 40 \\ \times\,100 \\ \hline \end{array}$

STOP • **STOP** • **STOP** • **STOP** • **STOP** • **STOP** • **STOP** • **STOP** • **STOP**

A21 a. 320   b. 720   c. 4,000

Q22 Determine the product:

a. $\begin{array}{r} 2400 \\ \times\,50 \\ \hline \end{array}$ b. $\begin{array}{r} 5600 \\ \times\,1200 \\ \hline \end{array}$

STOP • **STOP** • **STOP** • **STOP** • **STOP** • **STOP** • **STOP** • **STOP** • **STOP**

A22 a. 120,000: $\begin{array}{r} 2400 \\ \times\,50 \\ \hline 120000 \end{array}$ b. 6,720,000: $\begin{array}{r} 5600 \\ \times\,1200 \\ \hline 1120000 \\ 5600 \\ \hline 6720000 \end{array}$

Q23 Paul's car averages 15 miles per gallon of gasoline. If the gas tank holds 24 gallons, how far can he travel on a full tank?

STOP • **STOP** • **STOP** • **STOP** • **STOP** • **STOP** • **STOP** • **STOP** • **STOP**

A23 360 miles: because $15 \times 24 = 360$

Q24 If there are 26 rows of chairs in your classroom and 18 chairs in each row, how many chairs are there?

STOP • **STOP** • **STOP** • **STOP** • **STOP** • **STOP** • **STOP** • **STOP** • **STOP**

A24 468

10   *Division* of whole numbers can be thought of as finding how many equal groups are contained in a given number. For example, 12 divided by 4 means: How many 4s are contained in 12? This is pictured as follows:

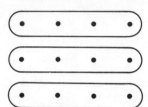

The 12 dots are divided into 3 equal groups, 4 in each group. Hence, 12 divided by 4 is 3. In symbols this can be stated: $12 \div 4 = 3$ (read "twelve divided by four equals three"). The symbol "$\div$" means "divided by."

Division can also be stated as the opposite of multiplication.

$12 \div 4 = 3$ because $3 \times 4 = 12$
$10 \div 2 = 5$ because $5 \times 2 = 10$
$20 \div 4 = 5$ because $5 \times 4 = 20$
$72 \div 9 = 8$ because $8 \times 9 = 72$

Q25   Complete the following division problem:

$63 \div 7 = \square$ because $\square \times 7 = 63$

STOP • STOP • STOP • STOP • STOP • STOP • STOP • STOP • STOP

A25   9, 9

Q26   (1) Complete the following division problems and (2) state them as multiplication problems:

a. (1) $72 \div 8 =$ _____   **b.** (1) $56 \div 7 =$ _____

(2) _____   (2) _____

STOP • STOP • STOP • STOP • STOP • STOP • STOP • STOP • STOP

A26   **a.** (1) 9   **b.** (1) 8
(2) $9 \times 8 = 72$ (2) $8 \times 7 = 56$

Q27   Complete the division $15 \div 3$ by dividing the dots into equal groups:

.   .   .   .   .
.   .   .   .   . $15 \div 3 =$ _____
.   .   .   .   .

STOP • STOP • STOP • STOP • STOP • STOP • STOP • STOP • STOP

A27 $15 \div 3 = 5$

Q28   Use the following dots to perform the division $8 \div 4$:

$$8 \div 4 = \underline{\hspace{2cm}}$$

STOP • STOP • STOP • STOP • STOP • STOP • STOP • STOP • STOP

A28

$$8 \div 4 = 2$$

---

11   The numbers 1 and 0 present interesting problems for division. Consider the following examples:

$$8 \div 1 = \square \quad \text{because} \quad \square \times 1 = 8$$
$$8$$

$$16 \div 1 = 16 \quad \text{because} \quad 16 \times 1 = 16$$

That is, (any number) $\div$ 1 = that same number.

$$0 \div 5 = \square \quad \text{because} \quad \square \times 5 = 0$$
$$0$$

$$0 \div 8 = 0 \quad \text{because} \quad 0 \times 8 = 0$$
$$0 \div 15 = 0 \quad \text{because} \quad 0 \times 15 = 0$$

Now consider

$$5 \div 0 = \square \quad \text{because} \quad \square \times 0 = 5$$
$$\uparrow \qquad\qquad\qquad \uparrow$$
$$\text{no such number}$$

$$8 \div 0 = \square \quad \text{because} \quad \square \times 0 = 8$$
$$\uparrow \qquad\qquad\qquad \uparrow$$
$$\text{no such number}$$

    These examples illustrate that *division by 0 is impossible*. Division involving zero may be summarized as follows:

1. 0 ÷ (any number except zero) = zero.
2. (any number) ÷ 0 is impossible. That is, division by zero is impossible.

Q29   (1) Complete the division problems and (2) state them as multiplication problems:

    a. (1) $0 \div 8 =$ _____   **b.** (1) $9 \div 1 =$ _____   **c.** (1) $4 \div 0 =$ _____

      (2) _____   (2) _____   (2) _____

STOP • STOP • STOP • STOP • STOP • STOP • STOP • STOP • STOP

A29   **a.** (1) 0   **b.** (1) 9   **c.** (1) impossible

        (2) $0 \times 8 = 0$ (2) $9 \times 1 = 9$ (2) $\square \times 0 = 4$
$$\uparrow$$
$$\text{no such number}$$

---

12   The number of equal groups in a division problem is called the *quotient* and the number you are dividing by is called the *divisor*. The number you are dividing into equal parts is called the *dividend*. That is,

$$\overset{\nearrow}{\underset{\text{dividend}}{}} \; 8 \div \underset{\text{divisor}}{2} = \underset{\text{quotient}}{4}$$

Division problems are often written

$$\begin{array}{r} 4 \leftarrow \text{quotient} \\ \text{divisor} \rightarrow 2\overline{)8} \leftarrow \text{dividend} \end{array}$$

Q30   Determine the quotient:

a. $36 \div 9$ _____   b. $0 \div 9$ _____   c. $7\overline{)42}$ _____   d. $7\overline{)56}$ _____

e. $9\overline{)63}$ _____   f. $9\overline{)81}$ _____   g. $5\overline{)45}$ _____   h. $13 \div 0$ _____

STOP • STOP • STOP • STOP • STOP • STOP • STOP • STOP • STOP

A30   a. 4   b. 0   c. 6   d. 8
       e. 7   f. 9   g. 9   h. impossible

13   When dividing, it is not always possible to divide the dividend into an equal number of groups exactly. There will often be a *remainder*, a certain amount left over. Consider $14 \div 4$.

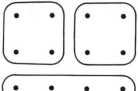

There are 3 equal groups of 4 and a remainder of 2.

That is, $14 \div 4 = 3\,\text{R}2$ (read "3 remainder 2"); or

$$\begin{array}{r} 3\,\text{R}2 \\ 4\overline{)14} \end{array}$$

It is not always possible to do this division process mentally. It can be performed as follows:

$$\begin{array}{r} 3 \\ 4\overline{)14} \\ \underline{12} \quad (3 \times 4) \\ 2 \quad \text{remainder} \end{array}$$

Division process:

Step 1: $14 \div 4$ ones $= 3 +$ remainder.

Step 2: Multiply the partial quotient by the divisor and position the product under the dividend.

Step 3: Subtract $14 - 12$. The remainder is 2.

Q31   Complete the division problem: $\begin{array}{r} 8 \\ 7\overline{)60} \\ \underline{} \quad (8 \times 7) \end{array}$

STOP • STOP • STOP • STOP • STOP • STOP • STOP • STOP • STOP

A31
$$\begin{array}{r} 8 \\ 7\overline{)60} \\ 56 \\ \hline 4 \end{array}$$

Q32   Determine the quotient and remainder (show the complete solution):

    a. $4\overline{)17}$ **b.** $7\overline{)45}$

STOP • **STOP** • **STOP** • **STOP** • **STOP** • **STOP** • **STOP** • **STOP** • **STOP**

A32   **a.** $\begin{array}{r} 4 \\ 4\overline{)17} \\ 16 \\ \hline 1 \end{array}$ (4 R1)   **b.** $\begin{array}{r} 6 \\ 7\overline{)45} \\ 42 \\ \hline 3 \end{array}$ (6 R3)

Q33   Determine the quotient and remainder:

    a. $9\overline{)67}$ **b.** $8\overline{)58}$ **c.** $6\overline{)35}$ **d.** $7\overline{)51}$

STOP • **STOP** • **STOP** • **STOP** • **STOP** • **STOP** • **STOP** • **STOP** • **STOP**

A33   **a.** 7 R4   **b.** 7 R2   **c.** 5 R5   **d.** 7 R2

---

14   For more difficult division problems, a method called *long division* is used.

$$\begin{array}{r} 23 \\ 3\overline{)69} \\ 6 \\ \hline 09 \\ 9 \\ \hline 0 \end{array}$$

    6 (2×3)

    9 (3×3)

Division process:

Step 1: 6 tens ÷ 3 = 2 tens.

Step 2: Write the 2 above the 6 in the tens column.

Step 3: 2 tens × 3 = 6 tens. Write the 6 under the 6.

Step 4: Subtract 6 − 6.

Step 5: Bring down the 9.

Step 6: Subtract 9 − 9, which equals 0.

Therefore, 69 ÷ 3 = 23 R0 = 23.

Q34 Complete the division problem:

$$\begin{array}{r} 2 \\ 4\overline{)87} \\ 8 \end{array}$$

STOP • **STOP** • **STOP** • **STOP** • **STOP** • **STOP** • **STOP** • **STOP** • **STOP**

A34

$$\begin{array}{r} 21 \\ 4\overline{)87} \\ 8 \\ \overline{07} \\ 4 \\ \overline{3} \end{array}$$

Q35 Show the complete solution:

a. $3\overline{)3711}$ b. $2\overline{)2838}$

STOP • **STOP** • **STOP** • **STOP** • **STOP** • **STOP** • **STOP** • **STOP** • **STOP**

A35

a.
$$\begin{array}{r} 1237 \\ 3\overline{)3711} \\ 3 \\ \overline{07} \\ 6 \\ \overline{11} \\ 9 \\ \overline{21} \\ 21 \\ \overline{0} \end{array}$$

b.
$$\begin{array}{r} 1419 \\ 2\overline{)2838} \\ 2 \\ \overline{08} \\ 8 \\ \overline{03} \\ 2 \\ \overline{18} \\ 18 \\ \overline{0} \end{array}$$

---

15 Often the divisor will not divide into the first digit of the dividend.

$$\begin{array}{r} 413 \\ 9\overline{)3717} \\ 36 \quad (4 \times 9) \\ \overline{11} \\ 9 \quad (1 \times 9) \\ \overline{27} \\ 27 \quad (3 \times 9) \\ \overline{} \end{array}$$

Division process:

Step 1: 3 is not divisible by 9; therefore, divide 37 by 9. $37 \div 9 = 4$ + remainder.

Step 2: Place the 4 above the 7 and continue the division process. Always place the first partial quotient above the digit in the dividend that indicates the number being divided. That is, place the 4 above the 7 because 37 is being divided by 9.

Q36   Show the complete solution:

a. $7\overline{)2058}$ b. $6\overline{)3516}$

STOP • **STOP** • **STOP** • **STOP** • **STOP** • **STOP** • **STOP** • **STOP** • **STOP**

A36   a. $\begin{array}{r} 294 \\ 7\overline{)2058} \\ \underline{14} \\ 65 \\ \underline{63} \\ 28 \\ \underline{28} \\ 0 \end{array}$ b. $\begin{array}{r} 586 \\ 6\overline{)3516} \\ \underline{30} \\ 51 \\ \underline{48} \\ 36 \\ \underline{36} \\ 0 \end{array}$

Q37   Determine the quotient and remainder:

a. $9\overline{)5863}$ b. $5\overline{)4938}$ c. $9\overline{)2937}$

STOP • **STOP** • **STOP** • **STOP** • **STOP** • **STOP** • **STOP** • **STOP** • **STOP**

A37   **a.** 651 R4   **b.** 987 R3   **c.** 326 R3

---

16   The solution of more difficult division problems will be aided through the use of *rounding* (to be discussed in greater detail in Section 1.4). When dividing by a two-digit divisor, it will be helpful to round the divisor to the nearest multiple of 10. The multiples of 10 are 10, 20, 30, 40, 50, 60, and so on. To round a number to a multiple of 10 is to find the multiple of 10 nearest to the number. For example, round 53 to the nearest multiple of 10.

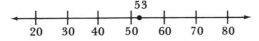

Notice that 53 is nearer 50 than 60; hence, 53 rounded to a multiple of 10 is 50. *A number halfway between two multiples is rounded to the higher multiple.* That is, 55 is rounded to 60; 45 is rounded to 50; 85 is rounded to 90; and so on.

Q38   Round the following numbers to the nearest multiple of 10:

         a. 22 _____   **b.** 15 _____   **c.** 67 _____   **d.** 83 _____

STOP • STOP • STOP • STOP • STOP • STOP • STOP • STOP • STOP

A38   **a.** 20   **b.** 20   **c.** 70   **d.** 80

Q39   When dividing by a three-digit divisor, it will be necessary to round the divisor to a multiple of _____ .

STOP • STOP • STOP • STOP • STOP • STOP • STOP • STOP • STOP

A39   100

Q40   List the next five multiples of 100:

         100, 200, _____, _____, _____, _____, _____ .

STOP • STOP • STOP • STOP • STOP • STOP • STOP • STOP • STOP

A40   300, 400, 500, 600, 700

Q41   Round the following numbers to the nearest multiple of 100:

         a. 123 _____   **b.** 489 _____   **c.** 450 _____   **d.** 349 _____

STOP • STOP • STOP • STOP • STOP • STOP • STOP • STOP • STOP

A41   **a.** 100   **b.** 500   **c.** 500   **d.** 300

---

17   Dividing by a two-digit divisor is accomplished as follows:

$$39\overline{)267} \quad \begin{array}{r} 6 \\ \hline 234 \\ \hline 33 \end{array}$$

Division process:

Step 1: Round the divisor to a multiple of 10: $40\overline{)267}$.

Step 2: *Approximate* the partial quotient by dividing the 4 into the first two digits of the dividend: $26 \div 4 = 6$ (approximate quotient).

Step 3: Multiply the approximated quotient by the divisor and subtract.

Step 4: Since the remainder is smaller than the divisor, the division is complete.

Q42   Approximate the partial quotient only:

         a. $37\overline{)625}$ **b.** $52\overline{)493}$ **c.** $48\overline{)352}$

STOP • STOP • STOP • STOP • STOP • STOP • STOP • STOP • STOP

A42   **a.** $37\overline{)625}$:   **b.** $52\overline{)493}$:   **c.** $48\overline{)352}$:

                      $40\overline{)625}$ $50\overline{)493}$ $50\overline{)352}$

       Think: $6 \div 4$ $49 \div 5$ $35 \div 5$

(partial quotients: **a.** 1   **b.** 9   **c.** 7)

Q43   Show the complete solution to the following division problems:

a. $25\overline{)632}$ b. $53\overline{)125}$ c. $78\overline{)563}$

STOP • STOP • STOP • STOP • STOP • STOP • STOP • STOP • STOP

A43   a.
$$
\begin{array}{r}
25 \\
25\overline{)632} \\
50 \\
\overline{132} \\
125 \\
\overline{7}
\end{array}
$$

b.
$$
\begin{array}{r}
2 \\
53\overline{)125} \\
106 \\
\overline{19}
\end{array}
$$

c.
$$
\begin{array}{r}
7 \\
78\overline{)563} \\
546 \\
\overline{17}
\end{array}
$$

---

18   The solution to A43, part a, is obtained as follows:

$$
\begin{array}{r}
25 \\
25\overline{)632} \\
50 \\
\overline{132} \\
125 \\
\overline{7}
\end{array}
$$

Division process:

Step 1: Think $30\overline{)632}$.

Step 2: $6 \div 3 = 2$ (first partial quotient).

Step 3: $2 \times 25 = 50$ and $63 - 50 = 13$. Bring down the 2.

Step 4: $13 \div 3 = 4$ (second partial quotient). $4 \times 25 = 100$ and $132 - 100 = 32$. Since 32 is larger than the divisor, the partial quotient was not large enough. Use 5 as the second partial quotient.

Step 5: $5 \times 25 = 125$ and $132 - 125 = 7$.

Remember that the partial quotient obtained by rounding the divisor is only an approximation. It may have to be adjusted upward or downward.

Q44   Show the complete solution to the following division problems:

a. $87\overline{)613}$ b. $37\overline{)1862}$

STOP • STOP • STOP • STOP • STOP • STOP • STOP • STOP • STOP

A44   a.
$$
\begin{array}{r}
7 \\
87\overline{)613} \\
609 \\
\overline{4}
\end{array}
$$

b.
$$
\begin{array}{r}
50 \\
37\overline{)1862} \\
185 \\
\overline{12}
\end{array}
$$

In both problems the approximated quotient had to be adjusted upward.

Q45   Determine the quotient and remainder:

        a. $48\overline{)352}$ **b.** $33\overline{)168}$

STOP • STOP • STOP • STOP • STOP • STOP • STOP • STOP • STOP

A45   **a.** 7 R16   **b.** 5 R3

Q46   Determine the quotient and remainder:

        a. $56\overline{)284}$ **b.** $547\overline{)2738}$

STOP • STOP • STOP • STOP • STOP • STOP • STOP • STOP • STOP

A46   **a.** 5 R4   **b.** 5 R3:

$$547\overline{)\begin{array}{l}2738\end{array}}\;\;\;\;\;5$$

b. 5 R3:
$$\begin{array}{r} 5 \\ 547\overline{)2738} \\ 2735 \\ \hline 3 \end{array}$$
Think:
$$500\overline{)2738} \quad 27 \div 5 = 5$$

---

19   A common error in division is the omission of zeros in the quotient. Examples:

1.
$$\begin{array}{r} 202 \\ 8\overline{)1623} \\ 16 \\ \hline 2 \\ 0 \quad (0 \times 8) \\ \hline 23 \\ 16 \\ \hline 7 \end{array}$$

Whenever a number is brought down, a division by the divisor is indicated, even though the partial quotient obtained is 0.

2.
$$\begin{array}{r} 200 \\ 67\overline{)13450} \\ 134 \\ \hline 5 \\ 0 \quad (0 \times 67) \\ \hline 50 \\ 0 \quad (0 \times 67) \\ \hline 50 \end{array}$$

The division is not complete until all the digits in the dividend have been brought down (one at a time) and divided by the divisor. The process is continued until the remainder is less than the divisor.

Q47   Complete the following division problem:
$$\begin{array}{r} 2 \\ 7\overline{)1421} \\ 14 \\ \hline 2 \end{array}$$

STOP • STOP • STOP • STOP • STOP • STOP • STOP • STOP • STOP

A47
```
        203
    7)1421
      14
      ‾
       2
       0
       ‾
      21
      21
      ‾
       0
```

Q48   Show the complete solution to the following division problems:

    a. 79)81304   **b.** 43)43197

STOP • **STOP** • **STOP** • **STOP** • **STOP** • **STOP** • **STOP** • **STOP** • **STOP**

A48 **a.**
```
          1029                    1004
      79)81304              43)43197
         79                    43
         ‾                     ‾
         23                     1
          0                     0
         ‾                     ‾
        230                    19
        158                     0
        ‾‾‾                    ‾‾
        724                   197
        711                   172
        ‾‾‾                   ‾‾‾
         13                    25
```
b.

Q49   Determine the quotient and remainder:

    a. 73)22058   **b.** 36)36140

STOP • **STOP** • **STOP** • **STOP** • **STOP** • **STOP** • **STOP** • **STOP** • **STOP**

A49   **a.** 302 R12   **b.** 1,003 R32

Q50   On a recent trip Ralph drove 1,173 miles and used 58 gallons of gasoline. What was his average number of miles per gallon on the trip (disregard the remainder).

STOP • *STOP* • *STOP* • *STOP* • *STOP* • *STOP* • *STOP* • *STOP* • *STOP*

A50   20

---

20   As with addition and subtraction, multiplication and division errors can be eliminated if you are aware of them and practice to avoid them. Some specific errors concerning multiplication and division are:

1. Lack of knowledge of basic facts.
2. Incorrect use of partial products and quotients.
3. Digits not in proper columns.
4. Lack of knowledge of multiplication and division procedures.

This completes the instruction for this section.

1.3   Exercises

1. Determine the product or quotient:
   a. 0×8
   b. $8 \div 0$
   c. 230×0
   d. $0 \div 8$
   e. 1×13
   f. $7 \div 1$
   g. $12 \div 0$
   h. $0 \div 100$

2. Determine the product:
   a. 52×100
   b. $37 \times 1,000$
   c. 100×100
   d. 50×60
   e. 800×50
   f. $30 \times 5,000$

3. Determine the product:
   a. 245×2
   b. 385×5
   c. 533×7
   d. 232×38

   e. 765×58
   f. 693×443
   g. 344×655
   h. $40 \times 2,700$

   i. 452×607
   j. 3070×409
   k. $8,006 \times 12$
   l. 507×804

4. Determine the quotient and remainder:
   a. $65 \div 8$
   b. $7 \overline{)378}$
   c. $26 \overline{)414}$
   d. $88 \overline{)5621}$

   e. $681 \overline{)2532}$
   f. $45 \overline{)478}$
   g. $742 \overline{)4237}$
   h. $1,609 \div 8$

   i. $49 \overline{)44628}$
   j. $24,805 \div 82$
   k. $67 \overline{)13407}$
   l. $70 \overline{)5880}$

1.3   Exercise Answers

1. **a.** 0 **b.** impossible **c.** 0 **d.** 0
   e. 13 **f.** 7 **g.** impossible **h.** 0
2. **a.** 5,200 **b.** 37,000 **c.** 10,000 **d.** 3,000
   e. 40,000 **f.** 150,000
3. **a.** 490 **b.** 1,925 **c.** 3,731 **d.** 8,816
   e. 44,370 **f.** 306,999 **g.** 225,320 **h.** 108,000
   i. 274,364 **j.** 1,255,630 **k.** 96,072 **l.** 407,628
4. **a.** 8 R1 **b.** 54 **c.** 15 R24 **d.** 63 R77
   e. 3 R489 **f.** 10 R28 **g.** 5 R527 **h.** 201 R1
   i. 910 R38 **j.** 302 R41 **k.** 200 R7 **l.** 84

1.4   Order of Operations, Rounding Off, and Exponential Notation

1   It will often be necessary to be concerned with the order in which we perform the operations of addition, subtraction, multiplication, and division. For example, consider the problem

$$5 + 2 \times 3$$

Is the answer

$7 \times 3 = 21$ (adding first)   *or*
$5 + 6 = 11$ (multiplying first)?

Since any given problem can have only one (unique) solution, the problem must be performed following a prescribed *order of operations*. When only addition and subtraction are involved, they must be carried out in order from left to right. For example,

$$8 - 3 + 2 = 5 + 2$$
$$= 7$$

Q1   Determine the solution: $12 - 7 + 3 - 4$

STOP • STOP • STOP • STOP • STOP • STOP • STOP • STOP • STOP

A1   4: $12 - 7 + 3 - 4 = 5 + 3 - 4$
$$= 8 - 4$$
$$= 4$$

Q2   Determine the solution:
   a. $37 - 13 + 12 - 15 + 21$ **b.** $16 - 8 + 5 - 3$

STOP • STOP • STOP • STOP • STOP • STOP • STOP • STOP • STOP

A2   **a.** 42   **b.** 10

2
When only multiplication and division are involved, they must be carried out in order from left to right. Examples of this are:

$$12 \div 3 \times 2 = 4 \times 2 \qquad \text{(divide first)}$$
$$= 8 \qquad \text{(multiply)}$$

$$4 \times 8 \div 2 = 32 \div 2 \qquad \text{(multiply first)}$$
$$= 16 \qquad \text{(divide)}$$

Q3
Determine the solution: $12 \div 3 \times 4 \div 2$

STOP • STOP • STOP • STOP • STOP • STOP • STOP • STOP • STOP

A3
$8: \quad 12 \div 3 \times 4 \div 2 = 4 \times 4 \div 2$
$$= 16 \div 2$$
$$= 8$$

Q4
Determine the solution:
a. $2 \times 3 \div 6 \times 8$ **b.** $42 \div 2 \times 3$

STOP • STOP • STOP • STOP • STOP • STOP • STOP • STOP • STOP

A4
a. 8 **b.** 63

3
Symbols of grouping should always be evaluated first. Symbols of grouping are usually parentheses, ( ), or brackets, [ ]. Consider the following example:

$$24 \div (3 \times 4) \times (6 \div 2) = 24 \div 12 \times 3$$
$$= 2 \times 3$$
$$= 6$$

Notice that the work in the parentheses is completed first and then the problem is rewritten, discarding the parentheses.

Q5
Determine the solution:
a. $15 - (8 + 3)$ **b.** $96 \div (8 \times 6) \div 2$

STOP • STOP • STOP • STOP • STOP • STOP • STOP • STOP • STOP

A5
a. $4: \quad 15 - (8 + 3) = 15 - 11$ **b.** $1: \quad 96 \div (8 \times 6) \div 2 = 96 \div 48 \div 2$
$$= 4 \qquad\qquad\qquad\qquad\qquad\qquad = 2 \div 2$$
$$= 1$$

4
The *order of operations* can be stated:

Step 1: Evaluate within symbols of grouping first.

Step 2: Next, multiplication and division are performed from left to right.

Step 3: Last, addition and subtraction are performed from left to right.

Problems must be worked following the above sequence of steps. For example,

	Steps
$24 \div 6 - 2 \times 6 \div 3 = 4 - 2 \times 6 \div 3$	$(24 \div 6)$
$= 4 - 12 \div 3$	(2×6)
$= 4 - 4$	$(12 \div 3)$
$= 0$	$(4 - 4)$

Q6 Determine the solution: $18 - 12 \div (4 + 2)$

STOP • STOP • STOP • STOP • STOP • STOP • STOP • STOP • STOP

A6 16: $18 - 12 \div (4 + 2) = 18 - 12 \div 6$
$= 18 - 2$
$= 16$

Q7 Determine the solution:

a. $24 + 4 \div 2$ **b.** $16 - 4 \times 2$

STOP • STOP • STOP • STOP • STOP • STOP • STOP • STOP • STOP

A7 **a.** 26: $24 + 4 \div 2 = 24 + 2 = 26$ **b.** 8: $16 - 4 \times 2 = 16 - 8 = 8$

Q8 Determine the solution:

a. $12 + 24 \div (2 \times 3)$ **b.** $18 - (9 - 6) \times 2$

STOP • STOP • STOP • STOP • STOP • STOP • STOP • STOP • STOP

A8 **a.** 16 **b.** 12

5 *Rounding numbers* was discussed briefly in Section 1.3. Rounded numbers are used a great deal in everyday life. When the newspaper reports that 50,000 people attended a baseball game, you should realize that it represents a rounded number. The actual attendance may have been 48,373. In this section, numbers will be rounded to a particular place value.

Example 1: Round 283 to the hundreds place.

Solution

To round 283 to the hundreds place means to find if 283 is closer to 200 or to 300. If the number to the right of the place you are rounding to (8) is 5 or larger, the place you are rounding to is increased by 1, and all digits to the right are replaced by 0.

283 rounded to the nearest hundred is 300
↑
5 or larger

Example 2: Round 329 to the hundreds place.

Solution

If the number to the right of the place you are rounding to (2) is 4 or smaller, the place you are rounding to is left as is, and all digits to the right are replaced by zero.

329 rounded to the nearest hundred is 300
↑
4 or smaller

Q9   Round to the nearest hundreds place:

    a. 481 _____   **b.** 267 _____   **c.** 450 _____   **d.** 849 _____

STOP • STOP • STOP • STOP • STOP • STOP • STOP • STOP • STOP

A9   **a.** 500   **b.** 300   **c.** 500   **d.** 800

Q10   Round to the nearest tens place:

    a. 289 _____   **b.** 144 _____   **c.** 367 _____   **d.** 85 _____

STOP • STOP • STOP • STOP • STOP • STOP • STOP • STOP • STOP

A10   **a.** 290   **b.** 140   **c.** 370   **d.** 90

Q11   Round to the nearest thousands place:
    a. 29,850   **b.** 12,576   **c.** 63,359   **d.** 3,500

    _____   _____   _____   _____

STOP • STOP • STOP • STOP • STOP • STOP • STOP • STOP • STOP

A11   **a.** 30,000   **b.** 13,000   **c.** 63,000   **d.** 4,000

Q12   Round to the nearest hundreds place:
    a. 5,876   **b.** 204   **c.** 192   **d.** 3,959

    _____   _____   _____   _____

STOP • STOP • STOP • STOP • STOP • STOP • STOP • STOP • STOP

A12   **a.** 5,900   **b.** 200   **c.** 200   **d.** 4,000

---

6   *Exponential notation* can be used as a short form of writing certain products. The product $5 \times 5 \times 5$ can be written 5^3. In the expression 5^3, the 3 is called the *exponent*, the 5 is called the *base*, and the entire expression is called a *power*.

base $\rightarrow 5^{3 \leftarrow \text{exponent}}$

The exponent indicates how many times the base is a factor. That is, there are 3 factors of 5 in the expression 5^3.

$$5^3 = \underbrace{5 \times 5 \times 5}_{\text{3 factors}}$$

Powers of numbers are read as follows:

3^2 is read "three to the second power," usually read "three squared."

4^3 is read "four to the third power" usually read "four cubed."

5^4 is read "five to the fourth power."

Q13   Write the following showing all factors (do not evaluate):

   a. 2^3_____   **b.** 3^2_____   **c.** 4^5_____

STOP • STOP • STOP • STOP • STOP • STOP • STOP • STOP • STOP

A13   **a.** $2 \times 2 \times 2$ **b.** 3×3
   c. $4 \times 4 \times 4 \times 4 \times 4$

Q14   Write the following expressions using exponential notation:

   a. $4 \times 4 \times 4 =$ _____   **b.** $8 \times 8 =$ _____

   c. $2 \times 2 \times 2 \times 2 \times 2 =$ _____

STOP • STOP • STOP • STOP • STOP • STOP • STOP • STOP • STOP

A14   **a.** 4^3 **b.** 8^2 **c.** 2^5

Q15 9^2 is read "_____."

STOP • STOP • STOP • STOP • STOP • STOP • STOP • STOP • STOP

A15   nine to the second power, or nine squared

Q16   Evaluate the following:
   a. 2^3 **b.** 4^2 **c.** 5^1 **d.** 3^4

STOP • STOP • STOP • STOP • STOP • STOP • STOP • STOP • STOP

A16   **a.** 8: because $2^3 = 2 \times 2 \times 2$ **b.** 16: because $4^2 = 4 \times 4$
   c. 5: because $5^1 = 5$ (Notice that in the expression 5^1, the exponent 1 indicates that 5
   is a factor one time.)   **d.** 81: because $3^4 = 3 \times 3 \times 3 \times 3$

Q17   In the expression 2^5:

   a. 2 is called the _____.   **b.** 5 is called the _____.

   c. The entire expression is called a _____.

STOP • STOP • STOP • STOP • STOP • STOP • STOP • STOP • STOP

A17   **a.** base   **b.** exponent   **c.** power

Q18   Evaluate the following:
    a. 3^3 **b.** 7^3 **c.** 20^2 **d.** 1^5

STOP • **STOP** • **STOP** • **STOP** • **STOP** • **STOP** • **STOP** • **STOP** • **STOP**

A18   **a.** 27   **b.** 343   **c.** 400   **d.** 1

Q19   Evaluate the following:
    a. 5^3 **b.** 11^2 **c.** $2^2 \times 3^3 \times 5^2$ **d.** 1^{15}

STOP • **STOP** • **STOP** • **STOP** • **STOP** • **STOP** • **STOP** • **STOP** • **STOP**

A19   **a.** 125   **b.** 121

       c. 2,700: $2^2 \times 3^3 \times 5^2 = 2 \times 2 \times 3 \times 3 \times 3 \times 5 \times 5$ **d.** 1
$$= 4 \times 27 \times 25$$
$$= 2,700$$

The remaining material in this section is optional.

---

7   Once you have mastered the basics of arithmetic you can practice and have fun at the same time. The following is an example of crypto-arithmetic.

$$
\begin{array}{r}
XXXX \\
YYYY \\
+ZZZZ \\
\hline
YXXXZ
\end{array}
$$

You must put a number in for each letter. All identical letters must have the same number, but all different letters must be replaced by different numbers.

$$
\begin{array}{rl}
9999 & X = 9 \\
1111 & Y = 1 \\
\underline{8888} & Z = 8 \\
19998 &
\end{array}
$$

Q20   Use the rules of Frame 7 to decipher the following problems:

    a.
$$
\begin{array}{r}
PNX \\
NX \\
\hline
RNX \\
NXS \\
\hline
ZPNX
\end{array}
\quad \text{(multiply)}
$$

    b.
$$
\begin{array}{r}
HIL \\
IL{\overline{)PHIL}} \\
IL \\
\hline
TI \\
LS \\
\hline
HIL \\
HIL \\
\hline
\end{array}
\quad \text{(divide)}
$$

STOP • **STOP** • **STOP** • **STOP** • **STOP** • **STOP** • **STOP** • **STOP** • **STOP**

A20   **a.** X = 5, N = 2, P = 1, S = 0, R = 6, Z = 3
           b. P = 3, H = 1, I = 2, L = 5, T = 6, S = 0

Q21   A group of soldiers is lined up as follows:

           2 soldiers in front of a soldier
           2 soldiers behind a soldier
           1 soldier in the middle

           How many soldiers are there?

STOP • STOP • STOP • STOP • STOP • STOP • STOP • STOP • STOP

A21   3

Q22   Select any three-digit number; reverse the digits and subtract the smaller number from the larger. The middle digit of the difference will always be 9 and the sum of the outer digits will always be 9. Try it.

STOP • STOP • STOP • STOP • STOP • STOP • STOP • STOP • STOP

A22   *Examples:*   721   341
                          -127 -143
                          ------   ------
                            594   198

                        5 + 4 = 9   1 + 8 = 9

Q23   Multiply any number by 9; the digits of the product will always sum to 9. Try it.

STOP • STOP • STOP • STOP • STOP • STOP • STOP • STOP • STOP

A23   *Examples:* $9 \times 2 = 18$; $1 + 8 = 9$
                        $9 \times 8 = 72$; $7 + 2 = 9$
                        $9 \times 23 = 207$; $2 + 0 + 7 = 9$
                        $9 \times 31 = 279$; $2 + 7 + 9 = 18$, and $1 + 8 = 9$

Q24   One last problem: Three men entered a hotel and asked for a room. The clerk said there was only one room available but that he would put dividers in the room. The room would cost $30. Each man paid $10. After the clerk thought the situation over, he decided that he had overcharged the men and called the bellboy and instructed him: "Take this $5 back to the three men. Tell them I overcharged them and that they should divide this $5 among themselves." The bellboy thought that dividing $5 among three men would be fairly difficult and, being dishonest, kept $2 for himself. Then he returned $1 to each man, so, actually, the cost to each man was $9. Now, 3 times 9 is 27 and the $2 the bellboy kept make $29. What happened to the other dollar?

STOP • STOP • STOP • STOP • STOP • STOP • STOP • STOP • STOP

A24   The problem is wrongly stated. The men did not actually pay $27 for the room, but $25. If the problem is stated this way, the amount will total $30 ($25 + $3 rebate + $2 bellboy kept = $30).

This completes the instruction for this section.

1.4   Exercises

1. Evaluate the following:
   a. $16 - 8 + 5 - 3$ b. $24 - 8 + 5$
   c. $12 - (8 + 3) + 3$ d. $14 \times 2 \div 7$
   e. $16 \div 4 \times 2$ f. $24 \div 3 \times 4 \div 2$
   g. $18 + 18 \div 6 + 3$ h. $18 - (9 - 6) \times 2$
   i. $24 - 6 \times (3 + 1)$
2. Round to the tens place:
   a. 5,916   b. 2,834   c. 516
3. Round to the hundreds place:
   a. 524   b. 4,327   c. 6,879
4. Round to the thousands place:
   a. 13,872   b. 21,504   c. 68,327
5. Write 81 as a power of 3.
6. Write 32 as a power of 2.
7. Evaluate the following:
   a. 10^3 b. 6^2 c. 8^2 d. 8^3
   e. 6^3 f. $5^3 \times 7^3$ g. $2^4 \times 3^4 \times 4^2$

1.4   Exercise Answers

1. a. 10   b. 21   c. 4   d. 4   e. 8   f. 16   g. 24
   h. 12   i. 0
2. a. 5,920   b. 2,830   c. 520
3. a. 500   b. 4,300   c. 6,900
4. a. 14,000   b. 22,000   c. 68,000
5. 3^4
6. 2^5
7. a. 1,000   b. 36   c. 64   d. 512   e. 216   f. 42,875   g. 20,736

Chapter 1 Sample Test

At the completion of Chapter 1 it is expected that you will be able to work the following problems.

1.1   Place Value

1. Write in words:
   a. 21,734   b. 210,053

2. Write in decimal notation:
   a. six thousand, seven hundred thirty-four
   b. twenty thousand, eighty-three

1.2 Addition and Subtraction

3. Add:

   a. 278
       325
      +82

   b. 2304
      +8927

4. Subtract:

   a. 2345
      −586

   b. 506 − 279

1.3 Multiplication and Division

5. Multiply:

   a. 674
      ×35

   b. 6,017 × 207

6. Divide:

   a. 8)4695

   b. 19)39015

1.4 Order of Operations, Rounding Off, and Exponential Notation

7. Evaluate:
   a. 6 × 3 − 8 ÷ 2 b. 5 + (12 − 7) × 2
8. Round 27,357 to the nearest:
   a. thousand   b. hundred
9. a. Evaluate 3^4.
   b. Use exponential notation to write 5 × 5 × 5 × 5 × 5 × 5.
10. How far can a car travel on a tankful of gas if it averages 14 miles on 1 gallon and the tank holds 23 gallons?

Chapter 1 Sample Test Answers

1. a. twenty-one thousand, seven hundred thirty-four
   b. two hundred ten thousand, fifty-three
2. a. 6,734   b. 20,083
3. a. 685   b. 11,231
4. a. 1,759   b. 227
5. a. 23,590   b. 1,245,519
6. a. 586 R7   b. 2,053 R8
7. a. 14   b. 15
8. a. 27,000   b. 27,400
9. a. 81   b. 5^6
10. 322 miles

Chapter 2

Operations with Fractions

2.1   Equivalent Fractions

1	A *fraction* is a number that indicates that some whole has been divided into a number of equal parts, and that a portion of the equal parts are counted or represented. For example, the fraction $\frac{2}{3}$ indicates that a whole has been divided into 3 equal parts and that 2 are represented. The bottom number is called the *denominator* and the top number is called the *numerator*. For the fraction $\frac{2}{3}$, the denominator, 3, indicates that a whole is divided into 3 equal parts.

the shaded area represents $\frac{2}{3}$

The numerator, 2, represents 2 of the 3 equal parts.

Q1   Into how many equal parts has the rectangle been divided?

_____

STOP • **STOP** • **STOP** • **STOP** • **STOP** • **STOP** • **STOP** • **STOP** • **STOP**

A1   8

Q2   The number of equal parts into which a whole is divided is indicated by the numerator/denominator (circle the correct answer).

STOP • **STOP** • **STOP** • **STOP** • **STOP** • **STOP** • **STOP** • **STOP** • **STOP**

A2   denominator

Q3   The number of equal parts represented (shaded) in Q1 is _____.

STOP • **STOP** • **STOP** • **STOP** • **STOP** • **STOP** • **STOP** • **STOP** • **STOP**

A3   6

Q4   The number of equal parts represented is called the numerator/denominator (circle the correct answer).

STOP • **STOP** • **STOP** • **STOP** • **STOP** • **STOP** • **STOP** • **STOP** • **STOP**

44

A4	numerator

Q5	What portion of the rectangle (Q1) is represented by the shaded area?_____

STOP • **STOP** • **STOP** • **STOP** • **STOP** • **STOP** • **STOP** • **STOP** • **STOP**

A5	$\dfrac{6}{8}$

Q6	What fraction is represented by the shaded area?

a.  _____ b.  _____

STOP • **STOP** • **STOP** • **STOP** • **STOP** • **STOP** • **STOP** • **STOP** • **STOP**

A6	a. $\dfrac{4}{8}$	b. $\dfrac{8}{8}$

Q7	What fraction is represented by the shaded area?

a. 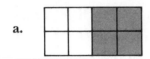 _____ b.  _____ c. 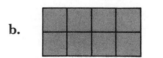 _____ d.  _____

STOP • **STOP** • **STOP** • **STOP** • **STOP** • **STOP** • **STOP** • **STOP** • **STOP**

A7	a. $\dfrac{1}{4}$	b. $\dfrac{1}{2}$	c. $\dfrac{2}{6}$	d. $\dfrac{2}{3}$

Q8	What fraction does the figure represent?_____

STOP • **STOP** • **STOP** • **STOP** • **STOP** • **STOP** • **STOP** • **STOP** • **STOP**

A8	$\dfrac{1}{7}$: because the line segment is divided into 7 equal parts (denominator) and 1 equal part is indicated or counted (numerator)

Q9	Shade the circle so that the fraction $\dfrac{5}{8}$ is indicated.

STOP • **STOP** • **STOP** • **STOP** • **STOP** • **STOP** • **STOP** • **STOP** • **STOP**

A9	or any combination of 5 of the 8 parts

2   The same area may be represented by different fractions.

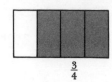   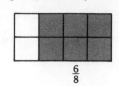

$$\tfrac{3}{4} \qquad\qquad \tfrac{6}{8}$$

The area on the left is represented by the fraction $\frac{3}{4}$; the area on the right is represented by $\frac{6}{8}$. However, both fractions represent the same part of the whole. Therefore, $\frac{3}{4} = \frac{6}{8}$ and they are called *equivalent fractions*.

Q10   What two equivalent fractions can be used to name the shaded area?_____

STOP • **STOP** • **STOP** • **STOP** • **STOP** • **STOP** • **STOP** • **STOP** • **STOP**

A10 $\dfrac{3}{9}, \dfrac{1}{3}$

3   Equivalent fractions can be obtained by multiplying the numerator and denominator of a fraction by the same number (not zero). For example,

$$\frac{3}{4} = \frac{3 \times 2}{4 \times 2} = \frac{6}{8} \qquad \frac{3}{4} = \frac{3 \times 3}{4 \times 3} = \frac{9}{12} \qquad \frac{3}{4} = \frac{3 \times 5}{4 \times 5} = \frac{15}{20}$$

The multiplication of the numerator and the denominator by the same number is usually done mentally. Thus,

$$\frac{2}{3} = \frac{8}{12}$$

multiplying the numerator and denominator of $\frac{2}{3}$ by 4.

Q11   Change $\dfrac{5}{7}$ to an equivalent fraction by multiplying numerator and denominator by 6.

STOP • **STOP** • **STOP** • **STOP** • **STOP** • **STOP** • **STOP** • **STOP** • **STOP**

A11 $\dfrac{30}{42}$: because $\dfrac{5 \times 6}{7 \times 6} = \dfrac{30}{42}$

Q12   Change $\frac{2}{3}$ to equivalent fractions by multiplying numerator and denominator by 2, 3, 4, and 5, respectively.

STOP   •   STOP   •   STOP   •   STOP   •   STOP   •   STOP   •   STOP   •   STOP   •   STOP

A12 $\frac{4}{6}, \frac{6}{9}, \frac{8}{12}, \frac{10}{15}$

---

4   To change $\frac{3}{4}$ to eighths, think

$$\frac{3}{4} = \frac{?}{8}$$

If the fraction with denominator 8 is to be equivalent to $\frac{3}{4}$, the numerator and denominator of $\frac{3}{4}$ must be multiplied by the same number. Since $8 \div 4 = 2$, the denominator, 4, was multiplied by 2. Likewise, multiply the numerator, 3, by 2, obtaining $\frac{3 \times 2}{4 \times 2} = \frac{6}{8}$.

Q13   What was the denominator multiplied by? $\frac{3}{4} = \frac{}{20}$

STOP   •   STOP   •   STOP   •   STOP   •   STOP   •   STOP   •   STOP   •   STOP   •   STOP

A13   5: because $20 \div 4 = 5$

Q14   Find the missing value: $\frac{3}{4} = \frac{}{20}$

STOP   •   STOP   •   STOP   •   STOP   •   STOP   •   STOP   •   STOP   •   STOP   •   STOP

A14   15: because $\frac{3 \times 5}{4 \times 5} = \frac{15}{20}$

Q15   Find the missing value: $\frac{5}{7} = \frac{}{21}$

STOP   •   STOP   •   STOP   •   STOP   •   STOP   •   STOP   •   STOP   •   STOP   •   STOP

A15   15: because $\frac{5 \times 3}{7 \times 3} = \frac{15}{21}$

Q16   Find the missing value: $\frac{8}{15} = \frac{}{60}$

STOP   •   STOP   •   STOP   •   STOP   •   STOP   •   STOP   •   STOP   •   STOP   •   STOP

A16   32: because $\frac{8 \times 4}{15 \times 4} = \frac{32}{60}$

Q17   Find the missing value:

   a. $\dfrac{2}{3} = \dfrac{}{24}$ **b.** $\dfrac{6}{11} = \dfrac{}{44}$ **c.** $\dfrac{3}{4} = \dfrac{}{28}$ **d.** $\dfrac{5}{12} = \dfrac{}{60}$

STOP • STOP • STOP • STOP • STOP • STOP • STOP • STOP • STOP

A17   **a.** 16   **b.** 24   **c.** 21   **d.** 25

---

5   It is sometimes necessary to represent a whole number as a fraction. First consider that $8 \div 2$ can be written as $2\overline{)8}$ or $\dfrac{8}{2}$. All these mean 8 divided by 2. To find the number of 20th's in 5,

$$5 = \dfrac{?}{20}$$

write 5 as $\dfrac{5^{*}}{1}$; hence,

$$\dfrac{5}{1} = \dfrac{?}{20}$$

Now, $1 \times 20 = 20$, so the numerator must also be multiplied by 20.

$$\dfrac{5 \times 20}{1 \times 20} = \dfrac{100}{20} \qquad \text{or} \qquad 5 = \dfrac{100}{20}$$

*Recall that (any number) $\div$ 1 = that number.

Q18   Find the missing value: $\dfrac{3}{1} = \dfrac{}{15}$

STOP • STOP • STOP • STOP • STOP • STOP • STOP • STOP • STOP

A18   45: because $\dfrac{3 \times 15}{1 \times 15} = \dfrac{45}{15}$

Q19   Find the missing value: $4 = \dfrac{}{8}$

STOP • STOP • STOP • STOP • STOP • STOP • STOP • STOP • STOP

A19   32: because $\dfrac{4 \times 8}{1 \times 8} = \dfrac{32}{8}$

Q20   Find the missing value:

   a. $5 = \dfrac{}{3}$ **b.** $8 = \dfrac{}{4}$

STOP • STOP • STOP • STOP • STOP • STOP • STOP • STOP • STOP

A20   **a.** 15   **b.** 32

---

6   The equivalent fractions obtained have all been of higher terms (numerator and denominator) than the original fraction. It is often necessary to find an equivalent fraction, but of lower terms. This will be called *reducing* the fraction to a lower term. An equivalent fraction is obtained if the numerator and the denominator of a fraction are divided by the same number (not zero):

Example 1: $\dfrac{4 \div 2}{8 \div 2} = \dfrac{2}{4}$;   hence, $\dfrac{4}{8} = \dfrac{2}{4}$

Example 2: Reduce $\dfrac{5}{10}$ by dividing numerator and denominator by 5.

Solution

$$\frac{5}{10} = \frac{5 \div 5}{10 \div 5} = \frac{1}{2}$$

Q21   Determine the equivalent fraction obtained by dividing the numerator and denominator of $\dfrac{6}{8}$ by 2.

STOP • STOP • STOP • STOP • STOP • STOP • STOP • STOP • STOP

A21 $\dfrac{3}{4}$: because $\dfrac{6 \div 2}{8 \div 2} = \dfrac{3}{4}$

Q22   Determine equivalent fractions, reducing by the common factor 3:

　　a. $\dfrac{3}{9}$　　　　　　b. $\dfrac{15}{30}$

STOP • STOP • STOP • STOP • STOP • STOP • STOP • STOP • STOP

A22   a. $\dfrac{1}{3}$: because $\dfrac{3 \div 3}{9 \div 3} = \dfrac{1}{3}$　　　　　b. $\dfrac{5}{10}$: because $\dfrac{15 \div 3}{30 \div 3} = \dfrac{5}{10}$

7   A *common factor* (divisor) of two numbers is a number that will divide both numbers evenly (remainder zero). A common factor of 9 and 12 is 3, because 3 divides both 9 and 12 evenly. A fraction is considered reduced to *lowest terms* if the numerator and denominator have no common factors (divisors). As an example, $\dfrac{5}{8}$ is reduced to lowest terms, because there are no common factors of 5 and 8.

Q23   Which fractions cannot be reduced (the numerator and denominator have no common factors)?

　　$\dfrac{21}{29}, \dfrac{15}{28}, \dfrac{12}{27}$ _____

STOP • STOP • STOP • STOP • STOP • STOP • STOP • STOP • STOP

A23 $\dfrac{21}{29}$ and $\dfrac{15}{28}$: $\dfrac{12}{27}$ could be reduced, because 12 and 27 have a common factor of 3.

Q24   Reduce $\dfrac{12}{27}$.

STOP • STOP • STOP • STOP • STOP • STOP • STOP • STOP • STOP

A24 $\dfrac{4}{9}$: because $\dfrac{12 \div 3}{27 \div 3} = \dfrac{4}{9}$

8   Often the terms of a fraction have many common factors. For example, $\frac{12}{18}$ is reduced as follows:

$$\frac{12 \div 2}{18 \div 2} = \frac{6}{9}$$

However, 6 and 9 have a common factor of 3. Thus, $\frac{6 \div 3}{9 \div 3} = \frac{2}{3}$. $\frac{2}{3}$ is reduced to lowest terms, because 2 and 3 have no common factors. Notice that $\frac{12}{18}$ could have been reduced to lowest terms by first reducing by the common factor 6:

$$\frac{12}{18} = \frac{12 \div 6}{18 \div 6} = \frac{2}{3}$$

Q25   Reduce $\frac{24}{36}$ to lowest terms.

STOP • STOP • STOP • STOP • STOP • STOP • STOP • STOP • STOP

A25 $\frac{2}{3}$: because $\frac{24 \div 12}{36 \div 12} = \frac{2}{3}$ or $\frac{24 \div 2}{36 \div 2} = \frac{12}{18}$

$$\text{(reducing by 2)} = \frac{6}{9}$$

$$\text{(reducing by 3)} = \frac{2}{3}$$

9   An aid to reducing fractions is to recall some simple *divisibility tests*. That is:

1. A number is divisible by 2 if the last digit is divisible by 2 (that is, if the digit is 0, 2, 4, 6, or 8).
Example: The following numbers are divisible by 2 because the last digit is divisible by 2:

14   38   54   1,026

2. A number is divisible by 3 if the sum of the digits is divisible by 3.
Example: 24 is divisible by 3 because the sum of the digits $(2 + 4 = 6)$ is divisible by 3. 54 is divisible by 3 because 9 $(5 + 4)$ is divisible by 3. The following numbers are divisible by 3.

15   126   102   390

3. A number is divisible by 5 if the last digit is 5 or 0.
Example: The following numbers are divisible by 5:

85   40   1,080   215

Q26   Show that the following numbers are divisible by 2: 14, 38, 54, 1,026.

STOP • STOP • STOP • STOP • STOP • STOP • STOP • STOP • STOP

A26   The last digit is divisible by 2. *Or:*

$14 \div 2 = 7$ $38 \div 2 = 19$
$54 \div 2 = 27$ $1,026 \div 2 = 513$

Q27   Show that the following numbers are divisible by 3: 15, 126, 102, 390.

STOP • STOP • STOP • STOP • STOP • STOP • STOP • STOP • STOP

A27   The sum of the digits is divisible by 3. *Or:*

$$15 \div 3 = 5 \qquad 126 \div 3 = 42$$
$$102 \div 3 = 34 \qquad 390 \div 3 = 130$$

Q28   Show that the following numbers are divisible by 5: 85, 40, 1,080, 215.

STOP • STOP • STOP • STOP • STOP • STOP • STOP • STOP • STOP

A28   The last digit is either 5 or 0. *Or:*

$$85 \div 5 = 17 \qquad 40 \div 5 = 8$$
$$1,080 \div 5 = 216 \qquad 215 \div 5 = 43$$

Q29   Reduce the fractions to lowest terms:

   a. $\dfrac{21}{42}$ b. $\dfrac{15}{39}$

STOP • STOP • STOP • STOP • STOP • STOP • STOP • STOP • STOP

A29   a. $\dfrac{1}{2}$ b. $\dfrac{5}{13}$

Q30   Reduce to lowest terms:

   a. $\dfrac{24}{36}$ b. $\dfrac{20}{30}$

STOP • STOP • STOP • STOP • STOP • STOP • STOP • STOP • STOP

A30   a. $\dfrac{2}{3}$ b. $\dfrac{2}{3}$

Q31   Reduce to lowest terms:

   a. $\dfrac{21}{28}$ b. $\dfrac{36}{40}$ c. $\dfrac{15}{25}$

STOP • STOP • STOP • STOP • STOP • STOP • STOP • STOP • STOP

A31   a. $\dfrac{3}{4}$ b. $\dfrac{9}{10}$ c. $\dfrac{3}{5}$

Q32   Reduce to lowest terms:

   a. $\dfrac{35}{49}$ b. $\dfrac{30}{45}$ c. $\dfrac{27}{36}$ d. $\dfrac{33}{54}$

STOP • STOP • STOP • STOP • STOP • STOP • STOP • STOP • STOP

A32   **a.** $\dfrac{5}{7}$ **b.** $\dfrac{2}{3}$ **c.** $\dfrac{3}{4}$ **d.** $\dfrac{11}{18}$

This completes the instruction for this section.

2.1   Exercises

1. What fraction is indicated by the shaded area?

a.
b.
c.
d.

2. Find the missing value:

a. $\dfrac{3}{4} = \dfrac{?}{12}$ **b.** $\dfrac{1}{3} = \dfrac{?}{9}$ **c.** $\dfrac{7}{8} = \dfrac{?}{24}$ **d.** $\dfrac{6}{7} = \dfrac{?}{42}$

e. $\dfrac{3}{8} = \dfrac{?}{16}$ **f.** $\dfrac{7}{9} = \dfrac{56}{?}$ **g.** $\dfrac{1}{2} = \dfrac{4}{?}$ **h.** $5 = \dfrac{?}{15}$

i. $8 = \dfrac{?}{9}$

3. Reduce to lowest terms:

a. $\dfrac{6}{8}$ **b.** $\dfrac{8}{12}$ **c.** $\dfrac{6}{9}$ **d.** $\dfrac{12}{16}$ **e.** $\dfrac{5}{6}$ **f.** $\dfrac{56}{63}$ **g.** $\dfrac{15}{18}$

h. $\dfrac{26}{39}$ **i.** $\dfrac{43}{86}$ **j.** $\dfrac{24}{28}$

2.1   Exercise Answers

1. a. $\dfrac{1}{4}$ **b.** $\dfrac{3}{8}$ **c.** $\dfrac{4}{4}$ or 1   **d.** $\dfrac{0}{4}$ or 0

2. a. 9   **b.** 3   **c.** 21   **d.** 36   **e.** 6   **f.** 72   **g.** 8
     h. 75   **i.** 72

3. a. $\dfrac{3}{4}$ **b.** $\dfrac{2}{3}$ **c.** $\dfrac{2}{3}$ **d.** $\dfrac{3}{4}$ **e.** $\dfrac{5}{6}$ **f.** $\dfrac{8}{9}$ **g.** $\dfrac{5}{6}$

     h. $\dfrac{2}{3}$ **i.** $\dfrac{1}{2}$ **j.** $\dfrac{6}{7}$

2.2   Mixed Numbers and Improper Fractions

---

1   A fraction can be defined as the quotient of two whole numbers (the denominator not zero). If the fraction is less than 1, it is called a *proper fraction*. That is, the numerator is less than the denominator. Examples of proper fractions are:

$$\frac{1}{2} \qquad \frac{1}{3} \qquad \frac{5}{8}$$

Fractions greater than or equal to 1 are called *improper fractions*. That is, the numerator is equal to or greater than the denominator. The following represents the improper fraction $\frac{4}{3}$:

 $+$ $= \frac{4}{3}$

If the numerator and denominator of a fraction are equal, the fraction is equal to 1. Examples are:

$$\frac{3}{3} = 1 \qquad \frac{8}{8} = 1 \qquad \frac{50}{50} = 1$$

Q1   **a.** A fraction greater than 1 is called a (an) _____ fraction.

 b. A fraction less than 1 is called a (an) _____ fraction.

 c. If the numerator is larger than the denominator, the fraction is a (an) _____ fraction.

 d. If the numerator is equal to the denominator, the fraction is equal to _____.

STOP • STOP • STOP • STOP • STOP • STOP • STOP • STOP • STOP

A1   **a.** improper   **b.** proper   **c.** improper   **d.** one

---

Q2   What improper fraction is represented by the following?

 $+$ 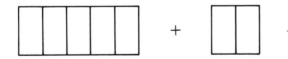 _____

STOP • STOP • STOP • STOP • STOP • STOP • STOP • STOP • STOP

A2 $\frac{6}{4}$

Q3   What improper fraction is represented by the following?

$+$ _____

STOP • STOP • STOP • STOP • STOP • STOP • STOP • STOP • STOP

A3 $\frac{7}{5}$

Q4   Which of the following are improper fractions?

   a. $\frac{3}{4}$ **b.** $\frac{7}{2}$ **c.** $\frac{5}{5}$ **d.** $\frac{8}{7}$ _____

STOP • STOP • STOP • STOP • STOP • STOP • STOP • STOP • STOP

A4   b, c, and d: because the numerator is larger than or equal to the denominator.

---

2   A *mixed number* is a whole number and a proper fraction. The following illustration represents the improper fraction $\frac{4}{3}$ and the mixed number $1\frac{1}{3}$.

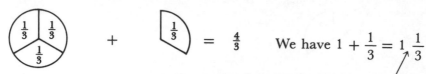

The plus sign, $+$, is usually omitted between the whole number and the fraction.

Q5   What mixed number is represented by this illustration?

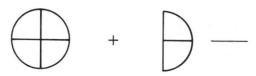   + _____

STOP • STOP • STOP • STOP • STOP • STOP • STOP • STOP • STOP

A5 $1\frac{2}{4}$ or $1\frac{1}{2}$

---

3   An improper fraction can be written as a mixed number by writing the improper fraction as the sum of a whole number and a common fraction. *Examples:*

1. $\dfrac{7}{4} = \dfrac{1}{4} + \dfrac{1}{4} + \dfrac{1}{4} + \dfrac{1}{4} + \dfrac{1}{4} + \dfrac{1}{4} + \dfrac{1}{4}$ 2. $\dfrac{6}{5} = \dfrac{5}{5} + \dfrac{1}{5}$ 3. $\dfrac{9}{4} = \dfrac{4}{4} + \dfrac{4}{4} + \dfrac{1}{4}$

      $= \dfrac{4}{4} + \dfrac{3}{4}$ $= 1 + \dfrac{1}{5}$ $= 1 + 1 + \dfrac{1}{4}$

      $= 1\dfrac{3}{4}$ $= 1\dfrac{1}{5}$ $= 2\dfrac{1}{4}$

Q6   Write the fraction $\frac{5}{4}$ as a mixed number by completing the following steps:

$$\frac{5}{4} = \frac{4}{4} + \frac{1}{4}$$

$$= \text{\_\_\_\_} + \text{\_\_\_\_}$$

$$= \text{\_\_\_\_}$$

STOP • STOP • STOP • STOP • STOP • STOP • STOP • STOP • STOP

A6 $= 1 + \dfrac{1}{4}$

       $= 1\dfrac{1}{4}$

Q7   Write $\frac{7}{5}$ as a mixed number.

STOP • STOP • STOP • STOP • STOP • STOP • STOP • STOP • STOP

A7 $\qquad 1\frac{2}{5}: \quad \frac{7}{5} = \frac{5}{5} + \frac{2}{5}$

$\qquad\qquad\qquad = 1\frac{2}{5}$

Q8 $\qquad$ Write $\dfrac{11}{4}$ as a mixed number.

STOP • **STOP** • **STOP** • **STOP** • **STOP** • **STOP** • **STOP** • **STOP** • **STOP**

A8 $\qquad 2\frac{3}{4}: \quad \frac{11}{4} = \frac{4}{4} + \frac{4}{4} + \frac{3}{4}$

$\qquad\qquad\qquad = 2\frac{3}{4}$

Q9 $\qquad$ Write as mixed numbers:

$\qquad$ **a.** $\dfrac{8}{3}$ $\qquad$ **b.** $\dfrac{5}{3}$ $\qquad$ **c.** $\dfrac{18}{7}$ $\qquad$ **d.** $\dfrac{9}{2}$

STOP • **STOP** • **STOP** • **STOP** • **STOP** • **STOP** • **STOP** • **STOP** • **STOP**

A9 $\qquad$ **a.** $2\frac{2}{3}$ $\quad$ **b.** $1\frac{2}{3}$ $\quad$ **c.** $2\frac{4}{7}$ $\quad$ **d.** $4\frac{1}{2}$

---

4 $\qquad$ An improper fraction can also be changed to a mixed number by dividing the numerator of the fraction by the denominator. The remainder (if any) is written over the denominator and added to the quotient. Two examples are:

$$\frac{5}{3} = 3\overline{)5} = 1\frac{2}{3} \qquad \frac{16}{5} = 5\overline{)16} = 3\frac{1}{5}$$

$$\frac{5}{3} = \begin{array}{r} 1 \\ 3\overline{)5} \\ \underline{3} \\ 2 \end{array} = 1\frac{2}{3} \qquad \frac{16}{5} = \begin{array}{r} 3 \\ 5\overline{)16} \\ \underline{15} \\ 1 \end{array} = 3\frac{1}{5}$$

The fractional part of the mixed number should be reduced, if possible. For example,

$$\frac{10}{4} = \begin{array}{r} 2 \\ 4\overline{)10} \\ \underline{8} \\ 2 \end{array} = 2\frac{2}{4} = 2\frac{1}{2}$$

Q10 $\qquad$ Change $\dfrac{15}{4}$ to a mixed number.

STOP • **STOP** • **STOP** • **STOP** • **STOP** • **STOP** • **STOP** • **STOP** • **STOP**

A10 $\quad 3\dfrac{3}{4}: \quad \dfrac{15}{4} = 4\overline{)15}\,^{3} = 3\dfrac{3}{4}$
$$\underline{12}$$
$$3$$

Q11 $\quad$ Change $\dfrac{42}{12}$ to a mixed number.

STOP • STOP • STOP • STOP • STOP • STOP • STOP • STOP • STOP

A11 $\quad 3\dfrac{1}{2}: \quad \dfrac{42}{12} = 12\overline{)42}\,^{3} = 3\dfrac{6}{12} = 3\dfrac{1}{2}$
$$\underline{36}$$
$$6$$

Q12 $\quad$ Change to mixed numbers:

$\quad$ **a.** $\dfrac{33}{2}$ $\qquad\qquad$ **b.** $\dfrac{73}{7}$ $\qquad\qquad$ **c.** $\dfrac{18}{8}$

STOP • STOP • STOP • STOP • STOP • STOP • STOP • STOP • STOP

A12 $\quad$ **a.** $16\dfrac{1}{2}$ $\quad$ **b.** $10\dfrac{3}{7}$ $\quad$ **c.** $2\dfrac{1}{4}$

Q13 $\quad$ Change to mixed or whole numbers:

$\quad$ **a.** $\dfrac{44}{4}$ $\qquad\qquad$ **b.** $\dfrac{17}{3}$ $\qquad\qquad$ **c.** $176 \div 24$

STOP • STOP • STOP • STOP • STOP • STOP • STOP • STOP • STOP

A13 $\quad$ **a.** 11 $\quad$ **b.** $5\dfrac{2}{3}$ $\quad$ **c.** $7\dfrac{1}{3}$

---

5 $\quad$ A mixed number can be changed to an improper fraction by writing the whole number as a fraction. For example, $1\dfrac{2}{3}$ is changed to an improper fraction by writing the 1 as 3 thirds and adding the $\dfrac{2}{3}$. That is,

$$1\dfrac{2}{3} = \dfrac{3}{3} + \dfrac{2}{3}$$
$$= \dfrac{5}{3}$$

Additional examples are:

$$3\frac{2}{3} = \frac{9}{3} + \frac{2}{3} \qquad 4\frac{1}{2} = \frac{8}{2} + \frac{1}{2}$$

$$= \frac{11}{3} \qquad\qquad = \frac{9}{2}$$

Q14   Change $2\frac{2}{3}$ to an improper fraction by completing the steps:

$$2\frac{2}{3} = \frac{6}{3} + \frac{2}{3}$$

$$= \underline{\hspace{1cm}}$$

STOP • **STOP** • **STOP** • **STOP** • **STOP** • **STOP** • **STOP** • **STOP** • **STOP**

A14 $\frac{8}{3}$

Q15   Change $1\frac{4}{5}$ to an improper fraction.

STOP • **STOP** • **STOP** • **STOP** • **STOP** • **STOP** • **STOP** • **STOP** • **STOP**

A15 $\frac{9}{5}$: $1\frac{4}{5} = \frac{5}{5} + \frac{4}{5}$

$$= \frac{9}{5}$$

Q16   Change to an improper fraction:

    a. $1\frac{3}{8}$ **b.** $1\frac{1}{7}$

STOP • **STOP** • **STOP** • **STOP** • **STOP** • **STOP** • **STOP** • **STOP** • **STOP**

A16   **a.** $\frac{11}{8}$: $1\frac{3}{8} = \frac{8}{8} + \frac{3}{8}$ **b.** $\frac{8}{7}$: $1\frac{1}{7} = \frac{7}{7} + \frac{1}{7}$

$$= \frac{11}{8} \qquad\qquad\qquad\qquad = \frac{8}{7}$$

---

6   Another method for changing a mixed number to an improper fraction is to multiply the whole number by the denominator and add the product to the existing numerator. This result is placed over the original denominator. Two examples are:

$$2\frac{3}{4} = \frac{2 \times 4 + 3}{4} \qquad\qquad 1\frac{3}{7} = \frac{1 \times 7 + 3}{7}$$

$$= \frac{11}{4} \qquad\qquad\qquad = \frac{10}{7}$$

Q17 Change $4\frac{2}{5}$ to an improper fraction by completing the following steps:

$$4\frac{2}{5} = \frac{4 \times 5 + 2}{5}$$

$$= \underline{}$$

STOP • STOP • STOP • STOP • STOP • STOP • STOP • STOP • STOP

A17 $\frac{22}{5}$

Q18 Change $5\frac{2}{3}$ to an improper fraction.

STOP • STOP • STOP • STOP • STOP • STOP • STOP • STOP • STOP

A18 $\frac{17}{3}$: because $5\frac{2}{3} = \frac{5 \times 3 + 2}{3}$

$$= \frac{17}{3}$$

Q19 Change to an improper fraction:

    a. $5\frac{3}{4}$ **b.** $4\frac{7}{8}$

STOP • STOP • STOP • STOP • STOP • STOP • STOP • STOP • STOP

A19   **a.** $\frac{23}{4}$: because $5\frac{3}{4} = \frac{5 \times 4 + 3}{4}$ **b.** $\frac{39}{8}$: because $4\frac{7}{8} = \frac{4 \times 8 + 7}{8}$

$$= \frac{23}{4} \hspace{6cm} = \frac{39}{8}$$

Q20 Change to an improper fraction:

    a. $4\frac{2}{3}$ **b.** $8\frac{7}{8}$ **c.** 8   **d.** $4\frac{3}{10}$

STOP • STOP • STOP • STOP • STOP • STOP • STOP • STOP • STOP

A20   **a.** $\frac{14}{3}$ **b.** $\frac{71}{8}$ **c.** $\frac{8}{1}$ **d.** $\frac{43}{10}$

This completes the instruction for this section.

2.2   Exercises

1. **a.** A fraction in which the numerator is equal to or larger than the denominator is called a (an) _____ fraction.
   b. A fraction less than 1 is called a (an) _____ fraction.
   c. A whole number and a fraction is called a _____ number.

2. Change to a mixed or whole number:

   a. $\dfrac{28}{6}$ **b.** $\dfrac{52}{8}$ **c.** $\dfrac{48}{9}$ **d.** $\dfrac{81}{9}$

   e. $222 \div 27$ **f.** $289 \div 51$ **g.** $\dfrac{102}{13}$ **h.** $\dfrac{13,218}{105}$

3. Change to an improper fraction:

   a. $2\dfrac{1}{3}$ **b.** $9\dfrac{2}{3}$ **c.** $6\dfrac{5}{12}$ **d.** $8\dfrac{9}{10}$

   e. $5\dfrac{7}{8}$ **f.** $10\dfrac{10}{11}$ **g.** $250\dfrac{7}{8}$ **h.** $42\dfrac{17}{19}$

2.2   Exercise Answers

1. **a.** improper   **b.** proper   **c.** mixed

2. **a.** $4\dfrac{2}{3}$ **b.** $6\dfrac{1}{2}$ **c.** $5\dfrac{1}{3}$ **d.** 9 **e.** $8\dfrac{2}{9}$ **f.** $5\dfrac{2}{3}$ **g.** $7\dfrac{11}{13}$

   h. $125\dfrac{31}{35}$

3. **a.** $\dfrac{7}{3}$ **b.** $\dfrac{29}{3}$ **c.** $\dfrac{77}{12}$ **d.** $\dfrac{89}{10}$ **e.** $\dfrac{47}{8}$ **f.** $\dfrac{120}{11}$ **g.** $\dfrac{2,007}{8}$

   h. $\dfrac{815}{19}$

2.3   Addition

Addition of fractions will be easier if you remember that the rules of addition for whole numbers also apply to fractions.

---

1 When adding whole numbers it was only possible to add numbers that represented the same unit of measure (place value). That is, tens were added to tens, hundreds added to hundreds, and so on. Similarily, when adding fractions it will be possible to add only those fractions expressed as the same unit of measure. Thirds must be added to thirds, fourths to fourths, and so on. Consider the following illustration:

Here 2 equal parts of seven are added to 3 equal parts of seven for a sum of 5 equal parts of seven:

$$\frac{2}{7} + \frac{3}{7} = \frac{5}{7}$$

Two fractions can be combined into a single fraction by addition only if they have the same denominator.

Q1   Find the sum of the fractions represented:

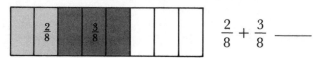 $\frac{2}{8} + \frac{3}{8}$ ——

STOP   •   STOP   •   STOP   •   STOP   •   STOP   •   STOP   •   STOP   •   STOP   •   STOP

A1 $\dfrac{5}{8}$

2   Notice that the sum of two fractions that have the same denominator can be expressed as a single fraction by writing the sum of the numerators over the common denominator:

$$\frac{2}{7} + \frac{3}{7} = \frac{2+3}{7} = \frac{5}{7}$$

$$\frac{2}{8} + \frac{3}{8} = \frac{2+3}{8} = \frac{5}{8}$$

When possible the sum should be reduced to lowest terms:

$$\frac{3}{12} + \frac{7}{12} = \frac{10}{12} = \frac{5}{6}$$

$$\frac{1}{8} + \frac{3}{8} = \frac{4}{8} = \frac{1}{2}$$

If the sum is an improper fraction, it should be expressed as a mixed number:

$$\frac{2}{5} + \frac{4}{5} = \frac{6}{5} = 1\frac{1}{5}$$

$$\frac{5}{6} + \frac{1}{6} = \frac{6}{6} = 1$$

Q2   Determine the sum: $\dfrac{1}{5} + \dfrac{3}{5}$

STOP   •   STOP   •   STOP   •   STOP   •   STOP   •   STOP   •   STOP   •   STOP   •   STOP

A2 $\dfrac{4}{5}$: $\dfrac{1}{5} + \dfrac{3}{5} = \dfrac{1+3}{5} = \dfrac{4}{5}$

Q3   Determine the sum: $\dfrac{1}{6} + \dfrac{3}{6}$

STOP   •   STOP   •   STOP   •   STOP   •   STOP   •   STOP   •   STOP   •   STOP   •   STOP

A3 $\quad \dfrac{2}{3}: \quad \dfrac{1}{6} + \dfrac{3}{6} = \dfrac{4}{6} = \dfrac{2}{3}$

Q4 $\qquad$ Determine the sum: $\dfrac{7}{3} + \dfrac{4}{3}$

STOP • STOP • STOP • STOP • STOP • STOP • STOP • STOP • STOP

A4 $\quad 3\dfrac{2}{3}: \quad \dfrac{7}{3} + \dfrac{4}{3} = \dfrac{11}{3} = 3\dfrac{2}{3}$

Q5 $\qquad$ Determine the sum:

$\qquad$ **a.** $\dfrac{5}{7} + \dfrac{6}{7}$ $\qquad$ **b.** $\dfrac{1}{15} + \dfrac{9}{15}$ $\qquad$ **c.** $\dfrac{11}{15} + \dfrac{9}{15}$

STOP • STOP • STOP • STOP • STOP • STOP • STOP • STOP • STOP

A5 $\qquad$ **a.** $1\dfrac{4}{7}$ $\qquad$ **b.** $\dfrac{2}{3}$ $\qquad$ **c.** $1\dfrac{1}{3}$

Q6 $\qquad$ Determine the sum:

$\qquad$ **a.** $\dfrac{3}{8} + \dfrac{5}{8}$ $\qquad$ **b.** $\dfrac{5}{11} + \dfrac{4}{11}$ $\qquad$ **c.** $\dfrac{5}{48} + \dfrac{11}{48}$

STOP • STOP • STOP • STOP • STOP • STOP • STOP • STOP • STOP

A6 $\qquad$ **a.** 1 $\qquad$ **b.** $\dfrac{9}{11}$ $\qquad$ **c.** $\dfrac{1}{3}$

---

3 $\qquad$ In an addition involving mixed numbers, it is usually better to add the whole numbers and fractions separately.

$\qquad$ *Examples:*

$\qquad$ **1.** $\quad 3\dfrac{2}{7}$ $\qquad$ **2.** $\quad 5\dfrac{7}{12}$

$\qquad\qquad \dfrac{+1\dfrac{4}{7}}{4\dfrac{6}{7}}$ $\qquad\qquad \dfrac{+3\dfrac{2}{12}}{8\dfrac{9}{12}} = 8\dfrac{3}{4}$

Q7 $\qquad$ Determine the sum: $\quad 3\dfrac{1}{3}$

$\qquad\qquad\qquad\qquad\qquad\qquad +2\dfrac{1}{3}$

STOP • STOP • STOP • STOP • STOP • STOP • STOP • STOP • STOP

A7 $5\dfrac{2}{3}$

Q8 Determine the sum:

a. $4\dfrac{1}{4}$ $\qquad\qquad$ b. $7\dfrac{3}{8}$

$+3\dfrac{2}{4}$ $\qquad\qquad\qquad$ $+2\dfrac{2}{8}$

$\overline{}$ $\qquad\qquad\qquad$ $\overline{}$

STOP • **STOP** • **STOP** • **STOP** • **STOP** • **STOP** • **STOP** • **STOP** • **STOP**

A8 a. $7\dfrac{3}{4}$ b. $9\dfrac{5}{8}$

---

4 In an addition involving mixed numbers, the sum of the fractions will often be an improper fraction that must be changed to a mixed number and added to the whole-number sum.

Examples:

1. $2\dfrac{5}{8}$ $\qquad\qquad\qquad$ 2. $2\dfrac{13}{16}$

$+3\dfrac{6}{8}$ $\qquad\qquad\qquad\qquad$ $+4\dfrac{15}{16}$

$\overline{}$ $\qquad\qquad\qquad\qquad$ $\overline{}$

$5\dfrac{11}{8} = 5 + 1\dfrac{3}{8} = 6\dfrac{3}{8}$ $\qquad$ $6\dfrac{28}{16} = 6 + 1\dfrac{12}{16} = 7\dfrac{12}{16} = 7\dfrac{3}{4}$

Q9 Determine the sum: $6\dfrac{1}{2}$

$+7\dfrac{1}{2}$

$\overline{}$

STOP • **STOP** • **STOP** • **STOP** • **STOP** • **STOP** • **STOP** • **STOP** • **STOP**

A9 14: $6\dfrac{1}{2} + 7\dfrac{1}{2} = 13\dfrac{2}{2} = 13 + 1 = 14$

Q10 Determine the sum:

a. $1\dfrac{3}{8}$ $\qquad\qquad$ b. $1\dfrac{5}{8}$

$+2\dfrac{1}{8}$ $\qquad\qquad\qquad$ $+2\dfrac{7}{8}$

$\overline{}$ $\qquad\qquad\qquad$ $\overline{}$

STOP • **STOP** • **STOP** • **STOP** • **STOP** • **STOP** • **STOP** • **STOP** • **STOP**

A10 a. $3\dfrac{1}{2}$ $\qquad\qquad$ b. $4\dfrac{1}{2}$: $1\dfrac{5}{8} + 2\dfrac{7}{8} = 3\dfrac{12}{8} = 3 + 1\dfrac{4}{8} = 4\dfrac{1}{2}$

Q11   Determine the sum:

      a. $15\dfrac{1}{12} + 8\dfrac{8}{12}$ **b.** $7\dfrac{1}{8} + 2$ **c.** $12 + 9\dfrac{1}{3}$

STOP • **STOP** • **STOP** • **STOP** • **STOP** • **STOP** • **STOP** • **STOP** • **STOP**

A11   **a.** $23\dfrac{3}{4}$ **b.** $9\dfrac{1}{8}$ **c.** $21\dfrac{1}{3}$

Q12   Determine the sum:

      a. $9\dfrac{3}{4} + 4\dfrac{2}{4} + 3\dfrac{1}{4}$ **b.** $4\dfrac{5}{8} + \dfrac{7}{8} + 2\dfrac{6}{8}$

STOP • **STOP** • **STOP** • **STOP** • **STOP** • **STOP** • **STOP** • **STOP** • **STOP**

A12   **a.** $17\dfrac{1}{2}$ **b.** $8\dfrac{1}{4}$

5   The sum $\dfrac{1}{2} + \dfrac{3}{4}$ cannot be written as one fraction as long as the two fractions are in this form. However, since the form of any fraction can be changed by multiplying both the numerator and denominator by the same number (not zero), the forms of the fractions in a sum can be changed so that they have the same denominator. For example, multiply numerator and denominator of $\dfrac{1}{2}$ by 2 so that both fractions will have the common denominator 4:

$$\frac{1}{2} = \frac{1 \times 2}{2 \times 2} = \frac{2}{4}$$

Hence,

$$\frac{1}{2} + \frac{3}{4} = \frac{2}{4} + \frac{3}{4}$$

$$= \frac{5}{4}$$

$$= 1\frac{1}{4}$$

Q13   Multiply numerator and denominator of $\dfrac{2}{3}$ and $\dfrac{3}{4}$ by 4 and 3, respectively, so that they have the common denominator 12.

STOP • **STOP** • **STOP** • **STOP** • **STOP** • **STOP** • **STOP** • **STOP** • **STOP**

A13 $\dfrac{2}{3} = \dfrac{2 \times 4}{3 \times 4} = \dfrac{8}{12}$ and $\dfrac{3}{4} = \dfrac{3 \times 3}{4 \times 3} = \dfrac{9}{12}$

Q14 The sum $\dfrac{2}{3} + \dfrac{3}{4}$ can be expressed as

$$\dfrac{}{12} + \dfrac{}{12}$$

STOP • STOP • STOP • STOP • STOP • STOP • STOP • STOP • STOP

A14 $\dfrac{8}{12} + \dfrac{9}{12}$

---

6 To add fractions with unlike denominators, it is necessary first to find a common denominator. Addition of fractions will be simpler if the common denominator is the smallest number possible. This number, the *least common denominator* (*LCD*), is the smallest number that is exactly divisible by each of the original denominators. The LCD of $\dfrac{2}{3}$ and $\dfrac{3}{4}$ is 12, because 12 is the smallest number divisible by 3 and 4.

Example: Find the least common denominator for $\dfrac{1}{9}$ and $\dfrac{2}{6}$.

Solution

$18 \div 9 = 2$ and $18 \div 6 = 3$

18 is the LCD, because 18 is the smallest number divisible by both 9 and 6.

Q15 Find the LCD for $\dfrac{1}{4}$ and $\dfrac{3}{8}$.

STOP • STOP • STOP • STOP • STOP • STOP • STOP • STOP • STOP

A15 8: because 8 is the smallest number divisible by both 4 and 8.

Q16 Find the LCD for $\dfrac{3}{4}$ and $\dfrac{1}{16}$.

STOP • STOP • STOP • STOP • STOP • STOP • STOP • STOP • STOP

A16 16: because 16 is the smallest number divisible by both 4 and 16.

Q17 Find the LCD for $\dfrac{1}{3}$ and $\dfrac{5}{6}$.

STOP • STOP • STOP • STOP • STOP • STOP • STOP • STOP • STOP

A17 6: because 6 is the smallest number divisible by both 3 and 6.

---

7   It is not always possible to determine the LCD by observation. One method for finding the LCD is to list the multiples of the larger denominator until a multiple is obtained that is divisible by the smaller denominator.

Example: Find the LCD for $\frac{1}{4}$ and $\frac{2}{5}$.

Solution

Multiples of 5 are obtained by multiplying 5 by 1, 2, 3, 4, and so on, respectively.

$5 \times 1 = 5$
$5 \times 2 = 10$ The smallest multiple of 5 divisible
$5 \times 3 = 15$ by 4 is 20; hence, 20 is the
$5 \times 4 = 20$
$5 \times 5 = 25$ LCD for $\frac{1}{4}$ and $\frac{2}{5}$.
$5 \times 6 = 30$

Q18   List the first five multiples of 8. _____

STOP • STOP • STOP • STOP • STOP • STOP • STOP • STOP • STOP

A18   8, 16, 24, 32, and 40: because $8 \times 1 = 8$, $8 \times 2 = 16$, $8 \times 3 = 24$, and so on.

Q19   Find the LCD for $\frac{1}{6}$ and $\frac{1}{8}$ (find the smallest multiple of 8 divisible by 6).

STOP • STOP • STOP • STOP • STOP • STOP • STOP • STOP • STOP

A19   24: because the multiples of 8 are 8, 16, 24, ..., and 24 is the smallest multiple of 8 divisible by 6. ("..." means "and so forth.")

Q20   Find the LCD for $\frac{7}{20}$ and $\frac{3}{15}$.

STOP • STOP • STOP • STOP • STOP • STOP • STOP • STOP • STOP

A20   60: because 60 is the smallest multiple of 20 divisible by 15 (20, 40, 60, 80, 100, ...) and $60 \div 15 = 4$.

---

8   The technique for finding the LCD for three or more fractions is the same as employed with two fractions. That is, find the smallest multiple of the largest denominator that is divisible by the other denominators.

As an example, the LCD for $\frac{1}{3}, \frac{1}{4}$, and $\frac{1}{8}$ is 24. Multiples of 8: 8, 16, 24, 32, ... Note that 16 is divisible by 4 and 8 but is not divisible by 3; hence, 16 is not the LCD. Also, 24 is divisible by 4 and 24 is divisible by 3 ($24 \div 4 = 6$ and $24 \div 3 = 8$); hence, 24 is the LCD for $\frac{1}{3}, \frac{1}{4}$, and $\frac{1}{8}$.

Q21   Find the LCD for $\frac{1}{8}$, $\frac{1}{6}$, and $\frac{1}{12}$ (determine the smallest multiple of 12 divisible by 8 and 6).

STOP • **STOP** • **STOP** • **STOP** • **STOP** • **STOP** • **STOP** • **STOP** • **STOP**

A21   24: because $24 \div 12 = 2$, $24 \div 8 = 3$, and $24 \div 6 = 4$.

Q22   Find the LCD for $\frac{1}{12}$ and $\frac{1}{16}$.

STOP • **STOP** • **STOP** • **STOP** • **STOP** • **STOP** • **STOP** • **STOP** • **STOP**

A22   48: because $48 \div 12 = 4$ and $48 \div 16 = 3$.

Q23   Find the LCD for the following fractions:

    a. $\frac{3}{4}$, $\frac{1}{5}$ **b.** $\frac{7}{15}$, $\frac{6}{45}$ **c.** $\frac{1}{4}$, $\frac{5}{6}$, $\frac{8}{16}$

STOP • **STOP** • **STOP** • **STOP** • **STOP** • **STOP** • **STOP** • **STOP** • **STOP**

A23   **a.** 20   **b.** 45   **c.** 48

---

9   To add fractions with unlike denominators it will be necessary to rewrite them as equivalent fractions with a common denominator. As an example, add $\frac{2}{3} + \frac{3}{4}$. The LCD is 12, because 12 is the smallest multiple of 4 divisible by 3. It is often easier to add fractions vertically.

$$\frac{2}{3} = \frac{}{12}$$
$$+\frac{3}{4} = \frac{}{12}$$

Write $\frac{2}{3}$ and $\frac{3}{4}$ as equivalent fractions with denominator 12.

$12 \div 3 = 4$; hence, $\frac{2}{3} = \frac{2 \times 4}{3 \times 4}$

$12 \div 4 = 3$; hence, $\frac{3}{4} = \frac{3 \times 3}{4 \times 3}$

$$\frac{2}{3} = \frac{2 \times 4}{3 \times 4} = \frac{8}{12}$$
$$+\frac{3}{4} = \frac{3 \times 3}{4 \times 3} = \frac{9}{12}$$
$$\frac{17}{12} = 1\frac{5}{12}$$

(this step is usually done mentally)

Therefore, $\frac{2}{3} + \frac{3}{4} = 1\frac{5}{12}$.

Q24   Rewrite the fractions as equivalent fractions with the least common denominator.

$$\frac{1}{4} = \underline{\hspace{1cm}}$$

$$+\frac{3}{8} = \underline{\hspace{1cm}}$$

STOP • STOP • STOP • STOP • STOP • STOP • STOP • STOP • STOP

A24 $\frac{1}{4} = \frac{2}{8}$: because 8 is the LCD and

$+\frac{3}{8} = \frac{3}{8}$ $8 \div 4 = 2$; multiply $\frac{1 \times 2}{4 \times 2}$

$8 \div 8 = 1$; multiply $\frac{3 \times 1}{8 \times 1}$

Q25   Complete the addition: $\frac{1}{4} = \frac{2}{8}$

$$+\frac{3}{8} = \frac{3}{8}$$

STOP • STOP • STOP • STOP • STOP • STOP • STOP • STOP • STOP

A25 $\frac{5}{8}$

Q26   Rewrite the fractions as equivalent fractions with the LCD:

$$\frac{3}{4} = \underline{\hspace{1cm}}$$

$$+\frac{1}{5} = $$

STOP • STOP • STOP • STOP • STOP • STOP • STOP • STOP • STOP

A26 $\frac{3}{4} = \frac{15}{20}$: because 20 is the LCD and

$+\frac{1}{5} = \frac{4}{20}$ $20 \div 4 = 5$; multiply $\frac{3 \times 5}{4 \times 5}$

$20 \div 5 = 4$; multiply $\frac{1 \times 4}{5 \times 4}$

Q27   Find the sum of the fractions in A26. _____

STOP • STOP • STOP • STOP • STOP • STOP • STOP • STOP • STOP

A27 $\frac{19}{20}$

Q28   Determine the sum: $\dfrac{1}{2}$

$+\dfrac{3}{4}$

STOP • STOP • STOP • STOP • STOP • STOP • STOP • STOP • STOP

A28 $1\dfrac{1}{4}$: $\dfrac{1}{2} = \dfrac{2}{4}$ $\left(\text{multiply } \dfrac{1 \times 2}{2 \times 2}\right)$

$+\dfrac{3}{4} = \dfrac{3}{4}$

$\dfrac{5}{4} = 1\dfrac{1}{4}$

Q29   Determine the sum: $\dfrac{2}{3}$

$+\dfrac{1}{9}$

STOP • STOP • STOP • STOP • STOP • STOP • STOP • STOP • STOP

A29 $\dfrac{7}{9}$: $\dfrac{2}{3} = \dfrac{6}{9}$ $\left(\text{multiply } \dfrac{2 \times 3}{3 \times 3}\right)$

$+\dfrac{1}{9} = \dfrac{1}{9}$

$\dfrac{7}{9}$

Q30   Determine the sum: $\dfrac{5}{6}$

$+\dfrac{7}{8}$

STOP • STOP • STOP • STOP • STOP • STOP • STOP • STOP • STOP

A30 $1\dfrac{17}{24}$: $\dfrac{5}{6} = \dfrac{20}{24}$ $\left(\text{multiply } \dfrac{5 \times 4}{6 \times 4}\right)$

$+\dfrac{7}{8} = \dfrac{21}{24}$ $\left(\text{multiply } \dfrac{7 \times 3}{8 \times 3}\right)$

$\dfrac{41}{24} = 1\dfrac{17}{24}$

Q31   Determine the sum: $\dfrac{3}{4}$

$+\dfrac{3}{5}$

STOP • STOP • STOP • STOP • STOP • STOP • STOP • STOP • STOP

A31 $1\dfrac{7}{20}$: $\dfrac{3}{4}=\dfrac{15}{20}$ $\left(\text{multiply } \dfrac{3\times5}{4\times5}\right)$

$+\dfrac{3}{5}=\dfrac{12}{20}$ $\left(\text{multiply } \dfrac{3\times4}{5\times4}\right)$

Q32   Determine the sum:

a. $\dfrac{5}{6}$ **b.** $\dfrac{2}{3}$

$+\dfrac{5}{12}$ $+\dfrac{1}{4}$

STOP • STOP • STOP • STOP • STOP • STOP • STOP • STOP • STOP

A32   **a.** $1\dfrac{1}{4}$: $\dfrac{5}{6}=\dfrac{10}{12}$ **b.** $\dfrac{11}{12}$: $\dfrac{2}{3}=\dfrac{8}{12}$

$+\dfrac{5}{12}=\dfrac{5}{12}$ $+\dfrac{1}{4}=\dfrac{3}{12}$

Q33   Determine the sum:

a. $\dfrac{3}{4}+\dfrac{1}{9}$ **b.** $\dfrac{3}{5}+\dfrac{5}{6}$

STOP • STOP • STOP • STOP • STOP • STOP • STOP • STOP • STOP

A33   **a.** $\dfrac{31}{36}$: $\dfrac{3}{4}=\dfrac{27}{36}$ **b.** $1\dfrac{13}{30}$: $\dfrac{3}{5}=\dfrac{18}{30}$

$+\dfrac{1}{9}=\dfrac{4}{36}$ $+\dfrac{5}{6}=\dfrac{25}{30}$

Q34   Determine the sum:

a. $\dfrac{1}{6} + \dfrac{4}{7}$ b. $\dfrac{2}{3} + \dfrac{5}{6}$ c. $\dfrac{1}{12} + \dfrac{1}{16}$

STOP • **STOP** • **STOP** • **STOP** • **STOP** • **STOP** • **STOP** • **STOP** • **STOP**

A34   a. $\dfrac{31}{42}$ b. $1\dfrac{1}{2}$ c. $\dfrac{7}{48}$

Q35   Determine the sum:

a. $\dfrac{3}{4} + \dfrac{2}{5} + \dfrac{3}{10}$ b. $\dfrac{2}{3} + \dfrac{5}{6} + \dfrac{4}{9}$

STOP • **STOP** • **STOP** • **STOP** • **STOP** • **STOP** • **STOP** • **STOP** • **STOP**

A35   a. $1\dfrac{9}{20}$: $\dfrac{3}{4} = \dfrac{15}{20}$ b. $1\dfrac{17}{18}$: $\dfrac{2}{3} = \dfrac{12}{18}$

$\dfrac{2}{5} = \dfrac{8}{20}$ $\dfrac{5}{6} = \dfrac{15}{18}$

$+\dfrac{3}{10} = \dfrac{6}{20}$ $+\dfrac{4}{9} = \dfrac{8}{18}$

10   The sum of mixed numbers with unlike fractions is determined as follows: Add the fractions and whole numbers separately, and then combine the separate sums. For example,

$2\dfrac{5}{6} = \quad 2\dfrac{15}{18}$

$8\dfrac{4}{9} = \quad 8\dfrac{8}{18}$

$= 10\dfrac{23}{18} = 10 + 1\dfrac{5}{18} = 11\dfrac{5}{18}$

Q36   Determine the sum: $1\dfrac{2}{3}$

$+ 3\dfrac{1}{4}$

STOP • **STOP** • **STOP** • **STOP** • **STOP** • **STOP** • **STOP** • **STOP** • **STOP**

A36 $4\dfrac{11}{12}$: $\quad 1\dfrac{2}{3} = 1\dfrac{8}{12}$

$\qquad\qquad +\, 3\dfrac{1}{4} = 3\dfrac{3}{12}$

$\qquad\qquad\qquad\quad = 4\dfrac{11}{12}$

Q37 Determine the sum: $\quad 1\dfrac{2}{3} + 3\dfrac{1}{4} + \dfrac{5}{8}$

STOP • STOP • STOP • STOP • STOP • STOP • STOP • STOP • STOP

A37 $5\dfrac{13}{24}$

Q38 A table top is $\dfrac{9}{16}$ inch thick. The legs of this table are $28\dfrac{3}{8}$ inches high. What is the distance from the floor to the top of the table?

STOP • STOP • STOP • STOP • STOP • STOP • STOP • STOP • STOP

A38 $28\dfrac{15}{16}$ inches

This completes the instruction for this section.

2.3 Exercises

1. Two fractions can be combined into a single fraction by addition only if they have the same _____.
2. The rule of adding fractions with the same denominator is to _____.
3. Addition of fractions with unlike denominators is performed by first finding the _____.
4. Determine the sum:

a. $\dfrac{3}{16}$
$+\dfrac{10}{16}$

b. $\dfrac{11}{14}$
$+\dfrac{12}{14}$

c. $\dfrac{4}{27}$
$+\dfrac{5}{27}$

d. $\dfrac{5}{6} + \dfrac{4}{6}$

e. $\begin{array}{r} 3\dfrac{2}{3} \\ +\dfrac{1}{3} \\ \hline \end{array}$ f. $\begin{array}{r} 2\dfrac{3}{7} \\ +4\dfrac{4}{7} \\ \hline \end{array}$

5. Find the sum:

a. $\begin{array}{r} \dfrac{1}{8} \\ +\dfrac{3}{4} \\ \hline \end{array}$ b. $\begin{array}{r} \dfrac{3}{4} \\ +\dfrac{7}{8} \\ \hline \end{array}$ c. $\begin{array}{r} \dfrac{9}{11} \\ +\dfrac{5}{6} \\ \hline \end{array}$ d. $\begin{array}{r} \dfrac{5}{8} \\ +\dfrac{7}{9} \\ \hline \end{array}$

e. $\dfrac{2}{3} + \dfrac{7}{9}$ f. $\dfrac{1}{8} + \dfrac{5}{6}$ g. $\dfrac{1}{2} + \dfrac{5}{9}$ h. $\dfrac{7}{14} + \dfrac{2}{7}$

i. $\dfrac{5}{8} + \dfrac{5}{7}$ j. $\dfrac{1}{6} + \dfrac{5}{9}$ k. $\dfrac{7}{12} + \dfrac{1}{18}$ l. $4\dfrac{3}{8} + \dfrac{7}{16}$

m. $\dfrac{2}{7} + \dfrac{1}{3} + \dfrac{1}{2}$ n. $3\dfrac{2}{3} + 2\dfrac{5}{6} + 1\dfrac{7}{12}$

2.3 Exercise Answers

1. denominator
2. Add the numerators and place the sum over the denominator.
3. LCD (least common denominator)
4. a. $\dfrac{13}{16}$ b. $1\dfrac{9}{14}$ c. $\dfrac{1}{3}$ d. $1\dfrac{1}{2}$ e. 4   f. 7

5. a. $\dfrac{7}{8}$ b. $1\dfrac{5}{8}$ c. $1\dfrac{43}{66}$ d. $1\dfrac{29}{72}$ e. $1\dfrac{4}{9}$ f. $\dfrac{23}{24}$ g. $1\dfrac{1}{18}$

   h. $\dfrac{11}{14}$ i. $1\dfrac{19}{56}$ j. $\dfrac{13}{18}$ k. $\dfrac{23}{36}$ l. $4\dfrac{13}{16}$ m. $1\dfrac{5}{42}$ n. $8\dfrac{1}{12}$

2.4 Prime and Composite Numbers

> **1**
>
> A *prime number* is a whole number greater than 1 divisible by exactly two different factors—itself and 1 only. All other whole numbers are *composite numbers*. For example,
>
> 1 is *not* prime, because 1 is not divisible by two different factors.
>
> 2 is prime, because 2 is divisible by itself and 1 only.
>
> 3 is prime, because 3 is divisible by itself and 1 only.
>
> 4 is *not* prime, because 4 is divisible by more than two factors. 4 is divisible by 4, 1, and 2. Therefore, 4 is composite.

Q1   What are the factors (divisors) of 5? _____

STOP • STOP • STOP • STOP • STOP • STOP • STOP • STOP • STOP

A1   5 and 1

Q2   Is 5 a prime number? _____ Why? _____

STOP • STOP • STOP • STOP • STOP • STOP • STOP • STOP • STOP

A2   yes; 5 has exactly two factors, 5 and 1.

Q3   What are the factors (divisors) of 6? _____

STOP • STOP • STOP • STOP • STOP • STOP • STOP • STOP • STOP

A3   6, 1, 2, and 3

Q4   Is 6 a prime number? _____ Why? _____

STOP • STOP • STOP • STOP • STOP • STOP • STOP • STOP • STOP

A4   no; 6 has more than two factors (6, 1, 2, and 3).

Q5   Circle the prime numbers:
      1 2 3 4 5 6 7 8 9 10 11 12 13 14 15 16 17 18 19 20 21
      22 23 24 25

STOP • STOP • STOP • STOP • STOP • STOP • STOP • STOP • STOP

A5   1 ② ③ 4 ⑤ 6 ⑦ 8 9 10 ⑪ 12 ⑬ 14 15 16 ⑰ 18 ⑲ 20 21
      22 ㉓ 24 25

Q6   Explain why 9 is not prime. _____

STOP • STOP • STOP • STOP • STOP • STOP • STOP • STOP • STOP

A6   9 has more than two factors (9, 1, and 3).

Q7   Explain why 11 is prime. _____

STOP • STOP • STOP • STOP • STOP • STOP • STOP • STOP • STOP

A7   11 has exactly two factors, itself and 1.

| 2 | The number 2 is the first prime number. The multiples of 2 are 2, 4, 6, 8, 10, 12, and so on. Any multiple of 2 is divisible by itself (that is, 2) and by 1; so the multiples of 2, except 2, are composite. For example, 8 is a multiple of 2 and is divisible by 8, 4, 2, and 1; hence, 8 is composite. |

Q8   List the first five multiples of 3. _____

STOP • STOP • STOP • STOP • STOP • STOP • STOP • STOP • STOP

A8   3, 6, 9, 12, 15

Q9   Which of the multiples of 3 are prime? _____

STOP • STOP • STOP • STOP • STOP • STOP • STOP • STOP • STOP

A9   3

Q10   List the first five multiples of 5._____

STOP   •   STOP   •   STOP   •   STOP   •   STOP   •   STOP   •   STOP   •   STOP   •   STOP

A10   5, 10, 15, 20, 25

Q11   Which of the multiples of 5 are prime?_____

STOP   •   STOP   •   STOP   •   STOP   •   STOP   •   STOP   •   STOP   •   STOP   •   STOP

A11   5

---

3   The preceding problems suggest a technique for finding the prime numbers. On the list given in Q12:

1. Circle the first prime, 2, and cross out all other multiples of 2 (every other number counting from 2).
2. Circle the second prime, 3, and cross out all other multiples of 3 (every third number counting from 3).
3. Circle 5 and cross out all other multiples of 5 (every fifth number counting from 5).
4. Circle 7 and cross out all other multiples of 7.

Continue in this manner until only primes remain on your list. This technique for locating primes is called the *sieve of Eratosthenes*.

Q12   Use the technique discussed in Frame 3 to locate all primes in the list below.

	2	3	4	5	6	7	8	9	10
11	12	13	14	15	16	17	18	19	20
21	22	23	24	25	26	27	28	29	30
31	32	33	34	35	36	37	38	39	40
41	42	43	44	45	46	47	48	49	50
51	52	53	54	55	56	57	58	59	60
61	62	63	64	65	66	67	68	69	70
71	72	73	74	75	76	77	78	79	80
81	82	83	84	85	86	87	88	89	90
91	92	93	94	95	96	97	98	99	100

STOP   •   STOP   •   STOP   •   STOP   •   STOP   •   STOP   •   STOP   •   STOP   •   STOP

A12

	(2)	(3)	4̶	(5)	6̶	(7)	8̶	9̶	1̶0̶
(11)	1̶2̶	(13)	1̶4̶	1̶5̶	1̶6̶	(17)	1̶8̶	(19)	2̶0̶
2̶1̶	2̶2̶	(23)	2̶4̶	2̶5̶	2̶6̶	2̶7̶	2̶8̶	(29)	3̶0̶
(31)	3̶2̶	3̶3̶	3̶4̶	3̶5̶	3̶6̶	(37)	3̶8̶	3̶9̶	4̶0̶
(41)	4̶2̶	(43)	4̶4̶	4̶5̶	4̶6̶	(47)	4̶8̶	4̶9̶	5̶0̶
5̶1̶	5̶2̶	(53)	5̶4̶	5̶5̶	5̶6̶	5̶7̶	5̶8̶	(59)	6̶0̶
(61)	6̶2̶	6̶3̶	6̶4̶	6̶5̶	6̶6̶	(67)	6̶8̶	6̶9̶	7̶0̶
(71)	7̶2̶	(73)	7̶4̶	7̶5̶	7̶6̶	7̶7̶	7̶8̶	(79)	8̶0̶
8̶1̶	8̶2̶	(83)	8̶4̶	8̶5̶	8̶6̶	8̶7̶	8̶8̶	(89)	9̶0̶
9̶1̶	9̶2̶	9̶3̶	9̶4̶	9̶5̶	9̶6̶	(97)	9̶8̶	9̶9̶	1̶0̶0̶

4	Often a number can be expressed as a product of factors in many different ways. For example,

$$20 = 4 \times 5 \qquad 20 = 10 \times 2 \qquad 20 = 2 \times 2 \times 5$$

However, a number may be expressed as a product of prime factors in only one way:

$$20 = 4 \times 5 \qquad\qquad 20 = 10 \times 2$$
$$ = 2 \times 2 \times 5 \qquad\qquad = 5 \times 2 \times 2$$

By expressing 4 and 10 as a product of primes, 20 has been expressed as a product of primes two times in exactly the same way (remember that the order of the factors is not important).

Q13   Express 8 as a product of primes. _____

STOP • STOP • STOP • STOP • STOP • STOP • STOP • STOP • STOP

A13 $8 = 2 \times 2 \times 2$

Q14   Express 30 as a product of primes. _____

STOP • STOP • STOP • STOP • STOP • STOP • STOP • STOP • STOP

A14 $30 = 2 \times 3 \times 5$

5	When expressing a composite number as a product of primes, it will be helpful to recall the following divisibility tests:

1. If the last digit of the number is divisible by 2, the number is divisible by 2.
Example: 304; 4 is divisible by 2, so 304 is divisible by 2. $304 \div 2 = 152$.
2. If the sum of the digits is divisible by 3, the number is divisible by 3.
Example: 411; the sum of the digits, $4 + 1 + 1 = 6$, is divisible by 3, so 411 is divisible by 3. $411 \div 3 = 137$.
3. If the last digit is 0 or 5, the number is divisible by 5.
Example: 85; the last digit is 5, so 85 is divisible by 5. $85 \div 5 = 17$.

The sieve of Eratosthenes can be used to help decide whether or not a number is prime.

Example: Express 45 as a product of primes.

Solution
The last digit is 5, so 45 is divisible by 5.

$$45 \div 5 = 9, \text{ hence } 45 = 9 \times 5$$
$$(\text{prime factors of 9}) = 3 \times 3 \times 5$$

Q15   Express 54 as a product of primes. _____

STOP • STOP • STOP • STOP • STOP • STOP • STOP • STOP • STOP

A15 $54 = 2 \times 3 \times 3 \times 3$; $54 = 2 \times 27$ (divide 54 by 2)
                             $= 2 \times 3 \times 9$ (divide 27 by 3)
                             $= 2 \times 3 \times 3 \times 3$

Other combinations of the same prime factors will give the same result.

Q16   Express 110 as a product of primes. _____

STOP • STOP • STOP • STOP • STOP • STOP • STOP • STOP • STOP

A16 $110 = 2 \times 5 \times 11$: because $110 = 2 \times 55$
                                                    $= 2 \times 5 \times 11$

Q17   Express as a product of primes:

                a. 90 _____   **b.** 53 _____

STOP • STOP • STOP • STOP • STOP • STOP • STOP • STOP • STOP

A17   **a.** $90 = 2 \times 3 \times 3 \times 5$ **b.** 53 is prime

---

6   Prime factors can be used to determine the least common denominator (LCD) for two or more fractions. The denominators are expressed as the product of prime factors. The LCD is the product of these prime factors; *each factor is used as many times as it is found in any one of the denominators.*

                Suppose that 12, 15, and 10 are the denominators for three fractions. First, list the prime factors of each:

                $12 = 2 \times 2 \times 3$
                $15 = 3 \times 5$
                $10 = 2 \times 5$

                Since 2 is used as a factor twice in the number 12, it is used twice in the LCD. The numbers 3 and 5 are used only once in each of the numbers 12, 10, and 15, so they are used only once in the LCD:

                $LCD = 2 \times 2 \times 3 \times 5$
                $\quad\quad = 60$

Q18   To find the sum $\dfrac{1}{54} + \dfrac{1}{20} + \dfrac{1}{12}$, first express the denominators as prime factors.

                54 = _____   20 = _____   12 = _____

STOP • STOP • STOP • STOP • STOP • STOP • STOP • STOP • STOP

A18 $54 = 2 \times 3 \times 3 \times 3$, $20 = 2 \times 2 \times 5$, $12 = 2 \times 2 \times 3$

Q19   In the LCD the factor

    a. 2 is used how many times? _____

    b. 3 is used how many times? _____

    c. 5 is used how many times? _____

STOP • **STOP** • **STOP** • **STOP** • **STOP** • **STOP** • **STOP** • **STOP** • **STOP**

A19   **a.** 2   **b.** 3   **c.** 1

Q20   Find the LCD for $\frac{1}{54}$, $\frac{1}{20}$, and $\frac{1}{12}$.

STOP • **STOP** • **STOP** • **STOP** • **STOP** • **STOP** • **STOP** • **STOP** • **STOP**

A20   540: because LCD $= 2 \times 2 \times 3 \times 3 \times 3 \times 5$
$$= 540$$

The factor 2 is used twice, because it is used twice in both 20 and 12. The factor 3 is used 3 times, because it is used 3 times in 54. The factor 5 is used once, because it is used only once in any of the denominators.

Q21   Use prime factors to find the LCD of the following denominators of fractions: 6, 8, and 9.

STOP • **STOP** • **STOP** • **STOP** • **STOP** • **STOP** • **STOP** • **STOP** • **STOP**

A21   72: because $6 = 2 \times 3$, $8 = 2 \times 2 \times 2$, $9 = 3 \times 3$. LCD $= 2 \times 2 \times 2 \times 3 \times 3$.

Q22   Find the LCD for $\frac{1}{5}$, $\frac{1}{7}$, and $\frac{1}{10}$.

STOP • **STOP** • **STOP** • **STOP** • **STOP** • **STOP** • **STOP** • **STOP** • **STOP**

A22   70: because 5 (prime), 7 (prime), $10 = 2 \times 5$; LCD $= 2 \times 5 \times 7$.

Q23   Find the LCD for $\frac{1}{12}$, $\frac{1}{5}$, and $\frac{1}{6}$.

STOP • **STOP** • **STOP** • **STOP** • **STOP** • **STOP** • **STOP** • **STOP** • **STOP**

A23   60: because $12 = 2 \times 2 \times 3$, 5, $6 = 2 \times 3$; LCD $= 2 \times 2 \times 3 \times 5$.

Q24   Find the LCD for $\frac{1}{8}$, $\frac{1}{12}$, and $\frac{1}{15}$.

STOP • **STOP** • **STOP** • **STOP** • **STOP** • **STOP** • **STOP** • **STOP** • **STOP**

A24   120: because $8 = 2 \times 2 \times 2$, $12 = 2 \times 2 \times 3$, $15 = 3 \times 5$; LCD $= 2 \times 2 \times 2 \times 3 \times 5$.

Q25   Find the sum: $\dfrac{5}{6} + \dfrac{11}{15} + \dfrac{17}{20}$

STOP • STOP • STOP • STOP • STOP • STOP • STOP • STOP • STOP

A25 $2\dfrac{5}{12}:$

$$\dfrac{5}{6} = \dfrac{50}{60}$$

$$\dfrac{11}{15} = \dfrac{44}{60}$$

$$+\dfrac{17}{20} = \dfrac{51}{60}$$

$$\dfrac{145}{60} = 2\dfrac{25}{60}$$

$6 = 2 \times 3,\ 15 = 3 \times 5,\ 20 = 2 \times 2 \times 5;\ \text{LCD} = 2 \times 2 \times 3 \times 5 = 60$

Q26   Find the sum: $\dfrac{7}{12} + \dfrac{12}{21} + \dfrac{18}{35}$

STOP • STOP • STOP • STOP • STOP • STOP • STOP • STOP • STOP

A26 $1\dfrac{281}{420}:$

$$\dfrac{7}{12} = \dfrac{245}{420}$$

$$\dfrac{12}{21} = \dfrac{240}{420}$$

$$\dfrac{18}{35} = \dfrac{216}{420}$$

$$\dfrac{701}{420}$$

$12 = 2 \times 2 \times 3,\ 21 = 3 \times 7;\ 35 = 5 \times 7;\ \text{LCD} = 2 \times 2 \times 3 \times 5 \times 7 = 420$

This completes the instruction for this section.

2.4 Exercises

1. A prime number is divisible by _____.
2. List the prime numbers 1–20.
3. If a whole number is not prime, it is _____.
4. Express the following as a product of primes:
   - **a.** 56
   - **b.** 78
   - **c.** 97
   - **d.** 185
   - **e.** 36
   - **f.** 100
   - **g.** 1,000
   - **h.** 13
5. Find the LCD for the following fractions:

   a. $\dfrac{1}{18}, \dfrac{1}{12}, \dfrac{1}{24}$ **b.** $\dfrac{1}{14}, \dfrac{1}{35}, \dfrac{1}{10}$ **c.** $\dfrac{1}{20}, \dfrac{1}{15}, \dfrac{1}{12}$ **d.** $\dfrac{1}{9}, \dfrac{1}{36}, \dfrac{1}{12}$

   e. $\dfrac{1}{6}, \dfrac{1}{8}, \dfrac{1}{3}$ **f.** $\dfrac{1}{15}, \dfrac{1}{14}, \dfrac{1}{35}$ **g.** $\dfrac{1}{48}, \dfrac{1}{75}, \dfrac{1}{27}$ **h.** $\dfrac{1}{4}, \dfrac{1}{36}, \dfrac{1}{27}$

6. Find the sum:

   a. $\dfrac{7}{9} + \dfrac{5}{8} + \dfrac{9}{14}$ **b.** $\dfrac{9}{11} + \dfrac{15}{33} + \dfrac{17}{30}$ **c.** $\dfrac{9}{20} + \dfrac{14}{15} + \dfrac{7}{8}$

2.4 Exercise Answers

1. itself and one only
2. 2, 3, 5, 7, 11, 13, 17, 19
3. composite
4. **a.** $2 \times 2 \times 2 \times 7$ **b.** $2 \times 3 \times 13$ **c.** prime
   d. 5×37 **e.** $2 \times 2 \times 3 \times 3$ **f.** $2 \times 2 \times 5 \times 5$
   g. $2 \times 2 \times 2 \times 5 \times 5 \times 5$ **h.** prime
5. **a.** 72   **b.** 70   **c.** 60
   d. 36   **e.** 24   **f.** 210
   g. 10,800   **h.** 108
6. **a.** $2\dfrac{23}{504}$ **b.** $1\dfrac{277}{330}$ **c.** $2\dfrac{31}{120}$

2.5 Subtraction

1

As in addition, you can subtract one fraction from another if they have the same denominator. Simply subtract numerators and write the difference over the common denominator:

$$\frac{9}{13} - \frac{4}{13} = \frac{9 - 4}{13}$$

$$= \frac{5}{13}$$

Q1   Find the difference:

a. $\dfrac{4}{5} - \dfrac{3}{5}$ b. $\dfrac{5}{8} - \dfrac{3}{8}$ c. $\dfrac{16}{24} - \dfrac{9}{24}$

STOP • STOP • STOP • STOP • STOP • STOP • STOP • STOP • STOP

A1   a. $\dfrac{1}{5}$: $\dfrac{4-3}{5}$ b. $\dfrac{1}{4}$: $\dfrac{5-3}{8} = \dfrac{2}{8}$ c. $\dfrac{7}{24}$: $\dfrac{16-9}{24}$

Q2   Find the difference:

a. $\dfrac{5}{6} - \dfrac{1}{6}$ b. $\dfrac{11}{12} - \dfrac{3}{12}$ c. $\dfrac{11}{15} - \dfrac{4}{15}$

STOP • STOP • STOP • STOP • STOP • STOP • STOP • STOP • STOP

A2   a. $\dfrac{2}{3}$ b. $\dfrac{2}{3}$ c. $\dfrac{7}{15}$

---

2   If two fractions do not have the same denominator, you must change the form of the fractions so that they have a common denominator. As in addition, it will be helpful if the common denominator is the LCD.

Example: Subtract $\dfrac{3}{4} - \dfrac{1}{3}$.

Solution

$$\begin{array}{r} \dfrac{3}{4} = \dfrac{9}{12} \\[2mm] -\dfrac{1}{3} = -\dfrac{4}{12} \\[2mm] \hline \dfrac{5}{12} \end{array}$$

LCD = 12

$\dfrac{3}{4} = \dfrac{3 \times 3}{4 \times 3} = \dfrac{9}{12}$

$\dfrac{1}{3} = \dfrac{1 \times 4}{3 \times 4} = \dfrac{4}{12}$

Q3   Find the difference:

$$\begin{array}{r} \dfrac{7}{8} \\[2mm] -\dfrac{3}{4} \\[2mm] \hline \end{array}$$

STOP • STOP • STOP • STOP • STOP • STOP • STOP • STOP • STOP

A3 $\dfrac{1}{8}$:

$$\begin{array}{r} \dfrac{7}{8} = \dfrac{7}{8} \\[2mm] -\dfrac{3}{4} = -\dfrac{6}{8} \\[2mm] \hline \dfrac{1}{8} \end{array}$$

Q4   Find the difference: $\dfrac{2}{3}$

$-\dfrac{1}{4}$

$\overline{}$

STOP • STOP • STOP • STOP • STOP • STOP • STOP • STOP • STOP

A4 $\dfrac{5}{12}$: $\dfrac{2}{3} = \dfrac{8}{12}$

$-\dfrac{1}{4} = -\dfrac{3}{12}$

$\overline{}\quad\overline{}$

Q5   Find the difference: $\dfrac{11}{16} - \dfrac{1}{3}$

STOP • STOP • STOP • STOP • STOP • STOP • STOP • STOP • STOP

A5 $\dfrac{17}{48}$: $\dfrac{11}{16} = \dfrac{33}{48}$

$-\dfrac{1}{3} = -\dfrac{16}{48}$

$\overline{}\quad\overline{}$

If you had difficulty finding the LCD, return to Section 2.3 or Section 2.4 and restudy the technique for finding the LCD.

Q6   Find the difference: $8\dfrac{2}{3}$

$-6\dfrac{3}{8}$

$\overline{}$

STOP • STOP • STOP • STOP • STOP • STOP • STOP • STOP • STOP

A6 $2\dfrac{7}{24}$: $8\dfrac{2}{3} = 8\dfrac{16}{24}$

$-6\dfrac{3}{8} = -6\dfrac{9}{24}$

$\overline{}\quad\overline{}$

$2\dfrac{7}{24}$

Q7 Find the difference:

a. $\dfrac{5}{7} - \dfrac{1}{3}$

b. $\dfrac{7}{8} - \dfrac{7}{10}$

STOP • STOP • STOP • STOP • STOP • STOP • STOP • STOP • STOP

A7 a. $\dfrac{8}{21}$ b. $\dfrac{7}{40}$

Q8 Find the difference:

a. $7\dfrac{2}{3} - 3\dfrac{1}{2}$

b. $4\dfrac{5}{16} - 2\dfrac{1}{4}$

STOP • STOP • STOP • STOP • STOP • STOP • STOP • STOP • STOP

A8 a. $4\dfrac{1}{6}$ b. $2\dfrac{1}{16}$

---

3 When subtracting fractions and mixed numbers, it will often be necessary to borrow to find the difference. Consider this example:

$$13\dfrac{1}{4} = \ 13\dfrac{2}{8}$$
$$-6\dfrac{7}{8} = -6\dfrac{7}{8}$$

Since $\dfrac{2}{8} - \dfrac{7}{8}$ does not have meaning at this time, it will be necessary to borrow 1 from the 13. The 1 is changed to $\dfrac{8}{8}$ (any nonzero number divided by itself is 1) and added to $\dfrac{2}{8}$. The subtraction is then completed.

$$13\dfrac{1}{4} = \ \overset{12}{\cancel{13}}\dfrac{2}{8} + \dfrac{8}{8} = \ 12\dfrac{10}{8}$$
$$-6\dfrac{7}{8} = -6\dfrac{7}{8} \qquad = -6\dfrac{7}{8}$$
$$6\dfrac{3}{8}$$

Q9 Express 1 with the following denominators:

a. $\dfrac{}{24}$

b. $\dfrac{}{19}$

c. $\dfrac{}{5}$

d. $\dfrac{}{101}$

STOP • STOP • STOP • STOP • STOP • STOP • STOP • STOP • STOP

A9   **a.** $\dfrac{24}{24}$ **b.** $\dfrac{19}{19}$ **c.** $\dfrac{5}{5}$ **d.** $\dfrac{101}{101}$

Q10   Complete the subtraction problem:

$$21\frac{3}{8} = \overset{20}{2\!\!\!1}\frac{9}{24} + \frac{24}{24}$$

$$-17\frac{5}{6} = -17\frac{20}{24}$$

STOP • **STOP** • **STOP** • **STOP** • **STOP** • **STOP** • **STOP** • **STOP** • **STOP**

A10 $3\dfrac{13}{24}:$

$$20\frac{33}{24}$$
$$-17\frac{20}{24}$$
$$\overline{}$$
$$3\frac{13}{24}$$

Q11   Find the difference:

$$3\frac{1}{4}$$
$$-1\frac{2}{3}$$
$$\overline{}$$

STOP • **STOP** • **STOP** • **STOP** • **STOP** • **STOP** • **STOP** • **STOP** • **STOP**

A11 $1\dfrac{7}{12}:$ $3\dfrac{1}{4} = \overset{2}{\cancel{3}}\dfrac{3}{12} + \dfrac{12}{12} = \quad 2\dfrac{15}{12}$

$$-1\frac{2}{3} = -1\frac{8}{12} \qquad = -1\frac{8}{12}$$
$$\overline{} \qquad \overline{} \qquad \qquad \overline{}$$
$$1\frac{7}{12}$$

Q12   Find the difference:

$$4\frac{1}{5}$$
$$-1\frac{3}{4}$$
$$\overline{}$$

STOP • **STOP** • **STOP** • **STOP** • **STOP** • **STOP** • **STOP** • **STOP** • **STOP**

A12 $2\dfrac{9}{20}$: $4\dfrac{1}{5} = 3\dfrac{24}{20}$

$-1\dfrac{3}{4} = -1\dfrac{15}{20}$

$2\dfrac{9}{20}$

---

4   The difference between a fraction and a whole number can be determined as follows:

$5 - \dfrac{2}{3}$ is written 5

$-\dfrac{2}{3}$

Borrow 1 from the 5 and change it to $\dfrac{3}{3}$:

$\overset{4}{\cancel{5}}\dfrac{3}{3}$

$-\dfrac{2}{3}$

$4\dfrac{1}{3}$

Q13   Complete the subtraction problem: $\overset{1}{\cancel{2}}(\ \ )$

$-\dfrac{1}{4}$

STOP • STOP • STOP • STOP • STOP • STOP • STOP • STOP • STOP

A13 $\overset{1}{\cancel{2}}\dfrac{4}{4}$

$-\dfrac{1}{4}$

$1\dfrac{3}{4}$

Q14   Find the difference: 8

$-1\dfrac{2}{3}$

STOP • STOP • STOP • STOP • STOP • STOP • STOP • STOP • STOP

A14 $6\frac{1}{3}$: $\overset{7}{\cancel{8}}\frac{3}{3}$

$-1\frac{2}{3}$
_____

$6\frac{1}{3}$

Q15   Find the difference: 7

$-3\frac{2}{5}$
_____

STOP • STOP • STOP • STOP • STOP • STOP • STOP • STOP • STOP

A15 $3\frac{3}{5}$: $\overset{6}{\cancel{7}}\frac{5}{5}$

$-3\frac{2}{5}$
_____

$3\frac{3}{5}$

Q16   Find the difference:

a. $3 - \frac{5}{8}$ b. $6\frac{1}{3} - 1\frac{2}{3}$

STOP • STOP • STOP • STOP • STOP • STOP • STOP • STOP • STOP

A16   a. $2\frac{3}{8}$ b. $4\frac{2}{3}$

Q17   Find the difference:

a. $3\frac{4}{5} - 2\frac{1}{2}$ b. $3\frac{5}{8} - 1\frac{11}{12}$

STOP • STOP • STOP • STOP • STOP • STOP • STOP • STOP • STOP

A17   a. $1\frac{3}{10}$ b. $1\frac{17}{24}$

Q18   Find the difference:

     a. $7\frac{5}{18} - 1\frac{11}{24}$ **b.** $4\frac{5}{15} - 2\frac{12}{25}$

STOP • **STOP** • **STOP** • **STOP** • **STOP** • **STOP** • **STOP** • **STOP** • **STOP**

A18   **a.** $5\frac{59}{72}$ **b.** $1\frac{64}{75}$

Q19   A board $1\frac{3}{4}$ feet long was sawed from a board $6\frac{1}{2}$ feet. How large was the piece of board left?

STOP • **STOP** • **STOP** • **STOP** • **STOP** • **STOP** • **STOP** • **STOP** • **STOP**

A19 $4\frac{3}{4}$ feet

This completes the instruction for this section.

2.5   Exercises

1. Find the difference:

     a. $\frac{5}{8} - \frac{3}{8}$ **b.** $\frac{5}{9} - \frac{2}{9}$ **c.** $\frac{3}{7} - \frac{3}{8}$ **d.** $\frac{3}{8} - \frac{1}{12}$

     e. $6\frac{3}{4} - 2\frac{1}{8}$ **f.** $4\frac{5}{9} - 3\frac{5}{12}$

2. Find the difference:

     a. $8 - 7\frac{7}{8}$ **b.** $3\frac{5}{12} - 1\frac{7}{12}$ **c.** $7\frac{5}{6} - 2\frac{7}{8}$ **d.** $3\frac{1}{3} - 2\frac{1}{2}$

     e. $1\frac{3}{10} - \frac{7}{15}$ **f.** $4\frac{1}{16} - 2\frac{1}{18}$

3. Subtract $\frac{1}{3}$ from $\frac{3}{4}$.

4. Subtract $\frac{2}{5}$ from $\frac{11}{12}$.

5. From $8\frac{1}{2}$ subtract $5\frac{1}{4}$.

6. From $27\frac{9}{16}$ subtract $13\frac{5}{8}$.

2.5   Exercise Answers

1. **a.** $\dfrac{1}{4}$ **b.** $\dfrac{1}{3}$ **c.** $\dfrac{3}{56}$ **d.** $\dfrac{7}{24}$ **e.** $4\dfrac{5}{8}$ **f.** $1\dfrac{5}{36}$

2. **a.** $\dfrac{1}{8}$ **b.** $1\dfrac{5}{6}$ **c.** $4\dfrac{23}{24}$ **d.** $\dfrac{5}{6}$ **e.** $\dfrac{5}{6}$ **f.** $2\dfrac{1}{144}$

3. $\dfrac{5}{12}$

4. $\dfrac{31}{60}$

5. $3\dfrac{1}{4}$

6. $13\dfrac{15}{16}$

2.6   Multiplication and Division

1   The product of whole numbers was first defined as repeated addition. That is, $4 \times 3 = 3 + 3 + 3 + 3$. The product of a whole number and a fraction can be defined in the same manner:

$$3 \times \frac{2}{7} = \frac{2}{7} + \frac{2}{7} + \frac{2}{7} \qquad 4 \times \frac{1}{2} = \frac{1}{2} + \frac{1}{2} + \frac{1}{2} + \frac{1}{2}$$

$$= \frac{6}{7} \qquad\qquad\qquad = \frac{4}{2}$$

$$\qquad\qquad\qquad\qquad = 2$$

Q1   Using repeated addition, find the product: $\;2 \times \dfrac{2}{3}$

STOP • **STOP** • **STOP** • **STOP** • **STOP** • **STOP** • **STOP** • **STOP** • **STOP**

A1 $1\dfrac{1}{3}$: $2 \times \dfrac{2}{3} = \dfrac{2}{3} + \dfrac{2}{3}$

$$= \frac{4}{3}$$

$$= 1\frac{1}{3}$$

Q2   Using repeated addition, find the product: $\dfrac{5}{12} \times 3$ $\left(\text{recall that } \dfrac{5}{12} \times 3 = 3 \times \dfrac{5}{12}\right)$.

STOP   •   STOP   •   STOP   •   STOP   •   STOP   •   STOP   •   STOP   •   STOP   •   STOP

A2 $1\dfrac{1}{4}$: $3 \times \dfrac{5}{12} = \dfrac{5}{12} + \dfrac{5}{12} + \dfrac{5}{12}$

$$= \dfrac{15}{12}$$

$$= 1\dfrac{3}{12}$$

---

2   The preceding examples suggest a shorter method for finding the product of a whole number and a fraction. In the problem

$$2 \times \dfrac{3}{7} = \dfrac{3}{7} + \dfrac{3}{7}$$

$$= \dfrac{3 + 3}{7}$$

the numerator of the sum could be written 2×3. That is, to find the product of a whole number and a fraction, multiply the whole number by the numerator of the fraction and place this product over the denominator:

$$2 \times \dfrac{3}{7} = \dfrac{2 \times 3}{7} \qquad 3 \times \dfrac{5}{12} = \dfrac{3 \times 5}{12}$$

$$= \dfrac{6}{7} \qquad\qquad\quad = \dfrac{15}{12}$$

$$\qquad\qquad\qquad\qquad = 1\dfrac{1}{4}$$

Q3   Find the product: $3 \times \dfrac{2}{7}$

STOP   •   STOP   •   STOP   •   STOP   •   STOP   •   STOP   •   STOP   •   STOP   •   STOP

A3 $\dfrac{6}{7}$: $3 \times \dfrac{2}{7} = \dfrac{3 \times 2}{7}$

$$= \dfrac{6}{7}$$

Q4   Find the product: $\dfrac{3}{16} \times 5$

STOP   •   STOP   •   STOP   •   STOP   •   STOP   •   STOP   •   STOP   •   STOP   •   STOP

A4 $\qquad \dfrac{15}{16}:\quad \dfrac{3}{16} \times 5 = \dfrac{3 \times 5}{16}$

$\qquad\qquad\qquad\qquad = \dfrac{15}{16}$

Q5 Find the product:

a. $6 \times \dfrac{2}{15}$
b. $4 \times \dfrac{1}{2}$
c. $\dfrac{1}{3} \times 7$

STOP • STOP • STOP • STOP • STOP • STOP • STOP • STOP • STOP

A5 a. $\dfrac{4}{5}:\quad 6 \times \dfrac{2}{15} = \dfrac{12}{15}$
b. $2:\quad 4 \times \dfrac{1}{2} = \dfrac{4}{2}$
c. $2\dfrac{1}{3}:\quad \dfrac{1}{3} \times 7 = \dfrac{7}{3}$

$\qquad\qquad\qquad = \dfrac{4}{5}$
$\qquad\qquad = 2$
$\qquad\qquad = 2\dfrac{1}{3}$

---

3 The product of two fractions can be illustrated as follows:

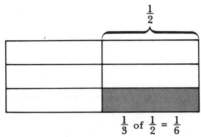

$\tfrac{1}{3}$ of $\tfrac{1}{2} = \tfrac{1}{6}$

The problem $\dfrac{1}{3}$ of $\dfrac{1}{2} = \dfrac{1}{6}$ can be restated as

$\dfrac{1}{3} \times \dfrac{1}{2} = \dfrac{1}{6}$

That is, "of" means to multiply.

Q6 In the following diagram illustrate $\dfrac{1}{2} \times \dfrac{2}{3} \cdot \left(\dfrac{1}{2} \text{ of } \dfrac{2}{3}\right)$:

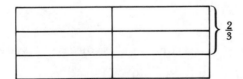

STOP • STOP • STOP • STOP • STOP • STOP • STOP • STOP • STOP

A6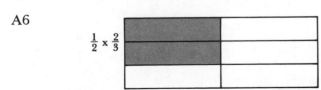

$\tfrac{1}{2} \times \tfrac{2}{3}$

Q7   Use the diagram in A6 to find how many sixths are represented by $\frac{1}{2} \times \frac{2}{3}$. _____

STOP   •   STOP   •   STOP   •   STOP   •   STOP   •   STOP   •   STOP   •   STOP   •   STOP

A7 $\frac{2}{6}$

---

4   The preceding examples illustrate the rule for multiplying fractions. The product of two fractions is the product of the numerators over the product of the denominators. For example:

$$1. \quad \frac{1}{2} \times \frac{2}{3} = \frac{1 \times 2}{2 \times 3} \qquad 2. \quad \frac{5}{7} \times \frac{3}{8} = \frac{5 \times 3}{7 \times 8}$$

$$= \frac{2}{6} \qquad\qquad\qquad = \frac{15}{56}$$

$$= \frac{1}{3}$$

Q8   Complete the multiplication process: $\dfrac{2}{3} \times \dfrac{4}{5} = \dfrac{2 \times (\ \ )}{(\ \ ) \times (\ \ )}$

STOP   •   STOP   •   STOP   •   STOP   •   STOP   •   STOP   •   STOP   •   STOP   •   STOP

A8 $\dfrac{2}{3} \times \dfrac{4}{5} = \dfrac{2 \times 4}{3 \times 5}$

$= \dfrac{8}{15}$

Q9   Find the product: $\dfrac{1}{5} \times \dfrac{1}{3}$

STOP   •   STOP   •   STOP   •   STOP   •   STOP   •   STOP   •   STOP   •   STOP   •   STOP

A9 $\dfrac{1}{15}$: $\dfrac{1}{5} \times \dfrac{1}{3} = \dfrac{1 \times 1}{5 \times 3}$

Q10   Find the product:

a. $\dfrac{5}{12} \times \dfrac{7}{9}$ b. $\dfrac{9}{5} \times \dfrac{3}{4}$

STOP   •   STOP   •   STOP   •   STOP   •   STOP   •   STOP   •   STOP •   STOP   •   STOP

A10   a. $\dfrac{35}{108}$: $\dfrac{5}{12} \times \dfrac{7}{9} = \dfrac{5 \times 7}{12 \times 9}$ b. $1\dfrac{7}{20}$: $\dfrac{9}{5} \times \dfrac{3}{4} = \dfrac{9 \times 3}{5 \times 4}$

Q11   Find the product:

a. $\dfrac{4}{5} \times \dfrac{3}{6}$ b. $\dfrac{8}{7} \times \dfrac{4}{7}$ c. $\dfrac{8}{5} \times \dfrac{3}{7}$

STOP   •   STOP   •   STOP   •   STOP   •   STOP   •   STOP   •   STOP   •   STOP   •   STOP

A11   **a.** $\dfrac{2}{5}$ **b.** $\dfrac{32}{49}$ **e.** $\dfrac{24}{35}$

---

5   The product of two fractions should be reduced to lowest terms when possible; however, it is often possible to reduce before multiplying. For example, $\dfrac{3}{4} \times \dfrac{2}{5}$ can be written as

$\dfrac{3 \times 2}{4 \times 5}$. Now,

$$\dfrac{3 \times 2}{4 \times 5} = \dfrac{2 \times 3}{4 \times 5}$$

Since $\dfrac{2}{4} = \dfrac{1}{2}$,

$$\dfrac{2 \times 3}{4 \times 5} = \dfrac{1 \times 3}{2 \times 5}$$

Hence,

$$\dfrac{3}{4} \times \dfrac{2}{5} = \dfrac{3}{10}$$

This reduction is usually shown as

$$\dfrac{3}{\overset{}{\underset{2}{4}}} \times \dfrac{\overset{1}{2}}{5} = \dfrac{3 \times 1}{2 \times 5} = \dfrac{3}{10}$$

The example illustrates the following rule: *In multiplying fractions, you can divide any numerator and any denominator by the same nonzero number.*

Q12   Find the product, reducing before multiplying (show all work): $\dfrac{3}{10} \times \dfrac{4}{9}$

STOP • STOP • STOP • STOP • STOP • STOP • STOP • STOP • STOP

A12 $\dfrac{\overset{1}{3}}{\underset{5}{10}} \times \dfrac{\overset{2}{4}}{\underset{3}{9}} = \dfrac{2}{15}$

Q13   Find the product: $\dfrac{11}{3} \times \dfrac{5}{11}$

STOP • STOP • STOP • STOP • STOP • STOP • STOP • STOP • STOP

A13 $1\dfrac{2}{3}:\ \dfrac{\overset{1}{11}}{3} \times \dfrac{5}{\underset{1}{11}} = \dfrac{5}{3}$

Q14   Find the product:

a. $\dfrac{7}{3} \times \dfrac{3}{8}$ b. $\dfrac{2}{3} \times \dfrac{5}{7}$

STOP • STOP • STOP • STOP • STOP • STOP • STOP • STOP • STOP

A14   a. $\dfrac{7}{8}$ b. $\dfrac{10}{21}$

Q15   Find the product:

a. $\dfrac{3}{4} \times \dfrac{1}{9}$ b. $\dfrac{7}{8} \times \dfrac{9}{7}$

STOP • STOP • STOP • STOP • STOP • STOP • STOP • STOP • STOP

A15   a. $\dfrac{1}{12}$ b. $1\dfrac{1}{8}$

---

6   The product of mixed numbers can be determined by changing the mixed numbers to improper fractions and multiplying. For example:

1. $2\dfrac{1}{3} \times \dfrac{2}{5} = \dfrac{7}{3} \times \dfrac{2}{5}$ 2. $8\dfrac{1}{2} \times 4\dfrac{2}{3} = \dfrac{17}{2} \times \dfrac{14}{3}$

$\qquad\qquad\quad = \dfrac{14}{15}$ $\qquad\qquad\quad = \dfrac{119}{3}$

$\qquad\qquad\qquad\qquad\qquad\qquad\quad = 39\dfrac{2}{3}$

Q16   Find the product: $2\dfrac{3}{4} \times \dfrac{5}{6}$

STOP • STOP • STOP • STOP • STOP • STOP • STOP • STOP • STOP

A16 $2\dfrac{7}{24}$: $2\dfrac{3}{4} \times \dfrac{5}{6} = \dfrac{11}{4} \times \dfrac{5}{6} = \dfrac{55}{24}$

Q17   Find the product:

a. $\dfrac{5}{8} \times 2\dfrac{1}{5}$ b. $6\dfrac{1}{4} \times \dfrac{4}{5}$

STOP • STOP • STOP • STOP • STOP • STOP • STOP • STOP • STOP

A17   a. $1\dfrac{3}{8}$ b. 5

Q18   Find the product:

  a. $1\frac{2}{3} \times 4\frac{1}{5}$ **b.** $1\frac{6}{7} \times 2\frac{1}{4}$

STOP • STOP • STOP • STOP • STOP • STOP • STOP • STOP • STOP

A18   **a.** 7   **b.** $4\frac{5}{28}$

Q19   John practices hockey every afternoon for $1\frac{3}{4}$ hours. How many hours does he practice in 5 days?

STOP • STOP • STOP • STOP • STOP • STOP • STOP • STOP • STOP

A19 $8\frac{3}{4}$ hours $\left(5 \times 1\frac{3}{4}\right)$

---

7   A division such as $3 \div \frac{1}{2}$ can be interpreted as: How many $\frac{1}{2}$ units are contained in 3 units? There are two $\frac{1}{2}$ units in each whole, so there are six $\frac{1}{2}$ units in 3 units.

Hence, $3 \div \frac{1}{2} = 6$.

Q20   How many $\frac{1}{3}$ units are contained in a whole?_____

STOP • STOP • STOP • STOP • STOP • STOP • STOP • STOP • STOP

A20   3

Q21   How many $\frac{1}{3}$ units are contained in 6 units?_____

STOP • STOP • STOP • STOP • STOP • STOP • STOP • STOP • STOP

A21   18

Q22 $6 \div \frac{1}{3} =$ _____

STOP • STOP • STOP • STOP • STOP • STOP • STOP • STOP • STOP

A22   18

Q23 How many $\frac{1}{4}$ units are contained in 5 units? _____

STOP • STOP • STOP • STOP • STOP • STOP • STOP • STOP • STOP

A23 20

Q24 $5 \div \frac{1}{4} =$ _____

STOP • STOP • STOP • STOP • STOP • STOP • STOP • STOP • STOP

A24 20

---

8 The quotient of two fractions can be determined by changing the division process into an equivalent multiplication process. Consider the example $\frac{3}{7} \div \frac{2}{3}$:

$$\frac{3}{7} \div \frac{2}{3} \text{ is equivalent to } \frac{\frac{3}{7}}{\frac{2}{3}}$$

This fraction can be simplified by multiplying numerator and denominator by $\frac{3}{2}$:

$$\frac{3}{7} \div \frac{2}{3} = \frac{\frac{3}{7}}{\frac{2}{3}} = \frac{\frac{3}{7} \times \frac{3}{2}}{\frac{2}{3} \times \frac{3}{2}} = \frac{\frac{3}{7} \times \frac{3}{2}}{1} = \frac{3}{7} \times \frac{3}{2}$$

The division problem $\frac{3}{7} \div \frac{2}{3}$ has been changed to an equivalent multiplication problem, $\frac{3}{7} \times \frac{3}{2}$.

Q25 Change the division $\frac{2}{3} \div \frac{5}{7}$ into a multiplication problem by completing the missing steps:

$$\frac{2}{3} \div \frac{5}{7} = \frac{\frac{2}{3} \times \frac{(\quad)}{(\quad)}}{\frac{5}{7} \times \frac{7}{5}} = \frac{\frac{(\quad)}{(\quad)} \times \frac{(\quad)}{(\quad)}}{(\quad)} = \frac{(\quad)}{(\quad)} \times \frac{(\quad)}{(\quad)}$$

STOP • STOP • STOP • STOP • STOP • STOP • STOP • STOP • STOP

A25 $\dfrac{\frac{2}{3} \times \frac{7}{5}}{\frac{5}{7} \times \frac{7}{5}} = \dfrac{\frac{2}{3} \times \frac{7}{5}}{1} = \frac{2}{3} \times \frac{7}{5}$

9   The preceding examples verify the rule for dividing fractions: *Invert the divisor and multiply.*
For example:

$$\frac{2}{3} \div \frac{5}{7}$$

(invert the divisor)

$\frac{2}{3} \div \frac{5}{7}$ is equivalent to $\frac{2}{3} \times \frac{7}{5} = \frac{14}{15}$

Hence, $\frac{2}{3} \div \frac{5}{7} = \frac{14}{15}$.

Q26   Which fraction is the divisor?

$$\frac{2}{3} \div \frac{7}{8} \underline{\hspace{2cm}}$$

STOP • STOP • STOP • STOP • STOP • STOP • STOP • STOP • STOP

A26 $\frac{7}{8}$

Q27   The rule for dividing fractions is to $\underline{\hspace{8cm}}$

$\underline{\hspace{12cm}}$

STOP • STOP • STOP • STOP • STOP • STOP • STOP • STOP • STOP

A27   invert the divisor and multiply

Q28   Write the division $\frac{2}{3} \div \frac{7}{8}$ as an equivalent multiplication problem.

STOP • STOP • STOP • STOP • STOP • STOP • STOP • STOP • STOP

A28 $\frac{2}{3} \times \frac{8}{7}$

Q29   Find the quotient: $\frac{2}{3} \div \frac{7}{8}$

STOP • STOP • STOP • STOP • STOP • STOP • STOP • STOP • STOP

A29 $\frac{16}{21}$: $\frac{2}{3} \div \frac{7}{8} = \frac{2}{3} \times \frac{8}{7}$

$$= \frac{16}{21}$$

Q30   Find the quotient: $\dfrac{4}{7} \div \dfrac{8}{9}$

STOP • STOP • STOP • STOP • STOP • STOP • STOP • STOP • STOP

A30 $\dfrac{9}{14}$: $\dfrac{4}{7} \div \dfrac{8}{9} = \dfrac{4}{7} \times \dfrac{9}{8}$

$$= \dfrac{36}{56}$$

Q31   Find the quotient: $\dfrac{4}{5} \div \dfrac{3}{7}$

STOP • STOP • STOP • STOP • STOP • STOP • STOP • STOP • STOP

A31 $1\dfrac{13}{15}$: $\dfrac{4}{5} \div \dfrac{3}{7} = \dfrac{4}{5} \times \dfrac{7}{3}$

$$= \dfrac{28}{15}$$

Q32   Find the quotient:

     a. $\dfrac{4}{7} \div \dfrac{3}{5}$ **b.** $\dfrac{3}{4} \div \dfrac{5}{8}$

STOP • STOP • STOP • STOP • STOP • STOP • STOP • STOP • STOP

A32   **a.** $\dfrac{20}{21}$ **b.** $1\dfrac{1}{5}$

Q33   Find the quotient:

     a. $\dfrac{6}{7} \div \dfrac{9}{10}$ **b.** $\dfrac{6}{7} \div \dfrac{12}{13}$

STOP • STOP • STOP • STOP • STOP • STOP • STOP • STOP • STOP

A33   **a.** $\dfrac{20}{21}$ **b.** $\dfrac{13}{14}$

---

10   The quotient of $\frac{3}{4} \div 6$ can be determined as follows: $6 = \frac{6}{1}$; hence,

$$\frac{3}{4} \div 6 \text{ is equivalent to } \frac{3}{4} \div \frac{6}{1}$$

$$\frac{3}{4} \div \frac{6}{1} = \frac{3}{4} \times \frac{1}{6}$$

$$= \frac{1}{8}$$

Q34 $\frac{2}{3} \div 5$ is equivalent to $\frac{2}{3} \div \left(\ \ \right)$.

STOP • **STOP** • **STOP** • **STOP** • **STOP** • **STOP** • **STOP** • **STOP** • **STOP**

A34 $\frac{5}{1}$

Q35 $\frac{2}{3} \div \frac{5}{1} = \frac{2}{3} \times \left(\ \ \right)$

STOP • **STOP** • **STOP** • **STOP** • **STOP** • **STOP** • **STOP** • **STOP** • **STOP**

A35 $\frac{1}{5}$

Q36   Find the quotient: $\frac{2}{3} \div 5$

STOP • **STOP** • **STOP** • **STOP** • **STOP** • **STOP** • **STOP** • **STOP** • **STOP**

A36 $\frac{2}{15}$: $\frac{2}{3} \div \frac{5}{1} = \frac{2}{3} \times \frac{1}{5}$

$$= \frac{2}{15}$$

Q37   Find the quotient: $\frac{2}{5} \div 10$

STOP • **STOP** • **STOP** • **STOP** • **STOP** • **STOP** • **STOP** • **STOP** • **STOP**

A37 $\frac{1}{25}$: $\frac{2}{5} \div \frac{10}{1} = \frac{2}{5} \times \frac{1}{10}$

$$= \frac{1}{25}$$

Q38 Find the quotient: $3 \div \dfrac{2}{3}$

STOP • STOP • STOP • STOP • STOP • STOP • STOP • STOP • STOP

A38 $4\dfrac{1}{2}$: $\quad 3 \div \dfrac{2}{3} = 3 \times \dfrac{3}{2}$

$\qquad\qquad = \dfrac{9}{2}$

Q39 Find the quotient: $3\dfrac{3}{4} \div \dfrac{3}{4}$

STOP • STOP • STOP • STOP • STOP • STOP • STOP • STOP • STOP

A39 5: $\quad 3\dfrac{3}{4} \div \dfrac{3}{4} = \dfrac{15}{4} \div \dfrac{3}{4}$

$\qquad\qquad = \dfrac{15}{4} \times \dfrac{4}{3}$

$\qquad\qquad = 5$

Q40 Find the quotient:

a. $\dfrac{6}{11} \div \dfrac{5}{11}$ 　　　　 b. $9 \div \dfrac{2}{3}$

STOP • STOP • STOP • STOP • STOP • STOP • STOP • STOP • STOP

A40 a. $1\dfrac{1}{5}$ 　　　　 b. $13\dfrac{1}{2}$

Q41 Find the quotient:

a. $\dfrac{6}{7} \div 3\dfrac{1}{4}$ 　　　　 b. $4\dfrac{2}{3} \div 3\dfrac{1}{4}$

STOP • STOP • STOP • STOP • STOP • STOP • STOP • STOP • STOP

A41 a. $\dfrac{24}{91}$ 　　　　 b. $1\dfrac{17}{39}$

Q42   Find the quotient:

   a. $2\frac{1}{2} \div 10$ **b.** $6\frac{1}{2} \div 4\frac{1}{3}$

STOP • STOP • STOP • STOP • STOP • STOP • STOP • STOP • STOP

A42   **a.** $\frac{1}{4}$ **b.** $1\frac{1}{2}$

Q43   Don went on a 500-mile trip this summer. He traveled $\frac{2}{5}$ of the way by train and the rest of the way by bus. How many miles did Don travel by train?_____ By bus?_____

STOP • STOP • STOP • STOP • STOP • STOP • STOP • STOP • STOP

A43   200 miles by train $\left(\frac{2}{5} \times 500\right)$, 300 miles by bus $\left(\frac{3}{5} \times 500\right)$

Q44   Mr. Pumford cut a large piece of meat weighing $18\frac{2}{3}$ pounds into pieces weighing $1\frac{3}{4}$ pounds each. How many smaller pieces did he get?_____

STOP • STOP • STOP • STOP • STOP • STOP • STOP • STOP • STOP

A44 $10\frac{2}{3}$ pieces: $\left(18\frac{2}{3} \div 1\frac{3}{4}\right)$

This completes the instruction for this section.

2.6   **Exercises**

   1. Find the products:

   a. $\frac{1}{2} \times \frac{3}{4}$ **b.** $\frac{1}{7} \times \frac{3}{11}$ **c.** $16 \times \frac{3}{4}$ **d.** $30 \times \frac{2}{15}$

   e. $\frac{1}{3} \times \frac{1}{6}$ **f.** $\frac{1}{8} \times 16$ **g.** $\frac{4}{9} \times 6$ **h.** $4 \times \frac{9}{10}$

   i. $\frac{4}{5} \times \frac{3}{8}$ **j.** $\frac{3}{4} \times \frac{12}{16}$ **k.** $\frac{4}{7} \times \frac{3}{8}$ **l.** $3\frac{1}{3} \times \frac{3}{5}$

   m. $2\frac{2}{5} \times \frac{5}{8}$ **n.** $2\frac{1}{7} \times 1\frac{2}{5}$ **o.** $1\frac{7}{8} \times 2\frac{1}{5}$

2. Find the quotients:

   a. $\dfrac{8}{9} \div \dfrac{3}{4}$ b. $\dfrac{5}{9} \div \dfrac{3}{4}$ c. $8 \div \dfrac{1}{4}$ d. $6 \div \dfrac{5}{6}$

   e. $5\dfrac{1}{4} \div 3\dfrac{1}{2}$ f. $6\dfrac{1}{4} \div 1\dfrac{3}{5}$ g. $8 \div 2\dfrac{1}{2}$ h. $6\dfrac{7}{8} \div \dfrac{11}{14}$

   i. $\dfrac{6}{7} \div 8$

3. A bag contains 25 pounds of fertilizer. How many $1\dfrac{1}{3}$-pound bags can be filled from the 25-pound bag?

4. A housing development requires $\dfrac{2}{5}$ of an acre for a house. How many acres will be needed to build 147 houses?

2.6 Exercise Answers

1. a. $\dfrac{3}{8}$ b. $\dfrac{3}{77}$ c. 12   d. 4   e. $\dfrac{1}{18}$ f. 2   g. $2\dfrac{2}{3}$

   h. $3\dfrac{3}{5}$ i. $\dfrac{3}{10}$ j. $\dfrac{9}{16}$ k. $\dfrac{3}{14}$ l. 2   m. $1\dfrac{1}{2}$ n. 3

   o. $4\dfrac{1}{8}$

2. a. $1\dfrac{5}{27}$ b. $\dfrac{20}{27}$ c. 32   d. $7\dfrac{1}{5}$ e. $1\dfrac{1}{2}$ f. $3\dfrac{29}{32}$ g. $3\dfrac{1}{5}$

   h. $8\dfrac{3}{4}$ i. $\dfrac{3}{28}$

3. $18\dfrac{3}{4}$ bags

4. $58\dfrac{4}{5}$ acres

Chapter 2 Sample Test

At the completion of Chapter 2 it is expected that you will be able to work the following problems.

2.1 Equivalent Fractions

1. Supply the missing value:

   a. $\dfrac{4}{5} = \dfrac{?}{35}$ b. $9 = \dfrac{?}{8}$

2. Reduce to lowest terms:

   a. $\dfrac{18}{24}$ **b.** $\dfrac{34}{51}$

2.2   Mixed Numbers and Improper Fractions

3. a. Change $5\dfrac{5}{8}$ to an improper fraction.

   b. Express $\dfrac{516}{8}$ as a mixed number.

2.3   Addition

4. Add:

  a. $\dfrac{1}{15} + \dfrac{9}{15}$ **b.** $3\dfrac{5}{12} + 2\dfrac{11}{12}$ **c.** $\dfrac{1}{2} + \dfrac{1}{7}$ **d.** $5\dfrac{5}{6} + 4\dfrac{7}{9}$

2.4   Prime and Composite Numbers

5. List the prime numbers between 20 and 30.

2.5   Subtraction

6. Subtract:

  a. $\dfrac{5}{6} - \dfrac{5}{12}$ **b.** $4\dfrac{5}{8} - 2\dfrac{1}{6}$ **c.** $7 - 2\dfrac{3}{5}$ **d.** $5\dfrac{7}{12} - 3\dfrac{7}{8}$

2.6   Multiplication and Division

7. Multiply:

  a. $\dfrac{3}{4} \times \dfrac{8}{11}$ **b.** $1\dfrac{1}{4} \times 1\dfrac{2}{3}$ **c.** $3\dfrac{1}{2} \times 6$ **d.** $8 \times \dfrac{1}{16}$

8. Divide:

  a. $3 \div \dfrac{1}{3}$ **b.** $\dfrac{2}{5} \div \dfrac{5}{6}$ **c.** $3\dfrac{1}{3} \div 2$ **d.** $2\dfrac{1}{2} \div \dfrac{5}{2}$

Chapter 2 Sample Test Answers

1. **a.** 28 **b.** 72

2. **a.** $\dfrac{3}{4}$ **b.** $\dfrac{2}{3}$

3. **a.** $\dfrac{45}{8}$ **b.** $64\dfrac{1}{2}$

4. **a.** $\dfrac{2}{3}$ **b.** $6\dfrac{1}{3}$ **c.** $\dfrac{9}{14}$ **d.** $10\dfrac{11}{18}$

5. 23, 29

6. **a.** $\dfrac{5}{12}$ **b.** $2\dfrac{11}{24}$ **c.** $4\dfrac{2}{5}$ **d.** $1\dfrac{17}{24}$

7. **a.** $\dfrac{6}{11}$ **b.** $2\dfrac{1}{12}$ **c.** 21 **d.** $\dfrac{1}{2}$

8. **a.** 9 **b.** $\dfrac{12}{25}$ **c.** $1\dfrac{2}{3}$ **d.** 1

Chapter 3

Operations with Decimal Fractions

A fraction was defined, in Chapter 2, as the quotient of two whole numbers with a nonzero denominator. These types of fractions could be called *common fractions*. The fractions discussed in this chapter are a type of fraction and are called *decimal fractions*.

3.1   Place Value

---

1   The Hindu–Arabic number system is called a *decimal system* because there are 10 digits and the place values are based on 10. The place values, from right to left, are ones, tens, hundreds, thousands, and so on.

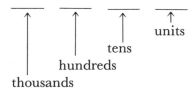

Each place value to the left of another has a value of 10 times the preceding value; that is, the thousands place is 10 times the hundreds place ($1,000 = 10 \times 100$), the hundreds place is 10 times the tens place ($100 = 10 \times 10$), and the tens place is 10 times the ones place ($10 = 10 \times 1$). Or, each place value to the right of another is one-tenth $\left(\frac{1}{10}\right)$ of the preceding place. Thus, the value to the right of the thousands place is $\frac{1}{10}$ of 1,000, or 100 (hundreds place, $\frac{1}{10} \times 1,000 = 100$).

Q1   The value of the place to the right of the hundreds place is $\frac{1}{10} \times$ _____ = _____.

STOP • STOP • STOP • STOP • STOP • STOP • STOP • STOP • STOP

A1   100, 10

Q2   The value of the place to the right of the tens place is $\frac{1}{10} \times$ _____ = _____.

STOP • STOP • STOP • STOP • STOP • STOP • STOP • STOP • STOP

A2   10, 1

Q3 Continuing with the above pattern, the value of the place to the right of the ones place is $\frac{1}{10} \times$ _____ = _____.

STOP • STOP • STOP • STOP • STOP • STOP • STOP • STOP • STOP

A3 $1, \frac{1}{10}$

Q4 The value of the place to the right of the tenths place is $\frac{1}{10} \times$ _____ = _____.

STOP • STOP • STOP • STOP • STOP • STOP • STOP • STOP • STOP

A4 $\frac{1}{10}, \frac{1}{100}$

---

2 Question 3 suggests place value to the right of the ones place. This idea can be extended with the introduction of decimal notation. A *decimal fraction* is a fraction whose denominator is 10, 100, 1,000, and so on (a power of 10). Decimal fractions are usually written, without denominators, with a *decimal point* representing place value:

$\frac{1}{10}$ is written 0.1 (one tenth)

$\frac{1}{100}$ is written 0.01 (one hundredth)

The place value one place to the right of the ones place is tenths. The digit 1 in 0.1 is one place to the right of the ones place.

The place value one place to the right of the tenths place is hundredths. The digit 1 in 0.01 is one place to the right of the tenths place:

$\frac{5}{10} = 0.5$ $\frac{8}{100} = 0.08$

$\frac{6}{10} = 0.6$ $\frac{3}{100} = 0.03$

Q5 Write the following decimal fractions using decimal notation:

a. $\frac{3}{10}$ _____   b. $\frac{9}{10}$ _____   c. $\frac{2}{100}$ _____

STOP • STOP • STOP • STOP • STOP • STOP • STOP • STOP • STOP

A5 a. 0.3   b. 0.9   c. 0.02

Q6 Write each of the following as a fraction:

a. 0.7 _____   b. 0.07 _____   c. 0.01 _____

STOP • STOP • STOP • STOP • STOP • STOP • STOP • STOP • STOP

A6 a. $\frac{7}{10}$ b. $\frac{7}{100}$ c. $\frac{1}{100}$

Q7   **a.** The place value one place to the right of the ones place is _____.

         b. The place value two places to the right of the ones place is _____.

STOP • STOP • STOP • STOP • STOP • STOP • STOP • STOP • STOP

A7   **a.** tenths   **b.** hundredths

---

3   Below is a diagram that shows some of the values of the places to the right of the decimal point. It should be used to determine the place value of numbers written in decimal notation. Decimal fractions written using decimal notation are simply called *decimals*.

decimal point

↓

. _ _ _ _ _ _ _

tenths
hundredths
thousandths
ten-thousandths
hundred-thousandths
millionths
ten-millionths
etc.

For the numeral 0.18793:

The place value of the 1 is tenths.

The place value of the 8 is hundredths.

The place value of the 7 is thousandths.

The place value of the 9 is ten-thousandths.

The place value of the 3 is hundred-thousandths.

Q8   In the numeral 21.0874, what digit is in the

      a. tenths place?   _____

      b. tens place?   _____

      c. hundredths place?   _____

      d. ones place?   _____

      e. ten-thousandths place?   _____

      f. thousandths place?   _____

STOP • STOP • STOP • STOP • STOP • STOP • STOP • STOP • STOP

A8   **a.** 0   **b.** 2   **c.** 8   **d.** 1   **e.** 4   **f.** 7

Q9   Determine the place value of each numeral in 93.1726.

      9 _____   3 _____   1 _____

      7 _____   2 _____   6 _____

STOP • STOP • STOP • STOP • STOP • STOP • STOP • STOP • STOP

A9   9 (tens), 3 (ones), 1 (tenths), 7 (hundredths), 2 (thousandths), 6 (ten-thousandths)

Q10   **a.** In the decimal 0.28, the 8 means _____.

b. In the decimal 3.107, the 7 means _____.

STOP • STOP • STOP • STOP • STOP • STOP • STOP • STOP • STOP • STOP

A10   **a.** $\dfrac{8}{100}$ **b.** $\dfrac{7}{1,000}$

---

4   The decimal 0.13 can be expressed as

$$0.13 = \frac{1}{10} + \frac{3}{100}$$
$$= \frac{1 \times 10}{10 \times 10} + \frac{3}{100}$$
$$= \frac{10}{100} + \frac{3}{100}$$
$$= \frac{13}{100}$$

The decimal 0.28 can be expressed as

$$0.28 = \frac{2}{10} + \frac{8}{100}$$
$$= \frac{2 \times 10}{10 \times 10} + \frac{8}{100}$$
$$= \frac{20}{100} + \frac{8}{100}$$
$$= \frac{28}{100}$$

The decimal 0.013 can be expressed as

$$0.013 = \frac{0}{10} + \frac{1}{100} + \frac{3}{1,000}$$
$$= 0 + \frac{1 \times 10}{100 \times 10} + \frac{3}{1,000}$$
$$= \frac{10}{1,000} + \frac{3}{1000}$$
$$= \frac{13}{1,000}$$

Q11   Express the decimal as a fraction by completing the following:

$$0.17 = \frac{1}{10} + \frac{7}{100}$$
$$= \frac{1 \times (\quad)}{10 \times (\quad)} + \frac{7}{100}$$
$$= (\qquad) + \frac{7}{100}$$
$$= \rule{2cm}{0.4pt}$$

STOP • STOP • STOP • STOP • STOP • STOP • STOP • STOP • STOP • STOP

A11
$$0.17 = \frac{1}{10} + \frac{7}{100}$$
$$= \frac{1 \times 10}{10 \times 10} + \frac{7}{100}$$
$$= \frac{10}{100} + \frac{7}{100}$$
$$= \frac{17}{100}$$

Q12   Express the decimals as fractions (do not reduce):

   a. 0.06 _____   **b.** 0.27 _____   **c.** 0.81 _____

STOP • STOP • STOP • STOP • STOP • STOP • STOP • STOP • STOP

A12   **a.** $\frac{6}{100}$ **b.** $\frac{27}{100}$ **c.** $\frac{81}{100}$

Q13   Express the decimal as a fraction by completing the following:

$$0.273 = \frac{2}{10} + \frac{7}{100} + \frac{3}{(\quad)}$$
$$= \frac{2 \times 100}{10 \times 100} + \frac{7 \times (\quad)}{100 \times (\quad)} + \frac{3}{1,000}$$
$$= (\qquad) + (\qquad) + (\qquad)$$
$$= \underline{\qquad}$$

STOP • STOP • STOP • STOP • STOP • STOP • STOP • STOP • STOP

A13
$$0.273 = \frac{2}{10} + \frac{7}{100} + \frac{3}{1,000}$$
$$= \frac{2 \times 100}{10 \times 100} + \frac{7 \times 10}{100 \times 10} + \frac{3}{1,000}$$
$$= \frac{200}{1,000} + \frac{70}{1,000} + \frac{3}{1,000}$$
$$= \frac{273}{1,000}$$

Q14   Express the decimals as fractions (do not reduce; do the problem mentally if possible):

   a. 0.213 _____   **b.** 0.085 _____   **c.** 0.372 _____

STOP • STOP • STOP • STOP • STOP • STOP • STOP • STOP • STOP

A14   **a.** $\frac{213}{1,000}$ **b.** $\frac{85}{1,000}$ **c.** $\frac{372}{1,000}$

Q15   Express the fractions as decimals:

   a. $\frac{3}{1,000}$ _____   **b.** $\frac{158}{1,000}$ _____   **c.** $\frac{29}{1,000}$ _____

STOP • STOP • STOP • STOP • STOP • STOP • STOP • STOP • STOP

A15   **a.** 0.003   **b.** 0.158   **c.** 0.029

---

5   A mixed number can be represented as a decimal if the fraction is a decimal fraction, and conversely.

Example: Write $2\frac{3}{10}$ as a decimal.

Solution

$2\frac{3}{10} = 2.3$

Two additional examples are

$2\frac{17}{100} = 2.17 \qquad 25\frac{3}{1,000} = 25.003$

Q16   Convert to decimals:

a. $17\frac{7}{10}$ _____   b. $12\frac{8}{100}$ _____   c. $201\frac{5}{1,000}$

STOP • **STOP** • **STOP** • **STOP** • **STOP** • **STOP** • **STOP** • **STOP** • **STOP**

A16   a. 17.7   b. 12.08   c. 201.005

Q17   Convert to fractions (do not reduce):

a. 213.7 _____   b. 2.002 _____   c. 9.16 _____

STOP • **STOP** • **STOP** • **STOP** • **STOP** • **STOP** • **STOP** • **STOP** • **STOP**

A17   a. $213\frac{7}{10}$ b. $2\frac{2}{1,000}$ c. $9\frac{16}{100}$

---

6   To read a decimal, first read the number involved and then give the place value in which the last digit of the number lies.

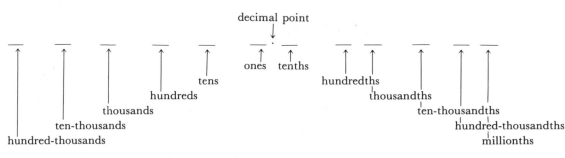

Thus,

0.0134 is read "one hundred thirty-four *ten-thousandths*."

0.85 is read "eighty-five *hundredths*."

0.00203 is read "two hundred three *hundred-thousandths*."

0.045 is read "forty-five *thousandths*."

To read a number consisting of a whole number and a decimal:

Step 1: Read the whole number.

Step 2: Read "and" for the decimal point.

Step 3: Read the decimal as above.

Examples:

1. 12.03 is read "twelve and three hundredths."
2. 5,280.123 is read "five thousand two hundred eighty and one hundred twenty-three thousandths."

Q18   Write the decimals in words:

    a. 1.03 _____

    b. 42.13 _____

    c. 12.0439 _____

    d. 10,800.876 _____

                  _____

    e. 400.002 _____

    f. 0.402 _____

STOP • STOP • STOP • STOP • STOP • STOP • STOP • STOP • STOP

A18   **a.** one and three hundredths
    b. forty-two and thirteen hundredths
    c. twelve and four hundred thirty-nine ten-thousandths
    d. ten thousand eight hundred and eight hundred seventy-six thousandths
    e. four hundred and two thousandths
    f. four hundred two thousandths

Q19   When reading decimals, the decimal point is read "_____."

STOP • STOP • STOP • STOP • STOP • STOP • STOP • STOP • STOP

A19   and

---

7   When reading a number, it is important to remember that *and* is used to connect a whole number and a decimal, but it is not used anywhere else in the number:

    two hundred thirty-six and fifty-two thousandths

    236.052

Q20   Write as decimals:

    a. two and six tenths _____

    b. fifteen and seventy-five hundredths _____

    c. one thousand fifty-seven and six thousandths _____

    d. four thousand five hundred seven and forty-three hundred-thousandths

    _____

STOP • STOP • STOP • STOP • STOP • STOP • STOP • STOP • STOP

A20   **a.** 2.6   **b.** 15.75   **c.** 1,057.006   **d.** 4,507.00043

Q21   Write the decimals in words:

    a. 0.3 _____

    b. 0.32 _____

    c. 6.35 _____

    d. 2.003 _____

    e. 50.052 _____

STOP • **STOP** • **STOP** • **STOP** • **STOP** • **STOP** • **STOP** • **STOP** • **STOP**

A21   **a.** three tenths
    b. thirty-two hundredths
    c. six and thirty-five hundredths
    d. two and three thousandths
    e. fifty and fifty-two thousandths

Q22   Write as decimals:

    a. five and six tenths _____

    b. twenty-five ten-thousandths _____

    c. two hundred eight and five hundredths _____

    d. two thousand fifteen ten-thousandths _____

    e. two thousand and fifteen ten-thousandths _____

STOP • **STOP** • **STOP** • **STOP** • **STOP** • **STOP** • **STOP** • **STOP** • **STOP**

A22   **a.** 5.6   **b.** 0.0025   **c.** 208.05   **d.** 0.2015
    e. 2,000.0015

This completes the instruction for this section.

3.1   Exercises

  1. What is the place value of the 7 in the following numbers?
    a. 13.872   **b.** 702.01   **c.** 0.0007

  2. In a number the word "and" represents the _____.

  3. Write the following decimals as a fraction or mixed number (do not reduce):
    a. 0.704   **b.** 0.0302   **c.** 2.07   **d.** 0.00045
    e. 17.108   **f.** 2.007

  4. Write as decimals:

    a. $\dfrac{19}{1,000}$ **b.** $5\dfrac{5}{100}$ **c.** $7\dfrac{9}{10}$ **d.** $\dfrac{72}{100}$

    e. $\dfrac{204}{10,000}$ **f.** $\dfrac{32}{100,000}$

  5. Write the decimals in words:
    a. 0.1   **b.** 0.07   **c.** 7.07   **d.** 21.4805

  6. Write as a decimal:
    a. three and fifty-seven hundredths
    b. six thousand four ten-thousandths
    c. six thousand and four ten-thousandths
    d. thirty-three and seventy-eight hundredths

3.1   Exercise Answers

1. **a.** hundredths   **b.** hundreds   **c.** ten-thousandths
2. decimal point

3. **a.** $\dfrac{704}{1,000}$ **b.** $\dfrac{302}{10,000}$ **c.** $2\dfrac{7}{100}$ **d.** $\dfrac{45}{100,000}$

   e. $17\dfrac{108}{1,000}$ **f.** $2\dfrac{7}{1,000}$

4. **a.** 0.019   **b.** 5.05   **c.** 7.9   **d.** 0.72
   e. 0.0204   **f.** 0.00032

5. **a.** one tenth
   b. seven hundredths
   c. seven and seven hundredths
   d. twenty-one and four thousand eight hundred five ten-thousandths

6. **a.** 3.57   **b.** 0.6004   **c.** 6,000.0004   **d.** 33.78

3.2   Rounding Off

---

1   Measurements are never exact. In measuring the length of something, you may measure to the nearest inch, the nearest 0.1 inch, the nearest 0.01 inch, or even closer. Each measurement is an approximation of the length. The measurement recorded is usually a number that has been rounded off.

Since a digit several places to the right of the decimal point represents a very small quantity, a number that has several decimal places is frequently "rounded off" to two or three decimal places. *Rounding off* a number is a way of saying that the number is correct to a specified number of decimal places. To round off a decimal number, look at the digit immediately to the right of the digit that you are rounding off to.

1. If this digit is 5 or larger, the digit in the place you are rounding off to is increased by 1 and all digits to the right are replaced by zeros.
2. If this digit is 4 or smaller, the digit in the place you are rounding off to is left as it is and all digits to the right are replaced by zeros. If these zeros are to the right of the decimal point, they are dropped.

Examples: Round off to the nearest hundredth:

1. 27.8463 to the nearest hundredth is 27.8500
      5 or larger   or   27.85
2. 0.7349 to the nearest hundredth is 0.7300
      4 or smaller   or   0.73

Q1   When rounding off 4.763 to the nearest tenth, which digit must be 5 or larger, or 4 or smaller? (Circle the answer.)

STOP • **STOP** • **STOP** • **STOP** • **STOP** • **STOP** • **STOP** • **STOP** • **STOP**

A1   4.7⑥3

Q2   Round off to the nearest tenth:

   a. 4.763 _____   **b.** 5.031 _____   **c.** 76.349 _____

STOP • **STOP** • **STOP** • **STOP** • **STOP** • **STOP** • **STOP** • **STOP** • **STOP**

A2	**a.** 4.8	**b.** 5.0	**c.** 76.3

Q3	Round off to the nearest tenth:

a. 76.352 _____   **b.** 89.015 _____   **c.** 7.555 _____

STOP • STOP • STOP • STOP • STOP • STOP • STOP • STOP • STOP

A3	**a.** 76.4	**b.** 89.0	**c.** 7.6

Q4	When rounding off 5.722 to the nearest hundredth, which digit must be 5 or larger, or 4 or smaller? (Circle the answer.)

STOP • STOP • STOP • STOP • STOP • STOP • STOP • STOP • STOP

A4	5.72②

Q5	Round off to the nearest hundredth:

a. 0.385 _____   **b.** 0.296 _____   **c.** 5.722 _____

STOP • STOP • STOP • STOP • STOP • STOP • STOP • STOP • STOP

A5	**a.** 0.39	**b.** 0.30	**c.** 5.72

Q6	Round off to the nearest hundredth:

a. 0.0782 _____   **b.** 6.0035 _____   **c.** 2.8347 _____

STOP • STOP • STOP • STOP • STOP • STOP • STOP • STOP • STOP

A6	**a.** 0.08	**b.** 6.00	**c.** 2.83

Q7	Round off to the nearest thousandth:

a. 0.0672 _____   **b.** 0.05550 _____   **c.** 0.6338 _____

STOP • STOP • STOP • STOP • STOP • STOP • STOP • STOP • STOP

A7	**a.** 0.067	**b.** 0.056	**c.** 0.634

Q8	Round off to the nearest thousandth:

a. 17.36371 _____   **b.** 0.3999 _____   **c.** 0.00191 _____

STOP • STOP • STOP • STOP • STOP • STOP • STOP • STOP • STOP

A8	**a.** 17.364	**b.** 0.400	**c.** 0.002

Q9	Round off to the nearest thousand:

a. 7,398 _____   **b.** 62,275 _____   **c.** 9,872.5 _____

STOP • STOP • STOP • STOP • STOP • STOP • STOP • STOP • STOP

A9	**a.** 7,000	**b.** 62,000	**c.** 10,000

Q10	Round off to the nearest whole number (ones place):

a. 479.23 _____   **b.** 6.872 _____   **c.** 17.50 _____

STOP • STOP • STOP • STOP • STOP • STOP • STOP • STOP • STOP

A10	**a.** 479	**b.** 7	**c.** 18

This completes the instruction for this section.

3.2   Exercises

Round off each number as indicated:

1. 4.54716
   a. nearest hundredth   **b.** nearest thousandth   **c.** nearest tenth
2. 8.58637
   a. nearest hundredth   **b.** nearest ten-thousandth
   c. nearest thousandth
3. 2.07462
   a. nearest hundredth   **b.** nearest thousandth   **c.** nearest tenth
4. 3.098736
   a. nearest hundredth   **b.** three decimals   **c.** one decimal
5. 1.70954
   a. two decimals   **b.** three decimals   **c.** one decimal

3.2   Exercise Answers

1. **a.** 4.55   **b.** 4.547   **c.** 4.5
2. **a.** 8.59   **b.** 8.5864   **c.** 8.586
3. **a.** 2.07   **b.** 2.075   **c.** 2.1
4. **a.** 3.10   **b.** 3.099   **c.** 3.1
5. **a.** 1.71   **b.** 1.710   **c.** 1.7

3.3   **Addition and Subtraction**

1   When adding two or more decimals it will be necessary to add digits that represent the same place value. This is illustrated in an example:

$$0.13 = \frac{1}{10} + \frac{3}{100}$$
$$+0.24 = \frac{2}{10} + \frac{4}{100}$$
$$0.37 = \frac{3}{10} + \frac{7}{100}$$

It is not necessary to write decimals as fractions to find their sum; however, it will be helpful to arrange the numbers in vertical columns so that tenths are added to tenths, hundredths are added to hundredths, and so on.

Q1   Find the sum: 0.34
                    +0.23

STOP • STOP • STOP • STOP • STOP • STOP • STOP • STOP • STOP

A1   0.57

Q2   Add 0.316 + 0.472 by arranging vertically.

STOP • STOP • STOP • STOP • STOP • STOP • STOP • STOP • STOP

A2   0.788:   0.316
                    +0.472
                     ‾‾‾‾‾
                     0.788

---

2   If the decimal numbers in an addition problem are arranged correctly the decimal points will lie on a vertical line.

Example: Find the sum 53.7 + 0.036 + 9.12 + 0.0005.

Solution

53.7
 0.036
 9.12
+0.0005
‾‾‾‾‾‾
62.8565   (decimal points aligned)

Q3   Align the decimals correctly: 273.1204 + 7.982

STOP • STOP • STOP • STOP • STOP • STOP • STOP • STOP • STOP

A3   273.1204
           +7.982
            ‾‾‾‾‾‾

Q4   Align the decimals correctly: 0.3 + 0.84 + 0.7 + 1.213

STOP • STOP • STOP • STOP • STOP • STOP • STOP • STOP • STOP

A4   0.3
           0.84
           0.7
          +1.213
           ‾‾‾‾‾

> **3** Zeros are often used to "square off" the columns of decimals as a means of keeping the column additions correct. The sum $53.7 + 0.036 + 9.12 + 0.0005$ could be written as follows by using zeros to "square off" the decimals.
>
> ```
>   53.7000
>    0.0360
>    9.1200
> + 0.0005
>   62.8565
> ```

Q5   Align and square off the decimals: $273.1204 + 7.982$

STOP • STOP • STOP • STOP • STOP • STOP • STOP • STOP • STOP

A5
```
  273.1204
+   7.9820
```

Q6   Align and square off the decimals: $0.3 + 0.84 + 0.7 + 1.213$

STOP • STOP • STOP • STOP • STOP • STOP • STOP • STOP • STOP

A6
```
  0.300
  0.840
  0.700
+ 1.213
```

Q7   Find the sum:

    a.   ```0.92```   **b.**   ```0.51```   **c.**   ```1.38```

```
a.   0.92          b.   0.51          c.   1.38
   +0.27                0.30               0.92
                      +0.84              +3.54
```

STOP • STOP • STOP • STOP • STOP • STOP • STOP • STOP • STOP

A7   **a.** 1.19   **b.** 1.65   **c.** 5.84

Q8   Find the sum:
    a. $4.09 + 3.987$ **b.** $7.56 + 2.3 + 5.879$

STOP • STOP • STOP • STOP • STOP • STOP • STOP • STOP • STOP

A8   **a.** 8.077:
$$\begin{array}{r} 4.090 \\ +3.987 \\ \hline 8.077 \end{array}$$

b. 15.739:
$$\begin{array}{r} 7.560 \\ 2.300 \\ +5.879 \\ \hline 15.739 \end{array}$$

Q9   Find the sum:
a. 8.632 + 0.234 + 0.81 + 0.065
b. 0.38 + 2.1 + 3.09 + 0.075 + 1.0004

STOP • STOP • STOP • STOP • STOP • STOP • STOP • STOP • STOP

A9   **a.** 9.741   **b.** 6.6454

Q10   Bob Fogg drove 681 miles on a hunting trip. He bought gas three times and paid $5.72, $6.10, and $5.34. His meals cost $9.60, and he paid $12 for a motel room. What was the total of his expenses?

STOP • STOP • STOP • STOP • STOP • STOP • STOP • STOP • STOP

A10 $38.76

Q11   Linda Fogg bought a dress for $24.63, a pair of shoes for $19.99, a handbag for $7, and a hat for $12.33. What was the total cost of these items?

STOP • STOP • STOP • STOP • STOP • STOP • STOP • STOP • STOP

A11 $63.95

---

4   When subtracting decimals it is especially important to use zeros to "square off" decimals.

Example: Find the difference 5.2 − 3.672.

Solution
$$\begin{array}{r} 5.200 \\ -3.672 \\ \hline 1.528 \end{array}$$

Q12   Find the difference:

       a.   4.9   **b.**   15.86

            −3.6   −12.78

STOP • **STOP** • **STOP** • **STOP** • **STOP** • **STOP** • **STOP** • **STOP** • **STOP**

A12   **a.**   1.3   **b.**   3.08

Q13   Find the difference:

       a.   2.63 − 1.7   **b.**   9.06 − 0.6

STOP • **STOP** • **STOP** • **STOP** • **STOP** • **STOP** • **STOP** • **STOP** • **STOP**

A13   **a.**   0.93   **b.**   8.46

Q14   Find the difference:

       a.   0.15 − 0.0367   **b.**   5 − 0.9163

STOP • **STOP** • **STOP** • **STOP** • **STOP** • **STOP** • **STOP** • **STOP** • **STOP**

A14   **a.**   0.1133   **b.**   4.0837

Q15   A block of copper 1 foot square and 1 inch thick weighs 45.835 pounds. A block of steel the same size weighs 40.809 pounds. How much heavier is the block of copper?

STOP • **STOP** • **STOP** • **STOP** • **STOP** • **STOP** • **STOP** • **STOP** • **STOP**

A15   5.026 pounds: 45.835

                            −40.809

This completes the instruction for this section.

3.3   Exercises

1. Find the sum:
     a.   6.237 + 1.986   **b.**   3.819 + 8.23 + 1.7
     c.   7.3 + 2.9186 + 1.79   **d.**   17.086 + 43.509 + 18.762
     e.   0.768 + 73.8 + 4.680   **f.**   453 + 289 + 387.6
     g.   0.003 + 600.01 + 10.1   **h.**   0.408 + 0.2976 + 0.34567

2. Find the difference:
     a.   43.62 − 37.96   **b.**   47 − 8.3   **c.**   4 − 0.68379   **d.**   97.8 − 0.4568
     e.   52.6 − 9.002   **f.**   4,006.1 − 969   **g.**   0.05 − 0.005   **h.** $45.90 − $3.86

3.3 Exercise Answers

1.	a. 8.223	b. 13.749	c. 12.0086	d. 79.357
	e. 79.248	f. 1129.6	g. 610.113	h. 1.05127
2.	a. 5.66	b. 38.7	c. 3.31621	d. 97.3432
	e. 43.598	f. 3,037.1	g. 0.045	h. $42.04

3.4 Multiplication and Division

---

1   The multiplication of decimals is carried out in the same way as multiplication of whole numbers, except for the placing of the decimal point in the product. Consider the product 0.2×0.3 in fraction form:

$$0.2 \times 0.3 = \frac{2}{10} \times \frac{3}{10} = \frac{6}{100} = 0.06$$

Notice that the product of a one-place decimal (tenths) by a one-place decimal (tenths) is a two-place decimal (hundredths). That is, tenths × tenths = hundredths.

In general, *the decimal places of the product will equal the sum of the decimal places of the factors.* For example,

$$
\begin{array}{r}
12.13 \quad \text{(2 decimal places)} \\
\times 0.212 \quad \text{(3 decimal places)} \\
\hline
2426 \\
1213 \\
2\,426 \\
\hline
2.57156 \quad \text{(5 decimal places)}
\end{array}
$$

Q1   How many decimal places will there be in the product 2.68×0.1703? _____

STOP • STOP • STOP • STOP • STOP • STOP • STOP • STOP • STOP

A1   6: because 2.68 has 2 decimal places, and 0.1703 has 4 decimal places. Hence, the product has $4 + 2 = 6$ decimal places.

Q2   How many decimal places will be in the product?

     a. 8×7.23 _____   **b.** 0.03×2.9 _____   **c.** 0.005×0.0002 _____

STOP • STOP • STOP • STOP • STOP • STOP • STOP • STOP • STOP

A2   **a.** 2   **b.** 3   **c.** 7

Q3   Place the decimal point in the product: $3.25 \times 1.6 = 5200$

STOP • STOP • STOP • STOP • STOP • STOP • STOP • STOP • STOP

A3   5.200

Q4   Place the decimal point in the product:

     a. $8.7 \times 0.36 = 3132$ **b.** $2.07 \times 0.308 = 63756$

STOP • STOP • STOP • STOP • STOP • STOP • STOP • STOP • STOP

A4	**a.** 3.132	**b.** 0.63756

Q5   Place the decimal point in the product:
           a. $37.5 \times 0.8 = 3000$ **b.** $60 \times 5.84 = 35040$

STOP • *STOP* • *STOP* • *STOP* • *STOP* • *STOP* • *STOP* • *STOP* • *STOP*

A5   **a.** 30.00   **b.** 350.40

Q6   Find the product: 27.3
                           $\times 6.5$

STOP • *STOP* • *STOP* • *STOP* • *STOP* • *STOP* • *STOP* • *STOP* • *STOP*

A6   177.45:   27.3   (1 decimal place)
                 $\times 6.5$ (1 decimal place)
                  1365
               1638
              177.45   (2 decimal places)

Q7   Find the product:
           a. 0.8×7 **b.** 0.3×0.8

STOP • *STOP* • *STOP* • *STOP* • *STOP* • *STOP* • *STOP* • *STOP* • *STOP*

A7   **a.** 5.6   **b.** 0.24

---

2   In some cases the product does not have as many digits as the required number of decimal places. When this happens, zeros are placed to the left of the first digit in the product to obtain the required number of places.

Example:

    0.028   (3 decimal places)
$\times 0.13$ (2 decimal places)
       84
      28
    364   (5 decimal places required)

From the right count 5 decimal places. Hence, $0.028 \times 0.13 = 0.00364$.

Q8   Complete the multiplication problem by placing zeros and the decimal point in the product: 0.09
                       $\times 0.7$
                       63

STOP • *STOP* • *STOP* • *STOP* • *STOP* • *STOP* • *STOP* • *STOP* • *STOP*

A8   0.063

Q9   Find the product: 0.12
                                  $\times$0.07

STOP • STOP • STOP • STOP • STOP • STOP • STOP • STOP • STOP

A9   0.0084:   0.12   (2 decimal places)
                       $\times$0.07   (2 decimal places)
                       ───────
                       0.0084   (4 decimal places)

Q10   Find the product:
              a. 0.7×0.07 **b.** 0.008×0.006

STOP • STOP • STOP • STOP • STOP • STOP • STOP • STOP • STOP

A10   **a.** 0.049   **b.** 0.000048

Q11   Find the product:
              a. 0.36×0.15 **b.** 3.6×0.15

STOP • STOP • STOP • STOP • STOP • STOP • STOP • STOP • STOP

A11   **a.** 0.054   **b.** 0.54

Q12   Find the product:
              a. 2.4×10 **b.** 0.08×10

STOP • STOP • STOP • STOP • STOP • STOP • STOP • STOP • STOP

A12   **a.** 24   **b.** 0.8

Q13   Find the product:
              a. 2.4×100 **b.** 0.08×100

STOP • STOP • STOP • STOP • STOP • STOP • STOP • STOP • STOP

A13   **a.** 240   **b.** 8

3   A shortcut can be observed for multiplying a power of 10 (10, 100, 1,000, etc.) by a decimal. Consider:

$$2.378 \times 10 = 23.78$$

$$2.378 \times 100 = 237.8$$

$$2.378 \times 1,000 = 2,378$$

$$2.378 \times 10,000 = 23,780$$

To multiply by a power of 10, the decimal point is moved to the *right* the same number of places as there are zeros in the power of 10. That is,

$$2.378 \times \underline{100} = 237.8$$
(2 zeros)   (decimal point moved 2 places)

Q14   Find the product mentally:

a. 0.072×10 _____ **b.** 0.072×100 _____ **c.** $0.072 \times 1,000$ _____

STOP • STOP • STOP • STOP • STOP • STOP • STOP • STOP • STOP

A14   **a.** 0.72   **b.** 7.2   **c.** 72

Q15   Find the product mentally:

a. 23.7×100 _____ **b.** 2.8×10 _____ **c.** $17 \times 1,000$ _____

STOP • STOP • STOP • STOP • STOP • STOP • STOP • STOP • STOP

A15   **a.** 2,370   **b.** 28   **c.** 17,000

Q16   Find the product mentally:

a. 0.0017×100 _____ **b.** 5×100 _____ **c.** $0.77 \times 1,000$ _____

STOP • STOP • STOP • STOP • STOP • STOP • STOP • STOP • STOP

A16   **a.** 0.17   **b.** 500   **c.** 770

4   The division of decimals is carried out in the same way as the division of whole numbers, except for the placing of the decimal point in the quotient.

Example: Consider the problem $0.8 \div 2$:

Solution

$$0.8 \div 2 = \frac{8}{10} \div 2$$

$$= \frac{8}{10} \times \frac{1}{2}$$

$$= \frac{4}{10}$$

$$= 0.4$$

The problem could have been written

$$
\begin{array}{r}
0.4 \\
2\overline{)0.8} \\
\underline{8}
\end{array}
$$

Notice that when the divisor is a whole number, *the decimal point in the quotient is directly above the decimal point in the dividend.*

Q17   Place the decimal point in the quotient (do not divide):

a. $23\overline{)19.218}$ b. $7\overline{)16.47}$ c. $19\overline{)5.612}$

STOP • STOP • STOP • STOP • STOP • STOP • STOP • STOP • STOP

A17   a. $23\overline{)19.218}$ b. $7\overline{)16.47}$ c. $19\overline{)5.612}$

(In all cases the decimal point is placed directly above the decimal point in the dividend.)

Q18   Find the quotient:

a. $3\overline{)1.8}$ b. $8\overline{)3.2}$ c. $15\overline{)30.45}$

STOP • STOP • STOP • STOP • STOP • STOP • STOP • STOP • STOP

A18   a. 0.6   b. 0.4   c. 2.03

5   When the divisor is a decimal, the division could be completed as follows: $0.168 \div 0.12$ means $\dfrac{0.168}{0.12}$. The fraction can now be multiplied by $\dfrac{100}{100}$, making the divisor a whole number.

$$
\frac{0.168}{0.12} \times \frac{100}{100} = \frac{16.8}{12} \quad \text{or} \quad
\begin{array}{r}
1.4 \\
12\overline{)16.8} \\
\underline{12} \\
4\,8 \\
\underline{4\,8}
\end{array}
$$

The same result could be obtained by using the following steps:

Step 1: Move the decimal point in the divisor to the right to make the divisor a whole number.

Step 2: Move the decimal point in the dividend the same number of places, using zeros as placeholders if necessary.

Step 3: The decimal point in the quotient will be directly above the new decimal point in the dividend.

Example: $0.12\overline{)0.168}$ (decimal point moved 2 places to the right)

Q19   Move the decimal point in the divisor and dividend, making the divisor a whole number.
       (Do no further computation.) $0.02\overline{)14.782}$

STOP • **STOP** • **STOP** • **STOP** • **STOP** • **STOP** • **STOP** • **STOP** • **STOP**

A19 $0.02.\overline{)14.782.}$

Q20   Place the decimal point in the quotient correctly:

$$\overset{5}{5.2\overline{)2.60}} \qquad\qquad \overset{205}{4.36\overline{)0.89380}}$$

       a. $5.2\overline{)2.60}$ b. $4.36\overline{)0.89380}$

STOP • **STOP** • **STOP** • **STOP** • **STOP** • **STOP** • **STOP** • **STOP** • **STOP**

A20   a. $5.2.\overline{)2.60}$ b. $4.36.\overline{)0.89380}$
       with quotients 0.5 and 0.205

Q21   Place the decimal point in the quotient correctly:

$$\overset{7391}{0.02\overline{)14.782}} \qquad\qquad \overset{13}{0.25\overline{)3.25}}$$

       a. $0.02\overline{)14.782}$ b. $0.25\overline{)3.25}$

STOP • **STOP** • **STOP** • **STOP** • **STOP** • **STOP** • **STOP** • **STOP** • **STOP**

A21   a. 739.1   b. 13.

6	Sometimes it may be necessary to place zeros in the quotient, between the decimal point and the first digit of the quotient, to obtain the required number of decimal places:

$$\overset{0.0117}{15\overline{)0.1755}}$$

Q22   Place the decimal point in the quotient correctly: $\overset{2}{9\overline{)0.18}}$

STOP • **STOP** • **STOP** • **STOP** • **STOP** • **STOP** • **STOP** • **STOP** • **STOP**

A22   0.02

Q23   Place the decimal point in the quotient correctly:

$$\overset{21}{25\overline{)0.525}} \qquad\qquad \overset{3}{90\overline{)0.0270}}$$

       a. $25\overline{)0.525}$ b. $90\overline{)0.0270}$

STOP • **STOP** • **STOP** • **STOP** • **STOP** • **STOP** • **STOP** • **STOP** • **STOP**

A23   a. 0.021   b. 0.0003

Q24   Find the quotient:

       a. $7\overline{)39.2}$ b. $0.9\overline{)23.4}$

STOP • **STOP** • **STOP** • **STOP** • **STOP** • **STOP** • **STOP** • **STOP** • **STOP**

A24   a. 5.6   b. 26

Q25   Find the quotient:

       a. $0.12\overline{)7.92}$ **b.** $0.016\overline{)11.68}$

STOP • STOP • STOP • STOP • STOP • STOP • STOP • STOP • STOP

A25   **a.** 66   **b.** 730: $0.016\overline{)11.680}$

---

7   It will often be necessary to round off quotients to a specified decimal place. One method often used when rounding off is to complete the division to one place beyond the place to which you are rounding off:

Example: $5.3 \div 8$ (round off to the nearest tenth)

```
       0.66
    8 ) 5.30
        4 8
        ‾‾‾
         50
         48
        ‾‾‾
```

Hence, $5.3 \div 8 = 0.7$ rounded off to the nearest tenth.

Q26   Round off the quotients to the nearest tenth:

       a. $0.14\overline{)31.9}$ **b.** $0.33\overline{)70.3}$

STOP • STOP • STOP • STOP • STOP • STOP • STOP • STOP • STOP

A26   **a.** 227.9:
```
          227.85
   0.14 ) 31.9000
          28
          ‾‾
          39
          28
          ‾‾
          110
           98
          ‾‾‾
          120
          112
          ‾‾‾
            80
            70
            ‾‾
```
       b. 213.0

Q27   Round off the quotient to the nearest hundredth:

a. $24\overline{)3.2}$ b. $0.73\overline{)5.88}$

STOP   •   **STOP**   •   **STOP**   •   **STOP**   •   **STOP**   •   **STOP**   •   **STOP**   •   **STOP**   •   **STOP**

A27   a. 0.13   b. 8.05

Q28   Find the quotient:
a. $27.63 \div 10$ b. $27.63 \div 100$

STOP   •   **STOP**   •   **STOP**   •   **STOP**   •   **STOP**   •   **STOP**   •   **STOP**   •   **STOP**   •   **STOP**

A28   a. 2.763   b. 0.2763

---

8   As was the case for multiplication, a shortcut can be observed when dividing a decimal by a power of 10:

$$2.3 \div 10 = 0.23$$

$$2.3 \div 100 = 0.023$$

$$2.3 \div 1,000 = 0.0023$$

When dividing by a power of 10, the decimal point is moved to the *left* the same number of places as there are zeros in the power of 10.

Q29   Find the quotient:

a. $4.76 \div 10$ _____   b. $16.5 \div 1,000$ _____

STOP   •   **STOP**   •   **STOP**   •   **STOP**   •   **STOP**   •   **STOP**   •   **STOP**   •   **STOP**   •   **STOP**

A29   a. 0.476   b. 0.0165

Q30   Find the quotient:

a. $0.0091 \div 100$ _____   b. $0.23 \div 10$ _____

STOP   •   **STOP**   •   **STOP**   •   **STOP**   •   **STOP**   •   **STOP**   •   **STOP**   •   **STOP**   •   **STOP**

A30   a. 0.000091   b. 0.023

---

9   Multiplying and dividing by powers of 10 can often be confusing. The decimal point is often moved the correct number of places but in the wrong direction. It may help to remember that multiplying by a power of 10 will make the number *larger;* hence, the decimal point is moved to the *right*. Dividing by a power of 10 will make the number *smaller;* hence, the decimal point is moved to the *left*.

Q31   Complete:

a. $2.713 \div 100 =$ _____   b. $2.713 \times 10 =$ _____

c. $0.018 \times 1,000 =$ _____   d. $0.018 \div 10 =$ _____

STOP • STOP • STOP • STOP • STOP • STOP • STOP • STOP • STOP

A31   a. 0.02713   b. 27.13   c. 18   d. 0.0018

Q32   Bill drove 603 miles and used 35.8 gallons of gasoline. What was his average number of miles per gallon to the nearest tenth? _____

STOP • STOP • STOP • STOP • STOP • STOP • STOP • STOP • STOP

A32   16.8 miles per gallon

This completes the instruction for this section.

3.4   Exercises

1. Dividing a decimal by 100 moves the decimal point _____ spaces to the _____.
2. Multiplying a decimal by 10,000 moves the decimal point _____ spaces to the _____.
3. Find the product:
   a. 0.0276×100 b. 100×0.0543 c. 3.04×10 d. $0.0063 \times 1,000$
4. Find the quotient:
   a. $0.0276 \div 10$ b. $0.0543 \div 100$ c. $3.04 \div 10$ d. $0.0063 \div 1,000$
5. Find the product:
   a. 0.08×0.03 b. 6×0.07 c. 0.5×8 d. 0.025×0.16
   e. 2.4×0.105 f. 3.14×2.5
6. Find the quotient:
   a. $0.3 \div 0.4$ b. $0.35 \div 0.05$ c. $0.084 \div 0.07$ d. $0.63 \div 0.7$
   e. $16.308 \div 0.36$ f. $1.5072 \div 0.628$
7. Round the quotient to the nearest hundredth:
   a. $1.05 \div 12$ b. $7.82 \div 0.047$

3.4   Exercise Answers

1. 2, left
2. 4, right
3. a. 2.76   b. 5.43   c. 30.4   d. 6.3
4. a. 0.00276   b. 0.000543   c. 0.304   d. 0.0000063
5. a. 0.0024   b. 0.42   c. 4   d. 0.004
   e. 0.252   f. 7.85
6. a. 0.75   b. 7   c. 1.2   d. 0.9
   e. 45.3   f. 2.4
7. a. 0.09   b. 166.38

3.5   **Fractions as Decimals**

---

1   Fractions whose denominators are a power of 10 are easily converted to a decimal. For example, $\frac{5}{10} = 0.5$, $\frac{7}{100} = 0.07$, and $\frac{23}{100} = 0.23$.

Fractions whose denominators are not powers of 10 can be converted to a decimal by dividing the numerator by the denominator. That is, to convert $\frac{3}{4}$ to a decimal, divide 3 by 4. This is accomplished by placing a decimal point and as many zeros as necessary after the 3.

$$
\begin{array}{r}
0.75 \\
4\overline{)3.00} \\
2\,8 \\
\hline
20 \\
20 \\
\hline
\end{array}
$$

Hence, $\frac{3}{4} = 0.75$.

Q1   Convert $\frac{1}{2}$ to a decimal by dividing 1 by 2.

STOP • **STOP** • **STOP** • **STOP** • **STOP** • **STOP** • **STOP** • **STOP** • **STOP**

A1   0.5: $\begin{array}{r} 0.5 \\ 2\overline{)1.0} \\ 1\,0 \\ \hline \end{array}$

Q2   Convert $\frac{1}{8}$ to a decimal (3 decimal places).

STOP • **STOP** • **STOP** • **STOP** • **STOP** • **STOP** • **STOP** • **STOP** • **STOP**

A2   0.125: $\begin{array}{r} 0.125 \\ 8\overline{)1.000} \\ 8 \\ \hline 20 \\ 16 \\ \hline 40 \\ 40 \\ \hline \end{array}$

2   If a zero remainder is obtained when converting a fraction to a decimal, the quotient is called a *terminating decimal*. It is given this name because there is a last digit. Often a zero remainder cannot be obtained and the digits of the quotient repeat themselves in a pattern. These decimals are called *repeating decimals*.

Example: Convert $\frac{7}{33}$ to a decimal.

Solution

$$
\begin{array}{r}
0.2121 = 0.212121\cdots \\
33\overline{)7.0000} \\
6\,6 \\
\hline
40 \\
33 \\
\hline
70 \\
66 \\
\hline
40 \\
33 \\
\hline
7 \\
\end{array}
$$

At these points the remainder is the same as the original dividend. This means that the same group of digits will repeat continuously. A bar is placed over the group of digits that repeat. Therefore, $\frac{7}{33} = 0.\overline{21}$.

Q3 $0.\overline{31}$ means _____

STOP • STOP • STOP • STOP • STOP • STOP • STOP • STOP • STOP

A3   0.313131 and so on (the digits 31 repeat continuously).

Q4   Convert $\frac{1}{3}$ to a repeating decimal.

STOP • STOP • STOP • STOP • STOP • STOP • STOP • STOP • STOP

A4 $0.\overline{3}$:
$$
\begin{array}{r}
0.3 = 0.333333\cdots \\
3\overline{)1.0} \\
9 \\
\hline
1 \\
\end{array}
$$
(the remainder is the same as the dividend)

Q5   Convert $\dfrac{2}{11}$ to a repeating decimal.

STOP • STOP • STOP • STOP • STOP • STOP • STOP • STOP • STOP

A5 $0.\overline{18}$:

$$
\begin{array}{r}
0.18 = 0.181818\cdots \\
11\overline{)2.00} \\
\underline{1\,1} \\
90 \\
\underline{88} \\
2
\end{array}
$$

Q6   Convert $\dfrac{3}{8}$ to a terminating decimal.

STOP • STOP • STOP • STOP • STOP • STOP • STOP • STOP • STOP

A6 0.375:

$$
\begin{array}{r}
0.375 \\
8\overline{)3.000} \\
\underline{2\,4} \\
60 \\
\underline{56} \\
40 \\
\underline{40} \\
\end{array}
$$

Q7   Convert $\dfrac{5}{8}$ to a decimal.

STOP • STOP • STOP • STOP • STOP • STOP • STOP • STOP • STOP

A7   0.625:
$$
\begin{array}{r}
0.625 \\
8\overline{)5.000} \\
4\,8 \\
\hline
20 \\
16 \\
\hline
40 \\
40 \\
\hline
\end{array}
$$

Q8   Convert $\dfrac{8}{9}$ to a decimal.

STOP • **STOP** • **STOP** • **STOP** • **STOP** • **STOP** • **STOP** • **STOP** • **STOP**

A8 $0.\overline{8}$:
$$
\begin{array}{r}
0.8 = 0.888888\cdots \\
9\overline{)8.0} \\
7\,2 \\
\hline
8 \\
\end{array}
$$

Q9   Convert to a decimal:

    a. $\dfrac{3}{20}$ **b.** $\dfrac{5}{32}$

STOP • **STOP** • **STOP** • **STOP** • **STOP** • **STOP** • **STOP** • **STOP** • **STOP**

A9   **a.** 0.15   **b.** 0.15625

Q10   Convert to a decimal:

    a. $\dfrac{4}{9}$ **b.** $\dfrac{1}{11}$

STOP • **STOP** • **STOP** • **STOP** • **STOP** • **STOP** • **STOP** • **STOP** • **STOP**

A10   **a.** $0.\overline{4}$ **b.** $0.\overline{09}$

3   When the dividend reappears as a remainder, the quotient will be a repeating decimal. In some cases, the dividend does not reappear as a remainder, but a remainder does appear a second time. This means that the digits in the quotient will repeat, starting with the one obtained when the remainder was first used as a new dividend.

Example: Convert $\frac{1}{6}$ to a decimal.

Solution

$$
\begin{array}{r}
0.16 = 0.166666\cdots \\
6\overline{)1.00} \\
\underline{6} \\
40 \\
\underline{36} \\
4
\end{array}
$$

(the remainder 4 repeats a second time)

Hence, $\frac{1}{6} = 0.1\overline{6}$.

Q11   Convert $\frac{7}{12}$ to a repeating decimal.

STOP • **STOP** • **STOP** • **STOP** • **STOP** • **STOP** • **STOP** • **STOP** • **STOP**

A11   0.58$\overline{3}$:

$$
\begin{array}{r}
0.583 = 0.5833333\cdots \\
12\overline{)7.000} \\
\underline{6\,0} \\
1\,00 \\
\underline{96} \\
40 \\
\underline{36} \\
4
\end{array}
$$

Q12   Convert $\frac{62}{495}$ to a decimal.

STOP • **STOP** • **STOP** • **STOP** • **STOP** • **STOP** • **STOP** • **STOP** • **STOP**

A12 $0.1\overline{25}$

---

4   Terminating decimals can be converted to fractions by determining the place value indicated and writing the digits over the appropriate power of 10. If possible, the fraction is reduced.

$$0.375 = \frac{375}{1,000} \quad \text{(dividing numerator and denominator by 5 three times)}$$

$$= \frac{75}{200}$$

$$= \frac{15}{40}$$

$$= \frac{3}{8}$$

Q13   Convert 0.12 to a fraction reduced to lowest terms.

STOP • STOP • STOP • STOP • STOP • STOP • STOP • STOP • STOP

A13 $\frac{3}{25}$: $0.12 = \frac{12}{100}$

$$= \frac{6}{50}$$

$$= \frac{3}{25}$$

Q14   Convert to a fraction reduced to lowest terms:
    a. 0.85   **b.** 0.024   **c.** 0.425

STOP • STOP • STOP • STOP • STOP • STOP • STOP • STOP • STOP

A14   **a.** $\frac{17}{20}$ **b.** $\frac{3}{125}$ **c.** $\frac{17}{40}$

Q15   Convert to a fraction reduced to lowest terms:
    a. 0.125   **b.** 0.1875   **c.** 0.203125

STOP • STOP • STOP • STOP • STOP • STOP • STOP • STOP • STOP

A15   **a.** $\frac{1}{8}$ **b.** $\frac{3}{16}$ **c.** $\frac{13}{64}$

---

5   In some occupations it is often necessary to convert from fractions to decimals. The following table may prove useful in making these conversions.

Fraction	Decimal	Fraction	Decimal	Fraction	Decimal	Fraction	Decimal
$\frac{1}{64}$	0.015625	$\frac{17}{64}$	0.265625	$\frac{33}{64}$	0.515625	$\frac{49}{64}$	0.765625
$\frac{1}{32}$	0.03125	$\frac{9}{32}$	0.28125	$\frac{17}{32}$	0.53125	$\frac{25}{32}$	0.78125
$\frac{3}{64}$	0.046875	$\frac{19}{64}$	0.296875	$\frac{35}{64}$	0.546875	$\frac{51}{64}$	0.796875
$\frac{1}{16}$	0.0625	$\frac{5}{16}$	0.3125	$\frac{9}{16}$	0.5625	$\frac{13}{16}$	0.8125
$\frac{5}{64}$	0.078125	$\frac{21}{64}$	0.328125	$\frac{37}{64}$	0.578125	$\frac{53}{64}$	0.828125
$\frac{3}{32}$	0.09375	$\frac{11}{32}$	0.34375	$\frac{19}{32}$	0.59375	$\frac{27}{32}$	0.84375
$\frac{7}{64}$	0.109375	$\frac{23}{64}$	0.359375	$\frac{39}{64}$	0.609375	$\frac{55}{64}$	0.859375
$\frac{1}{8}$	0.125	$\frac{3}{8}$	0.375	$\frac{5}{8}$	0.625	$\frac{7}{8}$	0.875
$\frac{9}{64}$	0.140625	$\frac{25}{64}$	0.390625	$\frac{41}{64}$	0.640625	$\frac{57}{64}$	0.890625
$\frac{5}{32}$	0.15625	$\frac{13}{32}$	0.40625	$\frac{21}{32}$	0.65625	$\frac{29}{32}$	0.90625
$\frac{11}{64}$	0.171875	$\frac{27}{64}$	0.421875	$\frac{43}{64}$	0.671875	$\frac{59}{64}$	0.921875
$\frac{3}{16}$	0.1875	$\frac{7}{16}$	0.4375	$\frac{11}{16}$	0.6875	$\frac{15}{16}$	0.9375
$\frac{13}{64}$	0.203125	$\frac{29}{64}$	0.453125	$\frac{45}{64}$	0.703125	$\frac{61}{64}$	0.953125
$\frac{7}{32}$	0.21875	$\frac{15}{32}$	0.46875	$\frac{23}{32}$	0.71875	$\frac{31}{32}$	0.96875
$\frac{15}{64}$	0.234375	$\frac{31}{64}$	0.484375	$\frac{47}{64}$	0.734375	$\frac{63}{64}$	0.984375
$\frac{1}{4}$	0.250	$\frac{1}{2}$	0.500	$\frac{3}{4}$	0.750	1	1.000

This completes the instruction for this section.

3.5   Exercises

1. Convert to a decimal:

a. $\frac{1}{10}$ **b.** $\frac{7}{100}$ **c.** $\frac{1}{2}$ **d.** $\frac{1}{4}$ **e.** $\frac{1}{5}$ **f.** $\frac{2}{5}$ **g.** $\frac{1}{3}$

h. $\frac{2}{3}$

2. Convert to a decimal:

a. $\frac{3}{8}$ b. $\frac{5}{6}$ c. $\frac{4}{7}$ d. $\frac{3}{11}$ e. $\frac{17}{20}$ f. $\frac{31}{40}$ g. $\frac{27}{20}$

h. $\frac{15}{64}$

3. Convert to a fraction reduced to lowest terms:
   a. 0.75   b. 0.25   c. 0.475   d. 0.724   e. 0.119   f. 0.648   g. 0.256
   h. 0.425

3.5   Exercise Answers

1. a. 0.1   b. 0.07   c. 0.5   d. 0.25
   e. 0.2   f. 0.4   g. $0.\overline{3}$ h. $0.\overline{6}$
2. a. 0.375   b. $0.8\overline{3}$ c. $0.\overline{571428}$ d. $0.\overline{27}$
   e. 0.85   f. 0.775   g. 1.35   h. 0.234375

3. a. $\frac{3}{4}$ b. $\frac{1}{4}$ c. $\frac{19}{40}$ d. $\frac{181}{250}$

   e. $\frac{119}{1,000}$ f. $\frac{81}{125}$ f. $\frac{32}{125}$ h. $\frac{17}{40}$

Chapter 3 Sample Test

At the completion of Chapter 3 it is expected that you will be able to work the following problems.

3.1   Place Value

1. a. Write in decimal form: two and seventeen thousandths
   b. Write in word form: 17.012
   c. Write in decimal form: 302/10,000
   d. What is the place value of the 4 in 2.143?

3.2   Rounding Off

2. Round off the numbers as indicated:
   a. 2.416 (nearest hundredth)   b. 137.054 (nearest *ten*)
   c. 60.37 (nearest whole number)   d. 0.6726 (nearest thousandth)

3.3   Addition and Subtraction

3. Add:
   a. 2.076 + 1.76 + 3.7   b. 0.003 + 22 + 3.5
4. Subtract:
   a. 5.06 − 4.982   b. 7 − 3.285

3.4   Multiplication and Division

5. Multiply:
   a. 3.27×0.79 **b.** 0.004×0.05 **c.** 0.46×10 **d.** 2.104×100
6. Divide:
   a. $0.0448 \div 0.032$ **b.** $1.216 \div 0.04$ **c.** $56.20 \div 100$ **d.** $0.215 \div 10$
7. Round the quotient to the nearest hundredth:
   a. $2.24 \div 43$ **b.** $0.4142 \div 0.35$

3.5   Fractions as Decimals

8. Convert to a decimal:

   a. $\dfrac{3}{8}$ **b.** $\dfrac{1}{11}$

9. Convert to a fraction in lowest terms:
   a. 0.025   **b.** 0.125

Chapter 3 Sample Test Answers

1. **a.** 2.017   **b.** seventeen and twelve thousandths
   c. 0.0302   **d.** hundredths
2. **a.** 2.42   **b.** 140   **c.** 60   **d.** 0.673
3. **a.** 7.536   **b.** 25.503
4. **a.** 0.078   **b.** 3.715
5. **a.** 2.5833   **b.** 0.0002   **c.** 4.6   **d.** 210.4
6. **a.** 1.4   **b.** 30.4   **c.** 0.5620   **d.** 0.0215
7. **a.** 0.05   **b.** 1.18
8. **a.** 0.375   **b.** $0.\overline{09}$

9. **a.** $\dfrac{1}{40}$ **b.** $\dfrac{1}{8}$

Chapter 4

Operations with Percents

4.1   Percents

---

1   In Chapter 3 fractions with a denominator of 100 were called decimal fractions. Fractions with a denominator of 100 are also called *percents*. The word percent means "hundredths." Thus, $\frac{7}{100}$ is equivalent to 7 percent. The symbol % is an abbreviation for the word "percent" and is derived from the denominator of 100. Therefore,

$$\frac{7}{100} \longrightarrow 7\%$$

The symbol 7% is read "7 percent" and means 7 hundredths.

Q1   Write the fractions using the % symbol:

    a. $\frac{23}{100} =$ _____   **b.** $\frac{3}{100} =$ _____   **c.** $\frac{25}{100} =$ _____   **d.** $\frac{50}{100} =$ _____

STOP • STOP • STOP • STOP • STOP • STOP • STOP • STOP • STOP

A1   **a.** 23%   **b.** 3%   **c.** 25%   **d.** 50%

Q2   Write the percents as fractions:

    a. 29% = _____   **b.** 9% = _____   **c.** 53% = _____   **d.** 97% = _____

STOP • STOP • STOP • STOP • STOP • STOP • STOP • STOP • STOP

A2   **a.** $\frac{29}{100}$ **b.** $\frac{9}{100}$ **c.** $\frac{53}{100}$ **d.** $\frac{97}{100}$

Q3   Write the percents as fractions and reduce to lowest terms:

    a. 25% = _____   **b.** 20% = _____   **c.** 75% = _____   **d.** 44% = _____

STOP • STOP • STOP • STOP • STOP • STOP • STOP • STOP • STOP

A3   **a.** $\frac{1}{4}$ **b.** $\frac{1}{5}$ **c.** $\frac{3}{4}$ **d.** $\frac{11}{25}$

2   Some percents can also be expressed as mixed numbers. For example,

$$225\% = \frac{225}{100} = 2\frac{25}{100} = 2\frac{1}{4}$$

Fractional percents can be expressed as a fraction as follows:

$$\frac{1}{2}\% = \frac{\frac{1}{2}}{100}$$

$$= \frac{1}{2} \div \frac{100}{1}$$

$$= \frac{1}{2} \times \frac{1}{100}$$

$$= \frac{1}{200}$$

Q4   Change 250% to a mixed number.

STOP • **STOP** • **STOP** • **STOP** • **STOP** • **STOP** • **STOP** • **STOP** • **STOP**

A4 $2\frac{1}{2}$: $250\% = \frac{250}{100}$

$$= 2\frac{50}{100}$$

$$= 2\frac{1}{2}$$

Q5   Change to a mixed number:
   a. 480%   **b.** 333%

STOP • **STOP** • **STOP** • **STOP** • **STOP** • **STOP** • **STOP** • **STOP** • **STOP**

A5   **a.** $4\frac{4}{5}$ **b.** $3\frac{33}{100}$

Q6   Change $\frac{1}{4}\%$ to a fraction.

STOP • **STOP** • **STOP** • **STOP** • **STOP** • **STOP** • **STOP** • **STOP** • **STOP**

A6 $\qquad \frac{1}{400}: \quad \frac{1}{4}\% = \dfrac{\frac{1}{4}}{100}$

$\qquad\qquad\qquad\qquad = \dfrac{1}{400}$

Q7 $\qquad$ Change $\frac{2}{3}\%$ to a fraction.

STOP • STOP • STOP • STOP • STOP • STOP • STOP • STOP • STOP

A7 $\qquad \frac{1}{150}: \quad \frac{2}{3}\% = \dfrac{\frac{2}{3}}{100}$

$\qquad\qquad\qquad\qquad = \dfrac{2}{300}$

---

3 $\qquad$ To change a mixed-number percent to a fraction it is necessary to write the mixed number as an improper fraction over 100. Consider this example:

$$66\frac{2}{3}\% = \frac{66\frac{2}{3}}{100}$$

$$= \frac{\frac{200}{3}}{100}$$

$$= \frac{200}{3} \div \frac{100}{1}$$

$$= \frac{200}{3} \times \frac{1}{100}$$

$$= \frac{200}{300}$$

$$= \frac{2}{3}$$

Q8 $\qquad$ Change $33\frac{1}{3}\%$ to a fraction.

STOP • STOP • STOP • STOP • STOP • STOP • STOP • STOP • STOP

A8 $\quad \dfrac{1}{3}: \quad 33\dfrac{1}{3}\% = \dfrac{33\dfrac{1}{3}}{100}$

$$= \dfrac{\dfrac{100}{3}}{100}$$

$$= \dfrac{100}{300}$$

$$= \dfrac{1}{3}$$

Q9 $\quad$ Change $12\dfrac{1}{2}\%$ to a fraction.

STOP • **STOP** • **STOP** • **STOP** • **STOP** • **STOP** • **STOP** • **STOP** • **STOP**

A9 $\quad \dfrac{1}{8}: \quad 12\dfrac{1}{2}\% = \dfrac{12\dfrac{1}{2}}{100}$

$$= \dfrac{\dfrac{25}{2}}{100}$$

$$= \dfrac{25}{200}$$

$$= \dfrac{1}{8}$$

Q10 $\quad$ Change to fractions:

$\quad$ **a.** $16\dfrac{2}{3}\%$ $\qquad\qquad$ **b.** $37\dfrac{1}{2}\%$

STOP • **STOP** • **STOP** • **STOP** • **STOP** • **STOP** • **STOP** • **STOP** • **STOP**

A10 $\quad$ **a.** $\dfrac{1}{6}$ $\quad$ **b.** $\dfrac{3}{8}$

Q11   Change $3\frac{1}{8}\%$ to a fraction.

STOP   •   STOP   •   STOP   •   STOP   •   STOP   •   STOP   •   STOP   •   STOP   •   STOP

A11 $\frac{1}{32}$: $3\frac{1}{8}\% = \dfrac{3\frac{1}{8}}{100}$

$$= \dfrac{\frac{25}{8}}{100}$$

$$= \dfrac{25}{800}$$

$$= \dfrac{1}{32}$$

---

4   In Chapter 3 it was shown that fractions with denominator 100 could be written as a decimal. Since percents are written as a fraction with denominator of 100, they can also be expressed as a decimal. For example,

$$18\% = \frac{18}{100} = 0.18$$

Q12   Express 23% as a fraction and a decimal. _____

STOP   •   STOP   •   STOP   •   STOP   •   STOP   •   STOP   •   STOP   •   STOP   •   STOP

A12 $23\% = \dfrac{23}{100} = 0.23$

---

5   A number can be divided by 100 by simply moving the decimal point two places to the left. Hence, to change a percent to a decimal, remove the percent sign and move the decimal point two places to the left. For example,

$$23\% = 0.23 \quad \text{because} \quad 23\% = 23. \div 100$$
$$7\% = 0.07 \quad \text{because} \quad 7\% = 7. \div 100$$
$$17.5\% = 0.175 \quad \text{because} \quad 17.5\% = 17.5 \div 100$$

Notice that the decimal point, for a number written without a decimal point, is placed after the last digit. Hence, 8 = 8.

Q13   Express 23.7% as a decimal. _____

STOP   •   STOP   •   STOP   •   STOP   •   STOP   •   STOP   •   STOP   •   STOP   •   STOP

A13   0.237:   23.7% = 0.237 (drop the percent sign and move the decimal point two places to the left).

Q14   Express as a decimal:

   a. 13% = _____   **b.** 75% = _____

STOP • STOP • STOP • STOP • STOP • STOP • STOP • STOP • STOP

A14   **a.** 0.13   **b.** 0.75

Q15   Express as a decimal:

   a. 19.8% = _____   **b.** 2.75% = _____   **c.** 250% = _____

STOP • STOP • STOP • STOP • STOP • STOP • STOP • STOP • STOP

A15   **a.** 0.198   **b.** 0.0275   **c.** 2.50 or 2.5

Q16   Express as a decimal:

   a. 700% = _____   **b.** 0.03% = _____

   c. 1% = _____   **d.** 5% = _____

STOP • STOP • STOP • STOP • STOP • STOP • STOP • STOP • STOP

A16   **a.** 7   **b.** 0.0003   **c.** 0.01   **d.** 0.05

6	To change a decimal to a percent, move the decimal point two places to the right and append the percent sign. For example,

$$0.28 = 28\% \quad \text{because} \quad 0.28 = \frac{28}{100} = 28\%$$

Examples:

   0.04 = 4%
   0.15 = 15%
   1.86 = 186%
   0.002 = 0.2%

Q17   Change 0.617 to a percent. _____

STOP • STOP • STOP • STOP • STOP • STOP • STOP • STOP • STOP

A17   61.7%:   0.617 = 61.7% (move the decimal point two places to the right and append the % sign).

Q18   Change to a percent:

   a. 0.93 = _____   **b.** 0.0015 = _____

   c. 1.72 = _____   **d.** 0.473 = _____

STOP • STOP • STOP • STOP • STOP • STOP • STOP • STOP • STOP

A18   **a.** 93%   **b.** 0.15%   **c.** 172%   **d.** 47.3%

7	It may be necessary to use zeros as placeholders when changing a decimal to a percent. For example, change 0.8 to a percent:

   0.8 = 0.80 = 80%

Q19   Change 0.7 to a percent. _____

STOP • STOP • STOP • STOP • STOP • STOP • STOP • STOP • STOP

A19   70%:   0.7 = 0.70
             = 70%

Q20   Change 17 to a percent._____

STOP • STOP • STOP • STOP • STOP • STOP • STOP • STOP • STOP

A20   1,700%:   17 = 17.00
                = 1,700%

Q21   Change to a percent:

   a. 0.2 = _____   **b.** 5 = _____   **c.** 0.1 = _____   **d.** 213 = _____

STOP • STOP • STOP • STOP • STOP • STOP • STOP • STOP • STOP

A21   **a.** 20%   **b.** 500%   **c.** 10%   **d.** 21,300%

---

8   To change a fraction to a percent, first change the common fraction to a decimal and then change the decimal to a percent. Consider these examples:

$$\frac{3}{4} = 0.75 = 75\%$$

$$\frac{1}{8} = 0.125 = 12.5\%$$

   Recall that to change a fraction to a decimal the numerator is divided by the denominator.

Q22   Change $\frac{3}{8}$ to a percent.

STOP • STOP • STOP • STOP • STOP • STOP • STOP • STOP • STOP

A22   37.5%: $8{\overline{)3.000}}$ with quotient 0.375

Q23   Change to a percent:

   a. $\frac{1}{4}$ = $\underset{\text{(decimal)}}{_____}$ = $\underset{\text{(percent)}}{_____}$ **b.** $\frac{5}{8}$ = $\underset{\text{(decimal)}}{_____}$ = $\underset{\text{(percent)}}{_____}$

STOP • STOP • STOP • STOP • STOP • STOP • STOP • STOP • STOP

A23   **a.** 0.25, 25%   **b.** 0.625, 62.5%

---

9   A fraction represented by a repeating decimal is usually changed to a percent by completing the division to two decimal places and writing the remainder in fraction form. That is, $\frac{1}{3} = 0.33\frac{1}{3} = 33\frac{1}{3}\%$. The problem could be stated $\frac{1}{3} = 0.33\overline{3} = 33.\overline{3}\%$; however, the first method is preferred.

   It should be noted that when moving the decimal point two places to the right, only digits count as places. Fractions do not hold a place. Hence, $0.3\frac{1}{3}$ *does not* equal $3\frac{1}{3}\%$.

The division must be carried out to at least two decimal places before the remainder is written as a fraction.

Q24   Express $\frac{1}{6}$ as a percent.

STOP • STOP • STOP • STOP • STOP • STOP • STOP • STOP • STOP

A24 $16\frac{2}{3}\%$: $6)\overline{1.00}$ with quotient $0.16\frac{2}{3}$

Q25   Change to a percent:

a. $\frac{2}{3}$ b. $\frac{2}{7}$

STOP • STOP • STOP • STOP • STOP • STOP • STOP • STOP • STOP

A25   a. $66\frac{2}{3}\%$ b. $28\frac{4}{7}\%$

This completes the instruction for this section.

4.1   Exercises

Fill in the blanks with the proper equivalents, as shown in problem 1:

	Fraction	Decimal	Percent
1.	$\frac{1}{2}$	0.5	50%
2.	_____	0.25	_____ %
3.	$\frac{1}{3}$	_____	_____ %
4.	_____	_____	10%
5.	_____	0.125	_____ %
6.	$\frac{2}{5}$	_____	_____ %
7.	_____	_____	$66\frac{2}{3}\%$
8.	_____	0.75	_____ %

	Fraction	Decimal	Percent
9.	$\frac{3}{10}$	_____	_____ %
10.	_____	_____	5%
11.	_____	0.0625	_____ %
12.	$\frac{3}{8}$	_____	_____ %
13.	_____	_____	80%
14.	_____	0.625	_____ %
15.	_____	0.90	_____ %
16.	_____	_____	100%
17.	$\frac{1}{6}$	_____	_____ %
18.	_____	_____	20%
19.	$\frac{11}{20}$	_____	_____ %
20.	_____	0.375	_____ %
21.	$\frac{3}{5}$	_____	_____ %
22.	_____	0.166	_____ %
23.	_____	_____	95%
24.	$\frac{7}{10}$	_____	_____ %
25.	_____	0.875	_____ %
26.	_____	_____	225%
27.	$1\frac{7}{8}$	_____	_____ %
28.	_____	$0.16\frac{2}{3}$	_____ %
29.	_____	_____	200%
30.	$\frac{1}{100}$	_____	_____ %
31.	_____	0.08	_____ %
32.	_____	_____	$16\frac{2}{3}$ %
33.	$4\frac{4}{5}$	_____	_____ %
34.	_____	5.00	_____ %
35.	_____	_____	180%
36.	$\frac{1}{50}$	_____	_____ %
37.	_____	6.166	_____ %
38.	_____	_____	110%
39.	$3\frac{3}{4}$	_____	_____ %
40.	_____	0.2	_____ %

	Fraction	Decimal	Percent
41.	$2\frac{1}{2}$	_____	_____ %
42.	_____	4.08	_____ %
43.	_____	_____	$166\frac{2}{3}$ %
44.	_____	2.125	_____ %
45.	$1\frac{5}{8}$	_____	_____ %
46.	_____	3.875	_____ %

4.1   Exercise Answers

1. given

2. $\frac{1}{4}$, 25%

3. $0.33\frac{1}{3}$, $33\frac{1}{3}$%

4. $\frac{1}{10}$, 0.1, or 0.10

5. $\frac{1}{8}$, 12.5%

6. 0.4 or 0.40, 40%

7. $\frac{2}{3}$, $0.66\frac{2}{3}$

8. $\frac{3}{4}$, 75%

9. 0.3 or 0.30, 30%

10. $\frac{1}{20}$, 0.05

11. $\frac{1}{16}$, 6.25%

12. 0.375, 37.5%

13. $\frac{4}{5}$, 0.8, or 0.80

14. $\frac{5}{8}$, 62.5%

15. $\frac{9}{10}$, 90%

16. 1, 1, or 1.00

17. $0.16\frac{2}{3}$, $16\frac{2}{3}$%

18. $\frac{1}{5}$, 0.2, or 0.20

19. 0.55, 55%

20. $\frac{3}{8}$, 37.5%

21. 0.6 or 0.60, 60%

22. $\frac{83}{500}$, 16.6%

23. $\frac{19}{20}$, 0.95

24. 0.7 or 0.70, 70%

25. $\frac{7}{8}$, 87.5%

26. $2\frac{1}{4}$, 2.25

27. 1.875, 187.5%

28. $\frac{1}{6}$, $16\frac{2}{3}$%

29. 2, 2, or 2.00

30. 0.01, 1%

31. $\frac{2}{25}$, 8%

32. $\frac{1}{6}$, $0.16\frac{2}{3}$

33. 4.8 or 4.80, 480%

34. 5, 500%

35. $1\frac{4}{5}$, 1.8, or 1.80

36. 0.02, 2%

37. $6\frac{83}{500}$, 616.6%

38. $1\frac{1}{10}$, 1.1, or 1.10

39. 3.75, 375%

40. $\frac{1}{5}$, 20%

41. 2.5 or 2.50, 250%

42. $4\frac{2}{25}$, 408%

43. $1\frac{2}{3}$, $1.66\frac{2}{3}$

44. $2\frac{1}{8}$, 212.5%

45. 1.625, 162.5%

46. $3\frac{7}{8}$, 387.5%

4.2   Writing and Solving Mathematical Sentences

---

1   When such a statement as "2% of 100 is 2" is written in the form $0.02 \times 100 = 2$, it is said to be written mathematically. Written mathematically, 6% of 20 is 1.2 would become: $0.06 \times 20 = 1.2$. Notice that "of" means "multiply" and is replaced by the multiplication symbol, $\times$. "Is" means "is equal to" and is replaced by the equal sign, $=$.

Q1   Write "10% of 20 is 2" mathematically. _____

STOP • **STOP** • **STOP** • **STOP** • **STOP** • **STOP** • **STOP** • **STOP** • **STOP**

A1 $0.10 \times 20 = 2$, or $\frac{1}{10} \times 20 = 2$

Q2   Write mathematically:

   a. 2% of 14 is 0.28   _____

   b. 20% of 50 is 10   _____

   c. 14% of 40 is 5.6   _____

   d. 200% of 4 is 8   _____

   e. $66\frac{2}{3}$% of 24 is 16   _____

STOP • **STOP** • **STOP** • **STOP** • **STOP** • **STOP** • **STOP** • **STOP** • **STOP**

A2   **a.** $0.02 \times 14 = 0.28$, or $\frac{1}{50} \times 14 = 0.28$

   b. $0.20 \times 50 = 10$, or $\frac{1}{5} \times 50 = 10$

   c. $0.14 \times 40 = 5.6$, or $\frac{7}{50} \times 40 = 5.6$

   d. $2 \times 4 = 8$

   e. $0.66\frac{2}{3} \times 24 = 16$, or $\frac{2}{3} \times 24 = 16$

> **2**   Consider the question: 5% of 12 is what number? Written mathematically: $0.05 \times 12 =$ what number?
>
>   When we use letter N to "hold the place" of the missing number, the statement becomes: $0.05 \times 12 = N$? Any letter can be used for the missing (unknown) number. N is a popular choice, because "number" begins with the letter n.

Q3   Write mathematically: 75% of 116 is what number?_____

STOP • **STOP** • **STOP** • **STOP** • **STOP** • **STOP** • **STOP** • **STOP** • **STOP**

A3 $0.75 \times 116 = N$? or $\frac{3}{4} \times 116 = N$?

Q4   Write mathematically:

     a. 420% of 7 is what number?   _____

     b. What number is 420% of 7?   _____

     c. 62% of 19 is what number?   _____

     d. What number is 62% of 19?   _____

STOP • **STOP** • **STOP** • **STOP** • **STOP** • **STOP** • **STOP** • **STOP** • **STOP**

A4   **a.** $4.2 \times 7 = N$?   **b.** $N = 4.2 \times 7$?   **c.** $0.62 \times 19 = N$?   **d.** $N = 0.62 \times 19$?

> **3**   In the following question a different part of the problem is missing:
>
>   4.1% of what number is 7.3?
>
> $0.041 \times N = 7.3$ (written mathematically)
>
>   Notice that "of" is replaced by "$\times$," "what number" by "N," and "is" by "$=$." (4.1% = 0.041.)

Q5   Write mathematically: 7% of what number is 0.09?_____

STOP • **STOP** • **STOP** • **STOP** • **STOP** • **STOP** • **STOP** • **STOP** • **STOP**

A5 $0.07 \times N = 0.09$?

Q6   Write mathematically:

     a. 22% of what number is 12?   _____

     b. 15 is 35% of what number?   _____

     c. 320% of what number is 400?   _____

     d. $37\frac{1}{2}$% of what number is 54?   _____

     e. 23.5 is 10% of what number?   _____

STOP • **STOP** • **STOP** • **STOP** • **STOP** • **STOP** • **STOP** • **STOP** • **STOP**

A6   **a.** $0.22 \times N = 12$?   **b.** $15 = 0.35 \times N$?   **c.** $3.2 \times N = 400$?

        d. $0.37\frac{1}{2} \times N = 54$?   or $\frac{3}{8} \times N = 54$?   **e.** $23.5 = 0.10 \times N$?

---

4   Consider still another type of percent question:

What percent of 90 is 7?
$$P \quad \times 90 = 7?$$

Notice that the percent is missing; hence, the letter P was used as a reminder that the missing number is a percent.

Q7   Write mathematically: What percent of 70 is 203? _____

STOP • STOP • STOP • STOP • STOP • STOP • STOP • STOP • STOP

A7 $P \times 70 = 203?$

Q8   Write mathematically:

    a. What percent of 12 is 0.6?   _____

    b. What percent of 9 is 24?   _____

    c. 23 is what percent of 19?   _____

    d. 14 is what percent of 30?   _____

    e. What percent of 30 is 14?   _____

STOP • STOP • STOP • STOP • STOP • STOP • STOP • STOP • STOP

A8   **a.** $P \times 12 = 0.6?$ **b.** $P \times 9 = 24?$ **c.** $23 = P \times 19?$ **d.** $14 = P \times 30?$
    e. $P \times 30 = 14?$

---

5   In a multiplication problem the numbers that are multiplied together to form the *product* are called the *factors*. In the example $4 \times 3 = 12$, 4 and 3 are the factors and 12 is the product.

      Two division problems can be formed from any multiplication problem. For example, $4 \times 3 = 12$ implies that $4 = 12 \div 3$ and $3 = 12 \div 4$.

Q9   Write two division problems that can be formed from $3 \times 5 = 15$.

    _____ and _____

STOP • STOP • STOP • STOP • STOP • STOP • STOP • STOP • STOP

A9 $3 = 15 \div 5, 5 = 15 \div 3$ (either order)

Q10   Write two division problems that can be formed from $5 \times 9 = 45$.

    _____ and _____

STOP • STOP • STOP • STOP • STOP • STOP • STOP • STOP • STOP

A10 $5 = 45 \div 9, 9 = 45 \div 5$ (either order)

---

6   Generally, a multiplication problem can be represented as: *factor* $\times$ *factor* = *product*. From the previous example this statement can be written *one factor* = *product* $\div$ *other factor*. For example, if $4 \times 8 = 32$, *what number* $= 32 \div 8$ or $N = 32 \div 8$? The answer is that $4 = 32 \div 8$; hence, $N = 4$.

Q11   Determine the missing factor represented by the letter N:

    a. If $4 \times 8 = 32$, then $N = 32 \div 4$; $N =$ _____

    b. If $2 \times 3 = 6$, then $N = 6 \div 3$; $N =$ _____

       c. If $45 = 5 \times 9$, then $N = 45 \div 5$; $N =$ ————

       d. If $25 \times N = 1,200$, then $N = 1,200 \div 25$; $N =$ ————

STOP • *STOP* • *STOP* • *STOP* • *STOP* • *STOP* • *STOP* • *STOP* • *STOP*

A11   **a.** 8   **b.** 2   **c.** 9   **d.** 48

7	Often one of the factors of a product will be unknown. That is, *unknown factor* × *known factor = product*. This statement could be written *unknown factor = product ÷ known factor*. For example, if $N \times 15 = 60$, then $N = 60 \div 15$. N represents the unknown factor, 15 the known factor, and 60 the product.

Q12   Write the unknown factor as the quotient of the product and the known factor:

       a. $N \times 7 = 42$ ———————   **b.** $7 \times N = 42$ ———————

STOP • *STOP* • *STOP* • *STOP* • *STOP* • *STOP* • *STOP* • *STOP* • *STOP*

A12   **a.** $N = 42 \div 7$ **b.** $N = 42 \div 7$

Q13   Solve for N by writing the unknown factor as a quotient of the product and the known factor:

       a. $0.03 \times N = 15$ **b.** $2.5 = 0.06 \times N$

STOP • *STOP* • *STOP* • *STOP* • *STOP* • *STOP* • *STOP* • *STOP* • *STOP*

A13   **a.** 500: $N = 15 \div 0.03$ **b.** $41\frac{2}{3}$: $N = 2.5 \div 0.06$

                     $= 500$ $= 41\frac{2}{3}$

Q14   Solve for N:

       a. $0.1 \times N = 2.4$ **b.** $\$1.62 = 0.06 \times N$

STOP • *STOP* • *STOP* • *STOP* • *STOP* • *STOP* • *STOP* • *STOP* • *STOP*

A14   **a.** 24: $N = 2.4 \div 0.1$ **b.** \$27: $N = \$1.62 \div 0.06$

Q15   Solve for N:

       a. $N \times \$21 = \$0.42$ **b.** $\$0.12 \times N = \$2.40$

STOP • *STOP* • *STOP* • *STOP* • *STOP* • *STOP* • *STOP* • *STOP* • *STOP*

A15   **a.** 0.02: $N = \$0.42 \div \$21$ **b.** 20: $N = \$2.40 \div \$0.12$

This completes the instruction for this section.

4.2 Exercises

1. Write mathematically:
   a. 6% of 10 is what number?
   b. What number is 17.3% of 92?
   c. 520% of what number is 62?
   d. What percent of 17 is 9?
   e. 17 is 43% of what number?
2. If factor $\times$ factor = product, one factor = _____ $\div$ _____ .
3. If known factor $\times$ unknown factor = product, _____ = product $\div$ _____ .
4. If $75 = N \times 3$, $N =$ _____ $\div$ _____ .
5. Solve for N or P:
   a. $5.2 \times N = 62$
   b. $17 = 0.43 \times N$
   c. $0.06 \times 10 = N$
   d. $\$13 = P \times \$52$
   e. $\$0.21 = 0.03 \times N$

4.2 Exercise Answers

1. **a.** $0.06 \times 10 = N$
   b. $N = 0.173 \times 92$
   c. $5.2 \times N = 62$
   d. $P \times 17 = 9$
   e. $17 = 0.43 \times N$
2. product, other factor
3. unknown factor, known factor
4. 75, 3
5. **a.** $11\dfrac{12}{13}$
   b. $39\dfrac{23}{43}$
   c. 0.6
   d. 0.25 or 25%
   e. $7

4.3 Solving Percent Problems

| 1 | Problems involving percents usually occur in three forms: |

1. Finding a percent of a number or quantity. *Example:* What number is 6% of 12?
2. Finding what percent one number is of another. *Example:* 18 is what percent of 36?
3. Finding a number when a certain percent of it is known. *Example:* 12 is 15% of what number?

These questions can be solved by first writing them mathematically.

Example: What number is 6% of 12?

Solution

$N = 0.06 \times 12$*
$N = 0.72$

$$\begin{array}{r} 12 \\ \times 0.06 \\ \hline 0.72 \end{array}$$

(show work
where necessary)

Therefore, 0.72 is 6% of 12.

*It is common to omit the "?" mark.

Q1   What number is 3% of 7?

STOP • STOP • STOP • STOP • STOP • STOP • STOP • STOP • STOP

A1   0.21: $N = 0.03 \times 7$

Q2   0.3% of 63 is what number?

STOP • STOP • STOP • STOP • STOP • STOP • STOP • STOP • STOP

A2   0.189: $0.003 \times 63 = N$

Q3   0.4% of 210 is what number?

STOP • STOP • STOP • STOP • STOP • STOP • STOP • STOP • STOP

A3   0.84: $0.004 \times 210 = N$

Q4   What number is 520% of 92?

STOP • STOP • STOP • STOP • STOP • STOP • STOP • STOP • STOP

A4   478.4: $N = 5.2 \times 92$

---

2   Recall that a mixed-number percent is simplified in the following manner: $23\frac{1}{3}\% =$

$\frac{70}{3} \times \frac{1}{100} = \frac{7}{30}$.

Example: $23\frac{1}{3}\%$ of 900 is what number?

Solution

$\frac{7}{30} \times 900 = N$

$210 = N$

Therefore, $23\frac{1}{3}\%$ of 900 is 210.

Q5   **a.** What number is 200% of 4?   **b.** 24% of 30 is what number?

        c. 0.5% of 18 is what number?   **d.** What is $15\frac{1}{3}\%$ of 600?

STOP • STOP • STOP • STOP • STOP • STOP • STOP • STOP • STOP

A5   **a.** 8   **b.** 7.2   **c.** 0.09   **d.** 92

---

3   The second type of percent problem is finding what percent one number is of another.

Example: 18 is what percent of 36?

Solution

$18 = P \times 36$ (written mathematically)

Recall that *factor* $\times$ *factor* = *product* and *unknown factor* = *product* $\div$ *known factor*. Hence,

$18 \div 36 = P$ [18 (product) $\div$ 36 (known factor)]
$$0.5 = P$$
$$50\% = P$$

Therefore, 18 is 50% of 36.

Q6   17 is what percent of 68?

STOP • STOP • STOP • STOP • STOP • STOP • STOP • STOP • STOP

A6   25%: $17 = P \times 68$
$$17 \div 68 = P$$
$$0.25 = P$$

Q7   24 is what percent of 30?

STOP • STOP • STOP • STOP • STOP • STOP • STOP • STOP • STOP

A7   80%: $24 = P \times 30$
$$24 \div 30 = P$$
$$0.8 = P$$

Q8   21 is what percent of 300?

STOP • STOP • STOP • STOP • STOP • STOP • STOP • STOP • STOP

A8   7%: $21 = P \times 300$
$$21 \div 300 = P$$
$$0.07 = P$$

Q9   80 is what percent of 64?

STOP • STOP • STOP • STOP • STOP • STOP • STOP • STOP • STOP

A9   125%: $80 = P \times 64$
$$80 \div 64 = P$$
$$1.25 = P$$

Q10   What percent of 6 is 4?

STOP • STOP • STOP • STOP • STOP • STOP • STOP • STOP • STOP

A10 $66\frac{2}{3}\%$: $P \times 6 = 4$

$$P = 4 \div 6$$

$$P = 0.66\frac{2}{3}$$

Q11   What percent of 25 is 30?

STOP • STOP • STOP • STOP • STOP • STOP • STOP • STOP • STOP

A11   120%: $P \times 25 = 30$
$$P = 30 \div 25$$
$$P = 1.2$$

Q12   **a.** What percent of 300 is 45?   **b.** 38 is what percent of 304?

         c. What percent of 12 is 4.6?   **d.** 66.6 is what percent of 74?

STOP • **STOP** • **STOP** • **STOP** • **STOP** • **STOP** • **STOP** • **STOP** • **STOP**

A12   **a.** 15%   **b.** 12.5%   **c.** $38\frac{1}{3}$% **d.** 90%

---

4   The third type of percent problem is finding a number when a certain percent of it is known.

Example: 12 is 15% of what number?

Solution

$$12 = 0.15 \times N$$
$$12 \div 0.15 = N$$
$$80 = N$$

Therefore, 12 is 15% of 80.

Q13   27 is 5% of what number?

STOP • **STOP** • **STOP** • **STOP** • **STOP** • **STOP** • **STOP** • **STOP** • **STOP**

A13   540: $27 = 0.05 \times N$
                $27 \div 0.05 = N$

Q14   17 is 4% of what number?

STOP • **STOP** • **STOP** • **STOP** • **STOP** • **STOP** • **STOP** • **STOP** • **STOP**

A14   425: $17 = 0.04 \times N$
                $17 \div 0.04 = N$

Q15   34 is 125% of what number?

STOP • **STOP** • **STOP** • **STOP** • **STOP** • **STOP** • **STOP** • **STOP** • **STOP**

A15   27.2: $34 = 1.25 \times N$
                         $34 \div 1.25 = N$

Q16   91 is 70% of what number?

STOP • **STOP** • **STOP** • **STOP** • **STOP** • **STOP** • **STOP** • **STOP** • **STOP**

A16   130: $91 = 0.7 \times N$
                   $91 \div 0.7 = N$

Q17   5% of what number is 8?

STOP • **STOP** • **STOP** • **STOP** • **STOP** • **STOP** • **STOP** • **STOP** • **STOP**

A17   160: $0.05 \times N = 8$
                        $N = 8 \div 0.05$

Q18   16% of what number is 10?

STOP • **STOP** • **STOP** • **STOP** • **STOP** • **STOP** • **STOP** • **STOP** • **STOP**

A18   62.5: $0.16 \times N = 10$
                         $N = 10 \div 0.16$

Q19   **a.** 78 is 15.6% of what number?   **b.** 8% of what number is 24?

           c. 5 is 0.2% of what number?   **d.** $33\frac{1}{3}$% of what number is 12.5?

STOP • **STOP** • **STOP** • **STOP** • **STOP** • **STOP** • **STOP** • **STOP** • **STOP**

A19   **a.** 500   **b.** 300   **c.** 2,500   **d.** 37.5

This completes the instruction for this section.

4.3   Exercises

1. 15 is what percent of 70?
2. 69% of $21 is what number?
3. 6 is 30% of what number?
4. What number is 19% of 20?
5. 152% of 5 is what number?
6. What percent of 320 is 500?
7. 8 is what percent of 15?
8. 32 is 64% of what number?
9. 75% of what number is 4.65?
10. 3.6 is 18% of what number?

11. What number is $33\frac{1}{3}$% of 60?
12. 12.5% of 28 is what number?

13. $63\frac{1}{3}$% of 3 is what number?
14. 122.5 is what percent of 98?

15. 2 is what percent of 400?

4.3   Exercise Answers

1. $21\frac{3}{7}$%
2. $14.49
3. 20
4. 3.8

5. 7.6
6. 156.25%
7. $53\frac{1}{3}$%
8. 50

9. 6.2
10. 20
11. 20
12. 3.5
13. 1.9
14. 125%
15. 0.5%

4.4   Story Problems that Involve Percents

1

The solution of story problems has always been an area of difficulty for many students. The difficulty arises because procedures for solving story problems vary with the problem and the difficulty students often have in stating a problem mathematically. Practice and an organized plan will aid in solving problems that involve percents. The following organization of a problem may be helpful.

Step 1: Decide what the problem asks you to find.

Step 2: List the useful information. Restate the problem.

Step 3: State the problem mathematically and solve.

Step 4: Check the solution.

Example: Jack sold his camera for $18.40. This was 120 percent of the cost. What was the cost of the camera?

Solution

Using the four steps:

Step 1: Cost of the camera?

Step 2: $18.40 (selling price)
$18.40 is 120% of the cost (C)

Step 3: $18.40 = 1.2 \times C$
$18.40 \div 1.2 = C$
$15.33 = C$ (rounded off to the nearest cent)

Step 4: Check: $18.40 = 120\%$ of $15.33
$18.40 = 1.2 \times \$15.33$
$18.40 = \$18.40$ (rounded off)

The cost, $15.33, checks and is the correct solution.

Q1 List the four steps for solving a problem:

a. _____

b. _____

c. _____

d. _____

STOP • STOP • STOP • STOP • STOP • STOP • STOP • STOP • STOP

A1 a. Decide what the problem asks you to find.
b. List the useful information.
c. State the problem mathematically and solve.
d. Check the solution.

Q2 Use the four steps to solve the following problem: Mr. Graves contributed 9% of a $180 Red Cross Fund. How much did he contribute?

STOP • STOP • STOP • STOP • STOP • STOP • STOP • STOP • STOP

A2 Step 1: The contribution?

Step 2: Contribution is 9% of $180

Step 3: $C = 0.09 \times 180$
$C = \$16.20$

Step 4: Check: $16.20 = 9\%$ of $180
$16.20 = 0.09 \times 180$
$16.20 = \$16.20$

Q3 Gordon worked 9 of 14 weeks of his vacation. What percent of his vacation did he work? (Use the four steps.)

STOP • STOP • STOP • STOP • STOP • STOP • STOP • STOP • STOP

A3   Step 1: Percent of vacation worked?

          Step 2: Vacation 14 weeks
                  Vacation worked 9 weeks
                  9 weeks is what percent of 14 weeks?

          Step 3: $9 = P \times 14$
                  $9 \div 14 = P$

                  $64\frac{2}{7}\% = P$

          Step 4: 9 weeks is $64\frac{2}{7}\%$ of 14 weeks

                  $9 = 0.64\frac{2}{7} \times 14$

                  $9 = 9$

2	The money that a salesperson receives for making a sale, usually a percent of the total sale, is called the *commission*.

Q4   An agent sold \$1,420 worth of sugar at a $1\frac{3}{4}\%$ commission. What was the commission?

STOP • STOP • STOP • STOP • STOP • STOP • STOP • STOP • STOP

A4   \$24.85:

          Step 1: The commission?

          Step 2: Commission is $1\frac{3}{4}\%$ of \$1,420

          Step 3: $C = 0.01\frac{3}{4} \times \$1,420$

                  $C = \$24.85$

          Step 4: $\$24.85 = 1\frac{3}{4}\%$ of \$1,420

                  $\$24.85 = \$24.85$

3	A certain amount of money is often subtracted from the list price of an article. This subtracted amount is generally called the *discount*. The selling price, after the discount has been subtracted, is the net price. That is,
	list price − discount = net price

Q5   A typewriter is listed to sell at $120. The owner is offering a 15% discount if cash is used to purchase the typewriter. What is the discount?

STOP   •   STOP   •   STOP   •   STOP   •   STOP   •   STOP   •   STOP   •   STOP   •   STOP

A5 $18:

Step 1: The discount?

Step 2: Discount is 15% of $120

Step 3: $d = 0.15 \times \$120$
$d = \$18$

Step 4: $\$18 = 15\%$ of $\$120$
$\$18 = \$18$

Q6   John's team won 14 games and lost 11 games. What is the percent of the total games that the team won?

STOP   •   STOP   •   STOP   •   STOP   •   STOP   •   STOP   •   STOP   •   STOP   •   STOP

A6   56%:

Step 1: Percent of games won?

Step 2: 14 games is what percent of 25 total games (total games played)?

Step 3: $14 = P \times 25$
$14 \div 25 = P$
$56\% = P$

Step 4: 14 is 56% of 25
$14 = 0.56 \times 25$
$14 = 14$

Q7   Sally bought a $17 dress. The state in which Sally lives has a $2\frac{1}{2}\%$ sales tax. What was the sales tax on the dress?

STOP • **STOP** • **STOP** • **STOP** • **STOP** • **STOP** • **STOP** • **STOP** • **STOP**

A7 $0.43:

Step 1: The sales tax?

Step 2: Sales tax is $2\frac{1}{2}\%$ of $17

Step 3: $T = 0.025 \times \$17$

$T = \$0.425$

$T = \$0.43$

Step 4: $\$17 \times 0.025 = \$0.425$

Q8   A real estate agent sold a house and received a commission of $1,787.50, which represented 5% of the selling price. How much did the house sell for?

STOP • **STOP** • **STOP** • **STOP** • **STOP** • **STOP** • **STOP** • **STOP** • **STOP**

A8 $35,750:

Step 1: Selling price of the house?

Step 2: $1,787.50 is 5% of the selling price

Step 3: $\$1,787.50 = 0.05 \times s$
$\$1,787.50 \div 0.05 = s$
$\$35,750 = s$

Step 4: Check: $\$1,787.50 = 0.05 \times \$35,750$

Q9   Mr. Belt says that a new automobile decreases (depreciates) $37\frac{1}{2}\%$ of its value the first year. At this rate, how much will the value of a $2,124 car drop in the first year?

STOP • STOP • STOP • STOP • STOP • STOP • STOP • STOP • STOP

A9 $796.50: partial solution $N = 37\frac{1}{2}\%$ of $2,124

Q10   When Mr. Dahlke's family moved to town, they rented a house for $285 a month. The landlord recently increased the rent $15 a month. What is the percent of increase in rent?

STOP • STOP • STOP • STOP • STOP • STOP • STOP • STOP • STOP

A10 $5\frac{5}{19}\%$: partial solution $15 = P \times \$285$

Q11   During the first 6 months, Mary gained 25% of her weight at birth. If she gained $2\frac{1}{2}$ pounds during this period, how much did she weigh at birth?

STOP • STOP • STOP • STOP • STOP • STOP • STOP • STOP • STOP

A11   10 pounds: partial solution $0.25 \times N = 2.5$ pounds

---

This completes the instruction for this section.

4.4   Exercises

Solve the following problems:

1. Mr. Lake left $36,000 to his family when he died. If he left $66\frac{2}{3}\%$ to his wife, how much did she receive?

2. Bob has traveled 250 miles of a 1,000-mile trip. What percent of the total distance has he traveled?

3. George earned $36.40 during 1 week of magazine selling. If this represented 40% of his total sales, what were his total sales for the week?

4. A baseball player made 18 hits in 57 times at bat. During what percent of the times at bat did he make a hit? Round off the answer to the nearest percent.

5. The price of a shirt was increased $0.40. If this represented $12\frac{1}{2}$% of the original price, what was the original price of the shirt?

6. At a sale, a $25 dress is marked "10% off." What is the amount of the discount on the dress?

7. The Pumford family had an income of $400 during a month. They spent 25% of the income for food and 20% for rent. During the month $140 was spent for clothes, recreation, and other expenses. The rest of the money was saved.
   a. How much was spent for food?
   b. How much was spent for rent?
   c. What was the total of all expenses?
   d. How much was saved?
   e. What percent of the income was saved?

8. In 1973 the school basketball team won 16 of its 20 games. In 1974 the team won 11 of its 15 games.
   a. What percent of its games did the 1973 team win?
   b. What percent of its games did the 1974 team win?
   c. Which team was more successful?

4.4   Exercise Answers

1. $24,000
2. 25%
3. $91
4. 32%
5. $3.20
6. $2.50
7. **a.** $100   **b.** $80   **c.** $320   **d.** $80   **e.** 20%
8. **a.** 80%   **b.** $73\frac{1}{3}$% **c.** 1973

Chapter 4 Sample Test

At the completion of Chapter 4 it is expected that you will be able to work the following problems.

4.1   Percents

1. Change to a decimal:
   a. 24%   **b.** 7.3%

2. Change to a percent:
   a. 0.8   **b.** 1.25
3. Change to a fraction:

   a. $16\frac{2}{3}\%$ **b.** $12\frac{1}{2}\%$
4. Change to a percent:

   a. $\dfrac{9}{40}$ **b.** $\dfrac{5}{6}$

4.2   Writing and Solving Mathematical Sentences

5. Write the following mathematically:
   a. 17% of 61 is 10.37   **b.** 5% of 30 is what number?
6. **a.** If $4 \times 5 = 20$, $5 =$ _____ ÷ _____.
   b. If $0.5 \times A = \$2.70$, $A =$ _____ ÷ _____.

4.3   Solving Percent Problems

7. **a.** 15 is 30% of what number? **b.** 336 is what percent of 16?

4.4   Story Problems that Involve Percents

8. Solve the following problems:
   a. Mr. Knox's rent during the past year amounted to $780 per year. His rent was 20% of his income. Find his income for last year.
   b. A store reduced by $35 the price of a $150 suit. Find the percent of reduction.

Chapter 4 Sample Test Answers

1. **a.** 0.24   **b.** 0.073
2. **a.** 80%   **b.** 125%

3. **a.** $\dfrac{1}{6}$ **b.** $\dfrac{1}{8}$

4. **a.** 22.5%   **b.** $83\frac{1}{3}\%$

5. **a.** $0.17 \times 61 = 10.37$ **b.** $0.05 \times 30 = N$
6. **a.** 20, 4   **b.** $2.70, 0.5
7. **a.** 50   **b.** 2,100%

8. **a.** $3,900   **b.** $23\frac{1}{3}\%$

Chapter 5

Sets and Properties of Whole Numbers

Section 5.1 is designed to provide the student with the necessary background for understanding the role of sets when developing number ideas throughout the text.

5.1   Set Ideas

---

1   A *set* is a well-defined collection of things. The things that make up a set are called *elements*. The collection of letters in the English alphabet is a set of 26 letters. The letter b is an element of this set. The collection of months in a year is a set of 12 months. July is an element of this set.

Q1   The collection of days in a week is a _____ of seven days.

STOP • STOP • STOP • STOP • STOP • STOP • STOP • STOP • STOP

A1   set

Q2   Name an element of the set in Q1._____

STOP • STOP • STOP • STOP • STOP • STOP • STOP • STOP • STOP

A2   Any day of the week: for example, Wednesday.

---

2   A set may contain elements that have a common property, such as the presidents of the United States or the residents of a particular city. A set may also consist of unrelated objects, such as a chair, a pencil, and a book. However, a set must be *well defined*. This means that membership in the set is clear. The set of men who have been President of the United States is a well-defined collection because it is clear exactly which men are included in this set. A collection of any three letters of the English alphabet is not a well-defined collection because many possibilities exist and you would not be sure which is intended.

Q3   Choose the sets from the following collections:
a.   The collection of states in the United States bordered by the Great Lakes
b.   The collection of three consecutive warm months
c.   The collection of beautiful people
d.   The collection of women who have been President of the United States

_____

STOP • STOP • STOP • STOP • STOP • STOP • STOP • STOP • STOP

A3   a and d: Part b is not well defined because temperature is dependent on location and the meaning of "warm" is debatable. Part c is not well defined because there would not be agreement as to who is beautiful.

---

3   Using braces, { }, to indicate a set is called *set notation*. The elements of the set are enclosed in braces and are separated by commas. For example, the set of one-digit numbers* between 1 and 7 is denoted as {2, 3, 4, 5, 6}. The braces are read "the set containing," so {1, 2, 5} is read "the set containing 1, 2, and 5."

*The "digits" are often referred to as 0, 1, 2, 3, 4, 5, 6, 7, 8, and 9.

Q4   Use set notation to indicate the set of digits between 2 and 9. _____

STOP • STOP • STOP • STOP • STOP • STOP • STOP • STOP • STOP

A4   {3, 4, 5, 6, 7, 8}

---

4   The use of set notation is not always convenient. A description is used when listing the elements would be difficult or impractical. Such an example would be the set of U.S. Congressmen.

Q5   Where convenient, use set notation to indicate the set:

     a. The set of vowels in the English alphabet   _____

     b. The set of different kinds of flowers found in Ontario   _____

     c. The set of digits in the numeral 100,537,375   _____

STOP • STOP • STOP • STOP • STOP • STOP • STOP • STOP • STOP

A5   a. {a, e, i, o, u} (*Note:* It is common to arrange letter elements of a set alphabetically.)
     b. Although this information could be listed, it would not be convenient to do so in set notation.
     c. {0, 1, 3, 5, 7} (*Note:* The order in which the elements are arranged does not affect the set. That is, {0, 1, 3, 5, 7} = {1, 0, 5, 3, 7}. Also, duplicate elements need be listed only once.)

---

5   The set that contains no elements is called the *empty set*. It is sometimes written using braces with no elements, { }. The symbol ∅ also denotes the empty set. The set of one-digit numbers between 4 and 5 is an example of the empty set.

Q6   The set of living 200-year-old people represents an _____ set.

STOP • STOP • STOP • STOP • STOP • STOP • STOP • STOP • STOP

A6   empty

Q7   Use set notation to indicate the set of two-digit numbers greater than 99. _____

STOP • STOP • STOP • STOP • STOP • STOP • STOP • STOP • STOP

A7   { } or ∅

---

6 How many men had walked on the moon by 1960? The answer to this question is zero. That is, the set of men who had walked on the moon by 1960 is empty. You should observe that the number of elements in the empty set is zero; that is, there are no elements in the empty set.

Q8   **a.** How many elements are there in {0}?   _____

        b. How many elements are there in { }?   _____

        c. Is {0} = { }?   _____

STOP • STOP • STOP • STOP • STOP • STOP • STOP • STOP • STOP

A8   **a.** one, the element 0   **b.** zero   **c.** no

---

7 If each element of one set is also a member of a second set, the first set is a *subset* of the second. The set of dogs is a subset of the set of animals.

Q9   The set of New England States is a _____ of the set of states in the United States.

STOP • STOP • STOP • STOP • STOP • STOP • STOP • STOP • STOP

A9   subset

Q10   Determine whether the first set is a subset of the second set:

        a. {Jane, Mary}   {June, Judy, Jane}   _____

        b. {1, 2, 5}   {0, 1, 2, 3, 4, 5}   _____

        c. {2, 4, 6}   {1, 2, 3, 4, 5}   _____

        d. {1, 2, 3}   {1, 2, 3}   _____

        e. set of vowels   set of all letters in the English alphabet   _____

STOP • STOP • STOP • STOP • STOP • STOP • STOP • STOP • STOP

A10   **a.** no   **b.** yes   **c.** no   **d.** yes   **e.** yes

---

8 The symbol $\in$ is used to abbreviate "is an element of." $x \in \{x, y, z\}$ is read "x is an element of the set containing $x, y,$ and z." The slant bar, /, is used as a negation symbol. $\notin$ means "is *not* an element of." $1 \notin \{0, 2, 4, 6\}$ is read "1 is not an element of the set containing 0, 2, 4, and 6."

Q11   Insert the correct symbol, $\in$ or $\notin$, in the following blanks:

        a. r _____ {a, e, i, o, u}

        b. 5 _____ {1, 3, 5, 7, 9}

STOP • STOP • STOP • STOP • STOP • STOP • STOP • STOP • STOP

A11   **a.** $\notin$ **b.** $\in$

---

9 Since the set of letters of the English alphabet are numerous, this set may be indicated as {a, b, c, . . . , z}. The three dots, read "and so on," indicate that not all elements of the set have been listed and that additional elements of the set can be found by following the pattern established in the elements preceding the dots.

Q12   Insert the correct symbol, $\in$ or $\notin$, in the following blanks:

    a. 17 _____ {1, 2, 3, ... , 20}

    b. 12 _____ {1, 3, 5, ... , 19}

    c. 10 _____ {2, 4, 6, ... , 20}

STOP • **STOP** • **STOP** • **STOP** • **STOP** • **STOP** • **STOP** • **STOP** • **STOP**

A12   **a.** $\in$ **b.** $\notin$ **c.** $\in$

> **10**   If the elements of a set can be counted and the count has a last number, the set is a *finite* set. A set whose count is unending is an *infinite* set. For example, {a, b, c, d} is a finite set, whereas {1, 2, 3, 4, ...} is an infinite set.

Q13   Describe each of the following sets as finite or infinite:

    a. The set of letters in the word "infinite" _____

    b. {1, 2, 3, 4, ... , 15} _____

    c. {5, 10, 15, ...} _____

    d. The set of provincial parks in Canada _____

    e. $\varnothing$ _____

STOP • **STOP** • **STOP** • **STOP** • **STOP** • **STOP** • **STOP** • **STOP** • **STOP**

A13   **a.** finite   **b.** finite   **c.** infinite   **d.** finite
    e. finite (the empty set has a count of zero)

This completes the instruction for this section.

5.1 Exercises

1. Give the word or phrase that best describes each of the following:
   a. well-defined collection
   b. set that contains no elements
   c. set in which the elements can be counted and the count has a last number
   d. set whose count is unending
2. Specify the members of each of the following sets:
   a. states in the United States bordered by the Pacific Ocean
   b. last three letters of the English alphabet
   c. {1, 2, 3}
3. Which of the following are not sets:
   a. {1, 3, 5, 7, 9, 11, ...}
   b. collection of ex-Presidents of the United States
   c. collection of consonants in the English alphabet
   d. collection of large numbers
4. Where convenient, use set notation to indicate the following sets:
   a. letters in the word "college"
   b. digits in the number 10,274
   c. five-letter words beginning with C
   d. days of the week beginning with R
   e. digits between 0 and 9

5. True or false:
   a. {a, c, d} is a subset of {a, b, c, d}.
   b. The set of cities that are provincial capitals in Canada is a subset of the set of all cities in Canada.
   c. {0} is a subset of {1, 2, 3, ...}.
   d. {Jane, Mary, Sue, Ann} is a subset of {Mary, Sue, Jane, Ann}.
6. Insert the correct symbol, $\in$ or $\notin$, in each of the following:
   a. h _____ {a, e, i, o, u}
   b. 0 _____ {1, 2, 3, ..., 12}
   c. 32 _____ {0, 2, 4, 6, ...}
   d. 2,000 _____ {0, 1, 2, 3, ...}
7. Describe each of the following sets as finite or infinite:
   a. consonants in the English alphabet
   b. {0, 1, 2, 3, 4, ...}
   c. digits between 5 and 6
   d. {0, 2, 4, 6, ..., 20}

5.1 Exercise Answers

1. a. set   b. empty set   c. finite set   d. infinite set
2. a. Alaska, California, Hawaii, Oregon, and Washington
   b. x, y, and z   c. 1, 2, and 3
3. d
4. a. {c, e, g, l, o}   b. {0, 1, 2, 4, 7}   c. not convenient   d. { } or Ø
   e. {1, 2, 3, 4, 5, 6, 7, 8} or {1, 2, 3, ..., 8}
5. a. true   b. true   c. false   d. true
6. a. $\notin$ b. $\notin$ c. $\in$ d. $\in$
7. a. finite   b. infinite   c. finite   d. finite

5.2 Evaluating Open Expressions

1   The evaluation of numerical expressions such as $432 - 178 + 92$, $3 + 5 \cdot 7$, $(13 - 2)(15 + 8)$, $12 \cdot 9 - 5 \cdot 6$, and $5(9 - 4)$ requires rules for order of operations:

Step 1: Perform operations within grouping symbols first.

Step 2: Next, perform all multiplications and divisions in the order in which they appear from left to right.

Step 3: Finally, perform all additions and subtractions in the order in which they appear from left to right.

     The raised dot, $\cdot$, indicates multiplication. Also, parentheses side by side or a number preceding parentheses indicates multiplication. That is,

$$(5)(7) = 5 \cdot 7 \qquad \text{or} \qquad 5(7) = 5 \cdot 7$$

The form of the evaluation of a numerical expression may vary. Two acceptable forms are

$$5 + 4 \cdot 7 = 5 + 28 \qquad \text{and} \qquad 5 + 4 \cdot 7$$
$$= \quad 33 \qquad\qquad\qquad 5 + \ 28$$
$$33$$

When steps are arranged vertically, as in the second example, each line is assumed to be equal to the previous line.

Q1   Evaluate the following numerical expressions (show your work):

a. $4 + 5 \cdot 6$ **b.** $3 \cdot 5 - 8$

c. $12(13 - 5)$ **d.** $15 \cdot 3 - 10 \cdot 2$

STOP • STOP • STOP • STOP • STOP • STOP • STOP • STOP • STOP

A1   **a.** 34: $4 + 5 \cdot 6 = 4 + 30$ **b.** 7: $3 \cdot 5 - 8 = 15 - 8$
$$= 34 \qquad\qquad\qquad\qquad\qquad = 7$$

    c. 96: $12(13 - 5)$ **d.** 25: $15 \cdot 3 - 10 \cdot 2$
$$12(8) \qquad\qquad\qquad\qquad\qquad 45 \ - \ 20$$
$$96 \qquad\qquad\qquad\qquad\qquad\qquad 25$$

Q2   Evaluate the following numerical expressions (show your work):

a. $2 + 4 \cdot 5 - 3$ **b.** $35 - 3(5 + 12 \div 3)$

c. $(32 - 13)(12 + 9)$ **d.** $45 - 15 \div 5$

STOP • STOP • STOP • STOP • STOP • STOP • STOP • STOP • STOP

A2   **a.** 19: $2 + 20 - 3$ **b.** 8: $35 - 3(5 + 4)$
$$22 - 3 \qquad\qquad\qquad\qquad\qquad 35 - 3 \cdot 9$$
$$19 \qquad\qquad\qquad\qquad\qquad\qquad 35 - 27$$
$$8$$

    c. 399: $19 \cdot 21$ **d.** 42: $45 - 3$
$$399 \qquad\qquad\qquad\qquad\qquad\qquad 42$$

---

2 $\square + 7$ is called an *open expression*. Other examples of open expressions are

$$15 + \square \qquad\qquad 2 + 3 \cdot \square - 4$$
$$5(3 \cdot \square + 7) \qquad 45 - 18 - \square$$

Open expressions may be evaluated when the $\square$ is replaced by some number.

Example: Evaluate $15 + \square$ when $\square$ is replaced by 7.

Solution

$15 + \square$ or $15 + \square = 15 + 7$
$15 + 7$ $\qquad = 22$
22

Q3   Evaluate each of the following open expressions when the $\square$ is replaced by the given value (show your work):

a. $(\square + 3) \cdot 5$ when $\square$ is 7

b. $(\square - 4)(\square + 8)$ when $\square$ is $12\frac{1}{2}$

c. $3(\square \div 3 - 2)$ when $\square$ is 9

d. $5(3 \cdot \square + 4 \cdot \square)$ when $\square$ is $\frac{1}{3}$

e. $\square \cdot (\square + 5 - 2 \cdot \square)$ when $\square$ is 5

STOP • **STOP** • **STOP** • **STOP** • **STOP** • **STOP** • **STOP** • **STOP** • **STOP**

A3   a. 50: $(7 + 3) \cdot 5$
              $10 \cdot 5$
              50

b. $174\frac{1}{4}$: $\left(12\frac{1}{2} - 4\right)\left(12\frac{1}{2} + 8\right)$

$8\frac{1}{2} \cdot 20\frac{1}{2}$

$\frac{17}{2} \cdot \frac{41}{2}$

$\frac{697}{4} = 174\frac{1}{4}$

c. 3: $3(9 \div 3 - 2)$
$3(3 - 2)$
$3 \cdot 1$
3

d. $11\frac{2}{3}$: $5\left(3 \cdot \frac{1}{3} + 4 \cdot \frac{1}{3}\right)$

$5\left(1 + 1\frac{1}{3}\right)$

$5 \cdot 2\frac{1}{3}$

$5 \cdot \frac{7}{3}$

$\frac{35}{3}$

$11\frac{2}{3}$

e. 0: $5 \cdot (5 + 5 - 2 \cdot 5)$
$5 \cdot (5 + 5 - 10)$
$5 \cdot 0$
0

3	Any letter of the alphabet may be used in place of the $\square$ in an open expression. Using x, $\square + 5$ may be written as $x + 5$. $x + 5$ is also called an *open expression*. Other examples of open expressions are $a - 3$, $5 \cdot y + 2 \cdot y$, $5(b + 7)$, $(x - 2)(x + 3)$, and $16 + 5 \cdot z$.

Q4 $2(y - 6)$ is an _____ expression.

STOP • STOP • STOP • STOP • STOP • STOP • STOP • STOP • STOP

A4   open

4	When a letter may be replaced by any one of a set of many numbers, the letter is called a *variable*. The set of permissible values of the variable is referred to as its *replacement set*. When variables are replaced by values from their replacement set, the resulting numerical expression may then be evaluated. (In this section, the replacement set will be whole numbers, common fractions, and decimals.)

Example: Evaluate $5(b + 7)$ when b is replaced by 4.

Solution

$5(b + 7)$ or $5(b + 7) = 5(4 + 7)$
$5(4 + 7)$ $= 5 \cdot 11$
$5 \cdot 11$ $= 55$
55

Q5   Evaluate:
a. $5x$ when x is 7 (*Note:* $5x = 5 \cdot x$.)   **b.** $3y - 12$ when y is 10

    c. $7(t - 8)$ when t is $9\frac{1}{2}$ **d.** $2a + 3a$ when a is 5

    e. $b(b + 6)$ when b is 2

STOP • **STOP** • **STOP** • **STOP** • **STOP** • **STOP** • **STOP** • **STOP** • **STOP**

A5   **a.** 35: $5x$ **b.** 18: $3y - 12$
                  $5 \cdot 7$ $3 \cdot 10 - 12$
                  35                               $30 - 12$
                                                  18

    c. $10\frac{1}{2}$: $7(t - 8)$ **d.** 25: $2a + 3a$
                              $7\left(9\frac{1}{2} - 8\right)$ $2 \cdot 5 + 3 \cdot 5$
                                                $10 + 15$
                              $7 \cdot 1\frac{1}{2}$ 25

                              $10\frac{1}{2}$

    e. 16: $b(b + 6)$
                  $2(2 + 6)$
                  $2 \cdot 8$
                  16

---

5   Open expressions may contain more than one variable.

    Example: Evaluate $3x + y - 2z$ when $x = 2$, $y = 5$, and $z = 4$.

    Solution

    $3x + y - 2z$
    $3 \cdot 2 + 5 - 2 \cdot 4$
    $6 + 5 - 8$
    $11 - 8$
    3

Q6   Evaluate (show your work):
    a. $x + 7y$ when $x = 5$ and $y = 0.2$

b. $(5 + 2a) + 3b$ when $a = 4$ and $b = 5$

c. $0.5x + (3y - 2)$ when $x = 10$ and $y = 4$

d. $2(a - 2b) + 3a$ when $a = 12$ and $b = \dfrac{1}{2}$

e. $x(y + z)$ when $x = 2, y = 3,$ and $z = 1.5$

f. $cd - 5c$ when $c = 7$ and $d = 5$ (*Note: cd = c · d.*)

g. $\dfrac{x}{y}$ when $x = \dfrac{2}{3}$ and $y = \dfrac{4}{5}$

STOP • **STOP** • **STOP** • **STOP** • **STOP** • **STOP** • **STOP** • **STOP** • **STOP**

A6   **a.** 6.4: $x + 7y$
$5 + 7 \cdot 0.2$
$5 + 1.4$
6.4

b. 28: $(5 + 2a) + 3b$
$(5 + 2 \cdot 4) + 3 \cdot 5$
$(5 + 8) + 15$
$13 + 15$
28

c. 15: $0.5x + (3y - 2)$
$0.5 \cdot 10 + (3 \cdot 4 - 2)$
$5 + (12 - 2)$
$5 + 10$
15

d. 58: $2(a - 2b) + 3a$
$2\left(12 - 2 \cdot \dfrac{1}{2}\right) + 3 \cdot 12$
$2(12 - 1) + 36$
$2 \cdot 11 + 36$
$22 + 36$
58

e. 9: $x(y + z)$
$2(3 + 1.5)$
$2(4.5)$
9

f. 0: $cd - 5c$
$7 \cdot 5 - 5 \cdot 7$
$35 - 35$
0

g. $\dfrac{5}{6}$: $\dfrac{x}{y}$

$$\dfrac{\dfrac{2}{3}}{\dfrac{4}{5}} = \dfrac{2}{3} \cdot \dfrac{5}{4} = \dfrac{5}{6}$$

---

6 The product of 3 and 4 is 12. 3 and 4 are said to be *factors* of 12. In such products as $2x$. $3y$, cd, and $7xyz$, the parts of the product are also called factors. For example, the factors of $2x$ are 2 and x. The factors of cd are c and d. The factors of $7xyz$ are 7, x, y, and z.

Q7 Indicate the factors of each open expression:

a. $5y$ _____

b. $3ab$ _____

c. rst _____

d. $0.6x$ _____

STOP • **STOP** • **STOP** • **STOP** • **STOP** • **STOP** • **STOP** • **STOP** • **STOP**

A7 **a.** 5 and y

b. 3, a, and b

c. r, s, and t

d. 0.6 and x

---

7 In an open expression such as $5x$, 5 is called the number factor and x is called the letter factor. Mathematicians often use *numerical coefficient* for number factor and *literal coefficient* for letter factor.

Q8 **a.** The numerical coefficient of $5x$ is   _____.

b. The literal coefficient of $5x$ is   _____.

c. The numerical coefficient of $3ab$ is   _____.

d. The literal coefficients of $3ab$ are   _____.

STOP • **STOP** • **STOP** • **STOP** • **STOP** • **STOP** • **STOP** • **STOP** • **STOP**

A8 **a.** 5   **b.** x **c.** 3   **d.** a and b

---

8 Parentheses are used for grouping expressions. $5 + (a + 3)$ is read "5 plus the *quantity* a plus 3." $2(x - 1)$ is read "2 times the *quantity* x minus 1." The word "quantity" indicates the expression that has been placed within the parentheses.

Q9 **a.** $2x - (3x + 5)$ is read "$2x$ minus the _____ $3x$ plus 5."

b. $h(h + 8)$ is read "h times the _____ h plus 8."

STOP • **STOP** • **STOP** • **STOP** • **STOP** • **STOP** • **STOP** • **STOP** • **STOP**

A9 **a.** quantity   **b.** quantity

This completes the instruction for this section.

5.2   Exercises

1. Evaluate each open expression for the value of the variable given:
   a. $x - 5$ when $x = 32.6$
   b. $3a + 7$ when $a = 4$

   c. $4(y + 3)$ when $y = \dfrac{1}{2}$

   d. $12 + 2(2x - 9)$ when $x = 5$
   e. $3b(b - 8)$ when $b = 12$

   f. $2x + 3y$ when $x = 9$ and $y = \dfrac{2}{3}$

   g. rst when $r = 0.2$, $s = 20$, and $t = 0.6$

   h. $\dfrac{2a}{b} + 6$ when $a = 10$ and $b = 5$

   i. $7x - 4x$ when $x = 4.3$
   j. $(2y - 3)(y + 6)$ when $y = 7$

2. Use the following words or phrases to correctly answer each of the following questions. If more than one applies, you need only select one. (variable, open expression, numerical coefficient, literal coefficient, quantity, factor)
   a. What is each of the following called? $x - 5$, $3a + 7$, $3b(b - 8)$, and $2x + 3y$
   b. What is x called in $x - 5$, xy, and $2x + 3y$?
   c. What is $b - 8$ called in $3b(b - 8)$?
   d. What is 2 called in $2x$ and $2(3 + x)$?
   e. What is x called in $2x$?
   f. What is $4x$ called in $(7x)(4x)$?
   g. What is $3b$ called in $3b(b - 8)$?
   h. What is $x + 2$ called in $5 + (x + 2)$?

5.2   Exercise Answers

1. **a.** 27.6   **b.** 19   **c.** 14   **d.** 14   **e.** 144   **f.** 20   **g.** 2.4
   h. 10   **i.** 12.9   **j.** 143
2. **a.** open expression   **b.** variable
   c. quantity or factor   **d.** factor or numerical coefficient
   e. variable, factor, or literal coefficient **f.** factor or quantity
   g. factor   **h.** quantity

5.3   Properties of Whole Numbers

---

1 | The *stated equality* of two *numerical* expressions is a *mathematical statement*. $5 + 3 = 8$ is a mathematical statement.

Q1   **a.** $4 \cdot 3 - 2$ and $12 - 2$ are _____ expressions.

          b. $4 \cdot 3 - 2 = 12 - 2$ is a _____ statement.

STOP • **STOP** • **STOP** • **STOP** • **STOP** • **STOP** • **STOP** • **STOP** • **STOP**

A1   **a.** numerical   **b.** mathematical

> **2**　　Mathematical statements may be judged as being true or false. $2 + (5 + 3) = 2 + 8$ is true. $(15 - 3) - 8 = 11 - 8$ is false.

Q2　　Answer true or false:

a. $5 \cdot 4 = 4 \cdot 5$　　_____

b. $2 + 3 = 3 + 2$　　_____

c. $\dfrac{12}{3} = \dfrac{3}{12}$　　_____

STOP　•　STOP　•　STOP　•　STOP　•　STOP　•　STOP　•　STOP　•　STOP　•　STOP

A2　　**a.** true　**b.** true　**c.** false

> **3**　　When two numerical expressions have the same evaluation, they are said to be *equivalent*. $3 + 8$ and $8 + 3$ are equivalent. $5 - 2$ and $1 + 2$ are equivalent. $12 - 2$ and $2 - 12$ are not equivalent.

Q3　　**a.** Are $12 + 5$ and $5 + 12$ equivalent?　　_____

　　　b. Are $12 - 5$ and $13 - 7$ equivalent?　　_____

STOP　•　STOP　•　STOP　•　STOP　•　STOP　•　STOP　•　STOP　•　STOP　•　STOP

A3　　**a.** yes　**b.** no

> **4**　　If two numerical expressions are equivalent, their stated equality is always true. The sums $12 + 5$ and $5 + 12$ are equivalent, so $12 + 5 = 5 + 12$ is true.

Answer yes or no to the following questions:

Q4　　**a.** Are $5 + (4 + 1)$ and $(5 + 4) + 1$ equivalent?　　_____

　　　b. Is $5 + (4 + 1) = (5 + 4) + 1$ a true statement?　　_____

STOP　•　STOP　•　STOP　•　STOP　•　STOP　•　STOP　•　STOP　•　STOP　•　STOP

A4　　**a.** yes: because both $5 + (4 + 1)$ and $(5 + 4) + 1$ equal 10
　　　b. yes

Q5　　**a.** Are $\dfrac{2}{5} \cdot 1$ and $1 \cdot \dfrac{2}{5}$ equivalent?　　_____

　　　b. Is $\dfrac{2}{5} \cdot 1 = 1 \cdot \dfrac{2}{5}$ a true statement?　　_____

STOP　•　STOP　•　STOP　•　STOP　•　STOP　•　STOP　•　STOP　•　STOP　•　STOP

A5　　**a.** yes　**b.** yes

Q6　　**a.** Are $17 \cdot 0$ and 0 equivalent?　　_____

　　　b. Is $17 \cdot 0 = 0$ a true statement?　　_____

STOP　•　STOP　•　STOP　•　STOP　•　STOP　•　STOP　•　STOP　•　STOP　•　STOP

A6　　**a.** yes　**b.** yes

Q7 **a.** Are $\dfrac{7}{8} + 0$ and $\dfrac{7}{8}$ equivalent? _____

    b. Is $\dfrac{7}{8} + 0 = \dfrac{7}{8}$ a true statement? _____

STOP • **STOP** • **STOP** • **STOP** • **STOP** • **STOP** • **STOP** • **STOP** • **STOP**

A7 **a.** yes **b.** yes

Q8 **a.** Are $2(3 \cdot 4)$ and $(2 \cdot 3)4$ equivalent? [*Note:* $(2 \cdot 3)4 = (2 \cdot 3) \cdot 4$] _____

    b. Is $2(3 \cdot 4) = (2 \cdot 3)4$ a true statement? _____

STOP • **STOP** • **STOP** • **STOP** • **STOP** • **STOP** • **STOP** • **STOP** • **STOP**

A8 **a.** yes **b.** yes

Q9 **a.** Are $20 - (12 - 3)$ and $(20 - 12) - 3$ equivalent? _____

    b. Is $20 - (12 - 3) = (20 - 12) - 3$ a true statement? _____

STOP • **STOP** • **STOP** • **STOP** • **STOP** • **STOP** • **STOP** • **STOP** • **STOP**

A9 **a.** no: because $20 - (12 - 3) = 11$, whereas $(20 - 12) - 3 = 5$
    b. no

Q10 **a.** Verify that $2(5 - 2)$ and $2 \cdot 5 - 2 \cdot 2$ are equivalent.

    b. Is $2(5 - 2) = 2 \cdot 5 - 2 \cdot 2$ a true statement? _____

STOP • **STOP** • **STOP** • **STOP** • **STOP** • **STOP** • **STOP** • **STOP** • **STOP**

A10 **a.** $2(5 - 2) = 2 \cdot 3 = 6$
       $2 \cdot 5 - 2 \cdot 2 = 10 - 4 = 6$
    b. yes

---

5 $\{1, 2, 3, 4, \ldots\}$ is called the set of *counting numbers*. This set is also called the set of *natural numbers*. It is common to denote this set by N:

$$N = \{1, 2, 3, 4, \ldots\}$$

The set of *whole numbers* is the set of natural numbers together with zero. It is common to denote this set by W:

$$W = \{0, 1, 2, 3, 4, \ldots\}$$

Q11 Answer true or false:

    a. The set of natural numbers is a subset of the set of whole numbers. _____

    b. $0 \in W$ but $0 \notin N$. _____

    c. $5 \in N$ and $5 \in W$. _____

    d. The set of whole numbers is an infinite set. _____

    e. $\{0, 1, 2, 3, \ldots\}$ is a well-defined collection. _____

STOP • **STOP** • **STOP** • **STOP** • **STOP** • **STOP** • **STOP** • **STOP** • **STOP**

A11   **a.** true: because each element of the set of natural numbers is also an element of the set of whole numbers.
b. true: zero belongs to the set of whole numbers, but zero does not belong to the set of natural numbers.
c. true: 5 belongs to both sets; hence, 5 is both a natural number and a whole number; every natural number is also a whole number.
d. true: because the count is unending.
e. true: membership is clear.

(*Note: In the remaining portion of this section, the replacement set for all variables will be the set of whole numbers.*)

---

6 $x + 2$ is called an *open expression*. $x + 2 = 5$ is called an *open sentence*. An open sentence cannot be judged as being true or false until a replacement is made for the variable. When x is replaced by 3, $x + 2 = 5$ becomes the true statement $3 + 2 = 5$. When x is replaced by 4, $x + 2 = 5$ becomes the false statement $4 + 2 = 5$.

Q12   In each open sentence, when the variable is replaced by the given value, determine whether the resulting statement is true or false:

a. $3x - 4 = 17$; $x = 7$ **b.** $x + 5 = 5$; $x = 0$

c. $x + 2 = 2 + x$; $x = 7$ **d.** $5x = x \cdot 5$; $x = 3$

e. $x + 2 = x$; $x = 0$

STOP • **STOP** • **STOP** • **STOP** • **STOP** • **STOP** • **STOP** • **STOP** • **STOP**

A12   **a.** true: because $3 \cdot 7 - 4 = 17$ **b.** true: because $0 + 5 = 5$
c. true: because $7 + 2 = 2 + 7$ **d.** true: because $5 \cdot 3 = 3 \cdot 5$
e. false: because $0 + 2 \neq 0$

---

7   The expressions $x + 7$ and $7 + x$ are open expressions. They are equivalent because they have the same evaluation for all replacements of the variable. Hence, $x + 7 = 7 + x$ is true for all whole-number replacements of x. For example,

$0 + 7 = 7 + 0$
$1 + 7 = 7 + 1$
$2 + 7 = 7 + 2$
$3 + 7 = 7 + 3$
       etc.

(*Note:* Although $x + 7 = 7 + x$ is an open sentence, it is true for all whole-number replacements of x.)

Q13   Answer yes or no:

  a. Are $a + b$ and $b + a$ equivalent for all whole-number replacements of a and b?_____ (Try to find a pair of whole numbers to replace a and b that do not give the same evaluation for both expressions.)

  b. Is $a + b = b + a$ true for all whole-number replacements of a and b?_____

STOP • STOP • STOP • STOP • STOP • STOP • STOP • STOP • STOP

A13   **a.** yes   (it is reasonable to assume that they are equivalent)
       b. yes

---

8   The assumption that $a + b = b + a$ is true for all whole-number replacements of a and b is called the *commutative property of addition*. It is concerned with the *order* in which two numbers are added. Addition is called a commutative operation because the order in which two numbers are added does not affect the sum. Examples of this property are:

$$0 + 5 = 5 + 0$$
$$2 + 7 = 7 + 2$$
$$52 + 1 = 1 + 52$$

Q14 $15 + 7 = 7 + 15$ is an example of the _____ property of addition.

STOP • STOP • STOP • STOP • STOP • STOP • STOP • STOP • STOP

A14   commutative

Q15   Use the commutative property of addition to complete a true statement:

  a. $17 + 8 =$ _____   **b.** $0 + 9 =$ _____

  c. $1 + 12 =$ _____   **d.** $92 + 18 =$ _____

STOP • STOP • STOP • STOP • STOP • STOP • STOP • STOP • STOP

A15   **a.** $8 + 17$ **b.** $9 + 0$ **c.** $12 + 1$ **d.** $18 + 92$

Q16   The commutative property of addition indicates that the _____ of two numbers in an addition problem may be reversed without affecting the sum.

STOP • STOP • STOP • STOP • STOP • STOP • STOP • STOP • STOP

A16   order

---

9   The commutative property of addition includes an infinite number of specific cases. The statement that $a + b = b + a$ is true for all possible whole-number replacements of a and b illustrates the power of making one statement that covers an infinite number of cases. A similar property is true for multiplication. You know that $3 \cdot 5 = 5 \cdot 3$. You realize that the *order* of the factors in a multiplication does not affect the product. The *commutative property of multiplication* assumes that $ab = ba$ is true for all whole-number replacements of a and b. Other examples of this property are:

$$2 \cdot 17 = 17 \cdot 2$$
$$93 \cdot 6 = 6 \cdot 93$$
$$0 \cdot 15 = 15 \cdot 0$$

Q17   Use the commutative property of multiplication to complete a true statement:

      a. $1 \cdot 15 =$ _____   **b.** $107 \cdot 49 =$ _____

      c. $7 \cdot 12 =$ _____   **d.** $50 \cdot 75 =$ _____

STOP • STOP • STOP • STOP • STOP • STOP • STOP • STOP • STOP

A17   **a.** $15 \cdot 1$ **b.** $49 \cdot 107$ **c.** $12 \cdot 7$ **d.** $75 \cdot 50$

Q18   **a.** Use the commutative property of addition to change $2x + 5$ into an equivalent expression. _____

      b. Use the commutative property of multiplication to change $3x \cdot 5$ into an equivalent expression. _____

STOP • STOP • STOP • STOP • STOP • STOP • STOP • STOP • STOP

A18   **a.** $5 + 2x$ **b.** $5 \cdot 3x$

Q19   Verify that $1 + (2 + 3) = (1 + 2) + 3$ is a true statement by showing that $1 + (2 + 3)$ and $(1 + 2) + 3$ are equivalent expressions.

STOP • STOP • STOP • STOP • STOP • STOP • STOP • STOP • STOP

A19 $1 + (2 + 3) = 1 + 5 = 6$
$(1 + 2) + 3 = 3 + 3 = 6$

Q20   **a.** Are $2 + (a + 3)$ and $(2 + a) + 3$ equivalent for all whole-number replacements of a? _____

      b. Is $2 + (a + 3) = (2 + a) + 3$ true for all whole-number replacements of a? _____

STOP • STOP • STOP • STOP • STOP • STOP • STOP • STOP • STOP

A20   **a.** yes   **b.** yes

Q21   **a.** Are $a + (b + c)$ and $(a + b) + c$ equivalent for all whole-number replacements of a, b, and c? _____

      b. Is $a + (b + c) = (a + b) + c$ true for all whole-number replacements of a, b, and c? _____

STOP • STOP • STOP • STOP • STOP • STOP • STOP • STOP • STOP

A21   **a.** yes   **b.** yes

---

10   The assumption that $a + (b + c) = (a + b) + c$ is true for all whole-number replacements of a, b, and c is called the *associative property of addition*. It indicates that in addition, the way three numbers are *grouped* does not affect the sum. Other examples of this property are:

$(5 + 6) + 2 = 5 + (6 + 2)$
$12 + (5 + 0) = (12 + 5) + 0$

Q22   **a.** $15 + (5 + 12) = (15 + 5) + 12$ is an example of the _____ property of addition.

     b. The associative property of addition indicates that the _____ of three numbers in an addition problem may be changed without affecting the sum.

STOP • STOP • STOP • STOP • STOP • STOP • STOP • STOP • STOP

A22   **a.** associative   **b.** grouping

Q23   Use the associative property of addition to complete the following in the form of true statements:

     a. $7 + (103 + 28) =$ _____

     b. $(9 + 12) + 8 =$ _____

     c. $(8 + 4) + 6 =$ _____

     d. $9 + (31 + 12) =$ _____

STOP • STOP • STOP • STOP • STOP • STOP • STOP • STOP • STOP

A23   **a.** $(7 + 103) + 28$ **b.** $9 + (12 + 8)$ **c.** $8 + (4 + 6)$ **d.** $(9 + 31) + 12$

11	A similar property is true for multiplication. In multiplication, the way three *factors* are grouped does not affect the product. This assumption is the *associative property of multiplication*, which states that $a(bc) = (ab)c$ is true for all whole-number replacements of a, b, and c. Examples of this property are

$$7(3 \cdot 2) = (7 \cdot 3)2$$
$$(3 \cdot 4)5 = 3(4 \cdot 5)$$

Q24   The associative property of multiplication indicates that the grouping of three _____ in a multiplication problem may be changed without affecting the product.

STOP • STOP • STOP • STOP • STOP • STOP • STOP • STOP • STOP

A24   factors

Q25   Use the associative property of multiplication to complete the following in the form of true statements:

     a. $(72 \cdot 5)2 =$ _____   **b.** $4(25 \cdot 7) =$ _____

     c. $125(8 \cdot 9) =$ _____   **d.** $(19 \cdot 8)5 =$ _____

STOP • STOP • STOP • STOP • STOP • STOP • STOP • STOP • STOP

A25   **a.** $72(5 \cdot 2)$ **b.** $(4 \cdot 25)7$ **c.** $(125 \cdot 8)9$ **d.** $19(8 \cdot 5)$

Q26   **a.** Use the associative property of addition to change $(x + 3) + 2$ into an equivalent expression. _____

     b. Use the associative property of multiplication to change $2(3x)$ into an equivalent expression. _____

STOP • STOP • STOP • STOP • STOP • STOP • STOP • STOP • STOP

A26   **a.** $x + (3 + 2)$ **b.** $(2 \cdot 3)x$

| 12 | You should note that subtraction and division are not associative operations. |

Q27 Verify the preceding true statement by showing that each of the following are false statements:

a. $(15 - 7) - 3 = 15 - (7 - 3)$ **b.** $60 \div (15 \div 3) = (60 \div 15) \div 3$

STOP • STOP • STOP • STOP • STOP • STOP • STOP • STOP • STOP

A27   **a.** $8 - 3 \neq 15 - 4$ **b.** $60 \div 5 \neq 4 \div 3$

Q28   True or false:

a. $1x = x \cdot 1$ _____

b. $2 + (x + 5) = 2 + (5 + x)$ _____

c. $(x - 2)5 = 5(x - 2)$ _____

d. $(3 - x) - 4 = 3 - (x - 4)$ _____

e. $(3x + 7) + 9 = 3x + (7 + 9)$ _____

f. $x + 0 = 0 + x$ _____

g. $24 \div (x \div 4) = (24 \div x) \div 4$ _____

h. $2(5x) = (2 \cdot 5)x$ _____

STOP • STOP • STOP • STOP • STOP • STOP • STOP • STOP • STOP

A28   **a.** true: commutative property of multiplication
b. true: commutative property of addition
c. true: commutative property of multiplication
d. false: (subtraction is not an associative operation)
e. true: associative property of addition
f. true: commutative property of addition
g. false: (division is not an associative operation)
h. true: associative property of multiplication

| 13 | The following properties are much easier to recognize and remember. The *addition property of zero* (additive identity property) states that $a + 0 = 0 + a = a$ is true for all whole-number replacements of a. If zero is added to any whole number, the result is always the identical whole number: |

$$5 + 0 = 0 + 5 = 5$$
$$17 + 0 = 0 + 17 = 17$$

The *multiplication property of zero* states that $a \cdot 0 = 0a = 0$ is true for all whole-number replacements of a. If zero is multiplied by any whole number, the result is always zero:

$$5 \cdot 0 = 0 \cdot 5 = 0$$
$$32 \cdot 0 = 0 \cdot 32 = 0$$

The *multiplication property of one* (multiplicative identity property) states that $a \cdot 1 = 1a = a$ is true for all whole-number replacements of a. If 1 is multiplied by any whole number, the result is always the identical whole number:

$$5 \cdot 1 = 1 \cdot 5 = 5$$
$$108 \cdot 1 = 1 \cdot 108 = 108$$

The *division property of one* states that $\frac{a}{1} = a$ is true for all whole-number replacements of a. If any whole number is divided by 1, the result is the same whole number:

$$\frac{5}{1} = 5$$

$$\frac{297}{1} = 297$$

$$\frac{0}{1} = 0$$

Q29   True or false:

a. $\dfrac{5x}{1} = 5x$ _____

b. $(2x + 1)1 = 2x + 1$ _____

c. $(5x + 0) + 3 = 5x + 3$ _____

d. $\dfrac{2}{3}x \cdot 0 = 0$ _____

STOP • **STOP** • **STOP** • **STOP** • **STOP** • **STOP** • **STOP** • **STOP** • **STOP**

A29   **a.** true: division property of one
b. true: multiplication property of one (multiplicative identity property)
c. true: addition property of zero (additive identity property)
d. true: multiplication property of zero

This completes the instruction for this section.

5.3   Exercises

1. Insert $=$ or $\neq$ in the blank to form a true statement (the replacement set for all variables is the set of whole numbers):
    a. $(xy)4$ _____ $4(xy)$ **b.** $7a + 0$ _____ $7a$
    c. $x - 8$ _____ $8 - x, x \neq 8$ **d.** $c \div 5$ _____ $5 \div c, c \neq 5$
    e. $1(x + y)$ _____ $x + y$ **f.** $(2a + b) + 3b$ _____ $2a + (b + 3b)$
    g. $(5 + b) + 3$ _____ $(b + 5) + 3$ **h.** $4(5x)$ _____ $(4 \cdot 5)x$

    i. $5x \cdot 0$ _____ 0 **j.** $\dfrac{2x - 1}{1}$ _____ $2x - 1$

    k. $14 \div (7 \div x)$ _____ $(14 \div 7) \div x, x \neq 1$
    l. $(x - 13) - 6$ _____ $x - (13 - 6)$
2. Indicate the number of the general statement that corresponds to its name. (Assume that general statements are true for all whole-number replacements of x, y, and z.)

a. commutative property of addition   1. $x \cdot 1 = 1x = x$
b. commutative property of multiplication   2. $x(yz) = (xy)z$
c. associative property of addition   3. $x \cdot 0 = 0x = 0$
d. associative property of multiplication   4. $x + y = y + x$

e. addition property of zero   5. $\frac{x}{1} = x$

f. multiplication property of zero   6. $(x + y) + z = x + (y + z)$
g. multiplication property of one   7. $xy = yx$
h. division property of one   8. $x + 0 = 0 + x = x$

3. Give one numerical example to illustrate that
    a. Subtraction of whole numbers is not a commutative operation.
    b. Division of whole numbers is not a commutative operation.
    c. Subtraction of whole numbers is not an associative operation.
    d. Division of whole numbers is not an associative operation.

4. Use the property given to write a true statement:
    a. By the associative property of addition, $2 + (3 + 4) =$ _____.
    b. By the multiplication property of one, $32 \cdot 1 =$ _____.
    c. By the commutative property of multiplication, $1 \cdot 15 =$ _____.
    d. By the addition property of zero, $(72 - 5) + 0 =$ _____.
    e. By the multiplication property of zero, $0(15 \cdot 3) =$ _____.
    f. By the associative property of multiplication, $1(4 \cdot 7) =$ _____.

    g. By the division property of one, $\frac{42}{1} =$ _____.

    h. By the commutative property of addition, $2(3 + 4) =$ _____.

5.3   Exercise Answers

1. a. $=$: commutative property of multiplication
    b. $=$: addition property of zero (additive identity property)
    c. $\neq$: (subtraction is not a commutative operation)
    d. $\neq$: (division is not a commutative operation)
    e. $=$: multiplication property of one (multiplicative identity property)
    f. $=$: associative property of addition
    g. $=$: commutative property of addition
    h. $=$: associative property of multiplication
    i. $=$: multiplication property of zero
    j. $=$: division property of one
    k. $\neq$: (division is not an associative operation)
    l. $\neq$: (subtraction is not an associative operation)
2. a. 4   b. 7   c. 6   d. 2   e. 8   f. 3   g. 1
    h. 5
3. a. $5 - 1 \neq 1 - 5$ (one of many examples)
    b. $6 \div 2 \neq 2 \div 6$ (one of many examples)
    c. $10 - (5 - 2) \neq (10 - 5) - 2$ (one of many examples)
    d. $20 \div (10 \div 2) \neq (20 \div 10) \div 2$ (one of many examples)
4. a. $(2 + 3) + 4$ b. 32   c. $15 \cdot 1$ d. $72 - 5$
    e. 0   f. $(1 \cdot 4)7$ g. 42   h. $2(4 + 3)$

5.4   Distributive Properties

1	Consider the following evaluations:
	$2(3 + 4)$ and $2 \cdot 3 + 2 \cdot 4$ $2 \cdot 7$ $6 + 8$ 14 14
	$2(3 + 4)$ and $2 \cdot 3 + 2 \cdot 4$ are equivalent expressions because both expressions have the same evaluation. Therefore, $2(3 + 4) = 2 \cdot 3 + 2 \cdot 4$ is a true statement.

Q1   Verify that $5 \cdot 6 + 5 \cdot 2 = 5(6 + 2)$ is a true statement by showing that $5 \cdot 6 + 5 \cdot 2$ and $5(6 + 2)$ are equivalent expressions.

STOP   •   STOP   •   STOP   •   STOP   •   STOP   •   STOP   •   STOP   •   STOP   •   STOP

A1 $5 \cdot 6 + 5 \cdot 2$ and $5(6 + 2)$
        $30 + 10$ $5 \cdot 8$
        40 40

Q2   Complete the following in order to form true statements:
a. $3(2 + 5)$ **b.** $7(8 + 6)$

       $3 \cdot 2 + 3 \cdot$ _____ $7 \cdot$ _____ $+ 7 \cdot 6$

STOP   •   STOP   •   STOP   •   STOP   •   STOP   •   STOP   •   STOP   •   STOP   •   STOP

A2   **a.** $3 \cdot 2 + 3 \cdot \underline{5}$ **b.** $7 \cdot \underline{8} + 7 \cdot 6$

2	The true statement $9(10 + 3) = 9 \cdot 10 + 9 \cdot 3$ is an example of the *left distributive property of multiplication over addition*. You should observe that $9(10 + 3)$ is a *product* of 9 times 13, whereas $9 \cdot 10 + 9 \cdot 3$ is a *sum* of 90 and 27. This property changes a product to a sum. It also changes a sum to a product. For example,
	$2 \cdot 6 + 2 \cdot 4 = 2(6 + 4)$

Q3   Use the left distributive property of multiplication over addition to change each sum to a product or each product to a sum (parts a and b are given as examples):

   a. $5 \cdot 3 + 5 \cdot 7 =$ $5(3 + 7)$ **b.** $6(4 + 5) =$ $6 \cdot 4 + 6 \cdot 5$

   c. $9(2 + 1) =$ _____   **d.** $17(5 + 6) =$ _____

   e. $10 \cdot 2 + 10 \cdot 3 =$ _____   **f.** $8 \cdot 12 + 8 \cdot 8 =$ _____

STOP   •   STOP   •   STOP   •   STOP   •   STOP   •   STOP   •   STOP   •   STOP   •   STOP

A3   **c.** $9 \cdot 2 + 9 \cdot 1$ **d.** $17 \cdot 5 + 17 \cdot 6$ **e.** $10(2 + 3)$ **f.** $8(12 + 8)$

3	In general, the left distributive property of multiplication over addition states that $a(b + c) = ab + ac$ is true for all whole-number replacements of a, b, and c.

Q4   Use the left distributive property of multiplication over addition to complete the evaluation of each numerical expression:

a. $4 \cdot 7 + 4 \cdot 3$

4(_____ + _____)

4 · _____

_____

b. $15(10 + 2)$

15 · _____ + 15 · _____

_____ + _____

_____

STOP • **STOP** • **STOP** • **STOP** • **STOP** • **STOP** • **STOP** • **STOP** • **STOP**

A4   **a.** $4(\underline{7} + \underline{3})$

$4 \cdot \underline{10}$

$\underline{40}$

b. $15 \cdot \underline{10} + 15 \cdot \underline{2}$

$\underline{150} + \underline{30}$

$\underline{180}$

---

4   Consider these evaluations:

$(3 + 4)5$ and $3 \cdot 5 + 4 \cdot 5$

$7 \cdot 5$ $15 + 20$

35   35

Hence, $(3 + 4)5 = 3 \cdot 5 + 4 \cdot 5$. This is an example of the *right distributive property of multiplication over addition*. This property also changes products to sums or sums to products.

Q5   Use the right distributive property of multiplication over addition to change each sum to a product or each product to a sum:

a. $(4 + 2)7 =$ _____

c. $18 \cdot 12 + 2 \cdot 12 =$ _____

b. $2 \cdot 3 + 21 \cdot 3 =$ _____

d. $(13 + 19)10 =$ _____

STOP • **STOP** • **STOP** • **STOP** • **STOP** • **STOP** • **STOP** • **STOP** • **STOP**

A5   **a.** $4 \cdot 7 + 2 \cdot 7$

c. $(18 + 2) \cdot 12$

b. $(2 + 21)3$

d. $13 \cdot 10 + 19 \cdot 10$

---

5   In general, the right distributive property of multiplication over addition states that $(a + b)c = ac + bc$ is true for all whole-number replacements of a, b, and c.

Q6   Complete the following evaluations:

a. $14 \cdot 9 + 6 \cdot 9$

(_____ + _____) · 9

_____ · 9

_____

b. $(17 + 8)12$

17 · _____ + 8 · _____

_____ + _____

_____

STOP • **STOP** • **STOP** • **STOP** • **STOP** • **STOP** • **STOP** • **STOP** • **STOP**

A6   **a.** $(\underline{14} + \underline{6}) \cdot 9$

$\underline{20} \cdot 9$

$\underline{180}$

b. $17 \cdot \underline{12} + 8 \cdot \underline{12}$

$\underline{204} + \underline{96}$

$\underline{300}$

---

6   Notice that:

$7(5 - 2)$ and $7 \cdot 5 - 7 \cdot 2$

$7 \cdot 3$ $35 - 14$

21   21

Hence, $7(5 - 2) = 7 \cdot 5 - 7 \cdot 2$. This is an example of the *left distributive property of multiplication over subtraction*. Observe that $7(5 - 2)$ is a product, whereas $7 \cdot 5 - 7 \cdot 2$ is a difference. This property also changes a difference to a product. For example,

$$7 \cdot 5 - 7 \cdot 2 = 7(5 - 2)$$

Q7   Use the left distributive property of multiplication over subtraction to change each product to a difference or each difference to a product (parts a and b are given as examples):

a. $5 \cdot 7 - 5 \cdot 3 =$ $5(7 - 3)$ b. $10(19 - 7) =$ $10 \cdot 19 - 10 \cdot 7$

c. $12(13 - 6) =$ _____   d. $8 \cdot 7 - 8 \cdot 3 =$ _____

e. $24 \cdot 12 - 24 \cdot 2 =$ _____   f. $3(36 - 27) =$ _____

STOP • **STOP** • **STOP** • **STOP** • **STOP** • **STOP** • **STOP** • **STOP** • **STOP**

A7   c. $12 \cdot 13 - 12 \cdot 6$ d. $8(7 - 3)$
e. $24(12 - 2)$ f. $3 \cdot 36 - 3 \cdot 27$

---

7   In general, the left distributive property of multiplication over subtraction states that $a(b - c) = ab - ac$ is true for all whole-number replacements of a, b, and c.

Q8   Complete the following evaluations:
a. $24(12 - 2)$ b. $7 \cdot 23 - 7 \cdot 3$

$24 \cdot$ _____ $- 24 \cdot$ _____ $7($ _____ $-$ _____ $)$

_____ $-$ _____ $7 \cdot$ _____

_____   _____

STOP • **STOP** • **STOP** • **STOP** • **STOP** • **STOP** • **STOP** • **STOP** • **STOP**

A8   a. $24 \cdot \underline{12} - 24 \cdot \underline{2}$ b. $7(\underline{23} - \underline{3})$
$\underline{288} - \underline{48}$ $7 \cdot \underline{20}$
$\underline{240}$ $\underline{140}$

---

8   The true statement $(12 - 9)2 = 12 \cdot 2 - 9 \cdot 2$ is an example of the *right distributive property of multiplication over subtraction*. This property also changes products to differences or differences to products.

Q9   Use the right distributive property of multiplication over subtraction to change each product to a difference or each difference to a product:

a. $(12 - 2)6 =$ _____

b. $7 \cdot 4 - 5 \cdot 4 =$ _____

STOP • **STOP** • **STOP** • **STOP** • **STOP** • **STOP** • **STOP** • **STOP** • **STOP**

A9   a. $12 \cdot 6 - 2 \cdot 6$ b. $(7 - 5)4$

---

9   In general, the right distributive property of multiplication over subtraction states that $(a - b)c = ac - bc$ is true for all whole-number replacements of a, b, and c.

Q10 Complete the following evaluations:

a. $(45 - 7)100$

  _____ $\cdot$ 100 $-$ _____ $\cdot$ 100

  _____ $-$ _____

  _____

b. $35 \cdot 9 - 5 \cdot 9$

  (_____ $-$ _____)9

  _____ $\cdot$ 9

  _____

STOP • STOP • STOP • STOP • STOP • STOP • STOP • STOP • STOP

A10

a. $\underline{45} \cdot 100 - \underline{7} \cdot 100$

 $\underline{4,500} - \underline{700}$

 $\underline{3,800}$

b. $(\underline{35} - \underline{5})9$

 $\underline{30} \cdot 9$

 $\underline{270}$

---

10 Consider the following examples of the distributive properties:

$$5(7 + 9)\ \ = 5 \cdot 7 + 5 \cdot 9$$
$$(12 - 5)6\ = 12 \cdot 6 - 5 \cdot 6$$
$$10(15 - 7) = 10 \cdot 15 - 10 \cdot 7$$
$$(32 + 11)8 = 32 \cdot 8 + 11 \cdot 8$$

In each example, a product was changed to a sum or difference. When the distribution properties are used in this manner, it is usually said that "the parentheses have been removed."

Q11 Use a distributive property to remove the parentheses from each of the following numerical expressions and complete the evaluation (part a is given as an example):

a. $2(7 - 3) =$ $2 \cdot 7 - 2 \cdot 3 = 14 - 6 = 8$

b. $18(5 + 3) =$ _____

c. $(12 - 8)6 =$ _____

d. $20(9 + 7) =$ _____

e. $35(2 - 1) =$ _____

STOP • STOP • STOP • STOP • STOP • STOP • STOP • STOP • STOP

A11

b. $18 \cdot 5 + 18 \cdot 3 = 90 + 54 = 144$
c. $12 \cdot 6 - 8 \cdot 6 = 72 - 48 = 24$
d. $20 \cdot 9 + 20 \cdot 7 = 180 + 140 = 320$
e. $35 \cdot 2 - 35 \cdot 1 = 70 - 35 = 35$

This completes the instruction for this section.

5.4 Exercises

1. Change each product to a sum or each sum to a product:
 a. $9(6 + 7)$ **b.** $2 \cdot 4 + 2 \cdot 7$ **c.** $(15 + 9)2$ **d.** $6 \cdot 8 + 10 \cdot 8$
2. Change each product to a difference or each difference to a product:
 a. $7 \cdot 9 - 7 \cdot 4$ **b.** $3(12 - 8)$ **c.** $(17 - 2)5$ **d.** $27 \cdot 8 - 7 \cdot 8$
3. Use one of the distributive properties to evaluate each expression (remove the parentheses):
 a. $2(31 - 6)$ **b.** $(15 + 7)3$ **c.** $(18 - 3)4$ **d.** $12(10 - 8)$

4. True or false:
   a. $3(x + 5) = 3x + 3 \cdot 5$ b. $4a - 1a = (4 - 1)a$
   c. $5y - 5 \cdot 3 = 5(y - 3)$ d. $7(b - 2) = 7b - 7 \cdot 2$
   e. $5x + 3x = (5 + 3)x$ f. $12(x + 5) = 12x + 5$
5. Complete the following general statements assumed true for all whole-number replacements of x, y, and z:
   a. $x(y - z) = $ _____   b. $(x + y)z = $ _____
   c. $(x - y)z = $ _____   d. $x(y + z) = $ _____

5.4   Exercise Answers

1. a. $9 \cdot 6 + 9 \cdot 7$ b. $2(4 + 7)$ c. $15 \cdot 2 + 9 \cdot 2$ d. $(6 + 10)8$
2. a. $7(9 - 4)$ b. $3 \cdot 12 - 3 \cdot 8$ c. $17 \cdot 5 - 2 \cdot 5$ d. $(27 - 7)8$
3. a. 50   b. 66   c. 60   d. 24
4. a. true: left distributive property of multiplication over addition
   b. true: right distributive property of multiplication over subtraction
   c. true: left distributive property of multiplication over subtraction
   d. true: left distributive property of multiplication over subtraction
   e. true: right distributive property of multiplication over addition
   f. false: 12 has not been multiplied times 5
5. a. $xy - xz$ b. $xz + yz$ c. $xz - yz$ d. $xy + xz$

Chapter 5 Sample Test

At the completion of Chapter 5 it is expected that you will be able to work the following problems.

5.1   Set Ideas

1. What word best describes each phrase?
   a. well-defined collection   b. things that belong to a set
   c. set whose count has a last number   d. set whose count is unending
2. Explain briefly why the set of people in the room in which you are located is a well-defined collection.
3. Use set notation where convenient to indicate the following sets:
   a. books in your city's public library
   b. digits in the number 1,041,257
4. Insert $\in$ or $\notin$ in the blank to form a true statement:
   a. 15 _____ $\{1, 3, 5, 7, 9, \ldots\}$ b. 12 _____ $\{1, 2, 3, 4, \ldots, 20\}$
5. Is the set on the left a subset of the set on the right?
   a. $\{1, 3, 6\}$ $\{1, 2, 3, \ldots\}$ b. $\{2, 10, 15\}$ $\{2, 4, 6, \ldots\}$
   c. $\{0, 7, 12\}$ $\{0, 7, 12\}$
6. a. Write two symbols that represent a set with no elements.
   b. What is the name of such a set?
   c. Explain briefly why $\{0\} \neq \{\ \ \}$.

5.2   Evaluating Open Expressions

7. Evaluate each open expression for the value of the variable given:
   a. $7(x - 3)$ when $x = 14$ b. $5a + 9$ when $a = 13$
   c. $15 - 3(y + 2)$ when $y = 3$ d. $7x - 9y$ when $x = 5$ and $y = 2$
   e. $(3b + 1)(b - 7)$ when $b = 9$

8. Use the following words or phrases to complete each open sentence: variable, open expression, numerical coefficient, literal coefficient, factor(s), quantity.
   a. In $5x$, 5 is called the _____.
   b. A _____ is a letter that may be replaced by any one of a set of many numbers.
   c. $2x + 3$, $2(x - 5)$, and $3x - 2y$ are each examples of a(n) _____.
   d. In $7x$ the letter factor x is called a _____ (two words).
   e. In $2(x - 2)$, the _____ $x - 2$ is a factor.
   f. The _____ of abc are a, b, and c.
   g. In $2 + (x + 2)$, the open expression $x + 2$ may be called a _____.

5.3   Properties of Whole Numbers

9. a. Use set notation to indicate the set of whole numbers.
   b. Complete: Two open expressions are _____ if they have the same evaluation for all replacements of the variable.
   c. The stated equality of two equivalent expressions forms a _____.
   d. $x + 2 = 5$ is an example of an _____ sentence.

10. The open sentence on the left is an application of which property listed on the right? Write the number associated with the property. (The replacement set is the set of whole numbers.)
   a. $(15y) \cdot 6 = 6(15y)$ 1. multiplication property of one
   b. $a \cdot 1 = a$ 2. associative property of addition
   c. $5(3z) = (5 \cdot 3)z$ 3. commutative property of addition
   d. $2x + 0 = 2x$ 4. associative property of multiplication
   e. $3 + 5c = 5c + 3$ 5. commutative property of multiplication

   f. $\dfrac{7d}{1} = 7d$ 6. multiplication property of zero

   g. $3 + (8 + 2m) = (3 + 8) + 2m$ 7. addition property of zero
   h. $0 \cdot 17t = 0$ 8. division property of one

5.4   Distributive Properties

11. The open sentence on the left is an application of which property on the right? Write the number associated with the property. (The replacement set is the set of whole numbers.)
   a. $(2x - 3)7 = 2x \cdot 7 - 3 \cdot 7$ 1. right distributive property of multiplication over addition
   b. $2(6y + 5) = 2 \cdot 6y + 2 \cdot 5$ 2. left distributive property of multiplication over addition
   c. $5a + 3a = (5 + 3)a$ 3. right distributive property of multiplication over subtraction
   d. $4(3z - 2) = 4 \cdot 3z - 4 \cdot 2$ 4. left distributive property of multiplication over subtraction

12. Change each product to a sum or difference. Change each sum or difference to a product:
    a. $2(3 + 7)$ b. $4 \cdot 5 - 2 \cdot 5$ c. $(12 - 8) \cdot 3$ d. $4 \cdot 7 + 4 \cdot 8$
    e. $12 \cdot 7 - 12 \cdot 2$

13. Use a distributive property to remove the parentheses (do not evaluate):
    a. $5(13 - 9)$ b. $(12 + 7)2$ c. $9(2 + 8)$ d. $(20 - 13)8$

Chapter 5 Sample Test Answers

1. **a.** set   **b.** elements   **c.** finite   **d.** infinite
2. membership is clear
3. **a.** not convenient   **b.** $\{0, 1, 2, 4, 5, 7\}$
4. **a.** $\in$ **b.** $\in$
5. **a.** yes   **b.** no   **c.** yes
6. **a.** $\{\ \}, \varnothing$ **b.** empty set
   c. $\{0\}$ has one element; $\{\ \}$ has zero elements
7. **a.** 77   **b.** 74   **c.** 0
   d. 17   **e.** 56
8. **a.** numerical coefficient   **b.** variable   **c.** open expression
   d. literal coefficient   **e.** quantity   **f.** factors
   g. quantity
9. **a.** $\{0, 1, 2, 3, \ldots\}$ **b.** equivalent   **c.** true statement   **d.** open
10. **a.** 5   **b.** 1   **c.** 4   **d.** 7
    e. 3   **f.** 8   **g.** 2   **h.** 6
11. **a.** 3   **b.** 2   **c.** 1   **d.** 4
12. **a.** $2 \cdot 3 + 2 \cdot 7$ **b.** $(4 - 2)5$ **c.** $12 \cdot 3 - 8 \cdot 3$ **d.** $4(7 + 8)$
    e. $12(7 - 2)$
13. **a.** $5 \cdot 13 - 5 \cdot 9$ **b.** $12 \cdot 2 + 7 \cdot 2$ **c.** $9 \cdot 2 + 9 \cdot 8$ **d.** $20 \cdot 8 - 13 \cdot 8$

Chapter 6

The Set of Integers and the Set of Rational Numbers

6.1   Introduction

---

1 In Chapter 5 we discussed the set of whole numbers and their properties. The natural numbers, $N = \{1, 2, 3, \ldots\}$, were presented as a subset of the set of whole numbers, $W = \{0, 1, 2, \ldots\}$. The set of whole numbers, W, can be thought of as an extension of the set of natural numbers, N.

Q1   Label each of the numbers below as a natural number, a whole number, or both a natural number and a whole number:

    a. 5 _____

    b. 0 _____

    c. 1 _____

STOP • **STOP** • **STOP** • **STOP** • **STOP** • **STOP** • **STOP** • **STOP** • **STOP**

A1   **a.** both a natural and a whole number
    b. whole number
    c. both a natural and a whole number

---

2 The set of whole numbers can be pictured using what is called a *number line* as follows:

Each dot on the number line indicates the point where a specific number occurs and is called the *graph* of the number. The number that corresponds with each dot is called the *coordinate* of the point. The arrow on the end of the number line and the three dots that follow the last number shown both serve to indicate that the set of whole numbers is an infinite set.

Q2   Graph the following whole numbers by placing dots at the appropriate points on the number line:
    a. 7   **b.** 2   **c.** 0

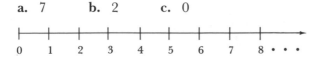

STOP • **STOP** • **STOP** • **STOP** • **STOP** • **STOP** • **STOP** • **STOP** • **STOP**

A2

3	The number line of Frame 2 pictures the whole numbers spaced an equal distance apart, with each number indicating the *number of units that it is to the right of zero*. That is, the number 2 indicates the point that is 2 units to the right of zero. A new set of numbers will now be defined by extending the number line to the *left* of zero and again marking off points an equal distance apart. It is called the set of *negative integers* and is written {⁻1, ⁻2, ⁻3, . . .}. (Read "negative one," "negative two," "negative three," etc.) In the same way that 2 represents 2 units to the *right of zero*, the negative integer ⁻2 (negative two) will represent a distance of 2 units to the *left of zero*. To emphasize the difference in direction between the negative integers {⁻1, ⁻2, ⁻3, . . .} and the natural numbers {1, 2, 3, . . .}, the set of natural numbers is sometimes written {⁺1, ⁺2, ⁺3, . . .}. (Read "positive one," "positive two," "positive three," etc.) This set is also correctly referred to as the set of *positive integers*. Thus any positive integer can be written either with or without the positive sign. The positive integer ⁺7, for example, is also correctly represented as 7.

Q3   Graph each of the following numbers by placing dots at the appropriate points on the number line:

a. ⁻5   **b.** ⁺3   **c.** 0   **d.** ⁻1   **e.** 6

STOP • STOP • STOP • STOP • STOP • STOP • STOP • STOP • STOP

A3

4	The set that contains the set of negative integers, zero, and the set of positive integers is called the set of *integers* and is represented by the letter *I*. That is, $$I = \{ \ldots, {}^{-}3, {}^{-}2, {}^{-}1, 0, {}^{+}1, {}^{+}2, {}^{+}3, \ldots \}$$ Each of the integers can be thought of as having two parts, *distance* and *direction*. The integer ⁺5, for example, represents a distance of 5 units to the *right* of zero. Similarly, the integer ⁻9 represents a distance of 9 units to the *left* of zero.

Q4   Write the distance and direction represented by each of the following integers:

		Distance	Direction
a.	⁻6	6 units	left of zero
b.	⁺42	_____	_____
c.	⁻42	_____	_____
d.	0	_____	_____

STOP • STOP • STOP • STOP • STOP • STOP • STOP • STOP • STOP

A4   **b.** 42 units, right of zero
     c. 42 units, left of zero
     d. 0 units, neither direction

5

The integers, . . . , $^-3$, $^-2$, $^-1$, 0, $^+1$, $^+2$, $^+3$, . . . , are also referred to as *signed numbers* because each one (except zero) has a direction designated by either a "$^-$" or a "$^+$" sign. Zero is an integer but is considered neither positive nor negative since its coordinate is neither to the right nor to the left of the zero point.

Q5   What is the sign of each of the following integers?

    a. $^+7$ _____   **b.** 6 _____

    c. $^-3$ _____   **d.** 0 _____

STOP • **STOP** • **STOP** • **STOP** • **STOP** • **STOP** • **STOP** • **STOP** • **STOP**

A5   **a.** positive   **b.** positive   **c.** negative
    d. none: zero has no sign

6

As a result of the definition of the set of integers, each integer can be paired with a second integer that is the same distance from zero but in a different direction. The paired integers are called *opposites* and are shown in the following drawing:

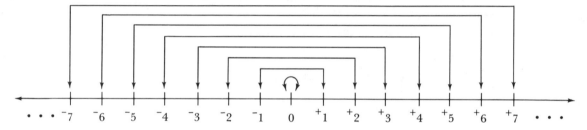

Notice that each of the integers has a unique opposite. The opposite of $^+4$ is $^-4$. The opposite of $^-6$ is $^+6$. The opposite of 0 is 0. That is, zero is the only integer that is its own opposite.

Q6   Write the opposite of each of the following integers:

    a. 5 _____   **b.** $^-7$ _____

    c. 0 _____   **d.** $^+1$ _____

    e. $^-95$ _____   **f.** 125 _____

STOP • **STOP** • **STOP** • **STOP** • **STOP** • **STOP** • **STOP** • **STOP** • **STOP**

A6   **a.** $^-5$ **b.** $^+7$ **c.** 0   **d.** $^-1$ **e.** $^+95$ **f.** $^-125$

7

Thus far, three sets have been defined. They are:

$N = \{1, 2, 3, \ldots\}$, the set of natural or counting numbers

$W = \{0, 1, 2, 3, \ldots\}$, the set of whole numbers

$I = \{\ldots, {}^-3, {}^-2, {}^-1, 0, {}^+1, {}^+2, {}^+3, \ldots\}$, the set of integers

Notice that some numbers are correctly named in more than one way. For example, 2 is a natural number, a whole number, and an integer, because it is an element of all three sets. Zero is both a whole number and an integer; $^-5$ is only an integer.

Q7　　　　Insert $\in$ or $\notin$ to form true statements:

　　　　a. 5 _____ N　　　　　　**b.** 5 _____ W　　　　　　**c.** 5 _____ I

　　　　d. 0 _____ N　　　　　　**e.** 0 _____ W　　　　　　**f.** 0 _____ I

　　　　g. ⁻12 _____ N　　　　　**h.** ⁻12 _____ W　　　　　**i.** ⁻12 _____ I

　　　　j. 100 _____ I　　　　　**k.** ⁻101 _____ N　　　　**l.** ⁻101 _____ W

STOP　•　STOP　•　STOP　•　STOP　•　STOP　•　STOP　•　STOP　•　STOP　•　STOP

A7　　　　**a.** $\in$　　**b.** $\in$　　**c.** $\in$　　**d.** $\notin$　　**e.** $\in$　　**f.** $\in$　　**g.** $\notin$
　　　　h. $\notin$　　**i.** $\in$　　**j.** $\in$　　**k.** $\notin$　　**l.** $\notin$

This completes the instruction for this section.

6.1　Exercises

1. The set $\{\ldots, {}^-2, {}^-1, 0, {}^+1, {}^+2, \ldots\}$ is called the set of _____ .
2. Each integer is thought of as having what two parts?
3. Write the distance and direction represented by each of the following integers:
　　a. 15　　**b.** ⁻3　　**c.** 0　　**d.** ⁺7
4. What is the sign of the integer 0?
5. If N, W, and I are defined as in Frame 7, insert $\in$ or $\notin$ to make each of the following a true statement.
　　a. ⁻4 _____ N　　　　**b.** 5 _____ I
　　c. 0 _____ W　　　　**d.** ⁺6 _____ I
6. Graph each of the following integers:
　　a. ⁻2　　**b.** 4　　**c.** 0　　**d.** ⁺3

```
 ←─┼──┼──┼──┼──┼──┼──┼──┼──┼──┼──┼──┼──┼──┼──┼──┼──┼─→
  ⁻8  ⁻7  ⁻6  ⁻5  ⁻4  ⁻3  ⁻2  ⁻1   0  ⁺1  ⁺2  ⁺3  ⁺4  ⁺5  ⁺6  ⁺7  ⁺8
```

7. Write the opposite of each of the following integers:
　　a. ⁺2　　**b.** ⁻3　　**c.** 11　　**d.** 0

6.1　Exercise Answers

1. integers
2. direction, distance
3. **a.** right of zero　　　　　　15 units
　　b. left of zero　　　　　　 3 units
　　c. neither right nor left　　0 units
　　d. right of zero　　　　　　 7 units
4. Zero has no sign, because it is neither right nor left of zero on the number line.
5. **a.** $\notin$　　**b.** $\in$　　**c.** $\in$　　**d.** $\in$
6.

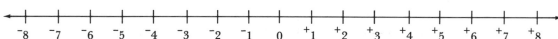

```
        a        c       d   b
 ←──┼──┼──┼──┼──┼──●──┼──●──┼──┼──●──●──┼──┼──→
 •••  ⁻7 ⁻6 ⁻5 ⁻4 ⁻3 ⁻2 ⁻1  0  ⁺1 ⁺2 ⁺3 ⁺4 ⁺5 ⁺6 ⁺7 •••
```

7. **a.** ⁻2　　**b.** ⁺3　　**c.** ⁻11　　**d.** 0

6.2   Addition of Integers

1   In Section 6.1 each integer was shown to have associated with it both a distance and a direction. For example, $^-5$ denotes a distance of 5 units to the left of zero, and $^+3$ denotes a distance of 3 units and a direction right of zero. To develop a procedure for the addition of integers it will be helpful to associate with each integer an arrow that pictures its distance and direction. For example, the integer $^+3$ can be represented by any of the arrows above the following number line:

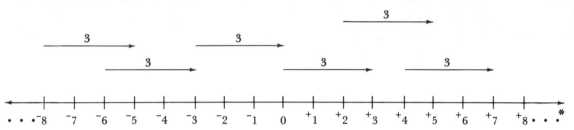

In this and the following sections the integer $^+3$ will be represented by any arrow that is *3 units long* and *points to the right*.

*When drawing the number line, it is common to omit the $\cdots$'s. It is understood that the line extends infinitely in both directions.

Q1   **a.**   What is the length of each arrow on the following number line?   _____

b.   What direction is indicated by each of the arrows?   _____

c.   What integer is represented by each of the arrows?   _____

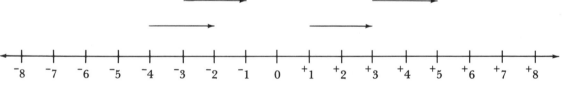

STOP • **STOP** • **STOP** • **STOP** • **STOP** • **STOP** • **STOP** • **STOP** • **STOP**

A1   **a.** 2 units   **b.** right   **c.** $^+2$

2   The integer $^-5$ is represented by each of the arrows above the following number line:

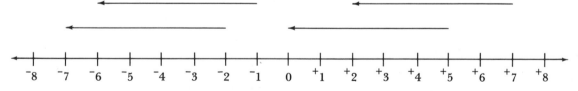

Each arrow represents $^-5$ since each is 5 *units in length* and each *points to the left*.

Q2   What integer is represented by each of the following arrows?   _____

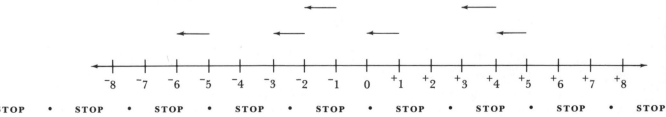

STOP • **STOP** • **STOP** • **STOP** • **STOP** • **STOP** • **STOP** • **STOP** • **STOP**

A2   ⁻1: each is 1 unit long and each points to the left.

Q3   Draw three arrows which each represent the integer ⁻4.

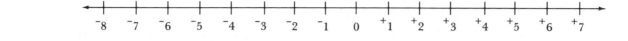

STOP • **STOP** • **STOP** • **STOP** • **STOP** • **STOP** • **STOP** • **STOP** • **STOP**

A3   The arrows shown must each be 4 units in length and point to the left. For example,

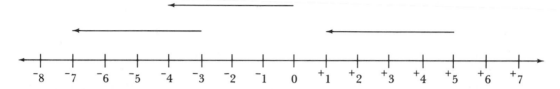

3   To add any two integers on the number line, use the following procedure:

Step 1: Represent the first addend* by an arrow that starts at zero.

Step 2: From the end of the first arrow, draw a second arrow to represent the second addend.

Step 3: Read the coordinate of the point at the end of the second arrow.

Examples:

1.

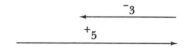

⁺5 + ⁻3

Dot denotes answer

Therefore, ⁺5 + ⁻3 = ⁺2.

2.

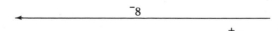

⁻4 + ⁻3

Therefore, ⁻4 + ⁻3 = ⁻7.

3.

⁺2 + ⁻8

Therefore, ⁺2 + ⁻8 = ⁻6.

*An addend is a value that is being added to another value.

Q4 Find the sum $^-3 + {}^+4$ by use of arrows for each of the addends on the following number line:

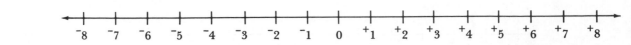

STOP • STOP • STOP • STOP • STOP • STOP • STOP • STOP • STOP

A4 $^+1$:

Q5 Find the sum $^+5 + {}^-7$ by use of arrows for each of the addends on the following number line:

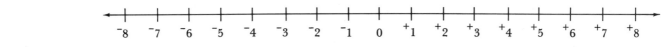

STOP • STOP • STOP • STOP • STOP • STOP • STOP • STOP • STOP

A5 -2:

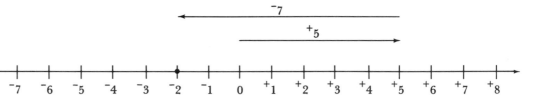

Q6 Find each of the following sums by use of the number line provided:

  a. $^-4 + {}^-5 =$ _____

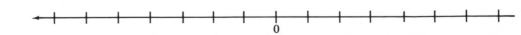

  b. $^+3 + {}^+2 =$ _____

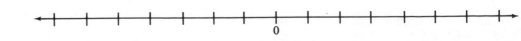

  c. $^-6 + 4 =$ _____

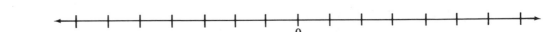

  d. $^-5 + {}^+9 =$ _____

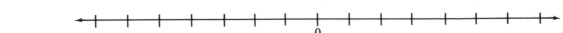

  e. $^-6 + {}^+6 =$ _____

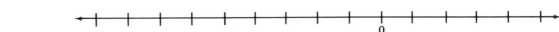

STOP • STOP • STOP • STOP • STOP • STOP • STOP • STOP • STOP

A6   **a.** $^-9$ **b.** $^+5$ **c.** $^-2$ **d.** $^+4$ **e.** 0

4

It is sometimes helpful to notice certain facts about the sum of two integers. Study each of the following three example sets to see if you can discover the three facts demonstrated.

Sum of two positives:	Sum of two negatives:	Sum of a positive and a negative:
$^+5 + {}^+3 = {}^+8$	$^-5 + {}^-3 = {}^-8$	$^-5 + {}^+3 = {}^-2$
$^+7 + {}^+6 = {}^+13$	$^-7 + {}^-6 = {}^-13$	$^+7 + {}^-6 = {}^+1$
$^+2 + {}^+9 = {}^+11$	$^-2 + {}^-9 = {}^-11$	$^+2 + {}^-2 = 0$

The three facts that correspond to the three preceding examples are:

1. The sum of two positive integers is a positive integer.
2. The sum of two negative integers is a negative integer.
3. The sum of a positive integer and a negative integer is sometimes positive, sometimes negative, and sometimes zero.

Q7

Use the facts of Frame 4 or a number line to find each of the following sums:

   a. $^-3 + {}^-6 = $ _____   **b.** $^+5 + {}^+8 = $ _____

   c. $^+4 + {}^-1 = $ _____   **d.** $^+2 + {}^-7 = $ _____

   e. $^+3 + {}^-3 = $ _____   **f.** $^-5 + {}^+5 = $ _____

STOP • **STOP** • **STOP** • **STOP** • **STOP** • **STOP** • **STOP** • **STOP** • **STOP**

A7   **a.** $^-9$ (fact 2)   **b.** $^+13$ (fact 1)   **c.** $^+3$ (fact 3)   **d.** $^-5$
        e. 0 (fact 3)   **f.** 0

5

Recall from Section 6.1 that every integer has a unique opposite, and that opposites represent the same distance but different directions. Consider the sum of $^-7$ and its opposite, $^+7$, shown on the following number line:

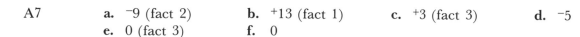

$^-7 + {}^+7 = 0$

If the order of the addends is reversed, the sum remains unchanged:

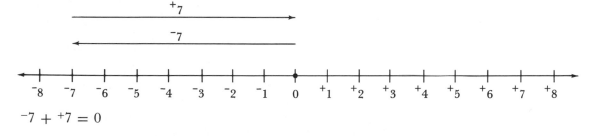

$^+7 + {}^-7 = 0$

Thus, regardless of the order of the addends, the sum of $^+7$ and its opposite, $^-7$, is 0:

$^+7 + {}^-7 = {}^-7 + {}^+7 = 0$

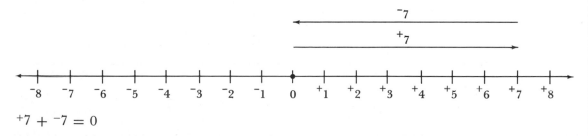

Q8 **a.** Find the sum $^-5 + {}^+5$.

b. Find the sum $^+5 + {}^-5$.

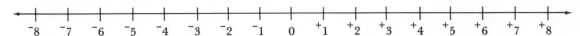

c. Is the sum in part a the same as the sum in part b?_____

STOP • STOP • STOP • STOP • STOP • STOP • STOP • STOP • STOP

A8 **a.** 0 **b.** 0 **c.** yes: $^-5 + {}^+5 = {}^+5 + {}^-5 = 0$

---

6 Because of the distance and direction relationships between opposites, the sum of any integer and its opposite is zero.

Examples:

Opposites   Sum

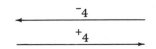

$^+4, {}^-4$ $^+4 + {}^-4 = 0$ 

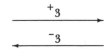

$^-3, {}^+3$ $^-3 + {}^+3 = 0$ 

$0, 0$ $0 + 0 = 0$ 

(Notice that the arrow representing zero is simply a dot at 0, because the integer zero has no length or direction.) The fact that the sum of any integer and its opposite is zero is generalized in the following statement: $a + {}^-a = {}^-a + a = 0$ for any integer replacement of a. In the above statement ^-a is read "the opposite of a." Thus, if $a = {}^+2$, ^-a is $^-({}^+2) = {}^-2$, the opposite of $^+2$. If $a = {}^-5$, ^-a is $^-({}^-5) = {}^+5$, the opposite of $^-5$. If $a = {}^-4$, the statement $a + {}^-a = 0$ is true, because

$a + {}^-a$
$^-4 + {}^-({}^-4)$
$^-4 + {}^+4$
0

Q9 **a.** What is ^-a if $a = {}^+6$? _____

b. What is ^-a if $a = {}^-9$? _____

STOP • STOP • STOP • STOP • STOP • STOP • STOP • STOP • STOP

A9 **a.** $^-6$ **b.** $^+9$

Q10   Show that $a + {}^-a = 0$ is true for $a = {}^-7$.

STOP • **STOP** • **STOP** • **STOP** • **STOP** • **STOP** • **STOP** • **STOP** • **STOP**

A10 $a + {}^-a$
        ${}^-7 + {}^-({}^-7)$
        ${}^-7 + {}^+7$
        0

Q11   Find each of the following sums:

        a. ${}^-3 + {}^-3 =$ _____   **b.** ${}^+5 + {}^-5 =$ _____

        c. ${}^+7 + {}^+7 =$ _____   **d.** ${}^-6 + {}^+6 =$ _____

        e. ${}^-8 + {}^+8 =$ _____   **f.** $0 + 0 =$ _____

STOP • **STOP** • **STOP** • **STOP** • **STOP** • **STOP** • **STOP** • **STOP** • **STOP**

A11   **a.** ${}^-6$: the sum of two negatives is a negative.
        b. 0: the sum of two opposites is zero.
        c. ${}^+14$: the sum of two positives is a positive.
        d. 0 **e.** 0 **f.** 0

Q12   Find the replacement for x that converts each of the following open sentences to a true statement:

        a. ${}^+2 + {}^-2 = x$ _____   **b.** $x + {}^+7 = 0$ _____

        c. ${}^-15 + x = 0$ _____   **d.** ${}^+13 + {}^-13 = x$ _____

STOP • **STOP** • **STOP** • **STOP** • **STOP** • **STOP** • **STOP** • **STOP** • **STOP**

A12   **a.** 0 **b.** ${}^-7$ **c.** ${}^+15$ **d.** 0

---

7   The sum of any natural number and zero is the natural number: for example, $0 + 2 = 2$ and $8 + 0 = 8$. The same is true of any integer and zero: for example,

        ${}^-3 + 0 = {}^-3$
        $0 + {}^+12 = {}^+12$
        ${}^-9 + 0 = {}^-9$

        This fact is called the *addition property of zero* and is generalized $a + 0 = 0 + a = a$ for any integer a.

Q13   Find each of the following sums:

        a. $0 + {}^-5 =$ _____   **b.** ${}^-4 + {}^-9 =$ _____

        c. ${}^+4 + {}^-4 =$ _____   **d.** ${}^+11 + 0 =$ _____

        e. ${}^+5 + {}^+7 =$ _____   **f.** ${}^-7 + {}^+3 =$ _____

STOP • **STOP** • **STOP** • **STOP** • **STOP** • **STOP** • **STOP** • **STOP** • **STOP**

A13   **a.** ${}^-5$ **b.** ${}^-13$ **c.** 0 **d.** ${}^+11$ **e.** ${}^+12$ **f.** ${}^-4$

---

8   To find the sum of more than two integers, use the methods presented earlier to add the integers two at a time. For example, the sum $^-3 + {}^+7 + {}^-6$ is found:

$$^-3 + {}^+7 + {}^-6 = (^-3 + {}^+7) + {}^-6$$
$$= {}^+4 + {}^-6$$
$$= {}^-2$$

Notice that each sum of two integers can be found by starting at zero and using the number-line procedure if necessary.

Q14   Find the sum $^+3 + {}^-2 + {}^+5$.

STOP • STOP • STOP • STOP • STOP • STOP • STOP • STOP • STOP

A14 $^+6$: $^+3 + {}^-2 + {}^+5 = (^+3 + {}^-2) + {}^+5$
$$= {}^+1 + {}^+5$$
$$= {}^+6$$

Q15   Find the sum $^-4 + {}^-7 + {}^+3$.

STOP • STOP • STOP • STOP • STOP • STOP • STOP • STOP • STOP

A15 $^-8$: $^-4 + {}^-7 + {}^+3 = (^-4 + {}^-7) + {}^+3$
$$= {}^-11 + {}^+3$$
$$= {}^-8$$

---

9   The sum $^-3 + {}^+5 + {}^-6 + {}^-2$ is found as follows:

$$^-3 + {}^+5 + {}^-6 + {}^-2 = (^-3 + {}^+5) + {}^-6 + {}^-2$$
$$= {}^+2 + {}^-6 + {}^-2$$
$$= (^+2 + {}^-6) + {}^-2$$
$$= {}^-4 + {}^-2$$
$$= {}^-6$$

Q16   Find each of the following sums:
a. $^-3 + {}^-5 + {}^+8$ b. $^-4 + {}^+5 + {}^-1 + {}^-3$

c. $^-2 + {}^-3 + {}^-6$ d. $^+1 + {}^+5 + {}^+7$

STOP • STOP • STOP • STOP • STOP • STOP • STOP • STOP • STOP

A16   **a.** 0   **b.** $^-3$ **c.** $^-11$ **d.** $^+13$

---

10   In Chapter 5 it was established that $a + b = b + a$ is true for all *whole-number replacements* of a and b (commutative property of addition). It is similarly true that $a + b = b + a$ for all *integer* replacements of a and b. An infinite number of examples can be used to demonstrate this fact. A few are:

$^-1 + {^+2} = {^+2} + {^-1}$ (both sums are $^+1$)
$^-7 + {^-4} = {^-4} + {^-7}$ (both sums are $^-11$)
$^+9 + {^-2} = {^-2} + {^+9}$ (both sums are $^+7$)

Q17   Verify that $a + b = b + a$ is true for $a = {^-5}$ and $b = {^+8}$.

STOP • STOP • STOP • STOP • STOP • STOP • STOP • STOP • STOP

A17 $^-5 + {^+8} = {^+8} + {^-5}$ (both sums are $^+3$)

---

11   From Chapter 5 it was also established that $(a + b) + c = a + (b + c)$ is true for all *whole-number* replacements of a, b, and c (associative property of addition). The associative property of addition is also true for all *integer* replacements of a, b, and c. For example,

1. $(^-3 + {^+4}) + {^-5} = {^-3} + ({^+4} + {^-5})$
    $^+1 + {^-5} = {^-3} + {^-1}$
    $^-4 = {^-4}$

2. $(^+7 + {^+5}) + {^-10} = {^+7} + ({^+5} + {^-10})$
    $^+12 + {^-10} = {^+7} + {^-5}$
    $^+2 = {^+2}$

Q18   Verify that $(a + b) + c = a + (b + c)$ is true for $a = {^-2}$, $b = {^-7}$, and $c = {^+3}$.

STOP • STOP • STOP • STOP • STOP • STOP • STOP • STOP • STOP

A18 $(^-2 + {^-7}) + {^+3} = {^-2} + ({^-7} + {^+3})$
    $^-9 + {^+3} = {^-2} + {^-4}$
    $^-6 = {^-6}$

This completes the instruction for this section.

6.2   Exercises

1. Find each of the following sums:

   a. $^-3 + {^-5}$ **b.** $^+5 + {^+9}$

   c. $^+7 + {^-6}$ **d.** $^-6 + {^+6}$

   e. $^-9 + 0$ **f.** $^-3 + {^+4} + {^-3}$

   g. $^-7 + {^+6} + {^+1}$ **h.** $0 + {^+4}$

   i. $^-4 + {^-3} + {^+7} + {^+4}$ **j.** $^-1 + {^+6} + {^-6} + {^+1}$

2. Find the replacement for x that converts each of the following open sentences to a true mathematical statement:

   a. $x + {}^-2 = 0$ **b.** ${}^-8 + {}^+8 = x$

   c. ${}^-12 + x = {}^-12$ **d.** ${}^-12 + x = 0$

3. Is the statement $a + b = b + a$ true for all integer replacements of a and b?

4. Match each of the following with the number of the statement that demonstrates it correctly.

   a. associative property of addition   **1.** ${}^-5 + {}^+3 = {}^+3 + {}^-5$

   b. addition property of zero   **2.** ${}^-2 + 0 = 0 + {}^-2 = {}^-2$

   c. the sum of two opposites is zero   **3.** ${}^-9 + {}^+9 = 0$

   d. commutative property of addition   **4.** $({}^-3 + {}^+7) + {}^-2 = {}^-3 + ({}^+7 + {}^-2)$

6.2   Exercise Answers

1. **a.** ${}^-8$ **b.** ${}^+14$ **c.** ${}^+1$ **d.** 0 **e.** ${}^-9$ **f.** ${}^-2$ **g.** 0

   h. ${}^+4$ **i.** ${}^+4$ **j.** 0

2. **a.** ${}^+2$ **b.** 0 **c.** 0 **d.** ${}^+12$

3. yes: commutative property of addition

4. **a.** 4   **b.** 2   **c.** 3   **d.** 1

6.3   Subtraction of Integers

1

Subtraction and addition can be thought of as opposite operations.* Consider, for example, the effect on any number x of first adding 5 and then performing the opposite operation of subtracting 5.

x

$x + 5$ add 5

$x + 5 - 5$ subtract 5

x

Notice that the operation of subtraction undoes what the operation of addition does, and the result is again the number x. The operation of addition also undoes an equal subtraction.

x

$x - 5$ subtract 5

$x - 5 + 5$ add 5

x

Thus, regardless of the order in which they are done, the operations of addition and subtraction are opposites. One operation undoes the other operation.

*Some mathematicians refer to addition and subtraction as "inverse operations."

Q1   **a.** What is the opposite operation of addition?   _____

       b. What is the opposite of adding 7 to any number?   _____

STOP • **STOP** • **STOP** • **STOP** • **STOP** • **STOP** • **STOP** • **STOP** • **STOP**

A1   **a.** subtraction   **b.** subtracting 7

Q2   **a.** What is the opposite operation of subtraction?   _____

   b. What is the opposite of subtracting 3 from any number?   _____

STOP • STOP • STOP • STOP • STOP • STOP • STOP • STOP • STOP

A2   **a.** addition   **b.** adding 3

---

2   Using the idea of addition and subtraction as opposite operations, the procedure for adding integers on the number line is easily modified for subtracting any two integers. Recall the procedure for finding the sum of any two integers on the number line.

$^-5 + {}^+7$:

Step 1: Draw the arrow for the first addend ($^-5$) from zero.

Step 2: From the tip of the first arrow, draw the arrow for the second addend ($^+7$).

Step 3: Read the answer below the tip of the arrow for the second addend.

Thus, $^-5 + {}^+7 = {}^+2$.

Q3   Find the sum $^+3 + {}^-9$ using the number line given.

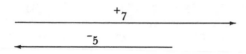

STOP • STOP • STOP • STOP • STOP • STOP • STOP • STOP • STOP

A3 $^-6$:

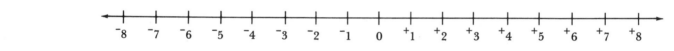

---

3   To find the difference $^-3 - {}^-5$, the following steps can be used:

Minuend   Subtrahend

$^-3 - {}^-5$

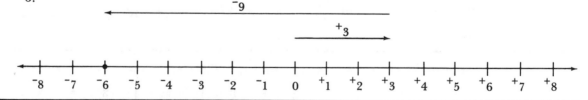

Step 1: Draw the arrow for the minuend ($^-3$).

Step 2: Since subtraction is the opposite of addition, draw the arrow for the *opposite* of the subtrahend ($^+5$) from the tip of the first arrow.

Step 3: Read the answer below the tip of the second arrow.

Thus, $^-3 - {^-5} = {^+2}$.

Q4   Use the following number line to find the difference $^-7 - {^-4}$:

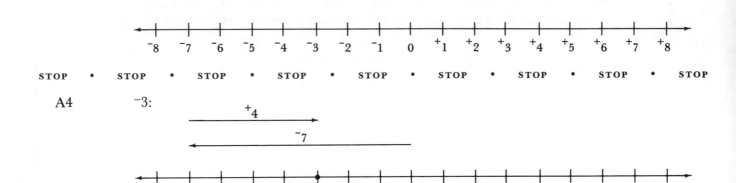

STOP   •   STOP   •   STOP   •   STOP   •   STOP   •   STOP   •   STOP   •   STOP   •   STOP

A4 $^-3$:

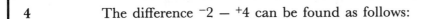

4   The difference $^-2 - {^+4}$ can be found as follows:

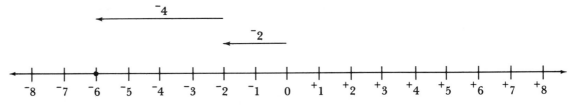

Step 1: Draw the arrow for the minuend ($^-2$).

Step 2: Since subtraction is the opposite of addition, draw the arrow for the *opposite* of the subtrahend ($^-4$).

Step 3: Read the answer below the tip of the second arrow.

Thus, $^-2 - {^+4} = {^-6}$.

Q5   Use the number line provided to find the difference $^+3 - {^+10}$.

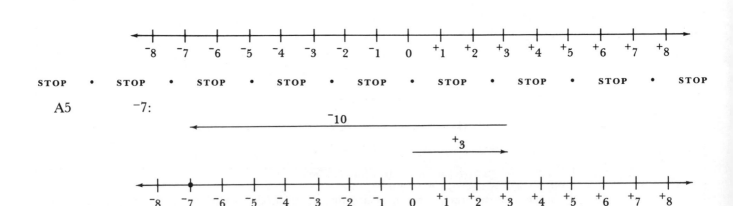

STOP   •   STOP   •   STOP   •   STOP   •   STOP   •   STOP   •   STOP   •   STOP   •   STOP

A5 $^-7$:

5 We repeat here the examples of Frames 3 and 4:

Minuend   Subtrahend   Difference

$$^-3 \;-\; ^-5 \;=\; ^+2$$
$$^-2 \;-\; ^+4 \;=\; ^-6$$

In each case the difference was found by first drawing the arrow for the minuend and then drawing the arrow for the *opposite* of the subtrahend.

    This procedure is the basis for defining subtraction in terms of addition: *To find the difference of two integers, add the minuend to the opposite of the subtrahend.*

Examples:

	Minuend	Subtrahend			Opposite of Subtrahend		Difference
1.	$^+7$	$-$	$\boxed{^+4}$	$= {}^+7 +$	$\boxed{^-4}$	$=$	$^+3$
2.	$^-3$	$-$	$\boxed{^+5}$	$= {}^-3 +$	$\boxed{^-5}$	$=$	$^-8$
3.	$^-4$	$-$	$\boxed{^-6}$	$= {}^-4 +$	$\boxed{^+6}$	$=$	$^+2$
4.	$^+2$	$-$	$\boxed{^+6}$	$= {}^+2 +$	$\boxed{^-6}$	$=$	$^-4$

Notice that in each example, two changes are involved: The operation sign for subtraction is changed to addition, and the subtrahend is changed to its opposite.

Q6 Find the difference $^-3 - {}^+1$ using the number line provided.

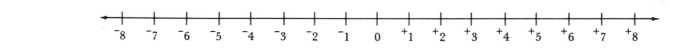

STOP • STOP • STOP • STOP • STOP • STOP • STOP • STOP • STOP

A6 $^-4$: $\;^-3 - {}^+1 = {}^-3 + {}^-1$

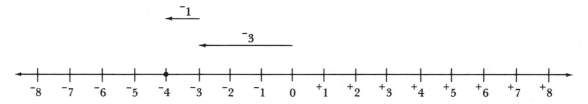

Q7 Find the difference by rewriting as a sum.

$$^-4 - {}^+7 = {}^-4 + \underline{\qquad} = \underline{\qquad}$$

STOP • STOP • STOP • STOP • STOP • STOP • STOP • STOP • STOP

A7 $^-7, {}^-11$

Q8 Find each of the following differences:

    a. $^-2 - {}^+5$ **b.** $^+7 - {}^-5$

    c. $^+1 - ^+9$ **d.** $^-5 - ^-3$

STOP • *STOP* • *STOP* • *STOP* • *STOP* • *STOP* • *STOP* • *STOP* • *STOP*

A8   **a.** $^-7$: $^-2 - ^+5 = ^-2 + ^-5 = ^-7$
             b. $^+12$: $^+7 - ^-5 = ^+7 + ^+5 = ^+12$
             c. $^-8$: $^+1 - ^+9 = ^+1 + ^-9 = ^-8$
             d. $^-2$: $^-5 - ^-3 = ^-5 + ^+3 = ^-2$

---

6   It is important to realize that the procedure of "adding the opposite" is done only with *subtraction*.

1. To find a *sum*, follow the procedure for adding integers directly.
2. To find a *difference*, *rewrite the problem as a sum* (by adding the opposite of the subtrahend to the minuend) and follow the procedure for adding integers.

Examples are:

$^-4 - ^+6 = ^-4 + ^-6 = ^-10$
$^+5 + ^-6 = ^-1$
$^-7 + ^-4 = ^-11$
$^-4 - ^-3 = ^-4 + ^+3 = ^-1$

Q9   Which of the following problems must be rewritten?
           a. $^-4 - ^-5$ **b.** $^-2 + ^-1$
           c. $^+6 + ^-9$ **d.** $^+6 - ^+15$ _____

STOP • *STOP* • *STOP* • *STOP* • *STOP* • *STOP* • *STOP* • *STOP* • *STOP*

A9   a and d: subtraction problems must be rewritten (b and c are addition problems and thus can be answered directly).

Q10   Complete the problems of Q9.

STOP • *STOP* • *STOP* • *STOP* • *STOP* • *STOP* • *STOP* • *STOP* • *STOP*

A10   **a.** $^-4 - ^-5 = ^-4 + ^+5 = ^+1$ **b.** $^-3$ **c.** $^-3$
             d. $^+6 - ^+15 = ^+6 + ^-15 = ^-9$

Q11   Complete each of the following as a sum or difference as indicated:
           a. $^-4 + ^-8$ **b.** $^+7 - ^+4$ **c.** $^-6 - ^+4$ **d.** $^-6 + ^+7$

           e. $^+2 - ^+5$ **f.** $^-4 - ^+8$ **g.** $^+4 + ^+7$ **h.** $^-1 - ^-1$

           i. $^+11 + ^-9$ **j.** $0 - ^-5$ **k.** $^-4 + 0$ **l.** $^+1 - ^+1$

m. $^+5 + {}^-5$	**n.** $^-2 + {}^-3$	**o.** $^-3 - {}^-3$	**p.** $^+5 + {}^-7$
q. $^+5 - {}^+9$	**r.** $^-1 + {}^+7$	**s.** $^-2 - 0$	**t.** $0 - {}^+1$

STOP • STOP • STOP • STOP • STOP • STOP • STOP • STOP • STOP

A11

a. $^-12$	**b.** $^+3$	**c.** $^-10$	**d.** $^+1$
e. $^-3$	**f.** $^-12$	**g.** $^+11$	**h.** 0
i. $^+2$	**j.** $^+5$	**k.** $^-4$	**l.** 0
m. 0	**n.** $^-5$	**o.** 0	**p.** $^-2$
q. $^-4$	**r.** $^+6$	**s.** $^-2$	**t.** $^-1$

---

7   When evaluating number sentences involving a combination of sums and differences, rewrite each of the differences as a sum and proceed as in Section 6.2, Frame 8. For example, the expression

$$^+3 + {}^-7 - {}^+5$$

$\underbrace{\phantom{^+3 + {}^-7}}_{\text{sum}}$

$\underbrace{\phantom{{}^-7 - {}^+5}}_{\text{difference}}$

involves a sum and a difference. First rewrite the difference as

$$^+3 + {}^-7 + {}^-5$$

The problem is now completed:

$$^-4 + {}^-5$$
$$^-9$$

Study the following examples:

1. $^-4 - {}^-2 + {}^+3$ 2. $^-3 - {}^+4 - {}^+6 + {}^+7$
    $^-4 + {}^+2 + {}^+3$ $^-3 + {}^-4 + {}^-6 + 7$
    $^-2 + {}^+3$ $^-13 + 7$
    $^+1$ $^-6$

Q12   Evaluate $^-3 + {}^+5 - {}^+7$.

STOP • STOP • STOP • STOP • STOP • STOP • STOP • STOP • STOP

A12 $^-5$: $^-3 + {}^+5 - 7$
         $^-3 + 5 + {}^-7$
         $^+2 + {}^-7$
         $^-5$

Q13   Evaluate $^+1 - {}^+5 - {}^+6$.

STOP • STOP • STOP • STOP • STOP • STOP • STOP • STOP • STOP

A13 $^-10$: $^+1 - {}^+5 - {}^+6$
                     $^+1 + {}^-5 + {}^-6$
                     $^-4 + {}^-6$
                     $^-10$

Q14   Evaluate $^+4 + {}^-3 - {}^+7 - {}^-6$.

STOP • **STOP** • **STOP** • **STOP** • **STOP** • **STOP** • **STOP** • **STOP** • **STOP**

A14   0: $^+4 + {}^-3 - {}^+7 - {}^-6$
                 $4 + {}^-3 + {}^-7 + {}^+6$
                 $^+1 + {}^-7 + {}^+6$
                 $^-6 + {}^+6$
                 0

Q15   Evaluate:
           a. $^-2 - {}^+5 + {}^+3$ **b.** $^+3 + {}^-5 + {}^-4$

           c. $^+4 - {}^-6 + {}^+7$ **d.** $^-6 - {}^+2 + {}^-7$

           e. $^-2 - {}^+3 - {}^+4$ **f.** $^+7 + {}^-2 - {}^+8 + {}^+3$

           g. $^+4 - 0 + {}^-4$ **h.** $0 - {}^+4 - {}^+3 + 2$

           i. $^+6 - {}^+3 - {}^+5$ **j.** $^-6 + {}^-3 - {}^+10$

STOP • **STOP** • **STOP** • **STOP** • **STOP** • **STOP** • **STOP** • **STOP** • **STOP**

A15   **a.** $^-4$: $^-2 - {}^+5 + 3 = {}^-2 + {}^-5 + {}^+3 = {}^-7 + {}^+3 = {}^-4$
           b. $^-6$
           c. $^+17$
           d. $^-15$: $^-6 - {}^+2 + {}^-7 = {}^-6 + {}^-2 + {}^-7 = {}^-15$
           e. $^-9$
           f. 0: $^+7 + {}^-2 - {}^+8 + {}^+3 = 7 + {}^-2 + {}^-8 + {}^+3 = 0$
           g. 0
           h. $^-5$: $0 - {}^+4 - {}^+3 + 2 = 0 + {}^-4 + {}^-3 + {}^+2 = {}^-7 + 2 = {}^-5$
           i. $^-2$: $^+6 - {}^+3 - {}^+5 = {}^+6 + {}^-3 + {}^-5 = {}^+3 + {}^-5 = {}^-2$
           j. $^-19$

> **8** Consider the following examples:
>
> $^+3 - {}^+4 = {}^+3 + {}^-4 = {}^-1$
> $^+4 - {}^+3 = {}^+1$
>
> The expressions $^+3 - {}^+4$ and $^+4 - {}^+3$ are not equivalent since they have different evaluations. Thus, $^+3 - {}^+4 = {}^+4 - {}^+3$ is a false statement. $^+3 - {}^+4 \neq {}^+4 - {}^+3$ is a true statement.

Q16 **a.** Are $^-2 - {}^+3$ and $^+3 - {}^-2$ equivalent expressions? _____

b. Is $^-2 - {}^+3 = {}^+3 - {}^-2$ a true statement? _____

STOP • STOP • STOP • STOP • STOP • STOP • STOP • STOP • STOP

A16 **a.** no: $^-2 - {}^+3 = {}^-5$, whereas $^+3 - {}^-2 = {}^+5$
b. no

Q17 Verify that $a - b$ and $b - a$ are not equivalent expressions for $a = {}^-5$ and $b = {}^-2$ by evaluating both expressions.

STOP • STOP • STOP • STOP • STOP • STOP • STOP • STOP • STOP

A17 $a - b = {}^-5 - {}^-2 = {}^-5 + {}^+2 = {}^-3$
$b - a = {}^-2 - {}^-5 = {}^-2 + {}^+5 = {}^+3$

> **9** $a - b$ and $b - a$ are not equivalent expressions since they do not have the same evaluation for all integer replacements of a and b.

Q18 For most integer replacements of a and b, $a - b = b - a$ is a _____ statement.
true/false

STOP • STOP • STOP • STOP • STOP • STOP • STOP • STOP • STOP

A18 false: $a - b = b - a$ is true when $a = b$

> **10** The fact that $a - b$ and $b - a$ are not equivalent expressions for all replacements of a and b demonstrates that subtraction is *not* a commutative operation.

Q19 Verify that $(^+3 - {}^+5) - {}^+7$ and $^+3 - (^+5 - {}^+7)$ do not have the same evaluation.

STOP • STOP • STOP • STOP • STOP • STOP • STOP • STOP • STOP

A19 $(^+3 - {}^+5) - {}^+7 = (^+3 + {}^-5) + {}^-7 = {}^-2 + {}^-7 = {}^-9$
$^+3 - (^+5 - {}^+7) = {}^+3 - (^+5 + {}^-7) = {}^+3 - (^-2) = {}^+3 + {}^+2 = {}^+5$

Q20 $(^+3 - {}^+5) - {}^+7 = {}^+3 - (^+5 - {}^+7)$ is a _____ statement.
true/false

STOP • STOP • STOP • STOP • STOP • STOP • STOP • STOP • STOP

A20　　　　false

Q21　　　　Verify that $(a - b) - c$ and $a - (b - c)$ are not equivalent by evaluating both expressions for $a = {}^+3$, $b = {}^-5$, and $c = {}^-4$.

STOP • **STOP** • **STOP** • **STOP** • **STOP** • **STOP** • **STOP** • **STOP** • **STOP**

A21　　　　

$(a - b) - c$	$a - (b - c)$
$({}^+3 - {}^-5) - {}^-4$	${}^+3 - ({}^-5 - {}^-4)$
$({}^+3 + {}^+5) + {}^+4$	${}^+3 - ({}^-5 + {}^+4)$
${}^+8 + {}^+4$	${}^+3 - ({}^-1)$
${}^+12$	${}^+3 + {}^+1$
	${}^+4$

11　　　　$(a - b) - c$ and $a - (b - c)$ are not equivalent expressions, because they do not have the same evaluation for all integer replacements of a, b, and c. This fact demonstrates that subtraction is *not* an associative operation.

Q22　　　　Answer true or false:

a. $a + b = b + a$ is true for all integer replacements of a and b.　　　　————

b. $a - b = b - a$ is true for all integer replacements of a and b.　　　　————

c. $a + (b + c) = (a + b) + c$ is true for all integer replacements of a, b, and c.

————

d. $a - (b - c) = (a - b) - c$ is true for all integer replacements of a, b, and c.

————

STOP • **STOP** • **STOP** • **STOP** • **STOP** • **STOP** • **STOP** • **STOP** • **STOP**

A22　　　　**a.** true　　**b.** false　　**c.** true　　**d.** false

This completes the instruction for this section.

6.3　　　Exercises

1. Find the following sums:
   a. ${}^+3 + {}^-7$　　　**b.** ${}^+12 + {}^-9$　　　**c.** ${}^-5 + {}^-6$　　　**d.** ${}^-9 + {}^+12$
   e. ${}^-4 + {}^+11$　　　**f.** ${}^+6 + {}^-6$　　　**g.** ${}^-2 + {}^+2$　　　**h.** ${}^+4 + 0$
   i. $0 + {}^-7$　　　　**j.** ${}^-3 + {}^-4$
2. Write the opposite for each of the following integers:
   a. ${}^-5$　　**b.** 0　　**c.** ${}^+6$　　**d.** ${}^-4$　　**e.** 2　　**f.** ${}^-1$

3. Find the following differences:
   a. $^-2 - 4$
   b. $^-5 - ^-2$
   c. $^-7 - ^-5$
   d. $^-3 - 0$
   e. $^+6 - ^+9$
   f. $^-4 - ^-4$
   g. $^+3 - ^+2$
   h. $0 - ^-1$
   i. $0 - ^+3$
   j. $^-2 - ^-8$

4. Complete each of the following:
   a. $^+7 + ^-3$
   b. $^-3 + 0$
   c. $^-6 - ^+4$
   d. $^+7 - ^+5$
   e. $^-5 + ^-9$
   f. $^+7 + ^-7$
   g. $0 + ^-4$
   h. $^-6 - ^-7$
   i. $^-1 - ^+1$
   j. $0 + ^+3$
   k. $^-3 - ^+4$
   l. $^-4 + ^+9$
   m. $^-5 - 0$
   n. $0 - ^-5$
   o. $^+6 + ^-9$
   p. $^+4 - ^+5$
   q. $^-3 + ^+12$
   r. $0 + ^-4$
   s. $^-5 + ^+5$
   t. $0 - ^-2$

5. Complete each of the following:
   a. $^+4 - ^+3 + ^-7$
   b. $^-2 - ^-5 - ^-6$
   c. $^-6 + ^+7 + ^-3$
   d. $^-4 + ^+6 - ^-4 - ^+6$
   e. $^+2 + ^-2 - 0$
   f. $^+3 - ^+2 - ^+4$
   g. $^+4 + ^-7 - ^+3$
   h. $0 - ^+4 + ^+7$
   i. $^-2 - ^+5 - ^+6$
   j. $^+6 - 0 + ^-3$

6. Answer true or false:
   a. $^-3 + ^+7 = ^+7 + ^-3$
   b. $^+2 - ^+1 = ^+1 - ^+2$
   c. $(^-3 + ^+5) + ^-7 = ^-3 + (^+5 + ^-7)$
   d. $^-5 + 0 = 0 + ^-5 = ^-5$
   e. $^+6 - (^+3 - ^-2) = (^+6 - ^+3) - ^-2$

6.3   Exercise Answers

1. a. $^-4$ b. $^+3$ c. $^-11$ d. $^+3$ e. $^+7$ f. 0 g. 0
   h. $^+4$ i. $^-7$ j. $^-7$

2. a. $^+5$ b. 0 c. $^-6$ d. $^+4$ e. $^-2$ f. $^+1$

3. a. $^-6$ b. $^-3$ c. $^-2$ d. $^-3$ e. $^-3$ f. 0 g. $^+1$
   h. $^+1$ i. $^-3$ j. $^+6$

4. a. $^+4$ b. $^-3$ c. $^-10$ d. $^+2$ e. $^-14$ f. 0 g. $^-4$
   h. $^+1$ i. $^-2$ j. $^+3$ k. $^-7$ l. $^+5$ m. $^-5$ n. $^+5$
   o. $^-3$ p. $^-1$ q. $^+9$ r. $^-4$ s. 0 t. $^+2$

5. a. $^-6$ b. $^+9$ c. $^-2$ d. 0 e. 0 f. $^-3$ g. $^-6$
   h. $^+3$ i. $^-13$ j. $^+3$

6. a. true: commutative property of addition
   b. false
   c. true: associative property of addition
   d. true: addition property of zero (additive identity property)
   e. false

6.4 Multiplication of Integers

1 The operation of multiplication was developed as a shortcut procedure for addition. For example, the product $2 \cdot 3$ can be represented either as the sum of 2 threes or as the sum of 3 twos:

$$2 \cdot 3 = \underbrace{3 + 3}_{\text{2 addends of 3}} = 6$$

or

$$3 \cdot 2 = \underbrace{2 + 2 + 2}_{\text{3 addends of 2}} = 6$$

Similarly, the product $7 \cdot 4$ can be represented as either 7 fours or 4 sevens.

$$7 \cdot 4 = 4 + 4 + 4 + 4 + 4 + 4 + 4 = 28$$

or

$$4 \cdot 7 = 7 + 7 + 7 + 7 = 28$$

Q1 Write $5 \cdot 6$ as a sum in two ways.

STOP • STOP • STOP • STOP • STOP • STOP • STOP • STOP • STOP

A1 $5 \cdot 6 = 6 + 6 + 6 + 6 + 6 = 30$

or

$$6 \cdot 5 = 5 + 5 + 5 + 5 + 5 + 5 = 30$$

Q2 Write $3 \cdot 9$ as a sum in two ways.

STOP • STOP • STOP • STOP • STOP • STOP • STOP • STOP • STOP

A2 $3 \cdot 9 = 9 + 9 + 9 = 27$

or

$$9 \cdot 3 = 3 + 3 + 3 + 3 + 3 + 3 + 3 + 3 + 3 = 27$$

2 The procedure of writing a product as a sum can also be used to find the product of two integers. For example, the product $^+4 \cdot {}^-2$ can be written as the sum of 4 negative twos:

$$^+4 \cdot {}^-2 = \underbrace{{}^-2 + {}^-2 + {}^-2 + {}^-2}_{\text{4 addends of } {}^-2}$$

Since the sum on the right is equal to $^-8$ the product $^+4 \cdot {}^-2$ is $^-8$:

$$^+4 \cdot {}^-2 = {}^-8$$

Similarly, the product $^+3 \cdot {}^-7$ is $^-21$, because

$$^+3 \cdot {}^-7 = {}^-7 + {}^-7 + {}^-7$$
$$= {}^-21$$

Q3   Write $^+2 \cdot {}^-6$ as a sum.   _____

STOP • STOP • STOP • STOP • STOP • STOP • STOP • STOP • STOP

A3 $^+2 \cdot {}^-6 = {}^-6 + {}^-6$

Q4   Find the product $^+2 \cdot {}^-6$.

STOP • STOP • STOP • STOP • STOP • STOP • STOP • STOP • STOP

A4 $^-12$

Q5   Find the product $^+3 \cdot {}^-5$ by writing it as a sum.

STOP • STOP • STOP • STOP • STOP • STOP • STOP • STOP • STOP

A5 $^-15$: $^+3 \cdot {}^-5 = {}^-5 + {}^-5 + {}^-5$
$$= {}^-15$$

Q6   Find the product $^+7 \cdot {}^-1$ by writing it as a sum.

STOP • STOP • STOP • STOP • STOP • STOP • STOP • STOP • STOP

A6 $^-7$: $7 \cdot {}^-1 = {}^-1 + {}^-1 + {}^-1 + {}^-1 + {}^-1 + {}^-1 + {}^-1$
$$= {}^-7$$

Q7   Find each of the following products:

    a. $^+5 \cdot {}^-2 = $ _____   **b.** $^+6 \cdot {}^-9 = $ _____

    c. $^+1 \cdot {}^-4 = $ _____   **d.** $0 \cdot {}^+4 = $ _____

    e. $^+3 \cdot {}^+3 = $ _____   **f.** $^+2 \cdot {}^-8 = $ _____

STOP • STOP • STOP • STOP • STOP • STOP • STOP • STOP • STOP

A7   **a.** $^-10$ **b.** $^-54$ **c.** $^-4$ **d.** 0 **e.** $^+9$ **f.** $^-16$

---

3   The product $^-5 \cdot {}^+4$ can be found by computing the sum of 4 negative fives:

$$^-5 \cdot {}^+4 = \underbrace{{}^-5 + {}^-5 + {}^-5 + {}^-5}_{4 \text{ addends of } {}^-5}$$
$$= {}^-20$$

Q8   Find the product $^-7 \cdot 5$ by writing it as a sum.

STOP • STOP • STOP • STOP • STOP • STOP • STOP • STOP • STOP

A8 -35: $^-7 \cdot 5 = ^-7 + ^-7 + ^-7 + ^-7 + ^-7$
                              $= ^-35$

Q9   Find the product $^+5 \cdot ^-7$.

STOP • STOP • STOP • STOP • STOP • STOP • STOP • STOP • STOP

A9 -35: $^+5 \cdot ^-7 = ^-7 + ^-7 + ^-7 + ^-7 + ^-7$
                              $= ^-35$

Q10   Find the product $^-4 \cdot ^+6$.

STOP • STOP • STOP • STOP • STOP • STOP • STOP • STOP • STOP

A10 -24: $^-4 \cdot ^+6 = ^-4 + ^-4 + ^-4 + ^-4 + ^-4 + ^-4$
                              $= ^-24$

Q11   Find the product $^-7 \cdot ^+9$.

STOP • STOP • STOP • STOP • STOP • STOP • STOP • STOP • STOP

A11 $^-63$

---

4   In each of the preceding products where the two factors had different signs (one positive and one negative), the product was negative. For example,

$$\underbrace{^+5 \cdot ^-9}_{} = ^-45 \qquad \underbrace{^-7 \cdot ^+8}_{} = ^-56$$

different   negative   different   negative
 signs   product   signs   product

These examples demonstrate the following rule for multiplying integers with different signs: *The product of two integers with different signs (one positive and one negative) is a negative integer.*

To make the multiplication of integers consistent with the multiplication of natural numbers, the rule for multiplying two positive integers is as follows: *The product of two positive integers is a positive integer.* For example,

$$^+2 \cdot ^+7 = ^+14 \qquad 5 \cdot 8 = 40$$

The "+" sign is frequently omitted from positive numbers such as in the second example above. A number written without a sign is assumed to be positive.

Q12   Find the product in each of the following:

   a. $^-2 \cdot ^+3 = $ _____   b. $^+6 \cdot ^-3 = $ _____   c. $4 \cdot 7 = $ _____

   d. $^+1 \cdot ^-5 = $ _____   e. $^+10 \cdot ^-8 = $ _____   f. $^+5 \cdot ^+3 = $ _____

   g. $^-7 \cdot ^+7 = $ _____   h. $^-8 \cdot ^+6 = $ _____

STOP • STOP • STOP • STOP • STOP • STOP • STOP • STOP • STOP

A12   a. $^-6$ b. $^-18$ c. 28   d. $^-5$ e. $^-80$ f. 15   g. $^-49$
          h. $^-48$

---

5   In Section 5.3 the multiplication property of zero was stated as being true for all whole-number replacements of the variable. This property is also true for all integer replacements of the variable; that is, $a \cdot 0 = 0a = 0$ for all integers a. Thus, any integer times zero is equal to zero. For example,

$$^-5 \cdot 0 = 0 \qquad 0 \cdot 3 = 0$$

Q13   Find each of the following products:

   a. $^-4 \cdot 0 =$ _____   b. $^+8 \cdot ^-1 =$ _____

   c. $^-3 \cdot ^+4 =$ _____   d. $0 \cdot ^+5 =$ _____

   e. $6 \cdot 7 =$ _____   f. $^+9 \cdot ^-9 =$ _____

STOP • STOP • STOP • STOP • STOP • STOP • STOP • STOP • STOP

A13   a. 0   b. $^-8$ c. $^-12$ d. 0   e. 42   f. $^-81$

---

6   The product of two integers with *different signs* is *negative*. The product of *two positive integers* is a *positive* integer. To discover the product of *two negative integers,* study the following series of products and notice the pattern that is present in the answers on the right.

$$^-2 \cdot ^+4 = ^-8$$
$$^-2 \cdot ^+3 = ^-6$$
$$^-2 \cdot ^+2 = ^-4$$
$$^-2 \cdot ^+1 = ^-2$$
$$^-2 \cdot 0 = 0$$
$$^-2 \cdot ^-1 = ?$$
$$^-2 \cdot ^-2 = ?$$

The pattern in the products on the right is that each answer increases by 2. Hence, to complete the pattern, the products are:

$$^-2 \cdot ^-1 = ^+2$$
$$^-2 \cdot ^-2 = ^+4$$

It is, thus, appropriate to state the rule for the product of two negative integers as follows: *The product of two negative integers is a positive integer.* For example,

$$^-7 \cdot ^-4 = ^+28 \qquad ^-11 \cdot ^-5 = ^+55$$

Q14   Find the product $^-4 \cdot ^-6$.

STOP • STOP • STOP • STOP • STOP • STOP • STOP • STOP • STOP

A14 $^+24$ (or simply 24)

Q15 $^-3 \cdot 5 =$ _____

STOP • STOP • STOP • STOP • STOP • STOP • STOP • STOP • STOP

A15 $^-15$

Q16 $^-9 \cdot ^-6 =$ _____

STOP • STOP • STOP • STOP • STOP • STOP • STOP • STOP • STOP

A16	54

Q17   Find the following products:

     a. $^-3 \cdot 4 = $ _____   **b.** $^-4 \cdot {}^-9 = $ _____

     c. $^-2 \cdot {}^-5 = $ _____   **d.** $9 \cdot {}^-8 = $ _____

     e. $0 \cdot {}^-6 = $ _____   **f.** $^-12 \cdot 0 = $ _____

     g. $5 \cdot {}^-1 = $ _____   **h.** $^-7 \cdot {}^-5 = $ _____

STOP • STOP • STOP • STOP • STOP • STOP • STOP • STOP • STOP

A17   **a.** $^-12$ **b.** 36 **c.** 10 **d.** $^-72$
         e. 0 **f.** 0 **g.** $^-5$ **h.** 35

---

7   Study the effect of multiplying any integer by $^-1$ in the following examples:

$$^-1 \cdot 4 = {}^-4 \qquad {}^-1 \cdot {}^-5 = {}^+5$$

In the first example, multiplying 4 by negative one changes it to $^-4$, its opposite. In the second example, the product of $^-5$ and negative one changes $^-5$ to $^+5$, its opposite. The fact that negative one times a number is the opposite of the number is often called the multiplication property of $^-1$. *The multiplication property of $^-1$ states that $^-1 \cdot a = {}^-a$ is true for all integer replacements of a.* (*Note:* ^-a is read "the *opposite* of a.")

Q18   Find each of the following products:

     a. $^-1 \cdot {}^-7 = $ _____   **b.** $1 \cdot {}^-7 = $ _____

     c. $^-1 \cdot 9 = $ _____   **d.** $^-1 \cdot 0 = $ _____

STOP • STOP • STOP • STOP • STOP • STOP • STOP • STOP • STOP

A18   **a.** 7 **b.** $^-7$ **c.** $^-9$ **d.** 0

---

8   It is important for later work that the student be able to read the multiplication property of $^-1$ both from left to right and from right to left. Reading from left to right it says that multiplying by $^-1$ gives the opposite of the integer involved:

The integer		Its opposite
$^-1 \cdot {}^+6$	$=$	$^-6$
$^-1 \cdot {}^-9$	$=$	$^+9$

Reading from right to left it says that any integer can be expressed as a product of $^-1$ and its opposite.

The integer			Its opposite
$^-5$	$=$	$^-1 \cdot$	$^+5$
$^-7$	$=$	$^-1 \cdot$	$^+7$
$^+6$	$=$	$^-1 \cdot$	$^-6$
$^+3$	$=$	$^-1 \cdot$	$^-3$

Q19   Complete the following statements using the multiplication property of $^-1$:

a. $^-1 \cdot {}^+7 =$ _____   b. $^-1 \cdot$ _____ $= {}^+5$

c. $^-3 = {}^-1 \cdot$ _____   d. _____ $\cdot {}^-8 = {}^+8$

STOP • STOP • STOP • STOP • STOP • STOP • STOP • STOP • STOP

A19   a. $^-7$ b. $^-5$ c. $^+3$ d. $^-1$

Q20   Find the product:

a. $^-3 \cdot {}^+2 =$ _____   b. $^+2 \cdot {}^-3 =$ _____

STOP • STOP • STOP • STOP • STOP • STOP • STOP • STOP • STOP

A20   a. $^-6$ b. $^-6$

Q21   Evaluate the expressions ab and ba for $a = {}^-7$ and $b = {}^-8$.

STOP • STOP • STOP • STOP • STOP • STOP • STOP • STOP • STOP

A21 $ab = {}^-7 \cdot {}^-8$ $ba = {}^-8 \cdot {}^-7$
       $= {}^+56$ $= {}^+56$

Q22   Do you think the statement $ab = ba$ is true for all integer replacements of a and b?

_____

STOP • STOP • STOP • STOP • STOP • STOP • STOP • STOP • STOP

A22   yes: the order of the factors when finding the product of two integers does not affect the product.

---

9   The set of integers is commutative with respect to the operation of multiplication. That is, the statement

$ab = ba$

is true for all integer replacements of a and b. For example,

1. $^-3 \cdot {}^+9 = {}^+9 \cdot {}^-3$ 2. $^-12 \cdot {}^-5 = {}^-5 \cdot {}^-12$
    $^-27 = {}^-27$ $^+60 = {}^+60$

Q23   Is $^-253 \cdot {}^+479 = {}^+479 \cdot {}^-253$ a true statement?   _____

STOP • STOP • STOP • STOP • STOP • STOP • STOP • STOP • STOP

A23   yes: the commutative property of multiplication is true for all integers.

---

10   Frame 9 makes it possible to state the following procedure for finding a product of more than two numbers. To find the product of more than two numbers:

Step 1: Do all work within parentheses first.

Step 2: Find the product of two numbers at a time in *any order desired.*

Examples:

1. $^-2 \cdot {}^+4 \cdot {}^-5$
   $^+10 \cdot {}^+4$ (since there are no parentheses, follow step 2)
   40

2. $^+3(^-6 \cdot {}^-5) \cdot {}^+9$
   $^+3(^+30)^+9$ (since parentheses are involved,
   $^+90 \cdot {}^+9$ use step 1 and then step 2)
   810

Q24   Find the product $^-2 \cdot {}^-3 \cdot {}^+5$.

STOP • **STOP** • **STOP** • **STOP** • **STOP** • **STOP** • **STOP** • **STOP** • **STOP**

A24   30

Q25   Find the product $(^-1 \cdot {}^+3)(^-4 \cdot {}^-7)$.

STOP • **STOP** • **STOP** • **STOP** • **STOP** • **STOP** • **STOP** • **STOP** • **STOP**

A25 $^-84$: $(^-1 \cdot {}^+3)(^-4 \cdot {}^-7)$
                  $^-3 \cdot {}^+28$
                  $^-84$

Q26   Find each of the following products:
            a. $^-2 \cdot {}^-4 \cdot 0$ b. $(^-3 \cdot {}^+4) \cdot {}^-5$

            c. $^-6 \cdot {}^+11 \cdot {}^+10$ d. $^-1(^+3 \cdot {}^-5)$

            e. $(^+4 \cdot 0)(^-7 \cdot {}^-6)$ f. $(^-2 \cdot {}^+3)(^-4 \cdot {}^-6)$

STOP • **STOP** • **STOP** • **STOP** • **STOP** • **STOP** • **STOP** • **STOP** • **STOP**

A26   **a.** 0   **b.** 60   **c.** $^-660$ **d.** 15   **e.** 0   **f.** $^-144$

11   In Chapter 5 it was established that the set of whole numbers for the operation of multiplication was associative, commutative, and distributive with respect to addition and subtraction. These properties are likewise true for the set of integers. Thus, the following statements are true for all integer replacements of *a*, *b*, and *c*:

            1. Associative property of multiplication

               $a(bc) = (ab)c$

2. Commutative property of multiplication

$$ab = ba$$

3. Distributive property of multiplication over addition

$$a(b + c) = ab + ac \quad \text{(left)}$$
$$(a + b)c = ac + bc \quad \text{(right)}$$

4. Distributive property of multiplication over subtraction

$$a(b - c) = ab - ac \quad \text{(left)}$$
$$(a - b)c = ac - bc \quad \text{(right)}$$

Q27   Verify that $a(bc) = (ab)c$ is true when $a = {}^-3$, $b = {}^+5$, and $c = {}^+7$.

$a(bc) =$ $(ab)c =$

STOP • **STOP** • **STOP** • **STOP** • **STOP** • **STOP** • **STOP** • **STOP** • **STOP**

A27
$$a(bc) = {}^-3({}^+5 \cdot {}^+7) \qquad\qquad (ab)c = ({}^-3 \cdot {}^+5) \cdot {}^+7$$
$$= {}^-3({}^+35) \qquad\qquad\qquad = ({}^-15) \cdot {}^+7$$
$$= {}^-105 \qquad\qquad\qquad\quad = {}^-105$$

Q28   Verify that $a(b - c) = ab - ac$ is true when $a = {}^+5$, $b = {}^-1$, and $c = {}^-4$.

$a(b - c) =$ $ab - ac =$

STOP • **STOP** • **STOP** • **STOP** • **STOP** • **STOP** • **STOP** • **STOP** • **STOP**

A28
$$a(b - c) = {}^+5({}^-1 - {}^-4) \qquad\qquad ab - ac = {}^+5 \cdot {}^-1 - {}^+5 \cdot {}^-4$$
$$= {}^+5({}^-1 + {}^+4) \qquad\qquad\qquad = {}^-5 - {}^-20$$
$$= {}^+5({}^+3) \qquad\qquad\qquad\quad = {}^-5 + {}^+20$$
$$= {}^+15 \qquad\qquad\qquad\qquad = {}^+15$$

This completes the instruction for this section.

6.4   Exercises

1. The product of two integers with different signs is a ＿＿＿＿＿＿＿ integer.
2. The product of any integer and zero is ＿＿＿＿＿＿.
3. The product of any two integers with the same sign is a ＿＿＿＿＿＿＿ integer.
4. Fill in the blanks so that a true statement results:
   a. ＿＿＿ $\cdot 5 = 0$ b. ${}^-1 \cdot 7 =$ ＿＿＿
   c. ${}^+8 =$ ＿＿＿ $\cdot {}^-8$ d. ${}^-1 \cdot {}^-5 =$ ＿＿＿

5. Find each of the following products:
   a. $^+4 \cdot ^-9$ b. $^-3 \cdot ^-4$
   c. $^-2 \cdot 0$ d. $^-1 \cdot ^+9$
   e. $^-1 \cdot ^+7$ f. $5 \cdot ^-7$
   g. $0 \cdot ^-9$ h. $^-6 \cdot ^-9$
   i. $11 \cdot ^-5$ j. $7 \cdot ^-8$

6. Find each of the following products:
   a. $^+2 \cdot ^-3 \cdot ^-4$ b. $(^+7 \cdot ^-1) \cdot ^-7$
   c. $^-1 \cdot ^-7 \cdot ^-8$ d. $^-1 \cdot ^-1 \cdot 0$
   e. $^-5 \cdot 0 \cdot ^-6 \cdot ^+3$ f. $4 \cdot 3 \cdot ^-5$
   g. $(^-4 \cdot ^-4)(^-2 \cdot ^+2)$ h. $^-7(9 \cdot ^-3) \cdot ^+2$
   i. $^+6(^-3 \cdot ^+2)$ j. $(^-2 \cdot 4)(3 \cdot ^-3)$
   k. $(^+6 \cdot ^-3) \cdot ^+2$ l. $^-4 \cdot 5(^-6 \cdot 0)$

7. Verify $a(b - c) = ab - ac$ when $a = ^-2$, $b = ^-3$, and $c = ^+5$.

8. True or false:
   a. $ab = ba$ for all integers a and b.
   b. $a - b = b - a$ for all integers a and b.
   c. $(ab)c = a(bc)$ for all integers a, b, and c.

6.4   Exercise Answers

1. negative
2. zero
3. positive
4. a. 0   b. $^-7$ c. $^-1$ d. 5
5. a. $^-36$ b. 12   c. 0   d. $^-9$ e. $^-7$
   f. $^-35$ g. 0   h. 54   i. $^-55$ j. $^-56$
6. a. 24   b. 49   c. $^-56$ d. 0   e. 0   f. $^-60$
   g. $^-64$ h. 378   i. $^-36$ j. 72   k. $^-36$ l. 0
7. $a(b - c) = ab - ac$
   $^-2(^-3 - ^+5) = ^-2 \cdot ^-3 - ^-2 \cdot ^+5$
   $^-2(^-3 + ^-5) = ^+6 - ^-10$
   $\qquad ^-2(^-8) = ^+6 + ^+10$
   $\qquad\qquad 16 = 16$
8. a. true   b. false   c. true

6.5   Division of Integers

1

The operation of division is closely related to that of multiplication. A division problem is often checked using the multiplication operation. As was the case with addition and subtraction, multiplication and division are also opposite or inverse operations; that is, one undoes the effect of the other. For example,

$10 \div 2 = 5$ because $5 \cdot 2 = 10$ or $8\overline{)56}^{\;7}$ because $7 \cdot 8 = 56$

Q1 $12 \div 3 = 4$, because $\underline{\qquad} \cdot \underline{\qquad} = \underline{\qquad}$.

STOP   •   STOP   •   STOP   •   STOP   •   STOP   •   STOP   •   STOP   •   STOP

A1 $4 \cdot 3 = 12$

Q2 $7\overline{)63}$, with 9 above, because _____ $\cdot$ _____ $=$ _____ .

STOP • STOP • STOP • STOP • STOP • STOP • STOP • STOP • STOP

A2 $9 \cdot 7 = 63$

Q3   Use multiplication to find the quotient and write the check: $45 \div 9 =$ _____ , because

     _____ .

STOP • STOP • STOP • STOP • STOP • STOP • STOP • STOP • STOP

A3 $45 \div 9 = \underline{5}$, because $\underline{5 \cdot 9 = 45}$

Q4   Use multiplication to find the quotient and check the result: $82 \div 2 =$ _____ , because

     _____ .

STOP • STOP • STOP • STOP • STOP • STOP • STOP • STOP • STOP

A4 $82 \div 2 = \underline{41}$, because $\underline{41 \cdot 2 = 82}$

> **2**   Consider the following quotient and check:
>
> $^-27 \div {}^+9 = \square$ because $\square \cdot {}^+9 = {}^-27$
>
>   Since $^-3$ makes the product statement true, it also satisfies the quotient statement:
>
> $^-27 \div {}^+9 = \boxed{^-3}$ because $\boxed{^-3} \cdot {}^+9 = {}^-27$

Q5   Place an integer in the $\square$ to form a true statement: $^+12 \div {}^-3 = \square$ because
   $\square \cdot {}^-3 = {}^+12$

STOP • STOP • STOP • STOP • STOP • STOP • STOP • STOP • STOP

A5 $^+12 \div {}^-3 = \boxed{^-4}$, because $\boxed{^-4} \cdot {}^-3 = {}^+12$

Q6   Place an integer in the $\square$ to form a true statement: $^-16 \div {}^+2 = \square$ because
   $\square \cdot {}^+2 = {}^-16$

STOP • STOP • STOP • STOP • STOP • STOP • STOP • STOP • STOP

A6 $^-16 \div {}^+2 = \boxed{^-8}$, because $\boxed{^-8} \cdot {}^+2 = {}^-16$

Q7   Find the quotient and write the check as a multiplication problem: $^+56 \div {}^-8 =$
   _____ because _____ $\cdot {}^-8 = {}^+56$

STOP • STOP • STOP • STOP • STOP • STOP • STOP • STOP • STOP

A7 $^+56 \div {}^-8 = \underline{^-7}$, because $\underline{^-7} \cdot {}^-8 = {}^+56$

Q8   Find the quotient and write the check as a multiplication problem: $^-30 \div {}^+10 =$ _____
   because _____

STOP • STOP • STOP • STOP • STOP • STOP • STOP • STOP • STOP

A8 $^-3$, because $^-3 \cdot {}^+10 = {}^-30$

Q9 $^+15 \div {}^-3 =$ _____

STOP • STOP • STOP • STOP • STOP • STOP • STOP • STOP • STOP

A9 $^-5$

---

3   Each of the problems in Q5 through Q9 involved the quotient of two integers with different signs. Study the problems and their answers below:

$$^+12 \div {}^-3 = {}^-4$$
$$^-16 \div {}^+2 = {}^-8$$
$$^+56 \div {}^-8 = {}^-7$$
$$^-30 \div {}^+10 = {}^-3$$
$$^+15 \div {}^-3 = {}^-5$$

Q10   What is true of all the answers in Frame 3?

_____

STOP • STOP • STOP • STOP • STOP • STOP • STOP • STOP • STOP

A10   Each answer is a negative integer.

---

4   The examples of Frame 3 demonstrate the following definition for the division of two integers with different signs: *The quotient of two integers with different signs is a negative integer.*
     A division problem is also correctly written as a fraction. Thus, $20 \div 4 = 5$ can also be written $\frac{20}{4} = 5$. Study the following examples:

$$\frac{^-36}{^+4} = {}^-9 \qquad {}^+27 \div {}^+9 = {}^+3$$
$$48 \div {}^-6 = {}^-8 \qquad \frac{^+14}{^-2} = {}^-7$$

Q11   Find the following quotients:

     a. $\dfrac{^-25}{^+5} =$ _____   b. $^-25 \div {}^+5 =$ _____

     c. $\dfrac{^-90}{^+10} =$ _____   d. $^+12 \div {}^-2 =$ _____

     e. $^+1 \div {}^-1 =$ _____   f. $\dfrac{^+45}{^-9} =$ _____

STOP • STOP • STOP • STOP • STOP • STOP • STOP • STOP • STOP

A11   a. $^-5$ b. $^-5$ c. $^-9$ d. $^-6$ e. $^-1$ f. $^-5$

---

5   To determine the sign of the *quotient* of two *integers* with the *same sign,* consider the integer that converts the following open sentence into a true statement.

     $^-20 \div {}^-5 = \square$ because $\square \cdot {}^-5 = {}^-20$

     The correct integer replacement is $^+4$. That is, $^-20 \div {}^-5 = {}^+4$, because $^+4 \cdot {}^-5 = {}^-20$.

Q12 Place an integer in the ☐ to form a true statement: $^-15 \div {}^-3 = $ ☐ because ☐ $\cdot {}^-3 = {}^-15$

STOP • **STOP** • **STOP** • **STOP** • **STOP** • **STOP** • **STOP** • **STOP** • **STOP**

A12 $^+5$: $^-15 \div {}^-3 = \boxed{^+5}$, because $\boxed{^+5} \cdot {}^-3 = {}^-15$.

Q13 Find the quotient and write the check as a multiplication problem: $^-63 \div {}^-7 = $ _____ because _____ $\cdot {}^-7 = {}^-63$

STOP • **STOP** • **STOP** • **STOP** • **STOP** • **STOP** • **STOP** • **STOP** • **STOP**

A13 $^+9$: $^-63 \div {}^-7 = \underline{^+9}$, because $\underline{^+9} \cdot {}^-7 = {}^-63$

Q14 $^-81 \div {}^-9 = $ _____

STOP • **STOP** • **STOP** • **STOP** • **STOP** • **STOP** • **STOP** • **STOP** • **STOP**

A14 $^+9$ (or simply 9)

6 The quotients of Q12 through Q14 involved integers with the same sign. The answer in each case was a positive integer. The rule suggested for quotients of this type is as follows: *The quotient of two integers with the same sign is a positive integer.* Some examples of the above rule are:

$$\frac{^-36}{^-4} = {}^+9 \qquad {}^-26 \div {}^-2 = {}^+13$$
$$^+42 \div {}^+21 = {}^+2 \qquad \frac{18}{6} = 3$$

Q15 Find the quotient $^-24 \div {}^-6$.

STOP • **STOP** • **STOP** • **STOP** • **STOP** • **STOP** • **STOP** • **STOP** • **STOP**

A15 4

Q16 $\dfrac{^-56}{^-7} = $ _____

STOP • **STOP** • **STOP** • **STOP** • **STOP** • **STOP** • **STOP** • **STOP** • **STOP**

A16 8

Q17 Find each of the following quotients:

a. $\dfrac{^-12}{^-3} = $ _____ b. $^+72 \div {}^-6 = $ _____

c. $^+4 \div {}^-4 = $ _____ d. $^-93 \div {}^-3 = $ _____

e. $\dfrac{^-19}{^-19} = $ _____ f. $14 \div 7 = $ _____

g. $\dfrac{^+24}{^-6} = $ _____ h. $\dfrac{^-24}{^+6} = $ _____

    i. $^-28 \div {}^-7 =$ _____   **j.** $^-1 \div {}^+1 =$ _____

    k. $^-1 \div {}^-1 =$ _____   **l.** $\dfrac{^-2}{^-2} =$ _____

STOP • STOP • STOP • STOP • STOP • STOP • STOP • STOP • STOP

A17   **a.** 4   **b.** $^-12$ **c.** $^-1$ **d.** 31   **e.** 1   **f.** 2   **g.** $^-4$
         h. $^-4$ **i.** 4   **j.** $^-1$ **k.** 1   **l.** 1

---

7   The *quotient* of two integers with *different signs* is a *negative* integer: for example, $^-12 \div {}^+6 = {}^-2$. The *product* of two integers with *different signs* is also a *negative* integer: for example, $^-4 \cdot {}^+6 = {}^-24$.

Q18   The quotient of two integers with the *same sign* is a _____ integer.

STOP • STOP • STOP • STOP • STOP • STOP • STOP • STOP • STOP

A18   positive

Q19   The product of two integers with the *same sign* is a _____ integer.

STOP • STOP • STOP • STOP • STOP • STOP • STOP • STOP • STOP

A19   positive

Q20   **a.** The quotient of two integers with *different signs* is a _____ integer.

         b. The product of two integers with *different signs* is a _____ integer.

STOP • STOP • STOP • STOP • STOP • STOP • STOP • STOP • STOP

A20   **a.** negative   **b.** negative

---

8   The rules for multiplication and division are the same. They may be summarized:

    1. *The product or quotient of two integers with the same sign is positive.*
    2. *The product or quotient of two integers with different signs is negative.*

    (*Note:* The above rules are sometimes abbreviated.)
        In multiplication or division of integers:

    1. Same sign—positive.
    2. Different signs—negative.

    This gives a quick and easy means to remember the signs in a multiplication or division problem.

Q21   Find the following products and quotients:

    a. $^-15 \div {}^-3 =$ _____   **b.** $\dfrac{^+18}{^-3} =$ _____

    c. $^-4 \cdot {}^-8 =$ _____   **d.** $\dfrac{^-14}{^-7} =$ _____

    e. $^+7 \cdot {}^-3 =$ _____   **f.** $^-5 \cdot {}^-9 =$ _____

    g. $\dfrac{^-12}{^+3} =$ _____   **h.** $^+7 \cdot {}^-6 =$ _____

   i. $^-8 \cdot 0 =$ _____   **j.** $\dfrac{^+27}{^-9} =$ _____

   k. $^+48 \div ^-12 =$ _____   **l.** $^-32 \div ^-4 =$ _____

STOP • STOP • STOP • STOP • STOP • STOP • STOP • STOP • STOP

A21   **a.** 5   **b.** $^-6$ **c.** 32   **d.** 2   **e.** $^-21$ **f.** 45   **g.** $^-4$
       h. $^-42$ **i.** 0   **j.** $^-3$ **k.** $^-4$ **l.** 8

---

9   In Section 6.4 the multiplication property of $^-1$ was stated:

$^-1 \cdot a = {}^-a$ is true for all integer replacements of a

According to this property, *negative one times a number is the opposite of the number.* For example, $^-1 \cdot {}^+5 = {}^-5$ and $^-1 \cdot {}^-3 = {}^+3$. Consider the result of dividing an integer by $^-1$ in the following examples:

$$\frac{^+3}{^-1} = {}^-3 \qquad \frac{^-5}{^-1} = {}^+5$$

Q22   When $^+3$ is divided by $^-1$, the quotient is _____.

STOP • STOP • STOP • STOP • STOP • STOP • STOP • STOP • STOP

A22 $^-3$: the opposite of $^+3$

Q23   When $^-5$ is divided by $^-1$, the quotient is _____.

STOP • STOP • STOP • STOP • STOP • STOP • STOP • STOP • STOP

A23 $^+5$: the opposite of $^-5$

---

10   The preceding examples demonstrate that the result of dividing an integer by $^-1$ is the same as when multiplying an integer by $^-1$. In each case, the answer is the opposite of the integer being multiplied or divided. The division property of $^-1$ states that $\dfrac{a}{^-1} = {}^-a$ is true for all integer replacements of a.

Q24   Complete each of the following:

   a. $^-1 \cdot {}^+7 =$ _____   **b.** $\dfrac{^+7}{^-1} =$ _____

   c. $^-1 \cdot {}^-5 =$ _____   **d.** $\dfrac{^-5}{^-1} =$ _____

   e. $^-x = {}^-1 \cdot$ _____   **f.** $\dfrac{x}{^-1} =$ _____

STOP • STOP • STOP • STOP • STOP • STOP • STOP • STOP • STOP

A24   **a.** $^-7$ **b.** $^-7$ **c.** 5   **d.** 5

   e. $^-x = {}^-1 \cdot x$ **f.** $\dfrac{x}{^-1} = {}^-x$

---

11   The number zero is frequently confusing when involved as a divisor or dividend in a division problem. Consider the open sentence

$$0 \div 5 = \square \quad \text{because} \quad \square \cdot 5 = 0$$

The integer that converts the open sentence to a true statement is zero:

$$0 \div 5 = \boxed{0} \quad \text{because} \quad \boxed{0} \cdot 5 = 0$$

Q25   What integer converts the open sentence to a true statement? $0 \div {}^-9 = \square$ because $\square \cdot {}^-9 = 0$

STOP • **STOP** • **STOP** • **STOP** • **STOP** • **STOP** • **STOP** • **STOP** • **STOP**

A25   0: $0 \div {}^-9 = \boxed{0}$, because $\boxed{0} \cdot {}^-9 = 0$

Q26 $0 \div {}^+2 = $ _____, because _____

STOP • **STOP** • **STOP** • **STOP** • **STOP** • **STOP** • **STOP** • **STOP** • **STOP**

A26 $\underline{0}$, because $\underline{0 \cdot {}^+2 = 0}$

Q27 $0 \div {}^-8 = $ _____

STOP • **STOP** • **STOP** • **STOP** • **STOP** • **STOP** • **STOP** • **STOP** • **STOP**

A27   0

---

12   When zero is the divisor, the quotient is said to be *undefined*. To understand why, consider the open sentence

$$4 \div 0 = \square \quad \text{because} \quad \square \cdot 0 = 4$$

There is no answer to the product $\square \cdot 0 = 4$, so there is no answer to the corresponding quotient $4 \div 0 = \square$. That is, $4 \div 0$ is undefined, because *no number* $\cdot 0 = 4$.

Q28 $7 \div 0 = $ _____, because _____

STOP • **STOP** • **STOP** • **STOP** • **STOP** • **STOP** • **STOP** • **STOP** • **STOP**

A28 $\underline{\text{undefined}}$, because $\underline{\text{no number} \cdot 0 = 7}$

Q29 ${}^-3 \div 0 = $ _____

STOP • **STOP** • **STOP** • **STOP** • **STOP** • **STOP** • **STOP** • **STOP** • **STOP**

A29   undefined

---

13   *When zero is the divisor* in a division problem, *the quotient is* said to be *undefined*. That is,

$\dfrac{x}{0} = $ undefined. For example,

$$\dfrac{{}^-3}{0} = \text{undefined} \qquad {}^+12 \div 0 = \text{undefined}$$

When zero is divided by any nonzero integer, the quotient is zero. That is, $\dfrac{0}{x} = 0$ *for* $x \in I$, $x \neq 0$.

For example,

$$\frac{0}{-7} = 0 \qquad 0 \div {}^+6 = 0$$

Q30   Complete each of the following quotients:

a. $0 \div {}^-2 = $ _____   b. ${}^-2 \div 0 = $ _____

c. $\dfrac{0}{+7} = $ _____   d. $\dfrac{7}{0} = $ _____

e. ${}^-3 \div $ _____   f. $10 \div {}^-1 = $ _____

g. $\dfrac{-4}{-2} = $ _____   h. $0 \div {}^-4 = $ _____

i. $\dfrac{42}{-6} = $ _____   j. $\dfrac{8}{0} = $ _____

STOP • STOP • STOP • STOP • STOP • STOP • STOP • STOP • STOP

A30   a. 0   b. undefined   c. 0
      d. undefined   e. undefined   f. $^-10$
      g. 2   h. 0   i. $^-7$
      j. undefined

Q31   Evaluate:

a. ${}^+4 \div {}^-2 = $ _____   b. ${}^-2 \div 4 = $ _____

STOP • STOP • STOP • STOP • STOP • STOP • STOP • STOP • STOP

A31   a. $^-2$ b. $\dfrac{-2}{4} = \dfrac{-1}{2}$

Q32   Is ${}^+4 \div {}^-2 = {}^-2 \div {}^+4$ a true statement?   _____

STOP • STOP • STOP • STOP • STOP • STOP • STOP • STOP • STOP

A32   no

14   The statement $a \div b = b \div a$ is false for most integer replacements of a and b. Therefore, the set of integers is not commutative with respect to the operation of division.

Q33   Evaluate:
a. $({}^-20 \div {}^-10) \div {}^+2$ b. ${}^-20 \div ({}^-10 \div {}^+2)$

STOP • STOP • STOP • STOP • STOP • STOP • STOP • STOP • STOP

A33   a. 1: $({}^-20 \div {}^-10) \div {}^+2 = {}^+2 \div {}^+2$
                                      $= 1$
      b. 4: ${}^-20 \div ({}^-10 \div {}^+2) = {}^-20 \div {}^-5$
                                      $= 4$

Q34 Is $(^-20 \div {}^-10) \div {}^+2 = {}^-20 \div (^-10 \div {}^+2)$ a true statement? _____

STOP • STOP • STOP • STOP • STOP • STOP • STOP • STOP • STOP

A34 no

---

15 The statement $(a \div b) \div c = a \div (b \div c)$ is false for most integer replacements of a, b, and c. Therefore, the set of integers is not associative with respect to the operation of division.

Q35 Complete the following:

a. The set of integers _____ commutative with respect to multiplication.
   is/is not

b. The set of integers _____ commutative with respect to division.
   is/is not

c. $(^-16 \div {}^+4) \div {}^+2 = {}^-16 \div (^+4 \div {}^+2)$ _____ a true statement.
   is/is not

STOP • STOP • STOP • STOP • STOP • STOP • STOP • STOP • STOP

A35 a. is   b. is not   c. is not

This completes the instruction for this section.

6.5 Exercises

1. Find each of the following quotients:

   a. $72 \div {}^-9$ b. $\dfrac{^-18}{2}$ c. $^-15 \div {}^-5$ d. $^-5 \div {}^-5$

   e. $0 \div {}^-7$ f. $\dfrac{^-24}{^-6}$ g. $\dfrac{10}{0}$ h. $\dfrac{0}{^-2}$

   i. $^-13 \div {}^-1$ j. $\dfrac{^-8}{^-8}$

2. Find each of the following products and quotients:

   a. $\dfrac{^-51}{3}$ b. $\dfrac{0}{^-12}$ c. $^-12 \cdot {}^-5$ d. $\dfrac{^-45}{^-3}$

   e. $\dfrac{5}{^-1}$ f. $\dfrac{4}{0}$ g. $^-5 \cdot 0$ h. $^-4 \cdot 0$

   i. $0 \div 3$ j. $^-1 \cdot 3$ k. $\dfrac{0}{^-5}$ l. $^-6 \cdot {}^-7$

   m. $\dfrac{^-75}{^-15}$ n. $7 \cdot {}^-9$

o. $^-3 \cdot 0$ **p.** $\dfrac{^-8}{^-2}$ **q.** $^-1 \cdot 6$ **r.** $^-32 \div 4$

s. $^-8 \cdot {}^-1$ **t.** $7 \cdot 0$

3. The product or quotient of two integers with the same signs is a _____ integer.
4. The product or quotient of two integers with different signs is a _____ integer.
5. Answer true or false: The set of integers is
   a. associative for division
   b. commutative for multiplication
   c. commutative for division
   d. associative for multiplication

6.5   Exercise Answers

1. **a.** $^-8$ **b.** $^-9$ **c.** 3   **d.** 1
   e. 0   **f.** 4   **g.** undefined   **h.** 0
   i. 13   **j.** 1
2. **a.** $^-17$ **b.** 0   **c.** 60   **d.** 15
   e. $^-5$ **f.** undefined   **g.** 0   **h.** 0
   i. 0   **j.** $^-3$ **k.** 0   **l.** 42
   m. 5   **n.** $^-63$ **o.** 0   **p.** 4
   q. $^-6$ **r.** $^-8$ **s.** 8   **t.** 0
3. positive
4. negative
5. **a.** false   **b.** true   **c.** false   **d.** true

6.6   Operations with Rational Numbers

1

The following number sets have been studied previously:

$N = \{1, 2, 3, \ldots\}$ natural numbers
$W = \{0, 1, 2, \ldots\}$ whole numbers
$I = \{\ldots, {}^-2, {}^-1, 0, 1, 2, \ldots\}$ integers

The set of rational numbers is an extension of the set of integers. Because it is impossible to list the set of rational numbers, a verbal description is usually given as a means of defining the set. A rational number is defined as follows: *A rational number is any number which can be written in the form* $\dfrac{p}{q}$, *where p and q are integers and* $q \neq 0$. Some examples of rational numbers are:

$$\frac{2}{3} \qquad \frac{^-5}{7} \qquad 4 \qquad ^-2 \qquad 0 \qquad 3\frac{1}{7}$$

$\dfrac{2}{3}$ and $\dfrac{^-5}{7}$ are rational numbers, because they are written in the form $\dfrac{p}{q}$ and their denominators (q) are not zero. Integers, such as 4, $^-2$, and 0, are rational numbers, because they can be written in the form $\dfrac{p}{q}$ by using 1 as their denominators. That is,

$4 = \frac{4}{1}$, $^-2 = \frac{^-2}{1}$, and $0 = \frac{0}{1}$. $3\frac{1}{7}$ is a rational number, because it can be written in the form $\frac{p}{q}$ using its improper fraction equivalent. That is, $3\frac{1}{7} = \frac{22}{7}$.

Q1   Which of the following are rational numbers?

$^-5, \frac{3}{0}, \frac{1}{2}, 2\frac{1}{6}$ _____

STOP • STOP • STOP • STOP • STOP • STOP • STOP • STOP • STOP

A1 $^-5, \frac{1}{2}$, and $2\frac{1}{6}$: $\frac{3}{0}$ is not a rational number, because its denominator is zero and its value is thus undefined.

Q2   Write in the $\frac{p}{q}$ form:

   a. $^-5 =$ ____   **b.** $\frac{1}{2} =$ ____   **c.** $2\frac{1}{6} =$ ____

STOP • STOP • STOP • STOP • STOP • STOP • STOP • STOP • STOP

A2   **a.** $\frac{^-5}{1}$ **b.** $\frac{1}{2}$ **c.** $\frac{13}{6}$

---

2   By its definition, the set of rational numbers includes many different types of numbers. It includes positive and negative *fractions*, both proper and improper, because they are already in the required $\frac{p}{q}$ form. It includes positive and negative *mixed numbers*, because any mixed number can be rewritten as an improper fraction in the required $\frac{p}{q}$ form. Finally, the set of rational numbers includes the natural numbers, the whole numbers, and the integers, because each of the elements of these sets can be written in the $\frac{p}{q}$ form, using the number 1 as the denominator q. As are the sets N, W, and I, the set of rational numbers is an infinite set.

Q3   Write each rational number in the $\frac{p}{q}$ form:

   a. $^-3 =$ ____   **b.** $4\frac{2}{7} =$ ____   **c.** $0 =$ ____

STOP • STOP • STOP • STOP • STOP • STOP • STOP • STOP • STOP

A3   **a.** $\frac{^-3}{1}$ **b.** $\frac{30}{7}$ **c.** $\frac{0}{1}$

---

3   A positive mixed number can be written as the sum of its whole-number and fraction parts. For example, $3\frac{2}{5} = 3 + \frac{2}{5}$. The improper-fraction form is found:

$$3\frac{2}{5} = \frac{3 \cdot 5 + 2}{5}$$

$$= \frac{15 + 2}{5}$$

$$= \frac{17}{5}$$

A similar procedure can be used with negative mixed numbers:

$$-3\frac{1}{2} = {}^{-}\left(3\frac{1}{2}\right)$$

$$= {}^{-}\left(\frac{3 \cdot 2 + 1}{2}\right)$$

$$= {}^{-}\left(\frac{7}{2}\right)$$

$$= \frac{-7}{2} \quad \text{(read ``negative seven halves'')}$$

Q4   Find the improper-fraction form:

   a. $-4\frac{1}{7}$

   b. $-2\frac{2}{9}$

STOP • STOP • STOP • STOP • STOP • STOP • STOP • STOP • STOP

A4   a. $\frac{-29}{7}$: $-4\frac{1}{7} = {}^{-}\left(4\frac{1}{7}\right)$

$$= {}^{-}\left(\frac{4 \cdot 7 + 1}{7}\right)$$

$$= {}^{-}\left(\frac{29}{7}\right)$$

$$= \frac{-29}{7}$$

   b. $\frac{-20}{9}$: $-2\frac{2}{9} = {}^{-}\left(2\frac{2}{9}\right)$

$$= {}^{-}\left(\frac{2 \cdot 9 + 2}{9}\right)$$

$$= {}^{-}\left(\frac{20}{9}\right)$$

$$= \frac{-20}{9}$$

Q5   Write each of the rational numbers in the $\frac{p}{q}$ form:

   a. $7 = $ _____

   b. $-5\frac{1}{4} = $ _____

   c. $\frac{2}{3} = $ _____

   d. $-4 = $ _____

   e. $4\frac{1}{2} = $ _____

   f. $-2\frac{4}{5} = $ _____

STOP • STOP • STOP • STOP • STOP • STOP • STOP • STOP • STOP

A5   a. $\frac{7}{1}$

   b. $\frac{-21}{4}$

   c. $\frac{2}{3}$

   d. $\frac{-4}{1}$

   e. $\frac{9}{2}$

   f. $\frac{-14}{5}$

4   The number line is used to graph rational numbers in a manner similar to that used with integers. The rational numbers $\frac{3}{4}$ and $^-4\frac{2}{3}$ and their opposites are graphed on the following number line:

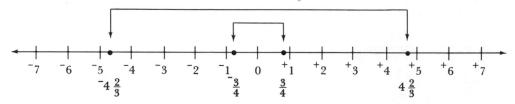

Q6   Graph the rational numbers and their opposites:

a. $^-2\frac{1}{2}$ b. $1\frac{2}{3}$ c. $^-4$ d. 0

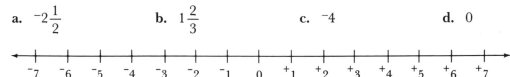

STOP   •   STOP   •   STOP   •   STOP   •   STOP   •   STOP   •   STOP   •   STOP   •   STOP

A6

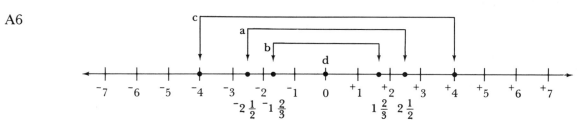

Note that zero is its own opposite.

5   In Section 3.5, procedures were developed for converting fractions into their decimal equivalents.* It is also possible to describe a *rational number* in terms of its decimal representation: *A rational number is any number whose decimal representation is either a terminating or a repeating decimal.*

Examples:

$\frac{1}{2} = 0.5$ terminating

$\frac{1}{3} = 0.\overline{3}\ (0.3333\cdots)$ repeating

$\frac{^-5}{8} = ^-0.625$ terminating

$\frac{2}{7} = 0.\overline{285714}\ (0.285714285714\cdots)$ repeating

$^-4 = ^-4.0$ terminating

The decimal representation of a rational number in $\frac{p}{q}$ form is found by dividing the numerator (p) by the denominator (q) and then applying the sign, if the rational number is negative.

*If necessary, see Section 3.5 for a review of these procedures.

Example 1: Find the decimal representation of $\frac{-3}{4}$.

Solution

```
  0.75
4)3.00
  2 8
  ──
    20
    20
    ──
     0
```

Thus, $\frac{-3}{4} = {}^-0.75$. The decimal is *terminating* since its division comes out evenly (0 remainder).

Example 2: Find the decimal representation of $2\frac{5}{33}$.

Solution

$$2\frac{5}{33} = \frac{71}{33}$$

```
      2.1515···(repeating 15s)
33)71.0000
   66
   ──
    50
    33
    ──
    170
    165
    ───
      50
      33
      ──
      170
      165
      ───
        5   etc.
```

Thus, $2\frac{5}{33} = 2.\overline{15}$. The decimal is infinite because its division does not come out evenly; it is *repeating* because it repeats itself in blocks of the digits 15. That is, $2.\overline{15} = 2.151515\cdots$ (repeating 15s).

Q7   Find the decimal representation and state whether each is terminating or repeating:

a. $\frac{-2}{3} =$ _____   _____

b. $2\dfrac{1}{5} =$ _____ _____

c. $5 =$ _____ _____

d. $\dfrac{^-7}{9} =$ _____ _____

e. $^-3\dfrac{1}{7} =$ _____ _____

f. $\dfrac{5}{8} =$ _____ _____

g. $\dfrac{^-7}{11} =$ _____ _____

STOP • **STOP** • **STOP** • **STOP** • **STOP** • **STOP** • **STOP** • **STOP** • **STOP**

A7   **a.** $^-0.\overline{6}$ repeating   **b.** 2.2, terminating
           c. 5.0, terminating   **d.** $^-0.\overline{7}$, repeating
           e. $^-3.\overline{142857}$, repeating   **f.** 0.625, terminating
           g. $^-0.\overline{63}$, repeating

---

6   Since the set of fractions is the set of nonnegative rationals, the procedure used in the addition of rational numbers is much the same as that used for the addition of fractions.

    To add two rational numbers:

Step 1: If the denominators are different, find a common denominator (see Section 2.4 for a review of LCDs).

Step 2: Write the sum of the numerators over the common denominator.

Step 3: Apply the rules for the addition of integers to the numerator.

Step 4: If possible, reduce the rational number to lowest terms.

Examples:

1. $\dfrac{3}{32} + \dfrac{^-11}{32} = \dfrac{3 + {^-11}}{32}$ Step 2   (step 1 is not necessary)

   $= \dfrac{^-8}{32}$ Step 3

   $= \dfrac{^-1}{4}$ Step 4

2. $\dfrac{^-5}{8} + \dfrac{3}{7} = \dfrac{^-35}{56} + \dfrac{24}{56}$ Step 1 $\left(\text{because } \dfrac{^-5 \times 7}{8 \times 7} = \dfrac{^-35}{56} \text{ and } \dfrac{3 \times 8}{7 \times 8} = \dfrac{24}{56}\right)$

   $= \dfrac{^-35 + 24}{56}$ Step 2

   $= \dfrac{^-11}{56}$ Step 3

Q8   Find the sum:

a. $\dfrac{^-7}{8} + \dfrac{3}{8}$ b. $\dfrac{^-1}{6} + \dfrac{4}{5}$

STOP • STOP • STOP • STOP • STOP • STOP • STOP • STOP • STOP

A8   a. $\dfrac{^-1}{2}$: $\dfrac{^-7}{8} + \dfrac{3}{8} = \dfrac{^-7 + 3}{8}$ b. $\dfrac{19}{30}$: $\dfrac{^-1}{6} + \dfrac{4}{5}$

   $= \dfrac{^-4}{8}$ $\dfrac{^-5}{30} + \dfrac{24}{30}$

   $= \dfrac{^-1}{2}$ $\dfrac{19}{30}$

Q9   Find the sum:

a. $\dfrac{^-2}{3} + \dfrac{^-5}{9}$ b. $\dfrac{5}{6} + \dfrac{^-7}{6}$

c. $\dfrac{^-2}{15} + \dfrac{19}{15}$ d. $\dfrac{7}{12} + \dfrac{4}{7}$

e. $\dfrac{5}{14} + \dfrac{-3}{4}$

f. $\dfrac{-7}{16} + \dfrac{-1}{4}$

g. $\dfrac{7}{9} + \dfrac{-4}{12}$

h. $\dfrac{-1}{8} + \dfrac{-4}{15}$

STOP • **STOP** • **STOP** • **STOP** • **STOP** • **STOP** • **STOP** • **STOP** • **STOP**

A9   a. $-1\dfrac{2}{9}$ b. $\dfrac{-1}{3}$ c. $1\dfrac{2}{15}$ d. $1\dfrac{13}{84}$

e. $\dfrac{-11}{28}$ f. $\dfrac{-11}{16}$ g. $\dfrac{4}{9}$ h. $\dfrac{-47}{120}$

7

To find a sum of two or more rational numbers where integers or mixed numbers are involved, the following procedures will be used:

1. Write all integers in $\dfrac{p}{q}$ form using a denominator of 1.

2. Write all mixed numbers as improper fractions in $\dfrac{p}{q}$ form.

3. Follow the rules for adding two rational numbers as previously stated, working from left to right.

Example 1: Find the sum $-1\dfrac{1}{5} + 3\dfrac{2}{5}$.

Solution

$$-1\dfrac{1}{5} + 3\dfrac{2}{5} = \dfrac{-6}{5} + \dfrac{17}{5}$$

$$= \dfrac{-6 + 17}{5}$$

$$= \dfrac{11}{5}$$

$$= 2\dfrac{1}{5}$$

Example 2: Find the sum $2\dfrac{4}{7} + {}^-5$.

Solution

$$2\dfrac{4}{7} + {}^-5 = \dfrac{18}{7} + \dfrac{-5}{1}$$

$$= \dfrac{18}{7} + \dfrac{-35}{7}$$

$$= \dfrac{18 + {}^-35}{7}$$

$$= \dfrac{-17}{7}$$

$$= {}^-2\dfrac{3}{7}$$

Q10   Find the sum:

a. $^-2\dfrac{3}{5} + 8$

b. $^-5\dfrac{2}{7} + \dfrac{3}{5}$

STOP • STOP • STOP • STOP • STOP • STOP • STOP • STOP • STOP

A10   a. $5\dfrac{2}{5}$: $^-2\dfrac{3}{5} + 8 = \dfrac{^-13}{5} + \dfrac{8}{1}$

$= \dfrac{^-13}{5} + \dfrac{40}{5}$

$= \dfrac{27}{5}$

$= 5\dfrac{2}{5}$

b. $^-4\dfrac{24}{35}$: $^-5\dfrac{2}{7} + \dfrac{3}{5} = \dfrac{^-37}{7} + \dfrac{3}{5}$

$= \dfrac{^-185}{35} + \dfrac{21}{35}$

$= \dfrac{^-185 + 21}{35}$

$= \dfrac{^-164}{35}$

$= ^-4\dfrac{24}{35}$

Q11   Find the sum:

a. $2\dfrac{7}{8} + ^-6\dfrac{3}{8}$

b. $^-7\dfrac{2}{3} + 4$

c. $6\dfrac{1}{5} + ^-3\dfrac{4}{7}$

d. $^-1\dfrac{2}{3} + ^-4\dfrac{5}{6}$

e. $\dfrac{5}{9} + {}^-3\dfrac{1}{2}$

f. $\dfrac{{}^-5}{7} + {}^-3$

g. ${}^-8 + \dfrac{3}{11}$

h. ${}^-4\dfrac{3}{8} + 5$

i. $\dfrac{2}{3} + \dfrac{{}^-5}{8} + \dfrac{{}^-5}{6}$

j. ${}^-3\dfrac{1}{7} + {}^-2\dfrac{5}{8}$

STOP • **STOP** • **STOP** • **STOP** • **STOP** • **STOP** • **STOP** • **STOP** • **STOP**

A11

a. ${}^-3\dfrac{1}{2}$

b. ${}^-3\dfrac{2}{3}$

c. $2\dfrac{22}{35}$

d. ${}^-6\dfrac{1}{2}$

e. ${}^-2\dfrac{17}{18}$

f. ${}^-3\dfrac{5}{7}$

g. ${}^-7\dfrac{8}{11}$

h. $\dfrac{5}{8}$

i. $\dfrac{{}^-19}{24}$

j. ${}^-5\dfrac{43}{56}$

Q12   Write the opposite for each rational number:

a. $\dfrac{2}{3}$ _____

b. $\dfrac{{}^-5}{8}$ _____

c.   7 _____

d. ${}^-4$ _____

e. $1\dfrac{2}{3}$ _____

f. ${}^-5\dfrac{4}{7}$ _____

STOP • **STOP** • **STOP** • **STOP** • **STOP** • **STOP** • **STOP** • **STOP** • **STOP**

A12

a. $\dfrac{-2}{3}$

b. $\dfrac{5}{8}$

c. $^-7$

d. 4

e. $^-1\dfrac{2}{3}$

f. $5\dfrac{4}{7}$

8

The definition of subtraction for integers is also true for the set of rational numbers. That is, $x - y = x +\ ^-y$ *for all rational-number replacements of x and y.*

To subtract any two rational numbers:

Step 1: Use the definition of subtraction to rewrite the difference as a sum.

Step 2: Apply the procedures previously developed to find the sum.

Example 1: Find the difference $\dfrac{4}{5} - \dfrac{7}{12}$.

Solution

$$\dfrac{4}{5} - \dfrac{7}{12} = \dfrac{4}{5} + \dfrac{-7}{12}$$

$$= \dfrac{48}{60} + \dfrac{-35}{60}$$

$$= \dfrac{13}{60}$$

Example 2: Find the difference $\dfrac{-4}{7} - \dfrac{-5}{6}$.

Solution

$$\dfrac{-4}{7} - \dfrac{-5}{6} = \dfrac{-4}{7} + \dfrac{5}{6}$$

$$= \dfrac{-24}{42} + \dfrac{35}{42}$$

$$= \dfrac{11}{42}$$

Example 3: Find the difference $^-7\dfrac{1}{3} -\ ^-2$.

Solution

$$^-7\dfrac{1}{3} -\ ^-2 =\ ^-7\dfrac{1}{3} + 2$$

$$= \dfrac{-22}{3} + \dfrac{2}{1}$$

$$= \dfrac{-22}{3} + \dfrac{6}{3}$$

$$= \dfrac{-16}{3}$$

$$=\ ^-5\dfrac{1}{3}$$

Q13 Rewrite each difference as a sum (do not evaluate):

a. $\dfrac{1}{3} - \dfrac{4}{9} =$ _____

b. $\dfrac{^{-}3}{5} - \dfrac{^{-}2}{3} =$ _____

c. $^{-}4\dfrac{2}{3} - 1\dfrac{2}{5} =$ _____

d. $3 - {}^{-}1\dfrac{3}{10} =$ _____

STOP • **STOP** • **STOP** • **STOP** • **STOP** • **STOP** • **STOP** • **STOP** • **STOP**

A13 a. $\dfrac{1}{3} + \dfrac{^{-}4}{9}$ b. $\dfrac{^{-}3}{5} + \dfrac{2}{3}$ c. $^{-}4\dfrac{2}{3} + {}^{-}1\dfrac{2}{5}$ d. $3 + 1\dfrac{3}{10}$

Q14 Find the difference:

a. $\dfrac{1}{3} - \dfrac{4}{9}$

b. $\dfrac{^{-}3}{5} - \dfrac{^{-}2}{3}$

c. $^{-}4\dfrac{2}{3} - 1\dfrac{2}{7}$

d. $3 - {}^{-}1\dfrac{3}{10}$

STOP • **STOP** • **STOP** • **STOP** • **STOP** • **STOP** • **STOP** • **STOP** • **STOP**

A14

a. $\dfrac{^{-}1}{9}$: $\dfrac{1}{3} - \dfrac{4}{9} = \dfrac{1}{3} + \dfrac{^{-}4}{9}$

$= \dfrac{3}{9} + \dfrac{^{-}4}{9}$

$= \dfrac{^{-}1}{9}$

b. $\dfrac{1}{15}$: $\dfrac{^{-}3}{5} - \dfrac{^{-}2}{3} = \dfrac{^{-}3}{5} + \dfrac{2}{3}$

$= \dfrac{^{-}9}{15} + \dfrac{10}{15}$

$= \dfrac{1}{15}$

c. $^{-}5\dfrac{20}{21}$: $^{-}4\dfrac{2}{3} - 1\dfrac{2}{7} = {}^{-}4\dfrac{2}{3} + {}^{-}1\dfrac{2}{7}$

$= \dfrac{^{-}14}{3} + \dfrac{^{-}9}{7}$

$= \dfrac{^{-}98}{21} + \dfrac{^{-}27}{21}$

$= \dfrac{^{-}125}{21}$

$= {}^{-}5\dfrac{20}{21}$

d. $4\dfrac{3}{10}$: $3 - {}^{-}1\dfrac{3}{10} = 3 + 1\dfrac{3}{10}$

$= \dfrac{3}{1} + \dfrac{13}{10}$

$= \dfrac{30}{10} + \dfrac{13}{10}$

$= \dfrac{43}{10}$

$= 4\dfrac{3}{10}$

Q15   Find the difference:

a. $\dfrac{^-5}{24} - \dfrac{7}{24}$ b. $2\dfrac{3}{16} - 1\dfrac{5}{16}$ c. $\dfrac{^-7}{21} - \dfrac{^-5}{7}$ d. $^-2\dfrac{1}{9} - 3\dfrac{2}{5}$

e. $5\dfrac{3}{8} - 7\dfrac{4}{9}$ f. $\dfrac{^-3}{16} - 4$ g. $18 - \dfrac{5}{17}$ h. $\dfrac{^-6}{13} - 4\dfrac{1}{2}$

i. $16\dfrac{3}{10} - 5\dfrac{7}{8}$ j. $^-15\dfrac{1}{6} - {}^-2\dfrac{7}{15}$

STOP • STOP • STOP • STOP • STOP • STOP • STOP • STOP • STOP

A15   a. $\dfrac{^-1}{2}$ b. $\dfrac{7}{8}$ c. $\dfrac{8}{21}$ d. $^-5\dfrac{23}{45}$

e. $^-2\dfrac{5}{72}$ f. $^-4\dfrac{3}{16}$ g. $17\dfrac{12}{17}$ h. $^-4\dfrac{25}{26}$

i. $10\dfrac{17}{40}$ j. $^-12\dfrac{7}{10}$

9

Let $\frac{a}{b}$ and $\frac{c}{d}$ stand for any two rational numbers. The product of $\frac{a}{b}$ and $\frac{c}{d}$ is defined as:

$$\frac{a}{b} \cdot \frac{c}{d} = \frac{ac}{bd}, \qquad bd \neq 0$$

where ac is the product of the numerators and bd is the product of the denominators.

Example 1: Find the product $\frac{-3}{7} \cdot \frac{-5}{4}$.

Solution

$$\frac{-3}{7} \cdot \frac{-5}{4} = \frac{-3 \cdot -5}{7 \cdot 4} = \frac{15}{28}$$

As with fractions, a common procedure when multiplying rational numbers is to divide any numerator and any denominator by a common factor. This procedure is called *reducing* (sometimes referred to as "canceling").

Example 2: Find the product $\frac{-3}{5} \cdot \frac{5}{12}$.

Solution

$$\frac{\overset{-1}{\cancel{-3}}}{\underset{1}{\cancel{5}}} \cdot \frac{\overset{1}{\cancel{5}}}{\underset{4}{\cancel{12}}} = \frac{-1}{4}$$

Example 3: Find the product $\frac{-9}{12} \cdot \frac{-5}{7}$.

Solution

$$\frac{\overset{-3}{\cancel{-9}}}{\underset{4}{\cancel{12}}} \cdot \frac{-5}{7} = \frac{15}{28}$$

As is the case with the sum and difference of two rational numbers, the product of two rational numbers is always reduced to lowest terms.

Q16   Find the product:

    a. $\dfrac{-6}{5} \cdot \dfrac{-2}{3}$ **b.** $\dfrac{15}{16} \cdot \dfrac{-4}{25}$

STOP • **STOP** • **STOP** • **STOP** • **STOP** • **STOP** • **STOP** • **STOP** • **STOP**

A16   **a.** $\dfrac{4}{5}$: $\dfrac{\overset{-2}{\cancel{-6}}}{5} \cdot \dfrac{-2}{\underset{1}{\cancel{3}}} = \dfrac{4}{5}$ **b.** $\dfrac{-3}{20}$: $\dfrac{\overset{3}{\cancel{15}}}{\underset{4}{\cancel{16}}} \cdot \dfrac{\overset{-1}{\cancel{-4}}}{\underset{5}{\cancel{25}}} = \dfrac{-3}{20}$

10 $\dfrac{^-2}{3}$ and $\dfrac{2}{^-3}$ are equivalent forms for the same rational number. Both represent a negative rational number since they involve a quotient of integers with unlike signs. When writing a negative rational number, it is customary to place the negative sign on the number in the numerator. For example, $\dfrac{5}{^-8}$ is usually written $\dfrac{^-5}{8}$.

Q17   Find the product:

a. $\dfrac{^-12}{15} \cdot \dfrac{3}{6}$

b. $\dfrac{^-4}{6} \cdot \dfrac{^-13}{18}$

c. $\dfrac{7}{12} \cdot \dfrac{9}{^-11}$

d. $\dfrac{^-5}{7} \cdot \dfrac{6}{^-13}$

e. $\dfrac{5}{^-28} \cdot \dfrac{^-16}{17}$

f. $\dfrac{4}{^-7} \cdot \dfrac{7}{4}$

STOP • **STOP** • **STOP** • **STOP** • **STOP** • **STOP** • **STOP** • **STOP** • **STOP**

A17   a. $\dfrac{^-2}{5}$ b. $\dfrac{13}{27}$ c. $\dfrac{^-21}{44}$ d. $\dfrac{30}{91}$ e. $\dfrac{20}{119}$ f. $^-1$

11   When finding a product of two or more rational numbers where integers or mixed numbers are involved:

1.   Write all integers in $\dfrac{p}{q}$ form with a denominator of 1.

2.   Write all mixed numbers in improper fractions in $\dfrac{p}{q}$ form.

3.   Use the rule of Frame 9 to find the product.

Example 1: Find the product $5\dfrac{2}{9} \cdot {}^-3$.

Solution

$$5\dfrac{2}{9} \cdot {}^-3 = \dfrac{47}{\cancel{9}_{3}} \cdot \dfrac{\cancel{^-3}^{^-1}}{1}$$

$$= \dfrac{^-47}{3}$$

$$= {}^-15\dfrac{2}{3}$$

Example 2: Find the product $^-5\frac{1}{3}\cdot{}^-3\frac{9}{16}$.

Solution

$$^-5\frac{1}{3}\cdot{}^-3\frac{9}{16}=\frac{\overset{-1}{\cancel{-16}}}{\underset{1}{\cancel{3}}}\cdot\frac{\overset{-19}{\cancel{-57}}}{\underset{1}{\cancel{16}}}$$

$$=19$$

Q18   Find the product:

a. $\dfrac{-3}{11}\cdot 3$

b. $\dfrac{4}{-5}\cdot\dfrac{-10}{12}$

c. $^-5\dfrac{2}{9}\cdot{}^-12$

d. $2\dfrac{1}{3}\cdot{}^-1\dfrac{3}{4}$

e. $^-4\cdot\dfrac{-1}{4}$

f. $3\dfrac{1}{5}\cdot{}^-1\dfrac{1}{2}$

STOP • STOP • STOP • STOP • STOP • STOP • STOP • STOP • STOP

A18   a. $\dfrac{-9}{11}$ b. $\dfrac{2}{3}$ c. $62\dfrac{2}{3}$ d. $^-4\dfrac{1}{12}$ e. 1   f. $^-4\dfrac{4}{5}$

12   Two rational numbers whose product is 1 are called *reciprocals*. Thus, 5 and $\dfrac{1}{5}$ are recipro-

cals, because $5\cdot\dfrac{1}{5}=\dfrac{\overset{1}{\cancel{5}}}{1}\cdot\dfrac{1}{\underset{1}{\cancel{5}}}=\dfrac{1}{1}=1$. Similarly, $\dfrac{-5}{8}$ and $\dfrac{8}{-5}$ are reciprocals, because

$$\frac{\overset{1}{\cancel{-5}}}{\underset{1}{\cancel{8}}}\cdot\frac{\overset{1}{\cancel{8}}}{\underset{1}{\cancel{-5}}}=\frac{1}{1}=1$$

In general, to find the reciprocal of any nonzero rational number, write the number in $\frac{p}{q}$ form and interchange the numerator and denominator to form $\frac{q}{p}$.

Example 1: Write the reciprocal for $^-2\frac{1}{3}$.

Solution

$-2\frac{1}{3} = \frac{-7}{3}$;   thus, the reciprocal of $^-2\frac{1}{3}$ is $\frac{3}{-7}$, which is written $\frac{-3}{7}$

Example 2: Write the reciprocal for 3.

Solution

$3 = \frac{3}{1}$;   thus, the reciprocal of 3 is $\frac{1}{3}$

Zero has no reciprocal, because $0 = \frac{0}{1}$ and $\frac{1}{0}$ is undefined.

Q19   Write the reciprocal:

a. $\frac{2}{3}$ _____   **b.** $^-1\frac{5}{9}$ _____   **c.** 12 _____   **d.** $\frac{1}{13}$ _____

e. $7\frac{3}{8}$ _____   **f.** $^-2$ _____

STOP • STOP • STOP • STOP • STOP • STOP • STOP • STOP • STOP

A19   **a.** $\frac{3}{2}$ **b.** $\frac{-9}{14}$ **c.** $\frac{1}{12}$ **d.** 13   **e.** $\frac{8}{59}$ **f.** $\frac{-1}{2}$

13   Let $\frac{a}{b}$ and $\frac{c}{d}$ stand for any two rational numbers with $\frac{c}{d} \neq 0$. Consider the simplification of the following quotient:

$$\frac{a}{b} \div \frac{c}{d} = \frac{\dfrac{a}{b}}{\dfrac{c}{d}} = \frac{\dfrac{a}{b} \cdot \dfrac{d}{c}}{\dfrac{c}{d} \cdot \dfrac{d}{c}} = \frac{\dfrac{a}{b} \cdot \dfrac{d}{c}}{1} = \frac{a}{b} \cdot \frac{d}{c}$$

Thus, the *definition of division for two rational numbers* (with a nonzero divisor) is stated:

$$\frac{a}{b} \div \frac{c}{d} = \frac{a}{b} \cdot \frac{d}{c}$$

$$= \frac{ad}{bc}, \qquad \frac{c}{d} \neq 0$$

In words, the quotient of two rational numbers with a nonzero divisor is equal to the product of the first (dividend) times the reciprocal of the second (divisor).

Example 1: Find the quotient $\frac{2}{3} \div \frac{4}{9}$.

Solution

$$\frac{2}{3} \div \frac{4}{9} = \frac{\overset{1}{\cancel{2}}}{\underset{1}{\cancel{3}}} \cdot \frac{\overset{3}{\cancel{9}}}{\underset{2}{\cancel{4}}} \qquad \left(\text{the reciprocal of } \frac{4}{9} \text{ is } \frac{9}{4}\right)$$

$$= \frac{3}{2}$$

$$= 1\frac{1}{2}$$

Example 2: Find the quotient $\frac{4}{5} \div \frac{^-6}{7}$.

Solution

$$\frac{4}{5} \div \frac{^-6}{7} = \frac{4}{5} \cdot \frac{^-7}{6} \qquad \left(\text{the reciprocal of } \frac{^-6}{7} \text{ is } \frac{^-7}{6}\right)$$

$$= \frac{^-14}{15}$$

14

When finding a quotient of two rational numbers where integers or mixed numbers are involved:

1. Write all integers in $\frac{p}{q}$ form with a denominator of 1.

2. Write all mixed numbers as improper fractions in $\frac{p}{q}$ form.

3. Use the rule of the previous frame to find the quotient.

Example 1: Find the quotient $\frac{^-5}{14} \div 3\frac{2}{7}$.

Solution

$$\frac{^-5}{14} \div 3\frac{2}{7} = \frac{^-5}{14} \div \frac{23}{7}$$

$$= \frac{^-5}{14} \cdot \frac{7}{23}$$

$$= \frac{^-5}{46}$$

Example 2: Find the quotient $^-2\frac{5}{8} \div 12$.

Solution

$$^-2\frac{5}{8} \div 12 = \frac{^-21}{8} \div \frac{12}{1}$$

$$= \frac{\overset{^-7}{\cancel{^-21}}}{8} \cdot \frac{1}{\underset{4}{\cancel{12}}}$$

$$= \frac{^-7}{32}$$

Q20   Find the quotient:

a. $\dfrac{^-11}{15} \div {}^-4\dfrac{2}{5}$ b. $^-8 \div 2\dfrac{2}{3}$

STOP • STOP • STOP • STOP • STOP • STOP • STOP • STOP • STOP

A20   a. $\dfrac{1}{6}$: $\dfrac{^-11}{15} \div {}^-4\dfrac{2}{5} = \dfrac{^-11}{15} \div \dfrac{^-22}{5}$

$$= \dfrac{^-\overset{-1}{\cancel{11}}}{\underset{3}{\cancel{15}}} \cdot \dfrac{^-\overset{-1}{\cancel{5}}}{\underset{2}{\cancel{22}}}$$

$$= \dfrac{1}{6}$$

b. $^-3$: $^-8 \div 2\dfrac{2}{3} = \dfrac{^-8}{1} \div \dfrac{8}{3}$

$$= \dfrac{^-8}{1} \cdot \dfrac{3}{8}$$

$$= {}^-3$$

Q21   Find the quotient:

a. $\dfrac{^-29}{50} \div 3\dfrac{1}{10}$ b. $^-4\dfrac{1}{5} \div {}^-3\dfrac{1}{3}$

c. $^-4\dfrac{1}{5} \div 3$ d. $\dfrac{1}{2} \div {}^-2$

e. $4 \div 4\dfrac{5}{8}$ f. $^-8 \div \dfrac{^-1}{8}$

g. $0 \div 3\frac{4}{7}$

h. $^{-}15 \div {}^{-}2\frac{5}{8}$

STOP • STOP • STOP • STOP • STOP • STOP • STOP • STOP • STOP

A21 **a.** $\dfrac{^{-}29}{155}$ **b.** $1\dfrac{13}{50}$ **c.** $^{-}1\dfrac{2}{5}$ **d.** $\dfrac{^{-}1}{4}$

e. $\dfrac{32}{37}$ **f.** 64 **g.** 0 **h.** $5\dfrac{5}{7}$

| 15 | By the definition of division, it is possible to find the quotient *only* if the divisor is not zero. Consider why this is so. The quotient of two rational numbers is found by multiplying the dividend by the reciprocal of the divisor. Since zero has no reciprocal, it is impossible to complete the quotient. Thus, the impossibility of division by zero in the set of rational numbers is consistent with the impossibility of division by zero in the set of integers. In other words, $\dfrac{a}{0}$ or $a \div 0$ is *undefined* for any rational-number replacement of a. |

Q22 **a.** What is the reciprocal of 0? _____

b. Is it possible to find $\dfrac{2}{3} \div 0$? _____

c. Why? _____

STOP • STOP • STOP • STOP • STOP • STOP • STOP • STOP • STOP

A22 **a.** There is none.
b. no

c. $\dfrac{2}{3} \div 0$ is undefined since division by zero is impossible (zero has no reciprocal).

Q23 Find the quotient:

a. $\dfrac{^{-}3}{0} =$ _____ **b.** $\dfrac{5}{8} \div 0 =$ _____

STOP • STOP • STOP • STOP • STOP • STOP • STOP • STOP • STOP

A23 *a* and *b* are undefined.

| 16 | It was established in previous sections that the set of integers for the operation of multiplication was associative and commutative. The distributive properties hold for multiplication over addition and subtraction. Similarly, the set of integers was associative and commutative for the operation of addition. These properties are also true for the set of *rational numbers*. In addition, the following statements are true for all rational-number replacements of a, b, and c. |

1. Addition property of zero:

$$a + 0 = 0 + a = a$$

2. Multiplication property of zero:

$a \cdot 0 = 0a = 0$

3. Multiplication property of one:

$a \cdot 1 = 1a = a$

Q24   The associative property of addition states that $(a + b) + c = a + (b + c)$. Verify that $(a + b) + c$ and $a + (b + c)$ are equivalent expressions when $a = \dfrac{3}{5}$, $b = \dfrac{7}{10}$, and $c = \dfrac{9}{20}$.

STOP • STOP • STOP • STOP • STOP • STOP • STOP • STOP • STOP

A24

$$(a + b) + c = \left(\frac{3}{5} + \frac{7}{10}\right) + \frac{9}{20}$$
$$= \left(\frac{6}{10} + \frac{7}{10}\right) + \frac{9}{20}$$
$$= \frac{13}{10} + \frac{9}{20}$$
$$= \frac{26}{20} + \frac{9}{20}$$
$$= \frac{35}{20}$$
$$= 1\frac{3}{4}$$

$$a + (b + c) = \frac{3}{5} + \left(\frac{7}{10} + \frac{9}{20}\right)$$
$$= \frac{3}{5} + \left(\frac{14}{20} + \frac{9}{20}\right)$$
$$= \frac{3}{5} + \frac{23}{20}$$
$$= \frac{12}{20} + \frac{23}{20}$$
$$= \frac{35}{20}$$
$$= 1\frac{3}{4}$$

Q25   The left distributive property of multiplication over subtraction states that $a(b - c) = ab - ac$. Verify that $a(b - c)$ and $ab - ac$ are equivalent expressions when $a = \dfrac{2}{3}$, $b = \dfrac{4}{5}$, and $c = \dfrac{3}{7}$.

STOP • STOP • STOP • STOP • STOP • STOP • STOP • STOP • STOP

A25

$$a(b - c) = \frac{2}{3}\left(\frac{4}{5} - \frac{3}{7}\right) \qquad ab - ac = \frac{2}{3} \cdot \frac{4}{5} - \frac{2}{3} \cdot \frac{3}{7}$$

$$= \frac{2}{3}\left(\frac{28}{35} - \frac{15}{35}\right) \qquad \qquad = \frac{8}{15} - \frac{2}{7}$$

$$= \frac{2}{3}\left(\frac{13}{35}\right) \qquad \qquad = \frac{56}{105} - \frac{30}{105}$$

$$= \frac{26}{105} \qquad \qquad = \frac{26}{105}$$

This completes the instruction for this section.

6.6   Exercises

1. Find the sum:

   a. $\dfrac{2}{3} + \dfrac{^-5}{3}$

   b. $\dfrac{^-4}{5} + \dfrac{3}{8}$

   c. $\dfrac{^-1}{6} + \dfrac{^-4}{18}$

   d. $\dfrac{^-6}{7} + 0$

   e. $^-3\dfrac{3}{4} + 1\dfrac{1}{4}$

   f. $15 + ^-2\dfrac{4}{11}$

   g. $3\dfrac{2}{5} + \dfrac{^-17}{5}$

   h. $^-1\dfrac{2}{3} + ^-7\dfrac{3}{5}$

   i. $\dfrac{3}{10} + \dfrac{^-7}{6} + \dfrac{5}{12}$

   j. $\dfrac{^-5}{24} + \dfrac{7}{8} + \dfrac{^-7}{12}$

2. Find the difference:

   a. $\dfrac{2}{3} - \dfrac{5}{3}$

   b. $\dfrac{3}{7} - \dfrac{7}{9}$

   c. $4 - 2\dfrac{4}{7}$

   d. $2\dfrac{1}{3} - 5\dfrac{1}{4}$

   e. $^-3 - 2\dfrac{3}{7}$

   f. $3\dfrac{2}{5} - 4\dfrac{1}{7}$

   g. $\dfrac{^-5}{8} - \dfrac{^-1}{8}$

   h. $^-6\dfrac{1}{5} - ^-1\dfrac{3}{7}$

   i. $^-2\dfrac{1}{3} - ^-2\dfrac{1}{3}$

   j. $12\dfrac{4}{9} - 3$

3. Find the product:

   a. $\dfrac{^-4}{15} \cdot \dfrac{9}{16}$

   b. $\dfrac{^-4}{5} \cdot \dfrac{^-2}{7}$

   c. $\dfrac{^-3}{4} \cdot 1\dfrac{1}{3}$

   d. $0 \cdot 5\dfrac{7}{17}$

   e. $\dfrac{^-6}{17} \cdot \dfrac{^-17}{6}$

   f. $^-1\dfrac{2}{5} \cdot \dfrac{^-5}{7}$

g. $5\frac{1}{4} \cdot 11\frac{1}{3}$

h. $\frac{1}{4} \cdot \frac{-2}{3} \cdot \frac{6}{7}$

i. $-2\frac{1}{3} \cdot \frac{6}{7} \cdot \frac{-1}{2}$

j. $-4\frac{2}{3} \cdot \frac{2}{5} \cdot \frac{-3}{14}$

4. Find the quotient:

   a. $\frac{-4}{15} \div \frac{3}{2}$

   b. $\frac{-7}{8} \div \frac{-3}{4}$

   c. $-2\frac{1}{2} \div 5$

   d. $-2 \div \frac{-6}{11}$

   e. $-3\frac{2}{3} \div -3\frac{1}{3}$

   f. $\frac{2}{5} \div \frac{5}{2}$

   g. $\frac{3}{17} \div -1$

   h. $0 \div 3$

   i. $3 \div 0$

   j. $\frac{-5}{11} \div \frac{0}{1}$

5. Perform the indicated operation:

   a. $1\frac{3}{7} + -2\frac{4}{7}$

   b. $\frac{-5}{8} - \frac{4}{5}$

   c. $\frac{8}{45} \cdot \frac{-5}{16}$

   d. $-3\frac{1}{3} \div \frac{-9}{10}$

   e. $1\frac{2}{11} + 3\frac{15}{11}$

   f. $\frac{-8}{15} \cdot 0 \cdot 8\frac{1}{4}$

   g. $-5 \div \frac{-7}{5}$

   h. $4\frac{2}{3} \div -8$

   i. $2\frac{1}{3} + -4 + 1\frac{2}{5}$

   j. $\left(\frac{3}{5} \div \frac{-2}{7}\right) \div \frac{-3}{5}$

6.6 Exercise Answers

1. a. -1
   b. $\frac{-17}{40}$
   c. $\frac{-7}{18}$
   d. $\frac{-6}{7}$

   e. $-2\frac{1}{2}$
   f. $12\frac{7}{11}$
   g. 0
   h. $-9\frac{4}{15}$

   i. $\frac{-9}{20}$
   j. $\frac{1}{12}$

2. a. -1
   b. $\frac{-22}{63}$
   c. $1\frac{3}{7}$
   d. $-2\frac{11}{12}$

   e. $-5\frac{3}{7}$
   f. $\frac{-26}{35}$
   g. $\frac{-1}{2}$
   h. $-4\frac{27}{35}$

   i. 0
   j. $9\frac{4}{9}$

3. a. $\frac{-3}{20}$
   b. $\frac{8}{35}$
   c. -1
   d. 0

   e. 1
   f. 1
   g. $59\frac{1}{2}$
   h. $\frac{-1}{7}$

   i. 1
   j. $\frac{2}{5}$

4. a. $\dfrac{-8}{45}$ b. $1\dfrac{1}{6}$ c. $\dfrac{-1}{2}$ d. $3\dfrac{2}{3}$

    e. $1\dfrac{1}{10}$ f. $\dfrac{4}{25}$ g. $\dfrac{-3}{17}$ h. 0

    i. undefined   j. undefined

5. a. $-1\dfrac{1}{7}$ b. $-1\dfrac{17}{40}$ c. $\dfrac{-1}{18}$ d. $3\dfrac{19}{27}$

    e. $5\dfrac{6}{11}$ f. 0   g. $3\dfrac{4}{7}$ h. $\dfrac{-7}{12}$

    i. $\dfrac{-4}{15}$ j. $3\dfrac{1}{2}$

Chapter 6 Sample Test

At the completion of Chapter 6 it is expected that you will be able to work the following problems.

6.1 Introduction

1. Label each of the following sets as the set of integers, whole numbers, or natural numbers:
    a. $\{1, 2, 3, \ldots\}$
    b. $\{0, 1, 2, 3, \ldots\}$
    c. $\{\ldots, -2, -1, 0, 1, 2, \ldots\}$
2. Each integer is thought of as having what two parts?
3. Graph each of the following integers on the number line:
    a. 2   b. -7 c. 5   d. 0

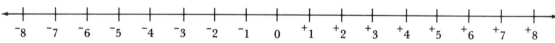

4. Write the opposite of each of the following integers:
    a. -5 b. 3   c. 0   d. 7

6.2 Addition of Integers

5. Find the sum:
    a. $-3 + -6$ b. $-5 + 8$
    c. $3 + -11$ d. $0 + -2$
    e. $5 + -5$ f. $-2 + 7 + -5$
    g. $-3 + -5 + 6$ h. $9 + -3 + 6$
6. The following properties are true for all integer replacements of a, b, and c. Match the property that each demonstrates with its name.
    a. $a + b = b + a$ 1. associative property of addition
    b. $a + 0 = 0 + a = a$ 2. commutative property of addition
    c. $(a + b) + c = a + (b + c)$ 3. addition property of zero

6.3   **Subtraction of Integers**

7. Find the difference:
   a. $^-2 - 5$ b. $3 - 7$ c. $6 - {}^-3$ d. $^-1 - {}^-2$
   e. $11 - 9$ f. $5 - 7 - {}^-3$
8. Complete each of the following:
   a. $^-5 + 7$ b. $^-5 - 7$ c. $0 - 5$ d. $^-3 + 0$
   e. $^-4 + {}^-9$ f. $6 - {}^-9$ g. $^-4 + 7 - 6$ h. $2 - 1 - {}^-1$

6.4   **Multiplication of Integers**

9. Find the product:
   a. $^-2 \cdot 5$ b. $^-4 \cdot {}^-6$ c. $^-3 \cdot 0$ d. $5 \cdot {}^-9$
   e. $^-1 \cdot 13$ f. $0 \cdot {}^-1$ g. $^-2 \cdot 4 \cdot {}^-5$ h. $(6 \cdot {}^-2)({}^-3 \cdot {}^-2)$
   i. $^-3(4 + {}^-7)$ j. $2({}^-4 + 4)$
10. The following properties are true for all integer replacements of a, b, and c. Match the property that each demonstrates with its name.
    a. $a \cdot 0 = 0a = 0$ 1. associative property of multiplication
    b. $1a = a \cdot 1 = a$ 2. commutative property of multiplication
    c. $a(bc) = (ab)c$ 3. multiplication property of one
    d. $ab = ba$ 4. left distributive property of multiplication over
    e. $a(b + c) = ab + ac$ addition
                                    5. multiplication property of zero

6.5   **Division of Integers**

11. Find the quotient:

    a. $^-36 \div 9$ b. $^-18 \div {}^-6$ c. $\dfrac{^-12}{4}$

    d. $\dfrac{^-25}{^-5}$ e. $10 \div 0$ f. $\dfrac{0}{^-6}$

    g. $0 \div {}^-2$ h. $\dfrac{4}{0}$

12. Answer true or false:
    a. $a \div b = b \div a$ is true for all integers a and b.
    b. $ab = ba$ is true for all integers a and b.
    c. $a \div (b \div c) = (a \div b) \div c$ is true for all integers a, b, and c.
    d. $a(bc) = (ab)c$ is true for all integers a, b, and c.

6.6   **Operations with Rational Numbers**

13. Find the sum:

    a. $\dfrac{^-4}{7} + \dfrac{7}{12}$

    b. $\dfrac{^-2}{3} + \dfrac{^-5}{8}$

    c. $^-3\dfrac{1}{3} + 2\dfrac{7}{9}$

    d. $\dfrac{^-3}{16} + 3$

14. Find the difference:

    a. $\dfrac{5}{16} - \dfrac{3}{16}$

    b. $1\dfrac{4}{5} - \dfrac{^-2}{3}$

    c. $4\dfrac{1}{3} - 5\dfrac{2}{5}$

    d. $\dfrac{^-7}{9} - 2$

15. Find the product:

    a. $\dfrac{^-4}{15} \cdot \dfrac{^-5}{16}$

    b. $3 \cdot {^-4\dfrac{2}{3}}$

    c. $^-2\dfrac{1}{3} \cdot 0$

    d. $^-5\dfrac{1}{3} \cdot {^-1\dfrac{2}{5}}$

16. Find the quotient:

    a. $\dfrac{^-9}{16} \div \dfrac{3}{8}$

    b. $\dfrac{5}{12} \div 7$

    c. $^-2\dfrac{1}{3} \div {^-4\dfrac{5}{16}}$

    d. $^-2 \div \dfrac{^-8}{9}$

17. Perform the indicated operation:

    a. $\dfrac{^-2}{3} - \dfrac{4}{7}$

    b. $^-1 \cdot \dfrac{3}{4}$

    c. $\dfrac{^-2}{7} \div {^-1\dfrac{4}{21}}$

    d. $3\dfrac{1}{5} \cdot {^-4\dfrac{2}{5}}$

    e. $\dfrac{^-5}{12} - \dfrac{^-5}{12}$

    f. $\dfrac{2}{7} \div \dfrac{24}{35}$

    g. $\dfrac{^-6}{51} \cdot \dfrac{17}{18} \cdot \dfrac{^-1}{3}$

    h. $^-1\dfrac{7}{8} + 9\dfrac{3}{5}$

Chapter 6 Sample Test Answers

1. a. the set of natural numbers
    b. the set of whole numbers
    c. the set of integers
2. distance and direction
3.

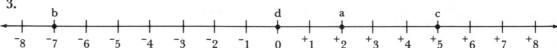

4. **a.** 5   **b.** ⁻3   **c.** 0   **d.** ⁻7
5. **a.** ⁻9   **b.** 3   **c.** ⁻8   **d.** ⁻2   **e.** 0   **f.** 0   **g.** ⁻2
    h. 12
6. **a.** 2, commutative property of addition
    b. 3, addition property of zero
    c. 1, associative property of addition
7. **a.** ⁻7   **b.** ⁻4   **c.** 9   **d.** 1   **e.** 2   **f.** 1
8. **a.** 2   **b.** ⁻12   **c.** ⁻5   **d.** ⁻3   **e.** ⁻13   **f.** 15   **g.** ⁻3
    h. 2
9. **a.** ⁻10   **b.** 24   **c.** 0   **d.** -45   **e.** ⁻13   **f.** 0   **g.** 40
    h. ⁻72   **i.** 9   **j.** 0
10. **a.** 5, multiplication property of zero
    b. 3, multiplication property of one
    c. 1, associative property of multiplication
    d. 2, commutative property of multiplication
    e. 4, left distributive property of multiplication over addition
11. **a.** ⁻4   **b.** 3   **c.** ⁻3   **d.** 5
    e. undefined   **f.** 0   **g.** 0   **h.** undefined
12. **a.** false   **b.** true   **c.** false   **d.** true

13. **a.** $\dfrac{1}{84}$ **b.** $-1\dfrac{7}{24}$ **c.** $\dfrac{-5}{9}$ **d.** $2\dfrac{13}{16}$

14. **a.** $\dfrac{1}{8}$ **b.** $2\dfrac{7}{15}$ **c.** $-1\dfrac{1}{15}$ **d.** $-2\dfrac{7}{9}$

15. **a.** $\dfrac{1}{12}$ **b.** ⁻14   **c.** 0   **d.** $7\dfrac{7}{15}$

16. **a.** $-1\dfrac{1}{2}$ **b.** $\dfrac{5}{84}$ **c.** $\dfrac{112}{207}$ **d.** $2\dfrac{1}{4}$

17. **a.** $-1\dfrac{5}{21}$ **b.** $\dfrac{-3}{4}$ **c.** $\dfrac{6}{25}$ **d.** $-14\dfrac{2}{25}$

    e. 0   **f.** $\dfrac{5}{12}$ **g.** $\dfrac{1}{27}$ **h.** $7\dfrac{29}{40}$

Chapter 7

Fundamental Operations with Algebraic Expressions

In Chapter 5 expressions such as $a + 3$, $5y - 2x$, $9(b + 7)$, and $(x - 2)(x + 3)$ were introduced and referred to as "open expressions." Expressions such as these are also correctly called *algebraic expressions*. Since letters (variables) in an algebraic expression can be replaced by rational numbers, all properties known to be valid for the set of rational numbers can be used when simplifying algebraic expressions.* The purpose of this chapter is to develop skill in the use of these properties for the simplification of algebraic expressions.

7.1 Simplifying Algebraic Expressions by Use of the Commutative and Associative Properties of Addition

---

1   The building blocks or components of algebraic expressions are called terms. When an algebraic expression shows only additions, the *terms* are the parts separated by plus signs. Subtraction signs appearing in an algebraic expression can be converted to addition signs using the definition of subtraction.

Examples:	*Terms*
$8y + 3$	$8y$ and 3
$\frac{1}{2}x - 4 + 2y = \frac{1}{2}x + {}^-4 + 2y$	$\frac{1}{2}x$, ${}^-4$, and $2y$
$x - 3y - 6 = x + {}^-3y + {}^-6$	x, ${}^-3y$, and ${}^-6$

Q1   Identify the terms in the algebraic expression $4a + 7ab - 12$. _____

STOP • STOP • STOP • STOP • STOP • STOP • STOP • STOP • STOP

A1 $4a$, $7ab$, and ${}^-12$

---

2   The *like terms* of an algebraic expression are terms that have exactly the same literal coefficients (letter factors). The like terms of the algebraic expression $7x - 3y - \frac{2}{3}x + 6$ are $\frac{-2}{3}x$ and $7x$, because each has the literal coefficient x.

In the expression $\frac{4}{5}y - 3xy + \frac{2}{3}x - 2y + 8xy$ there are two sets of like terms. They

---

*Unless otherwise stated, the replacement set for all variables will be the set of rational numbers.

258

are $\frac{4}{5}y$ and ^-2y, with the letter factor y, and ^-3xy and $8xy$, with the letter factors x and y.

Numbers without a literal factor are called *constants* and are considered to be like terms. That is, 5, $^-7$, $\frac{7}{8}$, 3, and so on, are constants and like terms.

Q2   Identify the like terms in each of the following algebraic expressions:

a. $4x - 2y + 3 - 6x$ _____

b. $6 - 2y + \frac{5}{9}$ _____

c. $5r - 6s + 2rs - \frac{1}{2}s$ _____

d. $6a - 2b + 3ab$ _____

STOP • STOP • STOP • STOP • STOP • STOP • STOP • STOP • STOP

A2   a. $4x, ^-6x$ b. $6, \frac{5}{9}$ c. $^-6s, \frac{^-1}{2}s$

d. There are no like terms in part d.

3   Algebraic expressions are simplified by combining (by addition or subtraction) like terms. The justification for combining like terms is the distributive property of multiplication over addition or the distributive property of multiplication over subtraction. For example, $5x - 7x$ is simplified:

Justification

$5x - 7x$
$(5 - 7)x$ right distributive property of multiplication over subtraction
^-2x number fact: $5 - 7 = ^-2$

Therefore, $5x - 7x$ simplifies to ^-2x.

Q3   Use a distributive property to simplify the following expressions:

a. $4a - 7a$ b. $\frac{2}{3}x + \frac{1}{4}x$

STOP • STOP • STOP • STOP • STOP • STOP • STOP • STOP • STOP

A3   a. ^-3a: $4a - 7a$ b. $\frac{11}{12}x$: $\frac{2}{3}x + \frac{1}{4}x$

$(4 - 7)a$ $\left(\frac{2}{3} + \frac{1}{4}\right)x$

^-3a $\frac{11}{12}x$

---

4   Consider the following simplification:

Justification

$4x - 3x$

$(4 - 3)x$ right distributive property of multiplication over subtraction

$1x$ number fact: $4 - 3 = 1$

x multiplication property of one

Therefore, $4x - 3x = x$.

Q4   Simplify $9y - 8y$.

STOP • STOP • STOP • STOP • STOP • STOP • STOP • STOP • STOP

A4 y: $9y - 8y$

         $(9 - 8)y$

         $1y$

         y

Q5   Simplify $7b - 6b$.

STOP • STOP • STOP • STOP • STOP • STOP • STOP • STOP • STOP

A5 b

---

5   By the multiplication property of one, x also equals $1x$. This idea is used to simplify expressions such as $4x + x$ or $3b - b$.

Example 1: Simplify $4x + x$.

Solution

Justification

$4x + x$

$4x + 1x$ multiplication property of one

$(4 + 1)x$ right distributive property of multiplication over addition

$5x$ number fact: $4 + 1 = 5$

Therefore, $4x + x = 5x$.

Example 2: Simplify $3b - b$.

Solution

$3b - b$

$3b - 1b$ multiplication property of one

$(3 - 1)b$ right distributive property of multiplication over subtraction

$2b$ number fact: $3 - 1 = 2$

Therefore, $3b - b = 2b$.

Q6   Simplify:

a. $7y + y$ **b.** $x + 2x$ **c.** $b - 5b$

d. $a - \dfrac{2}{3}a$

STOP • **STOP** • **STOP** • **STOP** • **STOP** • **STOP** • **STOP** • **STOP** • **STOP**

A6   **a.** $8y$: $7y + 7$ **b.** $3x$: $x + 2x$

$\qquad\qquad\quad 7y + 1y \qquad\qquad\qquad 1x + 2x$

$\qquad\qquad\quad (7 + 1)y \qquad\qquad\quad (1 + 2)x$

$\qquad\qquad\quad 8y \qquad\qquad\qquad\quad 3x$

c. ^-4b: $b - 5b$ **d.** $\dfrac{1}{3}a$: $a - \dfrac{2}{3}a$

$\qquad\qquad 1b - 5b \qquad\qquad\qquad 1a - \dfrac{2}{3}a$

$\qquad\qquad (1 - 5)b \qquad\qquad\qquad \left(1 - \dfrac{2}{3}\right)a$

$\qquad\qquad ^-4b \qquad\qquad\qquad\qquad \dfrac{1}{3}a$

---

6   By the multiplication property of negative one, $^-1x = {^-x}$. Consider its use in the simplification of $x - 2x$.

Justification

$x - 2x$

$1x - 2x$ multiplication property of one

$(1 - 2)x$ right distributive property of multiplication over subtraction

^-1x number fact: $1 - 2 = {^-1}$

^-x multiplication property of negative one

Therefore, $x - 2x = {^-x}$.

Q7   Simplify:

a. $7x - 8x$ **b.** $12y - 13y$

STOP • **STOP** • **STOP** • **STOP** • **STOP** • **STOP** • **STOP** • **STOP** • **STOP**

A7   **a.** ^-x: $7x - 8x$ **b.** ^-y

$\qquad\qquad\quad (7 - 8)x$

$\qquad\qquad\quad ^-1x$

$\qquad\qquad\quad ^-x$

7   Notice that using a distributive property to simplify algebraic expressions is merely the process of combining the numerical coefficients of the like terms. Thus, the simplification process can be shortened as follows:

$$^-5b + 3b = {}^-2b \quad \text{because} \quad {}^-5 + 3 = {}^-2$$

$$\frac{^-1}{2}z - \frac{2}{3}z = \frac{^-7}{6}z \quad \text{because} \quad \frac{^-1}{2} - \frac{2}{3} = \frac{^-7}{6}$$

In algebra, numbers are usually written in $\dfrac{p}{q}$ form rather than as mixed numbers. Thus, in the preceding example, the answer is written as $\dfrac{^-7}{6}z$ rather than as $^-1\dfrac{1}{6}z$.

Q8   Simplify the following algebraic expressions by combining the numerical coefficients of the like terms:

a. $7y - 12y =$ _____   **b.** $4b - b =$ _____

c. $\dfrac{^-2}{3}x + \dfrac{4}{5}x =$ _____   **d.** $y - y =$ _____

STOP • STOP • STOP • STOP • STOP • STOP • STOP • STOP • STOP

A8   **a.** ^-5y: because $7 - 12 = {}^-5$ **b.** $3b$: because $4 - 1 = 3$

c. $\dfrac{2}{15}x$: because $\dfrac{^-2}{3} + \dfrac{4}{5} = \dfrac{2}{15}$ **d.** 0: because $1 - 1 = 0$

8   To simplify expressions involving more than two terms, a similar procedure is followed. Study the following examples:

$$3y - 12y + 7 \qquad\qquad {}^-8x - 3x + 9 - 12$$
$$^-9y + 7 \qquad\qquad\qquad {}^-11x - 3$$

Q9   Simplify each of the following expressions:
a. $^-3m - 7m + 9$ **b.** $11 - 1 + 7x - 3x$

c. $5y + 7y - 6 + 3$ **d.** $4 - 10 + 8y - \dfrac{3}{2}y$

STOP • STOP • STOP • STOP • STOP • STOP • STOP • STOP • STOP

A9   **a.** $^-10m + 9$ **b.** $10 + 4x$

c. $12y - 3$ **d.** $^-6 + \dfrac{13}{2}y$

9   The associative property of addition is used to simplify $(3x - 8) + 4$ as follows:

Justification

$(3x - 8) + 4$	
$(3x + {}^-8) + 4$	definition of subtraction
$3x + ({}^-8 + 4)$	associative property of addition
$3x + {}^-4$	number fact
$3x - 4$	definition of subtraction

Recall that by the definition of subtraction,

$$a - b = a + {}^-b$$

Or, reversing the right and left sides,

$$a + {}^-b = a - b$$

The form $a - b$ is considered "simpler" than $a + {}^-b$. Therefore, when simplifying algebraic expressions, $5x - 3$ is written rather than $5x + {}^-3$.

Q10   Simplify $(2x - 5) + 3$.

STOP   •   STOP   •   STOP   •   STOP   •   STOP   •   STOP   •   STOP   •   STOP   •   STOP

A10 $2x - 2$: $(2x - 5) + 3$
$(2x + {}^-5) + 3$
$2x + ({}^-5 + 3)$
$2x + {}^-2$
$2x - 2$

10   The expression $(3 + 7y) - 16y$ is simplified as follows:

Justification

$(3 + 7y) - 16y$	
$(3 + 7y) + {}^-16y$	definition of subtraction
$3 + (7y + {}^-16y)$	associative property of addition
$3 + {}^-9y$	like terms combined
$3 - 9y$	definition of subtraction

Q11   Simplify $(7 - 4x) + 3x$.

STOP   •   STOP   •   STOP   •   STOP   •   STOP   •   STOP   •   STOP   •   STOP   •   STOP

A11 $7 - x$: $(7 - 4x) + 3x$
                    $(7 + {}^-4x) + 3x$
                    $7 + ({}^-4x + 3x)$
                    $7 + {}^-x$
                    $7 - x$

Q12   Simplify each of the following algebraic expressions using the associative property of addition where necessary:

a. $(x + 7) + 4$ **b.** $(3 + y) + 8y$

c. $(2b - 4b) + 6$ **d.** $4z + (3z - 5)$

e. $^-5y + (2y + 3)$ **f.** $^-5 + (2 + 4x)$

g. $\dfrac{4}{5} + \left(\dfrac{2}{3} - x\right)$ **h.** $\left(3x - \dfrac{2}{5}\right) + \dfrac{2}{5}$

i. $(^-5 - 2x) + 2x$ **j.** $(^-7x + 3) - 3$

STOP   •   STOP   •   STOP   •   STOP   •   STOP   •   STOP   •   STOP   •   STOP   •   STOP

A12   **a.** $x + 11$ **b.** $3 + 9y$
           c. $^-2b + 6$ **d.** $7z - 5$
           e. $^-3y + 3$ **f.** $^-3 + 4x$
           g. $\dfrac{22}{15} - x$ **h.** $3x$
           i. $^-5$ **j.** ^-7x

---

11   The commutative property of addition is used whenever there is a need to change the order of terms in addition as an aid in the simplification of algebraic expressions. For example, $(3x + 7) - 2x$ is simplified as follows:

Justification

$(3x + 7) - 2x$	
$(7 + 3x) - 2x$	commutative property of addition
$(7 + 3x) + {}^-2x$	definition of subtraction
$7 + (3x + {}^-2x)$	associative property of addition
$7 + x$	like terms combined

Q13   Use the commutative property of addition to change the order of the terms within the parentheses.

$(4y + 9) - 3y =$ _____

STOP • **STOP** • **STOP** • **STOP** • **STOP** • **STOP** • **STOP** • **STOP** • **STOP**

A13 $(9 + 4y) - 3y$

Q14   Use the associative property of addition to complete the simplification of A13.

STOP • **STOP** • **STOP** • **STOP** • **STOP** • **STOP** • **STOP** • **STOP** • **STOP**

A14 $(9 + 4y) - 3y$
$(9 + 4y) + {}^-3y$
$9 + (4y + {}^-3y)$
$9 + y$

---

12   The expression ${}^-5y + (3 - 2y)$ is simplified as follows:

Justification

${}^-5y + (3 - 2y)$	
${}^-5y + (3 + {}^-2y)$	definition of subtraction
${}^-5y + ({}^-2y + 3)$	commutative property of addition
$({}^-5y + {}^-2y) + 3$	associative property of addition
${}^-7y + 3$	like terms combined

Thus, ${}^-5y + (3 - 2y)$ simplifies to ${}^-7y + 3$.

Q15   Simplify $6t + (7 - 9t)$.

STOP • **STOP** • **STOP** • **STOP** • **STOP** • **STOP** • **STOP** • **STOP** • **STOP**

A15 $^-3t + 7$: $6t + (7 - 9t)$
$6t + (7 + {}^-9t)$
$6t + ({}^-9t + 7)$
$(6t + {}^-9t) + 7$
$^-3t + 7$

Q16 Simplify each of the following algebraic expressions:

    a. $4 + (3t - 9)$ **b.** $\left(\dfrac{2}{3}z - 5\right) - \dfrac{4}{7}z$

    c. $(3 + 5y) - 5y$ **d.** $4m + (2m - 7m)$

STOP • **STOP** • **STOP** • **STOP** • **STOP** • **STOP** • **STOP** • **STOP** • **STOP**

A16   **a.** $^-5 + 3t$ **b.** $^-5 + \dfrac{2}{21}z$ **c.** 3 **d.** ^-m

---

13 The expression $5x - 7$ may also be written $^-7 + 5x$. This fact is verified below:

Justification

$5x - 7$
$5x + {}^-7$ definition of subtraction
$^-7 + 5x$ commutative property of addition

Q17 Verify that $3x - 4 = {}^-4 + 3x$ by completing the following:

Justification

    $3x - 4$

    a. _____   definition of subtraction

    b. _____   commutative property of addition

STOP • **STOP** • **STOP** • **STOP** • **STOP** • **STOP** • **STOP** • **STOP** • **STOP**

A17   **a.** $3x + {}^-4$ **b.** $^-4 + 3x$

---

14 The expression $^-8 - 7x$ may also be written $^-7x - 8$. The verification is as follows:

Justification

$^-7x - 8$
$^-7x + {}^-8$ definition of subtraction
$^-8 + {}^-7x$ commutative property of addition
$^-8 - 7x$ definition of subtraction

Q18   Verify that $^-2x - 9 = {^-9} - 2x$ by completing the following:

Justification

$^-2x - 9$

a. _____   definition of subtraction

b. _____   commutative property of addition

c. _____   definition of subtraction

STOP • STOP • STOP • STOP • STOP • STOP • STOP • STOP • STOP

A18   a. $^-2x + {^-9}$ b. $^-9 + {^-2x}$ c. $^-9 - 2x$

---

15   Frame 13 showed that $5x - 7$ could be written $^-7 + 5x$. Frame 14 showed that $^-8 - 7x = {^-7x} - 8$. These frames demonstrate the following useful procedure: *In an algebraic expression, the terms may be rearranged in any order as long as the original sign of each term is left unchanged.* Some examples are:

$5 - 2x = {^-2x} + 5$

$3x - \dfrac{4}{5} = \dfrac{^-4}{5} + 3x$

$\dfrac{^-3}{4}x - 5 = {^-5} - \dfrac{3}{4}x$

$^-2x + 3 + 5x - 7 = {^-2x} + 5x + 3 - 7$

$7y - 4 - 3y - 5 + 2y + 6 = 7y - 3y + 2y - 4 - 5 + 6$

Notice that in the fourth and fifth examples the terms are rearranged so that the like terms are together.

Q19   Write equivalent expressions for each of the following:

a. $3x + 2 =$ _____   b. $^-5x - 4 =$ _____

c. $\dfrac{^-5}{8}x + 12 =$ _____   d. $^-4x + 3 - 2x =$ _____

e. $^-2x - 7 - 5x + 6 =$ _____

STOP • STOP • STOP • STOP • STOP • STOP • STOP • STOP • STOP

A19   a. $2 + 3x$ b. $^-4 - 5x$ c. $12 - \dfrac{5}{8}x$

d. $^-4x - 2x + 3$ e. $^-2x - 5x - 7 + 6$

---

16   The expression $5x - 2 - 3x - 9$ may be simplified by rearranging and combining the like terms as follows:

$5x - 2 - 3x - 9$

$5x - 3x - 2 - 9$

$2x - 11$

Q20   Simplify $7 - 3t + 9 - 6t$ by rearranging and combining the like terms.

STOP • STOP • STOP • STOP • STOP • STOP • STOP • STOP • STOP

A20 $^-9t + 16$ or $16 - 9t$: $7 - 3t + 9 - 6t$
$$^-3t - 6t + 7 + 9$$
$$^-9t + 16$$

Q21   Simplify $4x - 3 - 5x + 6 - x$.

STOP • STOP • STOP • STOP • STOP • STOP • STOP • STOP • STOP

A21 $^-2x + 3$ or $3 - 2x$: $4x - 3 - 5x + 6 - x$
$$4x - 5x - x - 3 + 6$$
$$^-2x + 3$$

Q22   Simplify the following algebraic expressions:

     a. $^-3 + 5x + 7$ **b.** $5y - 6 + 2y$

     c. $^-3x - 4 - x$ **d.** $x + 3x + 7 - 9$

     e. $^-2z + 3 - 5z - 1$ **f.** $6t + 3 - 7t - 6 + t$

STOP • STOP • STOP • STOP • STOP • STOP • STOP • STOP • STOP

A22   **a.** $5x + 4$ or $4 + 5x$ **b.** $7y - 6$ or $^-6 + 7y$
     c. $^-4x - 4$ or $^-4 - 4x$ **d.** $4x - 2$ or $^-2 + 4x$
     e. $^-7z + 2$ or $2 - 7z$ **f.** $^-3$

This completes the instruction for this section.

7.1   Exercises

1. Simplify the following algebraic expressions by combining like terms:

     a. $^-3x + 7x$ **b.** $2x - 5 - x$
     c. $5y - 7 - 6y$ **d.** $3 - 6 + 5z$
     e. $4b - 3b + 2$ **f.** $^-6a + 3 + 6a$

     g. $\dfrac{2}{3}x + \dfrac{7}{8}x$ **h.** $5y - \dfrac{3}{5}y + 7$

2. Simplify using the commutative and/or associative properties of addition:
   a. $(4y + 7) + 3y$
   b. $6 + (2x - 9)$

   c. $b + \left(\dfrac{1}{3}b + 2\right)$
   d. $\left(x - \dfrac{5}{8}x\right) + 3x$

   e. $(2x - 3) + 7x$
   f. $(5 - 4x) + 9$

3. Simplify:
   a. $3x + {}^-7$
   b. $5 - {}^-4y$

   c. $3 + {}^-7x + 5$
   d. ${}^-2y + 6 + \dfrac{{}^-2}{3}y$

4. Simplify:

   a. $3x - 4y + 7x - 6$
   b. $\dfrac{2}{5}z - \dfrac{3}{4}z + 8$

   c. $\dfrac{5}{2}b - 7a + \dfrac{3}{5}$
   d. $6 - 4r + 6s - 2r$

   e. ${}^-x + 6 - 3x + {}^-5$
   f. $7 - 2y + y$

   g. ${}^-4m + 6 + 4m$
   h. $\dfrac{{}^-2}{3}x + x - \dfrac{1}{3}x + 2$

   i. $4xy - 6y + 7 + 2y$
   j. $3m + 4n - 6m + n - 6$

7.1   Exercise Answers

1. **a.** $4x$
   b. $x - 5$
   c. ${}^-y - 7$
   d. $5z - 3$
   e. $b + 2$
   f. 3

   g. $\dfrac{37}{24}x$
   h. $\dfrac{22}{5}y + 7$

2. **a.** $7y + 7$
   b. $2x - 3$

   c. $\dfrac{4}{3}b + 2$
   d. $\dfrac{27}{8}x$

   e. $9x - 3$
   f. ${}^-4x + 14$ or $14 - 4x$

3. **a.** $3x - 7$
   b. $4y + 5$

   c. ${}^-7x + 8$ or $8 - 7x$
   d. $\dfrac{{}^-8}{3}y + 6$ or $6 - \dfrac{8}{3}y$

4. **a.** $10x - 4y - 6$
   b. $\dfrac{{}^-7}{20}z + 8$ or $8 - \dfrac{7}{20}z$

   c. $\dfrac{5}{2}b - 7a + \dfrac{3}{5}$ cannot be simplified because there are no like terms.

   d. $6s - 6r + 6$
   e. ${}^-4x + 1$ or $1 - 4x$
   f. $7 - y$
   g. 6
   h. 2
   i. $4xy - 4y + 7$
   j. $5n - 3m - 6$

7.2   Simplifying Algebraic Expressions by Use of the Commutative and Associative Properties of Multiplication

---

1   The associative and commutative properties of multiplication are used whenever there is a need to change the grouping or order of multiplication as an aid in the simplification of algebraic expressions. The associative property of multiplication is used to simplify $3(4x)$ as follows:

Justification

$3(4x)$

$(3 \cdot 4)x$ associative property of multiplication

$12x$ number fact

Thus, $3(4x) = 12x$.

Q1   Use the associative property of multiplication to simplify $^-5(7y)$.

STOP   •   STOP   •   STOP   •   STOP   •   STOP   •   STOP   •   STOP   •   STOP   •   STOP

A1 ^-35y: $^-5(7y)$

$(^-5 \cdot 7)y$

^-35y

---

2   The expression $7\left(\dfrac{-3}{7}t\right)$ is simplified:

Justification

$7\left(\dfrac{-3}{7}t\right)$

$\left(7 \cdot \dfrac{-3}{7}\right)t$ associative property of multiplication

^-3t number fact: $\dfrac{7}{1} \cdot \dfrac{-3}{7} = {}^-3$

Thus, $7\left(\dfrac{-3}{7}t\right) = {}^-3t$.

Q2   Simplify $^-9\left(\dfrac{-1}{9}x\right)$.

STOP   •   STOP   •   STOP   •   STOP   •   STOP   •   STOP   •   STOP   •   STOP   •   STOP

A2 x: $-9\left(\dfrac{-1}{9}x\right)$

$\left(-9\cdot\dfrac{-1}{9}\right)x$

$1x$

x

---

3   The expression $\dfrac{3}{5}\left(\dfrac{-4}{9}y\right)$ is simplified:

Justification

$\dfrac{3}{5}\left(\dfrac{-4}{9}y\right)$

$\left(\dfrac{3}{5}\cdot\dfrac{-4}{9}\right)y$ associative property of multiplication

$\dfrac{-4}{15}y$ number fact

Thus, $\dfrac{3}{5}\left(\dfrac{-4}{9}y\right)=\dfrac{-4}{15}y$.

Q3   Simplify $\dfrac{-3}{4}\left(\dfrac{-2}{7}z\right)$.

STOP   •   STOP   •   STOP   •   STOP   •   STOP   •   STOP   •   STOP   •   STOP   •   STOP

A3 $\dfrac{3}{14}z$: $\dfrac{-3}{4}\left(\dfrac{-2}{7}z\right)$

$\left(\dfrac{-3}{4}\cdot\dfrac{-2}{7}\right)z$

$\dfrac{3}{14}z$

Q4   Simplify each of the following algebraic expressions:

a. $4\left(\dfrac{3}{4}x\right)$ b. $\dfrac{2}{3}\left(\dfrac{3}{2}y\right)$

c. $-5\left(\dfrac{-2}{15}z\right)$ d. $\dfrac{-6}{7}\left(\dfrac{5}{12}t\right)$

e. $\dfrac{-5}{9}\left(\dfrac{-9}{5}m\right)$ f. $\dfrac{1}{7}(7x)$

STOP • **STOP** • **STOP** • **STOP** • **STOP** • **STOP** • **STOP** • **STOP** • **STOP**

A4   **a.** $3x$ **b.** y **c.** $\dfrac{2}{3}z$ **d.** $\dfrac{-5}{14}t$ **e.** m **f.** x

---

4   Both the commutative and associative properties of multiplication are used to simplify $\left(\dfrac{3}{4}x\right)\dfrac{2}{3}$ as follows:

Justification

$\left(\dfrac{3}{4}x\right)\dfrac{2}{3}$

$\dfrac{2}{3}\left(\dfrac{3}{4}x\right)$ commutative property of multiplication

$\left(\dfrac{2}{3}\cdot\dfrac{3}{4}\right)x$ associative property of multiplication

$\dfrac{1}{2}x$ number fact

Q5   Use the commutative and associative properties of multiplication to simplify $\left(\dfrac{2}{5}x\right)5$.

STOP • **STOP** • **STOP** • **STOP** • **STOP** • **STOP** • **STOP** • **STOP** • **STOP**

A5 $2x$: $\left(\dfrac{2}{5}x\right)5$

$5\left(\dfrac{2}{5}x\right)$

$\left(5\cdot\dfrac{2}{5}\right)x$

$2x$

---

5   The expression $\left(\dfrac{5}{8}y\right)\cdot\dfrac{-2}{3}$ is simplified as follows:

Justification

$\left(\dfrac{5}{8}y\right)\cdot\dfrac{-2}{3}$

$\dfrac{-2}{3}\left(\dfrac{5}{8}y\right)$ commutative property of multiplication

$$\left(\frac{-2}{3} \cdot \frac{5}{8}\right) y \qquad \text{associative property of multiplication}$$

$$\frac{-5}{12} y \qquad \text{number fact}$$

Thus, $\left(\frac{5}{8} y\right) \cdot \frac{-2}{3} = \frac{-5}{12} y.$

Q6 Simplify $\left(\frac{-3}{5} x\right) \cdot \frac{-5}{3}$ by use of the commutative and associative properties.

STOP • STOP • STOP • STOP • STOP • STOP • STOP • STOP • STOP

A6 $x:$ $\left(\frac{-3}{5} x\right) \cdot \frac{-5}{3}$

$$\frac{-5}{3}\left(\frac{-3}{5} x\right)$$

$$\left(\frac{-5}{3} \cdot \frac{-3}{5}\right) x$$

$1x$

x

6 The expression $\left(\frac{8}{13} z\right) 13$ is simplified:

Justification

$\left(\frac{8}{13} z\right) 13$

$13\left(\frac{8}{13} z\right) \qquad \text{commutative property of multiplication}$

$\left(13 \cdot \frac{8}{13}\right) z \qquad \text{associative property of multiplication}$

$8z \qquad \text{number fact}$

Thus, $\left(\frac{8}{13} z\right) 13 = 8z.$

Q7 Simplify the expression $\left(\frac{-4}{5} t\right) 5.$

STOP • STOP • STOP • STOP • STOP • STOP • STOP • STOP • STOP

A7 $\quad$ $^{-}4t$: $\quad \left(\dfrac{-4}{5}t\right)5$

$$5\left(\dfrac{-4}{5}t\right)$$

$$\left(5 \cdot \dfrac{-4}{5}\right)t$$

$$^{-}4t$$

Q8 $\quad$ Simplify each of the following algebraic expressions:

a. $\left(\dfrac{-3}{4}x\right) \cdot {}^{-}4$ $\qquad\qquad$ b. $\left(\dfrac{5}{7}y\right)\dfrac{7}{5}$

c. $(5z)\dfrac{1}{5}$ $\qquad\qquad$ d. $\left(\dfrac{-2}{3}t\right) \cdot \dfrac{-9}{12}$

e. $\left(\dfrac{-1}{7}x\right) \cdot {}^{-}7$ $\qquad\qquad$ f. $\left(\dfrac{3}{5}y\right)\dfrac{4}{7}$

STOP • **STOP** • **STOP** • **STOP** • **STOP** • **STOP** • **STOP** • **STOP** • **STOP**

A8 $\qquad$ a. $3x$ $\qquad$ b. y $\qquad$ c. z $\qquad$ d. $\dfrac{1}{2}t$ $\qquad$ e. x $\qquad$ f. $\dfrac{12}{35}y$

This completes the instruction for this section.

7.2 $\quad$ Exercises

1. Simplify each of the following algebraic expressions:

a. $2\left(\dfrac{3}{4}x\right)$ $\qquad$ b. $\dfrac{-1}{2}(8y)$ $\qquad$ c. $\dfrac{-4}{5}\left(\dfrac{-5}{4}z\right)$ $\qquad$ d. $\dfrac{-1}{5}({}^{-}5c)$

e. $15\left(\dfrac{4}{5}x\right)$ $\qquad$ f. $\dfrac{3}{4}\left(\dfrac{-2}{5}y\right)$

2. Simplify:

   a. $(2x)\dfrac{5}{8}$ **b.** $\left(\dfrac{-2}{3}x\right)\cdot\dfrac{-3}{2}$ **c.** $(^-7y)\cdot\dfrac{-1}{7}$ **d.** $\left(\dfrac{-2}{5}z\right)10$

   e. $\left(\dfrac{-7}{4}z\right)12$ **f.** $\left(\dfrac{-3}{25}x\right)25$ **g.** $\left(\dfrac{-2}{3}y\right)\dfrac{5}{7}$ **h.** $\left(\dfrac{-1}{9}z\right)9$

7.2   Exercise Answers

1. **a.** $\dfrac{3}{2}x$ **b.** ^-4y **c.** z **d.** c

   e. $12x$ **f.** $\dfrac{-3}{10}y$

2. **a.** $\dfrac{5}{4}x$ **b.** x **c.** y **d.** ^-4z

   e. ^-21z **f.** ^-3x **g.** $\dfrac{-10}{21}y$ **h.** ^-z

7.3   Using the Distributive Properties to Simplify Algebraic Expressions

---

1   The left distributive property of multiplication over addition and the left distributive property of multiplication over subtraction state that

$$a(b + c) = ab + ac$$

   and

$$a(b - c) = ab - ac$$

   for all rational-number replacements of a, b, and c. Two examples are:

   $3(7 + 5) = 3 \cdot 7 + 3 \cdot 5$
   $2(1 - 9) = 2 \cdot 1 - 2 \cdot 9$

Q1   Use the left distributive property of multiplication over addition to fill in the blanks.

   $5(2 + 9) = $ _____ $+$ _____

STOP • STOP • STOP • STOP • STOP • STOP • STOP • STOP • STOP

A1 $5(2 + 9) = 5 \cdot 2 + 5 \cdot 9$

Q2   Use the left distributive property of multiplication over subtraction to fill in the blanks.

   $4(8 - 5) = $ _____ $-$ _____

STOP • STOP • STOP • STOP • STOP • STOP • STOP • STOP • STOP

A2 $4(8 - 5) = 4 \cdot 8 - 4 \cdot 5$

2	When parentheses are removed from an algebraic expression, the same procedure is used. For example, to remove the parentheses from $3(x + 2)$, proceed as follows:

$$3(x + 2) = 3 \cdot x + 3 \cdot 2$$
$$= 3x + 6$$

Q3   Use the left distributive property of multiplication over addition to remove the parentheses from $7(y + 3)$.

STOP • STOP • STOP • STOP • STOP • STOP • STOP • STOP • STOP

A3 $7y + 21$: $7(y + 3) = 7 \cdot y + 7 \cdot 3$
                                    $= 7y + 21$

Q4   Remove the parentheses:
a. $5(x + 1)$ b. $8(2 + c)$

STOP • STOP • STOP • STOP • STOP • STOP • STOP • STOP • STOP

A4   a. $5x + 5$: $5(x + 1) = 5 \cdot x + 5 \cdot 1$ b. $16 + 8c$: $8(2 + c) = 8 \cdot 2 + 8 \cdot c$
                                         $= 5x + 5$ $= 16 + 8c$

3	The parentheses may be removed from $4(t - 3)$ by use of the left distributive property of multiplication over subtraction as follows:

$$4(t - 3) = 4 \cdot t - 4 \cdot 3$$
$$= 4t - 12$$

Q5   Use the left distributive property of multiplication over subtraction to remove the parentheses from $7(a - 5)$.

STOP • STOP • STOP • STOP • STOP • STOP • STOP • STOP • STOP

A5 $7a - 35$: $7(a - 5)$
                              $7 \cdot a - 7 \cdot 5$
                              $7a - 35$

Q6   Remove the parentheses:
a. $2(x - 9)$ b. $6(2 - y)$

STOP • STOP • STOP • STOP • STOP • STOP • STOP • STOP • STOP

A6   a. $2x - 18$ b. $12 - 6y$

4	To remove the parentheses from $(a + 3)$, recall that by the multiplication property of one:

$$(a + 3) = 1 \cdot (a + 3)$$

Thus,

$$(a + 3) = 1 \cdot (a + 3)$$
$$= 1 \cdot a + 1 \cdot 3$$
$$= a + 3$$

Notice that the result is exactly the expression within the parentheses. For example,

$$(b + 2) = b + 2$$
$$(7 - y) = 7 - y$$
$$(2x - 3) = 2x - 3$$

Q7   Remove the parentheses:

a. $(x - 9) = $ _____   b. $(3 - x) = $ _____

c. $(5x + 6) = $ _____

STOP • STOP • STOP • STOP • STOP • STOP • STOP • STOP • STOP

A7   a. $x - 9$ b. $3 - x$ c. $5x + 6$

Q8   Remove the parentheses from each of the following algebraic expressions:

a. $2(x + 7)$ b. $5(b - 1)$ c. $3(4 + y)$

d. $(x - 1)$ e. $8(7 - z)$ f. $4(m - 2)$

g. $9(3 + t)$ h. $(4 - a)$

STOP • STOP • STOP • STOP • STOP • STOP • STOP • STOP • STOP

A8   a. $2x + 14$ b. $5b - 5$ c. $12 + 3y$ d. $x - 1$
e. $56 - 8z$ f. $4m - 8$ g. $27 + 9t$ h. $4 - a$

5   The right distributive property of multiplication over addition and the right distributive property of multiplication over subtraction state that

$$(a + b)c = ac + bc$$

and

$$(a - b)c = ac - bc$$

for all rational-number replacements of a, b, and c. Two examples are:

$$(3 + 5)2 = 3 \cdot 2 + 5 \cdot 2$$
$$(7 - 4)\frac{3}{8} = 7 \cdot \frac{3}{8} - 4 \cdot \frac{3}{8}$$

Q9   Use the right distributive property of multiplication over addition to fill in the blanks.

$$(4 + 9)7 = \underline{\hspace{1cm}} + \underline{\hspace{1cm}}$$

STOP • STOP • STOP • STOP • STOP • STOP • STOP • STOP • STOP

A9 $\qquad$ $(4 + 9)7 = 4 \cdot 7 + 9 \cdot 7$

Q10 $\qquad$ Use the right distributive property of multiplication over subtraction to fill in the blanks.

$$\left(\frac{5}{6} - 1\right)6 = \underline{\qquad} - \underline{\qquad}$$

STOP • **STOP** • **STOP** • **STOP** • **STOP** • **STOP** • **STOP** • **STOP** • **STOP**

A10 $\qquad$ $\left(\frac{5}{6} - 1\right)6 = \frac{5}{6} \cdot 6 - 1 \cdot 6$

---

6 $\qquad$ The right distributive properties are also used to simplify algebraic expressions. Two examples are:

$(x - 2)3 = x \cdot 3 - 2 \cdot 3$
$\qquad\quad = 3x - 6$ $\quad$ (it is customary to rewrite $x \cdot 3$ as $3x$)

$(5 + r)3 = 5 \cdot 3 + r \cdot 3$
$\qquad\quad = 15 + 3r$

Q11 $\qquad$ Use the right distributive property of multiplication over addition to remove the parentheses from $(2 + x)7$.

STOP • **STOP** • **STOP** • **STOP** • **STOP** • **STOP** • **STOP** • **STOP** • **STOP**

A11 $\qquad$ $14 + 7x$: $\quad (2 + x)7 = 2 \cdot 7 + x \cdot 7$
$\qquad\qquad\qquad\qquad\qquad = 14 + 7x$

Q12 $\qquad$ Use the right distributive property of multiplication over subtraction to remove the parentheses from $(y - 4)5$.

STOP • **STOP** • **STOP** • **STOP** • **STOP** • **STOP** • **STOP** • **STOP** • **STOP**

A12 $\qquad$ $5y - 20$: $\quad (y - 4)5 = y \cdot 5 - 4 \cdot 5$
$\qquad\qquad\qquad\qquad\qquad = 5y - 20$

Q13 $\qquad$ Remove the parentheses from $(b - 7)9$.

STOP • **STOP** • **STOP** • **STOP** • **STOP** • **STOP** • **STOP** • **STOP** • **STOP**

A13 $\qquad$ $9b - 63$

---

7 $\qquad$ To remove the parentheses in the algebraic expression $3(4x - 7)$, the following steps are used:

$3(4x - 7) = 3 \cdot 4x - 3 \cdot 7$
$\qquad\qquad = 12x - 21$

Q14 $\qquad$ Remove the parentheses from $7(6x - 4)$.

STOP • **STOP** • **STOP** • **STOP** • **STOP** • **STOP** • **STOP** • **STOP** • **STOP**

A14 $42x - 28$: $7(6x - 4) = 7 \cdot 6x - 7 \cdot 4$
                            $= 42x - 28$

Q15   Remove the parentheses from $4\left(12 - \dfrac{3}{4}x\right)$.

STOP   •   STOP   •   STOP   •   STOP   •   STOP   •   STOP   •   STOP   •   STOP   •   STOP

A15 $48 - 3x$

| 8 | To remove the parentheses from $(3x - 2)5$, the procedure is as follows: |

$(3x - 2)5 = 3x \cdot 5 - 2 \cdot 5$
$\qquad\quad = 5 \cdot 3x - 2 \cdot 5$
$\qquad\quad = 15x - 10$

Q16   Remove the parentheses from $(7x - 4)9$.

STOP   •   STOP   •   STOP   •   STOP   •   STOP   •   STOP   •   STOP   •   STOP   •   STOP

A16 $63x - 36$: $(7x - 4)9 = 7x \cdot 9 - 4 \cdot 9$
                            $= 9 \cdot 7x - 4 \cdot 9$
                            $= 63x - 36$

Q17   Remove the parentheses from $(5a + 7)2$.

STOP   •   STOP   •   STOP   STOP   •   STOP   •   STOP   •   STOP   •   STOP   •   STOP

A17 $10a + 14$

| 9 | To remove the parentheses from $4(^-x + 7)$, recall that ^-x means ^-1x (multiplication property of negative one). Thus, |

$4(^-x + 7) = 4(^-1x + 7)$
$\qquad\quad = 4 \cdot {}^-1x + 4 \cdot 7$
$\qquad\quad = {}^-4x + 28$ or $28 - 4x$

Q18   Remove the parentheses from $3(^-x - 5)$.

STOP   •   STOP   •   STOP   •   STOP   •   STOP   •   STOP   •   STOP   •   STOP   •   STOP

A18 $^-3x - 15$: $3(^-x - 5) = 3(^-1x - 5)$
                            $= 3 \cdot {}^-1x - 3 \cdot 5$
                            $= {}^-3x - 15$

| 10 | When removing parentheses by use of the distributive properties, it is convenient to be able to find the result mentally (without showing work). For example, to remove parentheses from $5(3x - 6)$, write only $5(3x - 6) = 15x - 30$. |

Q19   Mentally remove the parentheses from $2(8 - 5t)$.

STOP • **STOP** • **STOP** • **STOP** • **STOP** • **STOP** • **STOP** • **STOP** • **STOP**

A19 $16 - 10t$

Q20   Mentally remove the parentheses from $3\left(7 - \dfrac{4}{3}y\right)$.

STOP • **STOP** • **STOP** • **STOP** • **STOP** • **STOP** • **STOP** • **STOP** • **STOP**

A20 $21 - 4y$

Q21   Remove the parentheses from $(6z + 1)9$.

STOP • **STOP** • **STOP** • **STOP** • **STOP** • **STOP** • **STOP** • **STOP** • **STOP**

A21 $54z + 9$

| 11 | To remove the parentheses from $^-5(3x - 4)$, the following steps can be used: $$^-5(3x - 4) = {}^-5 \cdot 3x - {}^-5 \cdot 4$$ $$= {}^-15x - {}^-20$$ $$= {}^-15x + 20$$ You should notice that $^-15x - {}^-20$ is equivalent to $^-15x + 20$ because of the definition of subtraction. The expression $^-15x + 20$ is considered to be in simplest form. The expression $20 - 15x$ may also be written. |

Q22   Remove the parentheses from $^-4(7x - 9)$ and write in simplest form.

STOP • **STOP** • **STOP** • **STOP** • **STOP** • **STOP** • **STOP** • **STOP** • **STOP**

A22 $^-28x + 36$ or $36 - 28x$: $^-4(7x - 9) = {}^-4 \cdot 7x - {}^-4 \cdot 9$
$$= {}^-28x - {}^-36$$
$$= {}^-28x + 36$$

Q23   Remove the parentheses from $^-1(7 + 3y)$ and write in simplest form.

STOP • **STOP** • **STOP** • **STOP** • **STOP** • **STOP** • **STOP** • **STOP** • **STOP**

A23 $^-7 - 3y$: $^-1(7 + 3y) = {}^-1 \cdot 7 + {}^-1 \cdot 3y$
$$= {}^-7 + {}^-3y$$
$$= {}^-7 - 3y$$

Q24   Remove the parentheses from $^-5(^-3x + 4)$ and write in simplest form.

STOP   •   **STOP**   •   **STOP**   •   **STOP**   •   **STOP**   •   **STOP**   •   **STOP**   •   **STOP**   •   **STOP**

A24 $15x - 20$: $^-5(^-3x + 4) = ^-5 \cdot ^-3x + ^-5 \cdot 4$
$$= 15x + ^-20$$
$$= 15x - 20$$

---

12   The expression $(^-4x - 7) \cdot ^-8$ is simplified:

$(^-4x - 7) \cdot ^-8 = ^-4x \cdot ^-8 - 7 \cdot ^-8$
$$= 32x - ^-56$$
$$= 32x + 56$$

Q25   Simplify $(3y + 9) \cdot ^-6$.

STOP   •   **STOP**   •   **STOP**   •   **STOP**   •   **STOP**   •   **STOP**   •   **STOP**   •   **STOP**   •   **STOP**

A25 $^-18y - 54$: $(3y + 9) \cdot ^-6 = 3y \cdot ^-6 + 9 \cdot ^-6$
$$= ^-18y + ^-54$$
$$= ^-18y - 54$$

Q26   Simplify $(^-a - 5) \cdot ^-2$.

STOP   •   **STOP**   •   **STOP**   •   **STOP**   •   **STOP**   •   **STOP**   •   **STOP**   •   **STOP**   •   **STOP**

A26 $2a + 10$: $(^-a - 5) \cdot ^-2 = ^-a \cdot ^-2 - 5 \cdot ^-2$
$$= 2a - ^-10$$
$$= 2a + 10$$

---

13   By the multiplication property of negative one, the expression $^-(3x + 7)$ is equivalent to $^-1(3x + 7)$. Thus $^-(3x + 7)$ is simplified:

$^-(3x + 7) = ^-1(3x + 7)$
$$= ^-1 \cdot 3x + ^-1 \cdot 7$$
$$= ^-3x + ^-7$$
$$= ^-3x - 7$$

An alternative method to the above is to notice that $^-(3x + 7)$ means the *opposite of* $(3x + 7)$, which can be found by taking the opposite of each term within the parentheses. Thus,

$^-(3x + 7) = ^-3x + ^-7$
$$= ^-3x - 7$$

Q27　　　　Write an equivalent expression for $^-(5 + 2y)$ by forming the opposite of each term within the parentheses.

STOP • STOP • STOP • STOP • STOP • STOP • STOP • STOP • STOP

A27　　　　$^-5 - 2y$:　$^-(5 + 2y) = {^-5} + {^-2y}$

Q28　　　　Simplify $^-1(5 + 2y)$ by removing the parentheses.

STOP • STOP • STOP • STOP • STOP • STOP • STOP • STOP • STOP

A28　　　　$^-5 - 2y$:　$^-1(5 + 2y) = {^-1} \cdot 5 + {^-1} \cdot 2y$
$$= {^-5} + {^-2y}$$
$$= {^-5} - 2y$$

Q29　　　　Simplify $^-(4b - 5)$.

STOP • STOP • STOP • STOP • STOP • STOP • STOP • STOP • STOP

A29　　　　$^-4b + 5$ or $5 - 4b$:　$^-(4b - 5) = {^-1}(4b - 5)$
$$= {^-1} \cdot 4b - {^-1} \cdot 5$$
$$= {^-4b} - {^-5}$$
$$= {^-4b} + 5 \text{ or } 5 - 4b$$

　　　　　　or　　　　　$^-(4b - 5) = {^-4b} - {^-5}$
$$= {^-4b} + 5 \text{ or } 5 - 4b$$

Q30　　　　Simplify $^-(x - 7)$.

STOP • STOP • STOP • STOP • STOP • STOP • STOP • STOP • STOP

A30　　　　$^-x + 7$ or $7 - x$

This completes the instruction for this section.

7.3　　Exercises

Simplify each of the following algebraic expressions:

1. $2(x + 3)$

2. $15\left(\dfrac{2}{3}y - \dfrac{4}{5}\right)$

3. $(4a - 3)$

4. $^-1(2x + 7)$

   5. $(3t + 1)2$ **6.** $(7 + 3b) \cdot {}^-2$

   7. $(4 - 9x) \cdot {}^-3$ **8.** ${}^-(9 + {}^-4c)$

   9. ${}^-2(7t - 4)$ **10.** $4\left({}^-3y + \dfrac{7}{4}\right)$

  11. ${}^-5(8x - 3)$ **12.** ${}^-3({}^-5 - 7z)$

  13. ${}^-(t - 1)$ **14.** ${}^-(3 - 5r)$

  15. ${}^-3({}^-x - 7)$

7.3   Exercise Answers

   1. $2x + 6$ **2.** $10y - 12$

   3. $4a - 3$ **4.** ${}^-2x - 7$

   5. $6t + 2$ **6.** ${}^-14 - 6b$

   7. ${}^-12 + 27x$ or $27x - 12$ **8.** ${}^-9 + 4c$ or $4c - 9$

   9. ${}^-14t + 8$ or $8 - 14t$ **10.** ${}^-12y + 7$ or $7 - 12y$

  11. ${}^-40x + 15$ or $15 - 40x$ **12.** $15 + 21z$

  13. ${}^-t + 1$ or $1 - t$ **14.** ${}^-3 + 5r$ or $5r - 3$

  15. $3x + 21$

7.4   Simplifying Algebraic Expressions by Use of the Associative, Commutative, and Distributive Properties

1   An algebraic expression is said to be in "simplest" form when:

1. All parentheses have been removed.
2. All like terms have been combined.
3. The definition of subtraction has been applied to remove all the "raised" negative signs that can be removed.

Examples:	*Simplest Form?*
$3x + 5 - 2x$	no, like terms not combined
$7x - 3 + 2y$	yes
$4(x - 2) + 3x$	no, parentheses not removed
$3 + {}^-5y$	no, raised negative sign can be removed by writing $3 - 5y$
$9t - {}^-4$	no, raised negative sign can be removed by writing $9t + 4$

Q1   Indicate whether each of the following expressions are in simplest form. If the expression is not in simplest form, briefly state why.

   a. $4x - 2y + 9$ _____ _____

   b. $3(2 - 4y) - 1$ _____ _____

   c. $5t - 6 + 9t$ _____ _____

   d. $8x - 5 + {}^-y$ _____ _____

   e. $5x + 7y - 3xy$ _____ _____

STOP • STOP • STOP • STOP • STOP • STOP • STOP • STOP • STOP

A1   **a.** yes
     b. no, parentheses not removed
     c. no, like terms not combined
     d. no, raised negative sign can be removed by writing $8x - 5 - y$
     e. yes

2   In Section 7.1 algebraic expressions such as $^-2x + 3 + 5x - 7$ were simplified by re-arranging and combining like terms as follows:

$^-2x + 3 + 5x - 7$
$^-2x + 5x + 3 - 7$
$3x - 4$

Q2   Simplify $5 - 6t + 3 - 9t$ by rearranging and combining like terms.

STOP • **STOP** • **STOP** • **STOP** • **STOP** • **STOP** • **STOP** • **STOP** • **STOP**

A2 $^-15t + 8$ or $8 - 15t$: $5 - 6t + 3 - 9t$
          $^-6t - 9t + 5 + 3$
          $^-15t + 8$ or $8 - 15t$

Q3   Simplify $^-3x + 7x - 5 + x$.

STOP • **STOP** • **STOP** • **STOP** • **STOP** • **STOP** • **STOP** • **STOP** • **STOP**

A3 $5x - 5$: $^-3x + 7x - 5 + x$
        $^-3x + 7x + x - 5$
        $5x - 5$

Q4   Simplify each of the following:

   a. $4y - 7 + 6y - 2$ **b.** $^-1 - \dfrac{2}{3}t + 7 - t$

   c. $10b + 6 - 3b$ **d.** $\dfrac{4}{5}z - \dfrac{1}{2}z + 9$

STOP • **STOP** • **STOP** • **STOP** • **STOP** • **STOP** • **STOP** • **STOP** • **STOP**

A4   **a.** $10y - 9$ **b.** $\dfrac{^-5}{3}t + 6$

   c. $7b + 6$ **d.** $\dfrac{3}{10}z + 9$

3   To simplify expressions that involve parentheses:

1. Use the distributive properties to remove parentheses.
2. Rearrange and combine the like terms.

Examples:

1. $2(3y - 7) + 4$
$6y - 14 + 4$
$6y - 10$

2. $^-2x + \dfrac{4}{9}(x - 4)$

$^-2x + \dfrac{4}{9}x - \dfrac{16}{9}$

$\dfrac{^-14}{9}x - \dfrac{16}{9}$

Q5   Simplify $5(2x - 1) + 7$.

STOP • STOP • STOP • STOP • STOP • STOP • STOP • STOP • STOP

A5 $10x + 2$: $5(2x - 1) + 7$
$10x - 5 + 7$
$10x + 2$

Q6   Simplify $3x + 7(^-x + 4)$.

STOP • STOP • STOP • STOP • STOP • STOP • STOP • STOP • STOP

A6 $^-4x + 28$ or $28 - 4x$: $3x + 7(^-x + 4)$
$3x + {}^-7x + 7 \cdot 4$
$^-4x + 28$ or $28 - 4x$

Q7   Simplify $\left(^-3y + \dfrac{1}{8}\right) - \dfrac{2}{5} + 2y$.

STOP • STOP • STOP • STOP • STOP • STOP • STOP • STOP • STOP

A7 $^-y - \dfrac{11}{40}$: $\left(^-3y + \dfrac{1}{8}\right) - \dfrac{2}{5} + 2y$

$^-3y + \dfrac{1}{8} - \dfrac{2}{5} + 2y$

$^-y - \dfrac{11}{40}$

Q8   Simplify each of the following:

a. $(7x + 8) - 3$ b. $3(2t - 4) + 7$

c. $8b + 2(b - 3)$ d. $(^-5 + 2x) - 3x$

e. $(^-3x + 4) - 9$ f. $^-7(4 - x) + 3x$

g. $\dfrac{5}{12}(z - 4) + 8$ h. $^-1(3 + y) + \dfrac{1}{2}y$

i. $4(^-3x + 2) + 12x$ j. $^-4y + 2(y + 8)$

k. $4a + {}^-3(a + 2)$ l. $\dfrac{-2}{7}(y + 7) - 8$

m. $x + {}^-1(3x - 5)$ n. $(5y + 4) - 4$

o. $^-2(4t - 8) - 5t$ p. $\dfrac{-1}{3}(3x + 7) - \dfrac{5}{6}x$

STOP • STOP • STOP • STOP • STOP • STOP • STOP • STOP • STOP

A8   a. $7x + 5$ b. $6t - 5$ c. $10b - 6$
      d. $^-x - 5$ e. $^-3x - 5$ f. $10x - 28$
      g. $\dfrac{5}{12}z + \dfrac{19}{3}$ h. $\dfrac{-1}{2}y - 3$ i. 8
      j. $^-2y + 16$ or $16 - 2y$ k. $a - 6$ l. $\dfrac{-2}{7}y - 10$

m. $^-2x + 5$ or $5 - 2x$ **n.** $5y$

o. $^-13t + 16$ or $16 - 13t$ **p.** $\dfrac{^-11}{6}x - \dfrac{7}{3}$

---

4   Recall that the expression $^-(x + 3)$ could be simplified in either of two ways:

$$^-(x + 3) = ^-1(x + 3) \qquad or \qquad ^-(x + 3) = ^-x + ^-3$$
$$= ^-1 \cdot x + ^-1 \cdot 3 \qquad\qquad\qquad\quad = ^-x - 3$$
$$= ^-x + ^-3$$
$$= ^-x - 3$$

$[^-(x + 3)$ means the opposite of $(x + 3)$, which is the same as the opposite of each term within the parentheses]

Q9   Simplify:

a. $^-(y + 9) =$ _____   **b.** $^-(2x + 7) =$ _____

c. $^-\left(4z - \dfrac{4}{3}\right) =$ _____   **d.** $^-(b - 12) =$ _____

e. $^-(5y + 7) =$ _____

STOP • STOP • STOP • STOP • STOP • STOP • STOP • STOP • STOP

A9   **a.** $^-y - 9$ **b.** $^-2x - 7$

c. $^-4z + \dfrac{4}{3}$ or $\dfrac{4}{3} - 4z$ **d.** $^-b + 12$ or $12 - b$

e. $^-5y - 7$

---

5   To simplify $4 - (x + 7)$ the following steps are used:

$$4 - (x + 7) = 4 + ^-(x + 7)$$
$$= 4 + ^-x + ^-7$$
$$= ^-x + ^-3$$
$$= ^-x - 3$$

Q10   Simplify $15 - (2x + 7)$.

STOP • STOP • STOP • STOP • STOP • STOP • STOP • STOP • STOP

A10 $^-2x + 8$ or $8 - 2x$: $15 - (2x + 7) = 15 + ^-(2x + 7)$
$$= 15 + ^-2x + ^-7$$
$$= ^-2x + 8 \text{ or } 8 - 2x$$

Q11   Simplify $^-4 - (2b - 3)$.

STOP • STOP • STOP • STOP • STOP • STOP • STOP • STOP • STOP

A11 $\quad$ $^-2b - 1$: $\quad$ $^-4 - (2b - 3) = {}^-4 + {}^-(2b - 3)$
$$= {}^-4 + {}^-2b - {}^-3$$
$$= {}^-4 - 2b + 3$$
$$= {}^-2b - 1$$

Q12 $\quad$ Simplify each of the following:

a. $4 - (y + 3)$ $\qquad$ b. $2z - (z - 5)$

c. $^-5 - \left(\dfrac{2}{3} + 4x\right)$ $\qquad$ d. $\dfrac{3}{2} + \left(2a - \dfrac{4}{5}\right)$

e. $9 - (2a - 4)$ $\qquad$ f. $6y - (2y + 7)$

STOP • STOP • STOP • STOP • STOP • STOP • STOP • STOP • STOP

A12 $\quad$ a. $^-y + 1$ or $1 - y$ $\qquad$ b. $z + 5$ $\qquad$ c. $^-4x - \dfrac{17}{3}$

d. $2a + \dfrac{7}{10}$ $\qquad$ e. $^-2a + 13$ or $13 - 2a$ $\qquad$ f. $4y - 7$

---

6 $\quad$ To simplify $(2x + 7) - (3x - 4)$, remove the parentheses and combine like terms as follows:

$$(2x + 7) - (3x - 4) = (2x + 7) + {}^-(3x - 4)$$
$$= (2x + 7) + {}^-3x - {}^-4$$
$$= 2x + 7 + {}^-3x + 4$$
$$= {}^-x + 11$$

Notice that parentheses preceded by no sign (or a "+" sign) are removed by just dropping them, whereas parentheses preceded by a "−" sign are removed by rewriting the subtraction problem as an addition problem and taking the opposite of the terms within the parentheses.

Q13 $\quad$ Remove the parentheses only (do not simplify): $\quad$ $(x - 5) - (2x + 7)$

STOP • STOP • STOP • STOP • STOP • STOP • STOP • STOP • STOP

A13 $\quad$ $x - 5 - 2x - 7$: $\quad$ $(x - 5) - (2x + 7) = (x - 5) + {}^-(2x + 7)$
$$= (x - 5) + {}^-2x + {}^-7$$
$$= x - 5 - 2x - 7$$

Q14   Simplify the result in A13.

STOP   •   STOP   •   STOP   •   STOP   •   STOP   •   STOP   •   STOP   •   STOP   •   STOP

A14 $^-x - 12$

Q15   Remove the parentheses only (do not simplify): $(1 - 3b) - (4b - 7)$

STOP   •   STOP   •   STOP   •   STOP   •   STOP   •   STOP   •   STOP   •   STOP   •   STOP

A15 $1 - 3b - 4b + 7$

Q16   Simplify the result in A15.

STOP   •   STOP   •   STOP   •   STOP   •   STOP   •   STOP   •   STOP   •   STOP   •   STOP

A16 $^-7b + 8$ or $8 - 7b$

Q17   Simplify $(x + 3) - (x - 3)$.

STOP   •   STOP   •   STOP   •   STOP   •   STOP   •   STOP   •   STOP   •   STOP   •   STOP

A17   6

Q18   Simplify $(^-z - 4) - (2z - 4)$.

STOP   •   STOP   •   STOP   •   STOP   •   STOP   •   STOP   •   STOP   •   STOP   •   STOP

A18. ^-3z

Q19   Simplify each of the following expressions:
           a. $(5 - 4x) - (x + 3)$ **b.** $(2x - 5) - (3 - 5x)$

           c. $(^-5 + b) - (b - 5)$ **d.** $^-(4y + 3) - (7y + 3)$

e. $(z - 9) - (^-z - 5)$

f. $^-(x + 5) - (x - 5)$

STOP • STOP • STOP • STOP • STOP • STOP • STOP • STOP • STOP

A19   a. $^-5x + 2$ or $2 - 5x$ b. $7x - 8$ c. 0
       d. $^-11y - 6$ e. $2z - 4$ f. ^-2x

---

7   To simplify expressions such as $3(x - 2) - 4(x + 3)$, the following steps are used:

$$3(x - 2) - 4(x + 3) = 3(x - 2) + ^-4(x + 3)$$
$$= 3(x - 2) + ^-4x + ^-12$$
$$= 3x - 6 - 4x - 12$$
$$= ^-x - 18$$

Note: The work is usually shortened to:

$$3(x - 2) - 4(x + 3) = 3x - 6 - 4x - 12$$
$$= ^-x - 18$$

Q20   Remove the parentheses only (do not simplify): $5(y + 7) - 2(y - 4)$

STOP • STOP • STOP • STOP • STOP • STOP • STOP • STOP • STOP

A20 $5y + 35 - 2y + 8$

Q21   Complete the simplification of A20.

STOP • STOP • STOP • STOP • STOP • STOP • STOP • STOP • STOP

A21 $3y + 43$

Q22   Remove the parentheses only (do not simplify): $2(2x - 3) - 4(x + 5)$

STOP • STOP • STOP • STOP • STOP • STOP • STOP • STOP • STOP

A22 $4x - 6 - 4x - 20$

Q23   Complete the simplification of A22.

STOP • STOP • STOP • STOP • STOP • STOP • STOP • STOP • STOP

A23 $^-26$

Q24   Remove the parentheses only (do not simplify): $^-3(^-2z + 7) - 4(z + 5)$

STOP • STOP • STOP • STOP • STOP • STOP • STOP • STOP • STOP

A24 $\qquad$ $6z - 21 - 4z - 20$

Q25 $\qquad$ Complete the simplification of A24.

STOP • STOP • STOP • STOP • STOP • STOP • STOP • STOP • STOP

A25 $\qquad$ $2z - 41$

Q26 $\qquad$ Simplify each of the following algebraic expressions:
 $\quad$ **a.** $(3x - 4) - 2(2x + 5)$ $\qquad$ **b.** $^-2(b + 7) - (3b - 5)$

 $\quad$ **c.** $2(y + 3) + 3(^-2y - 7)$ $\qquad$ **d.** $(y + 7) - 2(y + 5)$

 $\quad$ **e.** $^-(5x - 3) - \dfrac{4}{5}(3x - 1)$ $\qquad$ **f.** $^-4(t - 3) - 4(2t - 1)$

STOP • STOP • STOP • STOP • STOP • STOP • STOP • STOP • STOP

A26 $\qquad$ **a.** $^-x - 14$ $\qquad\qquad$ **b.** $^-5b - 9$
 $\qquad\qquad$ **c.** $^-4y - 15$ $\qquad\qquad$ **d.** $^-y - 3$
 $\qquad\qquad$ **e.** $\dfrac{-37}{5}x + \dfrac{19}{5}$ or $\dfrac{19}{5} - \dfrac{37}{5}x$ $\quad$ **f.** $^-12t + 16$ or $16 - 12t$

This completes the instruction for this section.

7.4 $\qquad$ Exercises

1. An algebraic expression is said to be in simplest form when what three conditions have been met?
2. Simplify each of the following by rearranging and combining like terms:
   - **a.** $^-3x - 5 - x$ $\qquad\qquad$ **b.** $4 - 2x + 7 + 3x$
   - **c.** $5y - 2y - 6 + 11$ $\qquad$ **d.** $^-7 - 5z + 6 - 3z$
   - **e.** $1 - t - 1 - t$ $\qquad\qquad$ **f.** $3t - 4t + 6$

g. $7 - 6b + 3 + \dfrac{15}{4}b$ **h.** $^-5 - 4 + 7z - 3z$

i. $\dfrac{3}{8}y - \dfrac{2}{3} - \dfrac{4}{7}y$ **j.** $4z - 5 - 4z + 5$

3. Remove the parentheses only (do not simplify):
   a. $(x + 8)$ **b.** $^-(x + 8)$
   c. $^-(y - 4)$ **d.** $(^-3x - 5)$
   e. $3(^-4y + 7)$ **f.** $^-(z + 3) + (2z - 4)$

   g. $(5b - 3) - (4 - 6b)$ **h.** $^-\left(\dfrac{1}{2}x + 3\right) - (4 + x)$

   i. $2(a + 7) - 3(4 - 3a)$ **j.** $^-4\left(\dfrac{1}{4}x - 5\right) + 7(3x - 4)$

4. Simplify each of the following algebraic expressions:
   a. $^-3(x + 7)$ **b.** $(4 + 3y) + (^-5y - 2)$
   c. $^-(a - 5) + (3a - 5)$ **d.** $2(3x - 5) - (^-7x + 5)$

   e. $\dfrac{^-3}{4}(y - 4) - 3(^-y - 7)$ **f.** $9(^-3 + z) - 3(^-3z - 9)$

   g. $^-3x - (4 + 7x) - 6$ **h.** $5y - 2(2 - 7y) - 3y$

   i. $4 - 7b - \dfrac{3}{4}(4 + b)$ **j.** $^-3(x + 5) + 5(^-4x - 7)$

7.4   Exercise Answers

1. **1.** All parentheses are removed.
   2. All like terms are combined.
   3. The definition of subtraction has been applied to remove "raised" negative signs.

2. **a.** $^-4x - 5$ **b.** $x + 11$ **c.** $3y + 5$

   d. $^-8z - 1$ **e.** ^-2t **f.** $^-t + 6$ or $6 - t$

   g. $\dfrac{^-9}{4}b + 10$ or $10 - \dfrac{9}{4}b$ **h.** $4z - 9$ **i.** $\dfrac{^-11}{56}y - \dfrac{2}{3}$

   j. 0

3. **a.** $x + 8$ **b.** $^-x - 8$ **c.** $^-y + 4$
   d. $^-3x - 5$ **e.** $^-12y + 21$ **f.** $^-z - 3 + 2z - 4$

   g. $5b - 3 - 4 + 6b$ **h.** $\dfrac{^-1}{2}x - 3 - 4 - x$ **i.** $2a + 14 - 12 + 9a$

   j. $^-x + 20 + 21x - 28$

4. **a.** $^-3x - 21$ **b.** $^-2y + 2$ **c.** $2a$

   d. $13x - 15$ **e.** $\dfrac{9}{4}y + 24$ **f.** $18z$

   g. $^-10x - 10$ **h.** $16y - 4$ **i.** $\dfrac{^-31}{4}b + 1$

   j. $^-23x - 50$

Chapter 7 Sample Test

At the completion of Chapter 7 it is expected that you will be able to work the following problems.

7.1

Simplifying Algebraic Expressions by Use of the Commutative and Associative Properties of Addition

1. Identify the like terms in each of the following algebraic expressions:
   a. $3x - 5y - 6x + 4$
   b. $3r - 2 + 7s - 6$
2. Simplify by combining like terms:
   a. $4x - 2x$
   b. $m - 3 - 7m$

   c. $7 - 4y - 6 - 3y$
   d. $\dfrac{3}{5}x - \dfrac{4}{7}x - 2$

3. Simplify by use of the commutative and/or associative properties of addition:
   a. $(3x - 2) + 5x$
   b. $^-15 + (7 - 8a)$

   c. $^-x + \left(7x - \dfrac{2}{5}x\right)$
   d. $\left(\dfrac{3}{5}b - \dfrac{2}{3}b\right) + b$

4. Simplify:
   a. $4y + {}^-9$
   b. $5 - 2r + 2s - 6r$

   c. $\dfrac{1}{3}x + y - 6xy - 4y$
   d. $4m + 7 - 3m + {}^-7 - m$

7.2

Simplifying Algebraic Expressions by Use of the Commutative and Associative Properties of Multiplication

5. Simplify:

   a. $2\left(\dfrac{3}{4}x\right)$
   b. $\dfrac{^-1}{5}\left(\dfrac{^-4}{9}y\right)$

   c. $\dfrac{^-1}{7}(7m)$
   d. $\dfrac{5}{7}\left(\dfrac{7}{5}x\right)$

   e. $\left(\dfrac{^-1}{9}z\right) \cdot {}^-9$
   f. $\left(\dfrac{^-3}{5}x\right)\dfrac{2}{3}$

   g. $(4y)\dfrac{5}{4}$
   h. $\left(\dfrac{^-3}{5}x\right)20$

7.3

Using the Distributive Properties to Simplify Algebraic Expressions

6. Simplify:
   a. $2(x - 3)$
   b. $(y + 7) \cdot {}^-3$
   c. $^-4(2x - 3)$
   d. $(4 - 2x)5$
   e. $(7x - 9)$
   f. $^-(7x - 9)$
   g. $^-(^-t + 2)$
   h. $^-5(^-7 - 3x)$
   i. $(6x - 5) \cdot {}^-8$
   j. $^-7(3y + {}^-9)$

7.4 **Simplifying Algebraic Expressions by Use of the Associative, Commutative, and Distributive Properties**

7. Simplify:

a. $^-4y + 9 + 2y + 5$

b. $\frac{4}{5}x - \frac{2}{3}x + 2 - 9$

c. $(7x + 8) - 8$

d. $^-2(x + 7) - 3$

e. $\frac{^-4}{5}(y + 10) + 7$

f. $4 - (z + 7)$

g. $3(x + 2) + 7(x - 5)$

h. $(y - 6) - (4y + 7)$

i. $^-7(x + 2) - 2(3x - 7)$

j. $^-\left(\frac{1}{3}x + 5\right) - (5 - x)$

Chapter 7 Sample Test Answers

1. a. $3x$ and ^-6x

   b. $^-2$ and $^-6$

2. a. $2x$

   b. $^-6m - 3$

   c. $^-7y + 1$ or $1 - 7y$

   d. $\frac{1}{35}x - 2$

3. a. $8x - 2$

   b. $^-8a - 8$

   c. $\frac{28}{5}x$

   d. $\frac{14}{15}b$

4. a. $4y - 9$

   b. $^-8r + 2s + 5$

   c. $\frac{1}{3}x - 3y - 6xy$

   d. 0

5. a. $\frac{3}{2}x$ b. $\frac{4}{45}y$ c. ^-m d. x e. z f. $\frac{^-2}{5}x$ g. $5y$

   h. ^-12x

6. a. $2x - 6$

   b. $^-3y - 21$

   c. $^-8x + 12$ or $12 - 8x$

   d. $20 - 10x$

   e. $7x - 9$

   f. $^-7x + 9$ or $9 - 7x$

   g. $t - 2$

   h. $15x + 35$

   i. $^-48x + 40$ or $40 - 48x$

   j. $^-21y + 63$ or $63 - 21y$

7. a. $^-2y + 14$ or $14 - 2y$

   b. $\frac{2}{15}x - 7$

   c. $7x$

   d. $^-2x - 17$

   e. $\frac{^-4}{5}y - 1$

   f. $^-z - 3$

   g. $10x - 29$

   h. $^-3y - 13$

   i. ^-13x

   j. $\frac{2}{3}x - 10$

Chapter 8

Solving Equations

Chapter 7 dealt with the procedures involved in simplifying open algebraic expressions. These included the use of the fundamental operations (addition, subtraction, multiplication, and division) in combining like terms and the application of the commutative, associative, and distributive properties. The skills developed in Chapter 7 will now be utilized in the study of equations.

8.1   Equations, Open Sentences, Replacement Set, and Solution Set

---

1   An equation is a statement that the expressions on opposite sides of an equal sign represent the same number. Some examples of equations are:

$$3 + 4 = 9 - 2$$
$$15 = y + 4$$
$$x - 2 = 6$$
$$2x - 3 = 5x + 9$$

The expressions on opposite sides of the equals sign are referred to as the left and right *sides* of the equation. For example,

$$\underbrace{4x - 7}_{\text{left side}} = \underbrace{2 - 6x}_{\text{right side}}$$

Q1   Identify the left and right sides of the following equations:

a. $14 - 9 = 5$ left side _____ ; right side _____

b. $3 = y + 1$ left side _____ ; right side _____

c. $3x - 7 = 8x + 13$ left side _____ ; right side _____

STOP • STOP • STOP • STOP • STOP • STOP • STOP • STOP • STOP

A1   a. left side, $14 - 9$; right side, 5
b. left side, 3; right side, $y + 1$
c. left side, $3x - 7$; right side, $8x + 13$

---

2   Equations that do not contain enough information to be judged as either true or false are often referred to as *open sentences*. Thus,

$$18 - y = 14 \qquad \text{and} \qquad x + 6 = 9$$

are examples of open sentences, because they cannot be judged true or false until numbers

---

are replaced for the unknown quantities y and x. If y is replaced by 10 in the open sentence $18 - y = 14$, the resulting statement $18 - 10 = 14$ is false. If x is replaced by 3 in the open sentence $x + 6 = 9$, the resulting statement $3 + 6 = 9$ is true.

Q2 **a.** If x is replaced by 7 in the open sentence $x - 2 = 5$, is the resulting statement true or false? _____

b. If y is replaced by 4 in the open sentence $12 + y = 15$, is the resulting statement true or false? _____

STOP • STOP • STOP • STOP • STOP • STOP • STOP • STOP • STOP

A2 **a.** true: $7 - 2 = 5$
b. false: $12 + 4 \neq 15$ ($\neq$ means "is not equal to")

3 The set of all numbers that may be used to replace the variable in an open sentence is called the *replacement set* of the variable. If the replacement set is not stated, it will be understood to be the set of rational numbers.

Consider the open sentence $x - 5 = 15$. If the replacement set for the variable x is the set $\{17, 18, 19, 20, 21\}$, the result of all possible replacements for x is four *false* statements,

$17 - 5 = 15$
$18 - 5 = 15$
$19 - 5 = 15$
$21 - 5 = 15$

and one *true* statement,

$20 - 5 = 15$

Q3 If the replacement set for the variable x is $\{0, 6, {}^-3, 7\}$, find the result of all possible replacements for x in the open sentence $x + 11 = 17$. Label each statement true or false.

STOP • STOP • STOP • STOP • STOP • STOP • STOP • STOP • STOP

A3 $0 + 11 = 17$ false ${}^-3 + 11 = 17$ false
$6 + 11 = 17$ true $7 + 11 = 17$ false

4 To solve an equation is to find all values of the variable from the replacement set which convert the open sentence into a true statement. The values are called the *solutions* of the equation. Thus the solution of the equation $x + 11 = 17$ is $x = 6$, because 6 is the only rational number that will convert $x + 11 = 17$ into a true statement, namely, $6 + 11 = 17$.

The *solution set* or *truth set* of an equation is the set of all values from the replacement set which converts the open sentence into a true statement. For example, the solution set for the equation $x + 11 = 17$ is $\{6\}$.

Q4 Use the given set as the replacement set to determine the solution (truth) set for each equation:

a. $\{4, 9, 13, 15\}$: $y + 4 = 17$ _____

b. $\{0, {}^-1, 2, {}^-3, 4, {}^-6, 7\}$: ${}^-3x = {}^-21$ _____

    c. $\{0, 5, {}^-4, {}^-7\}$: $9 + x = 9$ _____

    d. $\{6, {}^-3, 5, 1, 0\}$: $4x + 3 = 23$ _____

STOP • **STOP** • **STOP** • **STOP** • **STOP** • **STOP** • **STOP** • **STOP** • **STOP**

A4   **a.** $\{13\}$: because $13 + 4 = 17$ **b.** $\{7\}$: because ${}^-3(7) = {}^-21$
    c. $\{0\}$: because $9 + 0 = 9$ **d.** $\{5\}$: because $4(5) + 3 = 23$

This completes the instruction for this section.

8.1   Exercises

Use $\{0, 1, 2, 3, 4, 5, 6, 7, 8, 9, 10\}$ as the replacement set and find the truth set for each of the following equations:

1. $y + 7 = 15$ 2. $2 + x = 7$ 3. $6 = 2 + y$ 4. $x - 5 = 1$
5. $24 - y = 15$ 6. $6 = y - 2$ 7. $4y = 16$ 8. $2x = 0$
9. $56 = 8y$ 10. $2x + 5 = 7$ 11. $3 + 2x = 11$ 12. $3y - 1 = 5$
13. $8 + 7y = 8$ 14. $3x + 3 = x + 7$

8.1   Exercise Answers

1. $\{8\}$: $8 + 7 = 15$ 2. $\{5\}$: $2 + 5 = 7$
3. $\{4\}$: $6 = 2 + 4$ 4. $\{6\}$: $6 - 5 = 1$
5. $\{9\}$: $24 - 9 = 15$ 6. $\{8\}$: $6 = 8 - 2$
7. $\{4\}$: $4(4) = 16$ 8. $\{0\}$: $2(0) = 0$
9. $\{7\}$: $56 = 8(7)$ 10. $\{1\}$: $2(1) + 5 = 7$
11. $\{4\}$: $3 + 2(4) = 11$ 12. $\{2\}$: $3(2) - 1 = 5$
13. $\{0\}$: $8 + 7(0) = 8$ 14. $\{2\}$: $3(2) + 3 = 2 + 7$

8.2   Addition and Subtraction Principles of Equality

1

Since it is not always possible to guess the solution to an equation, it is necessary to study some basic procedures that can be used to solve equations. Recall that an equation is a statement that the expressions on opposite sides of the equal sign represent the same number. Since it is necessary to maintain this equality between sides, a basic rule in solving equations is that *whatever operation is performed on one side of an equation must also be performed on the other side of the equation.*

In general, solving an equation is like untying a knot, in that you always do the opposite of what has been done to form the equation. The equation has been solved when it has been changed to the form "a variable = a number" or "a number = a variable." If the variable is x, the form is "$x = a$ number" or "a number $= x$."

In the equation $x - 2 = 10$, for example, 2 has been subtracted from x to equal 10. Since *the opposite of subtracting 2 is adding 2,* the equation can be solved by adding 2 to both sides as follows:

$$x - 2 = 10 \quad \text{(add 2 to both sides)}$$
$$x - 2 + 2 = 10 + 2$$
$$x = 12$$

The solution can be checked by seeing if it converts the original open sentence into a true statement.

Check: $x - 2 = 10$
$$12 - 2 \overset{?}{=} 10$$
$$10 = 10$$

So 12 is the correct solution, because 12 converts $x - 2 = 10$ into the true statement $10 = 10$.

The solution of the preceding equation demonstrates the *addition principle of equality:* *If the same number is added to both sides of an equation, the result is another equation with the same truth set.* In general, if $a = b$, then $a + c = b + c$ for any numbers $a, b,$ and c.

Q1   Solve the following equations using the addition principle of equality and check the solutions:

a. $x - 3 = 5$ **b.** $y - 12 = {}^{-}7$ **c.** $3 = x - 6$

STOP • **STOP** • **STOP** • **STOP** • **STOP** • **STOP** • **STOP** • **STOP** • **STOP**

A1   **a.** $x - 3 = 5$ (3 was subtracted from x, so add 3 to both sides)
$$x - 3 + 3 = 5 + 3$$
$$x = 8*$$
Check: $x - 3 = 5$
$$8 - 3 \overset{?}{=} 5$$
$$5 = 5$$

b. $y - 12 = {}^{-}7$ (12 was subtracted from x, so add 12 to both sides)
$$y - 12 + 12 = {}^{-}7 + 12$$
$$y = 5$$
Check: $y - 12 = {}^{-}7$
$$5 - 12 \overset{?}{=} {}^{-}7$$
$${}^{-}7 = {}^{-}7$$

c. $3 = x - 6$ (6 was subtracted from x, so add 6 to both sides)
$$3 + 6 = x - 6 + 6$$
$$9 = x$$
Check: $3 = x - 6$
$$3 \overset{?}{=} 9 - 6$$
$$3 = 3$$

*The truth or solution set would be {8}. However, in this and the remaining sections of this chapter, solutions to equations will be left as $x = 5, y = 7,$ and so on.

2

Observe that in the equation $x + 4 = 13$, 4 has been added to x to equal 13. The opposite of adding 4 is subtracting 4, so the equation can be solved by subtracting 4 from both sides as follows:

$$x + 4 = 13$$
$$x + 4 - 4 = 13 - 4$$
$$x = 9$$

Check: $x + 4 = 13$
$$9 + 4 \overset{?}{=} 13$$
$$13 = 13$$

The solution of the preceding equation demonstrates a second principle useful in solving equations, the *subtraction principle of equality: If the same number is subtracted from both sides of an equation, the result is another equation with the same truth set.* In general, if $a = b$, then $a - c = b - c$ for any numbers a, b, and c.

Q2

Solve the following equations using the subtraction principle of equality and check each of the solutions:

a. $y + 1 = 16$ **b.** $x + 14 = {}^-29$ **c.** $34 = x + 11$

STOP • STOP • STOP • STOP • STOP • STOP • STOP • STOP • STOP

A2

a. $y + 1 = 16$ (1 was added to y, so subtract 1 from both sides)
$$y + 1 - 1 = 16 - 1$$
$$y = 15$$

Check: $y + 1 = 16$
$$15 + 1 \overset{?}{=} 16$$
$$16 = 16$$

b. $x + 14 = {}^-29$ (14 was added to x, so subtract 14 from both sides)
$$x + 14 - 14 = {}^-29 - 14$$
$$x = {}^-43$$

Check: $x + 14 = {}^-29$
$${}^-43 + 14 \overset{?}{=} {}^-29$$
$${}^-29 = {}^-29$$

c. $34 = x + 11$ (11 was added to x, so subtract 11 from both sides)
$$34 - 11 = x + 11 - 11$$
$$23 = x$$

Check: $34 = x + 11$
$$34 \overset{?}{=} 23 + 11$$
$$34 = 34$$

---

3   It is often necessary to simplify one or both sides of an equation by combining like terms before proceeding with the solution. For example,

$$x - 5 = 15 - 2 \quad \text{(simplify by combining like terms)}$$
$$x - 5 = 13$$
$$x - 5 + 5 = 13 + 5$$
$$x = 18$$

Q3   Solve the equation by first combining like terms: $y - 7 + 2 = 13$

STOP • **STOP** • **STOP** • **STOP** • **STOP** • **STOP** • **STOP** • **STOP** • **STOP**

A3
$$y - 7 + 2 = 13$$
$$y - 5 = 13$$
$$y - 5 + 5 = 13 + 5$$
$$y = 18$$

This completes the instruction for this section.

8.2   Exercises

Use the addition and subtraction principles of equality to solve the following equations and check each of the solutions:

1. $x - 2 = 5$
2. $y + 7 = 11$
3. $1 = x - 7$
4. $x + 3 = 3$
5. $12 = x + 8$
6. $x - 7 = -5$
7. $11 + y = -7$
8. $x + 6 = 5$
9. $^-3 = y + 9$
10. $7 + x = 2$
11. $^-1 = x + 1$
12. $9 - 7 = x + 5$
13. $y + 7 = 2 - 11$
14. $^-4 + x = 7 - 14$

8.2   Exercise Answers

1. $x = 7$
2. $y = 4$
3. $8 = x \, (x = 8)$
4. $x = 0$
5. $4 = x \, (x = 4)$
6. $x = 2$
7. $y = {}^-18$
8. $x = {}^-1$
9. $^-12 = y \, (y = {}^-12)$
10. $x = {}^-5$
11. $^-2 = x \, (x = {}^-2)$
12. $^-3 = x \, (x = {}^-3)$
13. $y = {}^-16$
14. $x = {}^-3$

8.3   Multiplication and Division Principles of Equality

1   In Section 8.2, equations such as $x - 2 = 5$ and $^-3 = x + 9$ were solved. In each of these equations the understood coefficient of the variable is 1. That is, x is understood to be the same as $1x$. The purpose of this section is to develop skill in solving equations in which the coefficient of the variable is a number other than 1. Some examples of this type of equation are $^-5x = 20$ and $\frac{3}{7}y = 27$.

Consider, first, the equation

$$4x = 24$$

Recall that the term $4x$ means 4 times x. Since the variable x has been multiplied by 4 and the opposite of multiplying by 4 is dividing by 4, the equation can be solved by dividing both sides of the equation by 4, as follows:

$$4x = 24 \qquad \text{Check:} \qquad 4x = 24$$
$$\frac{4x}{4} = \frac{24}{4} \qquad\qquad\qquad 4(6) \overset{?}{=} 24$$
$$1x = 6 \qquad\qquad\qquad\qquad 24 = 24$$
$$x = 6$$

The procedure used in the preceding equation demonstrates the *division principle of equality: If both sides of an equation are divided by the same nonzero number, the result is another equation with the same truth set.* (Zero is excluded since division by zero is impossible.) In general, if $a = b$, then $\frac{a}{c} = \frac{b}{c}$ for any numbers a, b, and c, $c \neq 0$.

Two examples of the division principle of equality are as follows:

$$^-3x = 75 \qquad (x \text{ was multiplied by } ^-3, \text{ so divide both sides by } ^-3)$$
$$\frac{^-3x}{^-3} = \frac{75}{^-3}$$
$$1x = ^-25$$
$$x = ^-25$$

Check:
$$^-3x = 75$$
$$^-3(^-25) \overset{?}{=} 75$$
$$75 = 75$$

$$^-17y = ^-29 \qquad (y \text{ was multiplied by } ^-17, \text{ so divide both sides by } ^-17)$$
$$\frac{^-17y}{^-17} = \frac{^-29}{^-17}$$
$$1y = \frac{29}{17}$$
$$y = \frac{29}{17}$$

Check:
$$^-17y = ^-29$$
$$^-17\left(\frac{29}{17}\right) \overset{?}{=} ^-29$$
$$^-29 = ^-29$$

Notice that the number used to divide both sides is exactly the same as the coefficient of the variable.

Q1   Solve the following equations using the division principle of equality and check each of the solutions:

a. $2x = 10$

b. $^-4y = 12$

c. $4y = ^-8$

d. $^-3x = ^-7$

STOP • STOP • STOP • STOP • STOP • STOP • STOP • STOP • STOP

A1

a. $2x = 10$

$\dfrac{2x}{2} = \dfrac{10}{2}$

$1x = 5$

$x = 5$

Check: $2x = 10$

$2(5) \overset{?}{=} 10$

$10 = 10$

b. $^-4y = 12$

$\dfrac{^-4y}{^-4} = \dfrac{12}{^-4}$

$1y = ^-3$

$y = ^-3$

Check: $^-4y = 12$

$^-4(^-3) \overset{?}{=} 12$

$12 = 12$

c. $4y = ^-8$

$\dfrac{4y}{4} = \dfrac{^-8}{4}$

$1y = ^-2$

$y = ^-2$

Check: $4y = ^-8$

$4(^-2) \overset{?}{=} ^-8$

$^-8 = ^-8$

d. $^-3x = ^-7$

$\dfrac{^-3x}{^-3} = \dfrac{^-7}{^-3}$

$1x = \dfrac{7}{3}$

$x = \dfrac{7}{3}$

Check: $^-3x = ^-7$

$\dfrac{^-3}{1}\left(\dfrac{7}{3}\right) \overset{?}{=} ^-7$

$^-7 = ^-7$

---

2   When the coefficient of the variable in an equation is a fraction, a similar procedure can be followed. For example, in the equation $\dfrac{3}{4}x = 12$, because the variable x has been multiplied by $\dfrac{3}{4}$, the equation can be solved by dividing both sides by $\dfrac{3}{4}$.

$\dfrac{3}{4}x = 12$ $\left(\text{divide by } \dfrac{3}{4}\right)$

$\dfrac{\frac{3}{4}x}{\frac{3}{4}} = \dfrac{12}{\frac{3}{4}}$

$1x = 12 \div \dfrac{3}{4}$

$x = 12 \cdot \dfrac{4}{3}$

$x = 16$

The above solution can be simplified if it is recalled that dividing by $\dfrac{3}{4}$ is the same as multiplying by its reciprocal, $\frac{4}{3}$. Thus, the value of $1x$ (or x) can be found by multiplying both sides of the equation by $\frac{4}{3}$ as follows:

$$\frac{3}{4}x = 12 \qquad\qquad \text{Check:} \qquad \frac{3}{4}x = 12$$

$$\frac{4}{3}\left(\frac{3}{4}x\right) = \frac{4}{3}(12) \qquad\qquad \frac{3}{4}(16) \overset{?}{=} 12$$

$$1x = \frac{4}{3}\left(\frac{12}{1}\right) \qquad\qquad\qquad 12 = 12$$

$$x = 16$$

The procedure used to solve the preceding equation demonstrates the fourth principle useful in solving equations, the *multiplication principle of equality*: *If both sides of an equation are multiplied by the same nonzero number, the result is another equation with the same truth set.* In general, if $a = b$, then $ac = bc$ for any numbers a, b, and c, $c \neq 0$.

Study the following examples of the multiplication principle of equality before proceeding to the problems of Q2.

$$\frac{5}{7}y = 10 \qquad\qquad\qquad \frac{^{-}2}{3}x = \frac{4}{5}$$

$$\frac{7}{5}\left(\frac{5}{7}y\right) = \frac{7}{5}(10) \qquad\qquad \frac{^{-}3}{2}\left(\frac{^{-}2}{3}x\right) = \frac{^{-}3}{2}\left(\frac{4}{5}\right)$$

$$1y = \frac{7}{5}\cdot\frac{10}{1} \qquad\qquad\qquad 1x = \frac{^{-}6}{5}$$

$$y = 14 \qquad\qquad\qquad\qquad x = \frac{^{-}6}{5} \text{ or } ^{-}1\frac{1}{5}$$

$$\text{Check:} \quad \frac{5}{7}y = 10 \qquad \text{Check:} \quad \frac{^{-}2}{3}x = \frac{4}{5}$$

$$\frac{5}{7}(14) \overset{?}{=} 10 \qquad\qquad \frac{^{-}2}{3}\left(\frac{^{-}6}{5}\right) \overset{?}{=} \frac{4}{5}$$

$$10 = 10 \qquad\qquad\qquad \frac{4}{5} = \frac{4}{5}$$

Q2   Use the multiplication principle of equality to solve the following equations and check each of the solutions:

a. $\frac{1}{2}x = 12$ **b.** $\frac{4}{5}y = ^{-}40$

c. $\dfrac{-3}{7}x = \dfrac{5}{12}$

d. $\dfrac{-6}{7}y = {}^-3$

STOP • STOP • STOP • STOP • STOP • STOP • STOP • STOP • STOP

A2   a. $\dfrac{1}{2}x = 12$

$\dfrac{2}{1}\left(\dfrac{1}{2}x\right) = \dfrac{2}{1}(12)$

$1x = \dfrac{2}{1}\left(\dfrac{12}{1}\right)$

$x = 24$

Check: $\dfrac{1}{2}x = 12$

$\dfrac{1}{2}(24) \overset{?}{=} 12$

$12 = 12$

b. $\dfrac{4}{5}y = {}^-40$

$\dfrac{5}{4}\left(\dfrac{4}{5}y\right) = \dfrac{5}{4}(-40)$

$1y = \dfrac{5}{4}\left(\dfrac{-40}{1}\right)$

$y = {}^-50$

Check: $\dfrac{4}{5}y = {}^-40$

$\dfrac{4}{5}({}^-50) \overset{?}{=} {}^-40$

${}^-40 = {}^-40$

c. $\dfrac{-3}{7}x = \dfrac{5}{12}$

$\dfrac{-7}{3}\left(\dfrac{-3}{7}x\right) = \dfrac{-7}{3}\left(\dfrac{5}{12}\right)$

$1x = \dfrac{-35}{36}$

$x = \dfrac{-35}{36}$

Check: $\dfrac{-3}{7}x = \dfrac{5}{12}$

$\dfrac{-3}{7}\left(\dfrac{-35}{36}\right) \overset{?}{=} \dfrac{5}{12}$

$\dfrac{5}{12} = \dfrac{5}{12}$

d. $\dfrac{-6}{7}y = {}^-3$

$\dfrac{-7}{6}\left(\dfrac{-6}{7}y\right) = \dfrac{-7}{6}({}^-3)$

$1y = \dfrac{7}{2}$

$y = \dfrac{7}{2}$

Check: $\dfrac{-6}{7}y = {}^-3$

$\dfrac{-6}{7}\left(\dfrac{7}{2}\right) \overset{?}{=} {}^-3$

${}^-3 = {}^-3$

3   The equation $\dfrac{x}{7} = 3$ can be solved in a manner similar to that used with the preceding equations. Using the understood coefficient 1 for x, the solution proceeds as follows:

$$\frac{x}{7} = 3$$

$$\frac{1x}{7} = 3$$

$$\frac{1}{7}x = 3$$

$$\frac{7}{1}\left(\frac{1}{7}x\right) = \frac{7}{1}(3)$$

$$1x = 21 \qquad \text{(this step is usually omitted)}$$

$$x = 21$$

Q3   Solve the following equations using the understood coefficient 1 for the variable:

a. $\dfrac{x}{5} = 2$ b. $\dfrac{^-y}{8} = 2$

STOP • STOP • STOP • STOP • STOP • STOP • STOP • STOP • STOP

A3   a. $\dfrac{x}{5} = 2$ b. $\dfrac{^-y}{8} = 2$

$$\frac{1x}{5} = 2 \qquad\qquad\qquad \frac{^-1y}{8} = 2$$

$$\frac{1}{5}x = 2 \qquad\qquad\qquad \frac{^-1}{8}y = 2$$

$$\frac{5}{1}\cdot\frac{1}{5}x = \frac{5}{1}(2) \qquad\qquad \frac{^-8}{1}\cdot\frac{^-1}{8}y = \frac{^-8}{1}(2)$$

$$x = 10 \qquad\qquad\qquad\qquad y = {}^-16$$

---

4   It is often necessary to solve equations of the form $^-x = a$ for some number a, that is, equations with a coefficient of $^-1$ on the variable. These can be solved using the multiplication principle of equality. For example,

$$^-x = 9$$
$$^-1x = 9$$
$$(^-1)(^-1x) = (^-1)(9)$$
$$x = {}^-9$$

Q4   Use the procedure of Frame 4 to solve each of the following equations:
a. $^-x = 5$ b. $^-y = {}^-7$

STOP • STOP • STOP • STOP • STOP • STOP • STOP • STOP • STOP

A4   **a.**

$$^-x = 5$$
$$^-1x = 5$$
$$(^-1)(^-1x) = (^-1)(5)$$
$$x = ^-5$$

b.

$$^-y = ^-7$$
$$^-1y = ^-7$$
$$(^-1)(^-1y) = ^-1(^-7)$$
$$y = 7$$

Q5   Solve the following equations (do step 2 mentally):

a. $^-x = \dfrac{^-3}{5}$ **b.** $^-y = 0$

STOP • **STOP** • **STOP** • **STOP** • **STOP** • **STOP** • **STOP** • **STOP** • **STOP**

A5   **a.**

$$^-x = \frac{^-3}{5}$$
$$(^-1)(^-x) = (^-1)\left(\frac{^-3}{5}\right)$$
$$x = \frac{3}{5}$$

b.

$$^-y = 0$$
$$(^-1)(^-y) = (^-1)(0)$$
$$y = 0$$

This completes the instruction for this section.

8.3   Exercises

1. Use the multiplication and division principles of equality to solve the following equations, and check each of the solutions:

a. $2x = 10$ **b.** $\dfrac{2}{5}x = 20$ **c.** $^-4y = 12$ **d.** $\dfrac{3}{8}x = 24$

e. $\dfrac{x}{5} = ^-3$ **f.** $5y = ^-15$ **g.** $^-4x = ^-8$ **h.** $^-y = 12$

i. $\dfrac{x}{4} = ^-4$ **j.** $\dfrac{^-5}{7}y = 10$ **k.** $5y = ^-6$ **l.** $\dfrac{^-3}{11}x = \dfrac{2}{3}$

m. $^-x = ^-1$ **n.** $13x = ^-26$ **o.** $\dfrac{x}{4} = \dfrac{3}{4}$ **p.** $^-7y = ^-5$

q. $\dfrac{9}{16} = \dfrac{^-3}{4}x$ **r.** $12 = \dfrac{x}{2}$ **s.** $32y = ^-4$ **t.** $\dfrac{7}{8}y = 0$

u. $\dfrac{x}{5} = \dfrac{1}{5}$ **v.** $\dfrac{^-5}{6}x = \dfrac{^-2}{3}$ **w.** $8 = ^-y$ **x.** $3y = ^-4$

y. $^-7x = \dfrac{3}{5}$ **z.** $\dfrac{^-4}{9} = ^-3y$

2. Use the addition, subtraction, multiplication, and division principles of equality to solve the following equations, and check each of the solutions:

a. $x - 3 = 7$ **b.** $4y = {}^-12$ **c.** $\dfrac{{}^-2}{3}x = \dfrac{{}^-4}{5}$ **d.** $y + 9 = 2$

e. $3 + x = 5$ **f.** ${}^-5y = 25$ **g.** $\dfrac{1}{2}x = 7$ **h.** $\dfrac{3}{4}y = {}^-12$

i. $15 = x - 7$ **j.** $42 = {}^-6y$ **k.** $10 = \dfrac{{}^-2}{5}y$ **l.** $0 = \dfrac{4}{7}x$

m. $x - 3 = {}^-5$ **n.** ${}^-12 + x = 5$ **o.** $x + 7 = 11 - 5$ **p.** $\dfrac{{}^-4}{3}x = 2$

q. $5y = \dfrac{3}{7}$ **r.** $13 = 4 + x$ **s.** $5 + x = {}^-3$ **t.** ${}^-7y = \dfrac{14}{15}$

8.3   Exercise Answers

1. a. $x = 5$ **b.** $x = 50$ **c.** $y = {}^-3$ **d.** $x = 64$

  e. $x = {}^-15$ **f.** $y = {}^-3$ **g.** $x = 2$ **h.** $y = {}^-12$

  i. $x = {}^-16$ **j.** $y = {}^-14$ **k.** $y = \dfrac{{}^-6}{5}$ **l.** $x = \dfrac{{}^-22}{9}$

  m. $x = 1$ **n.** $y = {}^-2$ **o.** $x = 3$ **p.** $y = \dfrac{5}{7}$

  q. $x = \dfrac{{}^-3}{4}$ **r.** $x = 24$ **s.** $y = \dfrac{{}^-1}{8}$ **t.** $y = 0$

  u. $x = 1$ **v.** $x = \dfrac{4}{5}$ **w.** $y = {}^-8$ **x.** $y = \dfrac{{}^-4}{3}$

  y. $x = \dfrac{{}^-3}{35}$ **z.** $y = \dfrac{4}{27}$

2. a. $x = 10$ **b.** $y = {}^-3$ **c.** $x = \dfrac{6}{5}$ **d.** $y = {}^-7$

  e. $x = 2$ **f.** $x = {}^-5$ **g.** $x = 14$ **h.** $y = {}^-16$

i. $x = 22$ j. $y = {}^-7$ k. $y = {}^-25$ l. $x = 0$

m. $x = {}^-2$ n. $x = 17$ o. $x = {}^-1$ p. $x = \dfrac{{}^-3}{2}$

q. $y = \dfrac{3}{35}$ r. $x = 9$ s. $x = {}^-8$ t. $y = \dfrac{{}^-2}{15}$

8.4   Solving Equations by the Use of Two or More Steps

1

Frequently it is necessary to use more than one step in solving an equation. Recall that if the variable is x, an equation is solved when it is changed to the form "$x =$ a number" or "a number $= x$." Thus the aim is to *isolate all terms involving variables on one side of the equation and all numbers on the opposite side.* By "isolate" it is meant that the *only* terms involving variables are alone on one side of the equation and *only* number terms are alone on the other side of the equation.

In the equation $3x - 4 = 11$ the objective is to isolate the x term on the left side. Since 4 has been subtracted from $3x$, add 4 (the opposite of subtracting 4) to both sides of the equation.

$$3x - 4 = 11$$
$$3x - 4 + 4 = 11 + 4$$
$$3x = 15$$

With the number and variable terms isolated on opposite sides, reduce the $3x$ to $1x$ by use of the division principle of equality.

$$3x = 15 \qquad \text{Check:} \quad 3x - 4 = 11$$
$$\frac{3x}{3} = \frac{15}{3} \qquad\qquad 3(5) - 4 \overset{?}{=} 11$$
$$x = 5 \qquad\qquad\qquad 11 = 11$$

Q1   **a.** Solve the equation $5x - 3 = 7$ by first adding 3 to both sides.

b. Check the solution to the equation.

STOP • STOP • STOP • STOP • STOP • STOP • STOP • STOP • STOP

A1   **a.** $5x - 3 = 7$ (3 was subtracted, so add 3 to both sides)

$$5x - 3 + 3 = 7 + 3$$

$$5x = 10$$ (x was multiplied by 5, so divide both sides by 5)

$$\frac{5x}{5} = \frac{10}{5}$$

$$x = 2$$

b. Check: $5x - 3 = 7$

$$5(2) - 3 \stackrel{?}{=} 7$$

$$7 = 7$$

---

2   In the equation $^-9 = 5x + 6$, since 6 has been added to $5x$, do the opposite and subtract 6 from both sides.

$$^-9 = 5x + 6$$

$$^-9 - 6 = 5x + 6 - 6$$

$$^-15 = 5x$$

With the number and variable terms isolated on opposite sides, reduce the $5x$ to $1x$ by use of the division principle of equality.

$$^-15 = 5x$$ Check: $^-9 = 5x + 6$

$$\frac{^-15}{5} = \frac{5x}{5}$$ $^-9 \stackrel{?}{=} 5(^-3) + 6$

$$^-3 = x$$ $^-9 \stackrel{?}{=} ^-15 + 6$

$$^-9 = ^-9$$

Q2   **a.** Solve $12 = ^-4x + 8$.

b. Check the solution to the equation.

STOP • STOP • STOP • STOP • STOP • STOP • STOP • STOP • STOP

A2 **a.** $12 = {}^-4x + 8$
$12 - 8 = {}^-4x + 8 - 8$
$4 = {}^-4x$
$$\frac{4}{{}^-4} = \frac{{}^-4x}{{}^-4}$$
${}^-1 = x$

b. Check: $12 = {}^-4x + 8$
$12 \overset{?}{=} {}^-4({}^-1) + 8$
$12 \overset{?}{=} 4 + 8$
$12 = 12$

---

3 In the equation $3 + 2x = 7$, the variable term will be isolated on the left side if 3 is subtracted from both sides:

$3 + 2x = 7$
$3 + 2x - 3 = 7 - 3$
$2x = 4$

The solution can now be completed by use of the division principle of equality:

$$\frac{2x}{2} = \frac{4}{2}$$
$x = 2$

Q3 Solve $7 - 5x = 12$ by first isolating the variable term.

STOP • **STOP** • **STOP** • **STOP** • **STOP** • **STOP** • **STOP** • **STOP** • **STOP**

A3 $7 - 5x = 12$
$7 - 5x - 7 = 12 - 7$
${}^-5x = 5$
$$\frac{{}^-5x}{{}^-5} = \frac{5}{{}^-5}$$
$x = {}^-1$

---

4 In the equation ${}^-17 = {}^-8 - 3x$, the variable term will be isolated on the right side if 8 is added to both sides:

${}^-17 = {}^-8 - 3x$
${}^-17 + 8 = {}^-8 - 3x + 8$
${}^-9 = {}^-3x$

The solution can now be completed by use of the division principle of equality:

$$\frac{{}^-9}{{}^-3} = \frac{{}^-3x}{{}^-3}$$
$3 = x$

Q4   Solve $0 = {}^-5 + 7x$ by first isolating the variable term.

STOP • STOP • STOP • STOP • STOP • STOP • STOP • STOP • STOP

A4 $$0 = {}^-5 + 7x$$
          $$0 + 5 = {}^-5 + 7x + 5$$
          $$5 = 7x$$
          $$\frac{5}{7} = \frac{7x}{7}$$
          $$\frac{5}{7} = x$$

5   In the preceding equations, it is important to notice that the *addition or subtraction principles of equality* are used *first* with the *multiplication or division principles of equality* used in the *final step* of the problem. Study the following two examples before proceeding to Q5.

$$\frac{-2}{3}x + 7 = 1 \qquad \text{Check:} \quad \frac{-2}{3}x + 7 = 1$$

$$\frac{-2}{3}x + 7 - 7 = 1 - 7 \qquad \frac{-2}{3}(9) + 7 \overset{?}{=} 1$$

$$\frac{-2}{3}x = {}^-6 \qquad {}^-6 + 7 = 1$$

$$\qquad\qquad 1 = 1$$

$$\frac{-3}{2} \cdot \frac{-2}{3}x = \frac{-3}{2}({}^-6)$$

$$x = 9$$

$$^-3 = {}^-3 + 4x \qquad \text{Check:} \quad {}^-3 = {}^-3 + 4x$$

$$^-3 + 3 = {}^-3 + 4x + 3 \qquad {}^-3 \overset{?}{=} {}^-3 + 4(0)$$

$$0 = 4x \qquad\qquad {}^-3 \overset{?}{=} {}^-3 + 0$$

$$\frac{0}{4} = \frac{4x}{4} \qquad\qquad {}^-3 = {}^-3$$

$$0 = x$$

Q5   Solve the following equations by first applying the addition or subtraction principles of equality and then the multiplication or division principles of equality:

    a. $3x - 7 = {}^-19$ **b.** $5 = \frac{4}{5}y - 3$

c. $0 = {}^-6 + 12x$ **d.** $25 = 7 - 2x$

STOP • **STOP** • **STOP** • **STOP** • **STOP** • **STOP** • **STOP** • **STOP** • **STOP**

A5

a.
$$3x - 7 = {}^-19$$
$$3x - 7 + 7 = {}^-19 + 7$$
$$3x = {}^-12$$
$$\frac{3x}{3} = \frac{{}^-12}{3}$$
$$x = {}^-4$$

b.
$$5 = \frac{4}{5}y - 3$$
$$5 + 3 = \frac{4}{5}y - 3 + 3$$
$$8 = \frac{4}{5}y$$
$$\frac{5}{4}(8) = \frac{5}{4} \cdot \frac{4}{5}y$$
$$10 = y$$

c.
$$0 = {}^-6 + 12x$$
$$0 + 6 = {}^-6 + 12x + 6$$
$$6 = 12x$$
$$\frac{6}{12} = \frac{12x}{12}$$
$$\frac{1}{2} = x$$

d.
$$25 = 7 - 2x$$
$$25 - 7 = 7 - 2x - 7$$
$$18 = {}^-2x$$
$$\frac{18}{{}^-2} = \frac{{}^-2x}{{}^-2}$$
$${}^-9 = x$$

---

6   In the preceding equations a variable term occurred on only one side of the equation. If variable terms occur on both sides of the equation, the general procedure used in solving the equation is the same. That is, *isolate all terms involving variables on one side of the equation and all numbers on the opposite side.*

For example, consider the equation

$$3x - 2 = x + 6$$

To isolate the variable terms on the left side, subtract x from both sides:

$$3x - 2 - x = x + 6 - x$$
$$2x - 2 = 6$$

To isolate the numbers on the right, add 2 to both sides:

$$2x - 2 + 2 = 6 + 2$$
$$2x = 8$$

The final step involves the division principle of equality:

$$\frac{2x}{2} = \frac{8}{2}$$
$$x = 4$$

Q6   **a.** Solve $5x - 4 = 2x + 5$ by isolating the terms that involve the variables on the left side and the numbers on the right side.

        b. Check the solution.

STOP • **STOP** • **STOP** • **STOP** • **STOP** • **STOP** • **STOP** • **STOP** • **STOP**

A6   **a.**
$$5x - 4 = 2x + 5$$
$$5x - 4 - 2x = 2x + 5 - 2x$$
$$3x - 4 = 5$$
$$3x - 4 + 4 = 5 + 4$$
$$3x = 9$$
$$\frac{3x}{3} = \frac{9}{3}$$
$$x = 3$$

        b.
$$5x - 4 = 2x + 5$$
$$5(3) - 4 \stackrel{?}{=} 2(3) + 5$$
$$15 - 4 \stackrel{?}{=} 6 + 5$$
$$11 = 11$$

---

7   Consider the equation $x + 3 = 15 - 5x$. If you decide to isolate the terms involving variables on the left side and the numbers on the right side, you must add $5x$ and subtract 3 from both sides.

$$x + 3 = 15 - 5x$$
$$x + 3 + 5x = 15 - 5x + 5x$$
$$6x + 3 = 15$$
$$6x + 3 - 3 = 15 - 3$$
$$6x = 12$$
$$\frac{6x}{6} = \frac{12}{6}$$
$$x = 2$$

Check:
$$x + 3 = 15 - 5x$$
$$2 + 3 \stackrel{?}{=} 15 - 5(2)$$
$$5 = 5$$

Q7   **a.** Solve $4x + 7 = 2x + 8$ by isolating the terms involving variables on the left and the numbers on the right.

b. Check the solution.

STOP • STOP • STOP • STOP • STOP • STOP • STOP • STOP • STOP

A7   **a.** $4x + 7 = 2x + 8$ **b.** $4x + 7 = 2x + 8$

$4x + 7 - 2x = 2x + 8 - 2x$ $4 \cdot \dfrac{1}{2} + 7 \overset{?}{=} 2 \cdot \dfrac{1}{2} + 8$

$2x + 7 = 8$

$2x + 7 - 7 = 8 - 7$ $2 + 7 \overset{?}{=} 1 + 8$

$2x = 1$ $9 = 9$

$\dfrac{2x}{2} = \dfrac{1}{2}$

$x = \dfrac{1}{2}$

---

8   The preceding problems have been solved by isolating the terms that involve variables on the left side. However, equations can be solved by isolating the variable on either side. You may wish to isolate the variable on the side that makes the coefficient of the variable positive.

Example 1: Solve $3x - 1 = x + 5$ by isolating the variable on the left side.

Solution

Subtract x from and add 1 to both sides:

$3x - 1 = x + 5$
$3x - 1 - x = x + 5 - x$
$2x - 1 = 5$
$2x - 1 + 1 = 5 + 1$
$2x = 6$
$x = 3$

Example 2: Solve $3x - 1 = x + 5$ by isolating the variable on the right side.

Solution

Subtract $3x$ and 5 from both sides:

$3x - 1 = x + 5$
$3x - 1 - 3x = x + 5 - 3x$
$^-1 = {}^-2x + 5$
$^-1 - 5 = {}^-2x + 5 - 5$
$^-6 = {}^-2x$
$3 = x$

The same solution was obtained by isolating the variable on either side.

Q8   Solve $5x - 3 = 6x + 9$ by isolating the variable on the left side.

STOP   •   STOP   •   STOP   •   STOP   •   STOP   •   STOP   •   STOP   •   STOP   •   STOP

A8
$$5x - 3 = 6x + 9$$
$$5x - 3 - 6x = 6x + 9 - 6x$$
$$^-x - 3 = 9$$
$$^-x - 3 + 3 = 9 + 3$$
$$^-x = 12$$
$$x = ^-12$$

Q9   Solve $5x - 3 = 6x + 9$ by isolating the variable on the right side.

STOP   •   STOP   •   STOP   •   STOP   •   STOP   •   STOP   •   STOP   •   STOP   •   STOP

A9
$$5x - 3 = 6x + 9$$
$$5x - 3 - 5x = 6x + 9 - 5x$$
$$^-3 = x + 9$$
$$^-3 - 9 = x + 9 - 9$$
$$^-12 = x$$

Q10   Solve by isolating the variable on the side that makes the coefficient of the variable positive:

a. $2x + 1 = ^-x - 2$ **b.** $^-4x + 2 = 7x + 3$

STOP   •   STOP   •   STOP   •   STOP   •   STOP   •   STOP   •   STOP   •   STOP   •   STOP

A10   **a.**
$$2x + 1 = ^-x - 2$$
$$2x + 1 + x = ^-x - 2 + x$$
$$3x + 1 = ^-2$$
$$3x + 1 - 1 = ^-2 - 1$$
$$3x = ^-3$$
$$x = ^-1$$

b.
$$^-4x + 2 = 7x + 3$$
$$^-4x + 2 + 4x = 7x + 3 + 4x$$
$$2 = 11x + 3$$
$$2 - 3 = 11x + 3 - 3$$
$$^-1 = 11x$$
$$\frac{^-1}{11} = x$$

9   To solve equations that involve parentheses, first simplify both sides of the equation wherever possible, and then proceed as before.

Examples:

1. $2x - (x + 2) = 7$
    $2x - x - 2 = 7$ (remove parentheses)
    $x - 2 = 7$ (combine like terms)
    $x - 2 + 2 = 7 + 2$
    $x = 9$

2. $x + 5 = x + (2x - 3)$
        $x + 5 = x + 2x - 3$ (remove parentheses)
        $x + 5 = 3x - 3$ (combine like terms)
    $x + 5 - 3x = 3x - 3 - 3x$
        $^-2x + 5 = ^-3$
    $^-2x + 5 - 5 = ^-3 - 5$
        $^-2x = ^-8$

$$\frac{^-2x}{^-2} = \frac{^-8}{^-2}$$

        $x = 4$

Q11   Solve by first simplifying both sides of the equation:
    a. $5x - (2x + 7) = 8$ **b.** $6x = 8 + (2x - 4)$

STOP • STOP • STOP • STOP • STOP • STOP • STOP • STOP • STOP

A11   **a.** $5x - (2x + 7) = 8$ **b.** $6x = 8 + (2x - 4)$
        $5x - 2x - 7 = 8$ $6x = 8 + 2x - 4$
        $3x - 7 = 8$ $6x = 4 + 2x$
    $3x - 7 + 7 = 8 + 7$ $6x - 2x = 4 + 2x - 2x$
        $3x = 15$ $4x = 4$

$$\frac{3x}{3} = \frac{15}{3} \qquad\qquad \frac{4x}{4} = \frac{4}{4}$$

        $x = 5$ $x = 1$

---

10   To solve $2(5x + 3) - 3(x - 5) = 7$, the following steps are used:

$$2(5x + 3) - 3(x - 5) = 7$$
$$10x + 6 - 3x + 15 = 7$$
$$7x + 21 = 7$$
$$7x + 21 - 21 = 7 - 21$$
$$7x = {}^-14$$
$$\frac{7x}{7} = \frac{{}^-14}{7}$$
$$x = {}^-2$$

Q12   Solve:

    a. $5(x + 6) = 45$ **b.** $3(x - 2) = x - 2(3x - 1)$

STOP • **STOP** • **STOP** • **STOP** • **STOP** • **STOP** • **STOP** • **STOP** • **STOP**

A12   **a.**

$$5(x + 6) = 45$$
$$5x + 30 = 45$$
$$5x + 30 - 30 = 45 - 30$$
$$5x = 15$$
$$\frac{5x}{5} = \frac{15}{5}$$
$$x = 3$$

b.

$$3(x - 2) = x - 2(3x - 1)$$
$$3x - 6 = x - 6x + 2$$
$$3x - 6 = {}^-5x + 2$$
$$3x - 6 + 5x = {}^-5x + 2 + 5x$$
$$8x - 6 = 2$$
$$8x - 6 + 6 = 2 + 6$$
$$8x = 8$$
$$\frac{8x}{8} = \frac{8}{8}$$
$$x = 1$$

---

11   When solving equations such as $5 - 4(x + 3) = 1$, it is again important to first simplify both sides of the equation by removing parentheses and combining like terms. For example,

$$5 - 4(x + 3) = 1$$
$$5 - 4x - 12 = 1$$
$${}^-4x - 7 = 1$$
$${}^-4x - 7 + 7 = 1 + 7$$
$${}^-4x = 8$$
$$\frac{{}^-4x}{{}^-4} = \frac{8}{{}^-4}$$
$$x = {}^-2$$

Q13 Solve:
a. $5 - 3(x - 2) = 14$ **b.** $x + 8 = {}^{-}10 - 6(2x - 3)$

STOP • STOP • STOP • STOP • STOP • STOP • STOP • STOP • STOP

A13

a.
$$5 - 3(x - 2) = 14$$
$$5 - 3x + 6 = 14$$
$${}^{-}3x + 11 = 14$$
$${}^{-}3x + 11 - 11 = 14 - 11$$
$$\frac{{}^{-}3x}{{}^{-}3} = \frac{3}{{}^{-}3}$$
$$x = {}^{-}1$$

b.
$$x + 8 = {}^{-}10 - 6(2x - 3)$$
$$x + 8 = {}^{-}10 - 12x + 18$$
$$x + 8 = 8 - 12x$$
$$x + 8 + 12x = 8 - 12x + 12x$$
$$13x + 8 = 8$$
$$13x + 8 - 8 = 8 - 8$$
$$13x = 0$$
$$\frac{13x}{13} = \frac{0}{13}$$
$$x = 0$$

This completes the instruction for this section.

8.4 Exercises

1. Solve and check each of the following:
 a. $2x - 3 = 9$ **b.** $15 = 4y + 7$
 c. $6x + 11 = {}^{-}7$ **d.** ${}^{-}5y + 1 = {}^{-}4$
 e. $\frac{2}{5}x - 3 = 11$ **f.** $6 = 6 - 7y$
 g. $5 - y = 12$ **h.** $6 = 2x + 5$
 i. $7 - \frac{1}{2}x = 3$ **j.** ${}^{-}3y = {}^{-}5$
 k. $4 + 3y = 0$ **l.** $2 = {}^{-}3 + \frac{5}{7}y$
 m. $6 = 9 - x$ **n.** $\frac{{}^{-}3}{4}x - 3 = 5$

2. Solve the following equations by isolating the terms that involve variables on one side and the numbers on the opposite side:
 a. $2x + 3 = x + 4$ **b.** $7 + 3x = 4x - 2$
 c. $12 + x = 2x + 3$ **d.** $5x + 3 = x - 1$
 e. $4y = y + 9$ **f.** $2 - 3y = 2y + 2$
 g. $x + 11 = {}^{-}x + 5$ **h.** ${}^{-}2y + 3 = {}^{-}5y + 5$

    i. $^{-}7x + 3 - 5x = 0$ **j.** $9 = 3x - 15$

    k. $4x - 11 = 2 - x$ **l.** $3 + 4y = {}^{-}2y + 1$

3. Solve:

    a. $2(x - 4) = 10$ **b.** $2x + 3 = 4 + (x - 6)$

    c. $6x - (x - 7) = 22$ **d.** $3(x - 5) = {}^{-}7 + 2(x - 4)$

    e. $2(x - 3) - 3(2x + 5) = {}^{-}25$ **f.** $4 - 2(x + 1) = 2x + 4$

    g. $3x + 10 = 7 - 5(2x + 15)$ **h.** $2(3x - 6) + 4 = 22 - 2(x - 1)$

8.4   Exercise Answers

1.   **a.** $x = 6$ **b.** $y = 2$ **c.** $x = {}^{-}3$ **d.** $y = 1$

     e. $x = 35$ **f.** $y = 0$ **g.** $y = {}^{-}7$ **h.** $x = \dfrac{1}{2}$

     i. $x = 8$ **j.** $y = \dfrac{5}{3}$ **k.** $y = \dfrac{{}^{-}4}{3}$ **l.** $y = 7$

     m. $x = 3$ **n.** $x = \dfrac{{}^{-}32}{3}$

2.   **a.** $x = 1$ **b.** $x = 9$ **c.** $x = 9$ **d.** $x = {}^{-}1$

     e. $y = 3$ **f.** $y = 0$ **g.** $x = {}^{-}3$ **h.** $y = \dfrac{2}{3}$

     i. $x = \dfrac{1}{4}$ **j.** $x = 8$ **k.** $x = \dfrac{13}{5}$ **l.** $y = \dfrac{{}^{-}1}{3}$

3.   **a.** $x = 9$ **b.** $x = {}^{-}5$ **c.** $x = 3$ **d.** $x = 0$

     e. $x = 1$ **f.** $x = \dfrac{{}^{-}1}{2}$ **g.** $x = {}^{-}6$ **h.** $x = 4$

8.5   **Equations with Fractions**

1

Many times it is necessary to solve equations that involve several rational numbers (fractions). This is usually done by eliminating the fractions from the equation by use of the multiplication property of equality, and solving the resulting equation using the procedures of previous sections.

The fractions can be eliminated from the equation $\dfrac{x}{2} - 4 = \dfrac{x}{3}$ by multiplying both sides of the equation by each of the denominators, 2 and 3. For example,

$$\frac{x}{2} - 4 = \frac{x}{3}$$

$$2\left(\frac{x}{2} - 4\right) = 2\left(\frac{x}{3}\right)$$

$$2\left(\frac{x}{2}\right) - 2(4) = 2\left(\frac{x}{3}\right)$$

$$x - 8 = \frac{2x}{3}$$

The first fraction has now been eliminated.

$$3(x - 8) = 3\left(\frac{2x}{3}\right)$$

$$3(x) - 3(8) = 3\left(\frac{2x}{3}\right)$$

$$3x - 24 = 2x$$

All fractions have now been eliminated and the solution can be completed using the procedures already studied.

$$3x - 24 = 2x \qquad \text{Check:} \quad \frac{x}{2} - 4 = \frac{x}{3}$$
$$3x - 24 - 2x = 2x - 2x$$
$$x - 24 = 0 \qquad\qquad \frac{24}{2} - 4 \stackrel{?}{=} \frac{24}{3}$$
$$x - 24 + 24 = 0 + 24$$
$$x = 24 \qquad\qquad 12 - 4 \stackrel{?}{=} 8$$
$$8 = 8$$

Q1   **a.** Eliminate the fractions from the equation $\dfrac{x}{2} = 5 + \dfrac{x}{3}$ by first multiplying both sides of the equation by 2 and then by 3.

b. Solve the resulting equation.   **c.** Check the solution.

A1 **a.** $\dfrac{x}{2} = 5 + \dfrac{x}{3}$ **b.** $3x - 2x = 30 + 2x - 2x$ **c.** $\dfrac{x}{2} = 5 + \dfrac{x}{3}$

$$2\left(\dfrac{x}{2}\right) = 2\left(5 + \dfrac{x}{3}\right)$$

$$x = 30$$

$$\dfrac{30}{2} \overset{?}{=} 5 + \dfrac{30}{3}$$

$$2\left(\dfrac{x}{2}\right) = 2(5) + 2\left(\dfrac{x}{3}\right)$$

$$15 \overset{?}{=} 5 + 10$$

$$15 = 15$$

$$x = 10 + \dfrac{2x}{3}$$

$$3(x) = 3\left(10 + \dfrac{2x}{3}\right)$$

$$3(x) = 3(10) + 3 \cdot \dfrac{2x}{3}$$

$$3x = 30 + 2x$$

2

A much shorter procedure for eliminating several fractions from an equation is to multiply both sides by just *one* number. For example, rather than to multiply the equation $\dfrac{x}{4} - 3 = \dfrac{x}{5}$ by the two numbers 4 and 5, the fractions can be eliminated by multiplying both sides by *one* number which has both factors 4 and 5. The smallest number with both factors 4 and 5 is 20. Therefore, multiply both sides of the equation by 20.

$$20\left(\dfrac{x}{4} - 3\right) = 20\left(\dfrac{x}{5}\right)$$

$$20\left(\dfrac{x}{4}\right) - 20(3) = 20\left(\dfrac{x}{5}\right)$$

$$5x - 60 = 4x$$

With the fractions eliminated, the solution of the equation can now be completed:

$$5x - 60 = 4x$$
$$5x - 60 - 4x = 4x - 4x$$
$$x - 60 = 0$$
$$x - 60 + 60 = 0 + 60$$
$$x = 60$$

Check: $\dfrac{x}{4} - 3 = \dfrac{x}{5}$

$$\dfrac{60}{4} - 3 \overset{?}{=} \dfrac{60}{5}$$

$$15 - 3 \overset{?}{=} 12$$

$$12 = 12$$

Q2 **a.** Eliminate the fractions from the equation $\dfrac{x}{3} - 4 = \dfrac{x}{5}$ by multiplying both sides by the smallest number with the factors 3 and 5.

b. Solve the resulting equation.   **c.** Check the solution.

STOP • **STOP** • **STOP** • **STOP** • **STOP** • **STOP** • **STOP** • **STOP** • **STOP**

A2   **a.** The smallest number is 15.   **b.** $5x - 60 - 3x = 3x - 3x$

$$15\left(\frac{x}{3} - 4\right) = 15\left(\frac{x}{5}\right)$$

$$2x - 60 = 0$$

$$2x - 60 + 60 = 0 + 60$$

$$15\left(\frac{x}{3}\right) - 15(4) = 15\left(\frac{x}{5}\right)$$

$$2x = 60$$

$$\frac{2x}{2} = \frac{60}{2}$$

$$5x - 60 = 3x$$

$$x = 30$$

c. Check: $\dfrac{x}{3} - 4 = \dfrac{x}{5}$

$$\frac{30}{3} - 4 \stackrel{?}{=} \frac{30}{5}$$

$$10 - 4 \stackrel{?}{=} 6$$

$$6 = 6$$

---

3   The procedure of eliminating the fractions from an equation is called "clearing an equation of fractions." An equation can be cleared of fractions by multiplying both sides by the smallest number that all the denominators in the equation will divide into evenly. This number is called the least common denominator (LCD) for the fractions in the equation.

Q3   Solve the following equations by first clearing them of fractions using the LCD.

a. $\dfrac{2x}{3} - 5 = \dfrac{x}{4}$ **b.** $\dfrac{2}{3}x - \dfrac{2}{5} = \dfrac{2}{5}x$ **c.** $\dfrac{x}{2} + \dfrac{5}{2} = \dfrac{2x}{3}$

STOP • **STOP** • **STOP** • **STOP** • **STOP** • **STOP** • **STOP** • **STOP** • **STOP**

A3   **a.** The LCD is 12.   **b.** The LCD is 15.

$$12\left(\frac{2x}{3} - 5\right) = 12\left(\frac{x}{4}\right)$$ $$15\left(\frac{2}{3}x - \frac{2}{5}\right) = 15\left(\frac{2}{5}x\right)$$

$$12\left(\frac{2x}{3}\right) - 12(5) = 12\left(\frac{x}{4}\right)$$ $$15\left(\frac{2}{3}x\right) - 15\left(\frac{2}{5}\right) = 15\left(\frac{2}{5}x\right)$$

$$8x - 60 = 3x$$ $$10x - 6 = 6x$$

$$8x - 60 - 3x = 3x - 3x$$ $$10x - 6 - 6x = 6x - 6x$$

$$5x - 60 = 0$$ $$4x - 6 = 0$$

$$5x - 60 + 60 = 0 + 60$$ $$4x - 6 + 6 = 0 + 6$$

$$5x = 60$$ $$4x = 6$$

$$\frac{5x}{5} = \frac{60}{5}$$ $$\frac{4x}{4} = \frac{6}{4}$$

$$x = 12$$ $$x = \frac{3}{2}$$

c. The LCD is 6. The solution is $x = 15$.

---

4   The steps used when solving an equation with fractions are:

Step 1: Determine the LCD.

Step 2: Multiply both sides of the equation by the LCD. This step "clears" the equation of all fractions.

Step 3: Solve the resulting equation.

Step 4: Check the solution.

Q4   Solve each of the following equations using the preceding four-step procedure:

a. $\dfrac{x}{2} = 7 - \dfrac{2x}{3}$ **b.** $\dfrac{7x}{8} + \dfrac{5}{6} = \dfrac{1}{12}$

c. $\dfrac{9}{10} = \dfrac{^-3}{4}x + \dfrac{2}{5}$

d. $\dfrac{3y}{5} + \dfrac{5}{2} = \dfrac{^-y}{5} - \dfrac{3}{2}$

STOP • STOP • STOP • STOP • STOP • STOP • STOP • STOP • STOP

A4   **a.** The LCD is 6.

$$6\left(\dfrac{x}{2}\right) = 6\left(7 - \dfrac{2x}{3}\right) \qquad \text{Check:} \quad \dfrac{x}{2} = 7 - \dfrac{2x}{3}$$

$$6\left(\dfrac{x}{2}\right) = 6(7) - 6\left(\dfrac{2x}{3}\right) \qquad\qquad \dfrac{6}{2} \overset{?}{=} 7 - \dfrac{2(6)}{3}$$

$$3x = 42 - 4x \qquad\qquad\qquad 3 \overset{?}{=} 7 - \dfrac{12}{3}$$

$$3x + 4x = 42 - 4x + 4x \qquad\qquad 3 = 3$$

$$7x = 42$$

$$\dfrac{7x}{7} = \dfrac{42}{7}$$

$$x = 6$$

b. The LCD is 24. The solution is $x = \dfrac{^-6}{7}$.

c. The LCD is 20. The solution is $x = \dfrac{^-2}{3}$.

d. The LCD is 10.

$$10\left(\dfrac{3y}{5} + \dfrac{5}{2}\right) = 10\left(\dfrac{^-y}{5} - \dfrac{3}{2}\right) \qquad \text{Check:} \quad \dfrac{3y}{5} + \dfrac{5}{2} = \dfrac{^-y}{5} - \dfrac{3}{2}$$

$$10\left(\dfrac{3y}{5}\right) + 10\left(\dfrac{5}{2}\right) = 10\left(\dfrac{^-y}{5}\right) - 10\left(\dfrac{3}{2}\right) \qquad \dfrac{3(^-5)}{5} + \dfrac{5}{2} \overset{?}{=} \dfrac{^-(^-5)}{5} - \dfrac{3}{2}$$

$$6y + 25 = {^-2}y - 15 \qquad\qquad\qquad ^-3 + \dfrac{5}{2} \overset{?}{=} 1 - \dfrac{3}{2}$$

$$6y + 25 + 2y = {^-2}y - 15 + 2y \qquad\qquad \dfrac{^-1}{2} = \dfrac{^-1}{2}$$

$$8y + 25 = {^-15}$$

$$8y + 25 - 25 = {^-15} - 25$$

$$8y = {^-40}$$

$$\dfrac{8y}{8} = \dfrac{^-40}{8}$$

$$y = {^-5}$$

This completes the instruction for this section.

8.5   Exercises

Solve each of the following equations:

1. $\dfrac{4x}{5} - 2 = 10$

2. $\dfrac{2x}{3} + 3 = \dfrac{4}{5}$

3. $\dfrac{3y}{4} - \dfrac{2}{3} = \dfrac{5}{12}$

4. $\dfrac{x}{2} - 6 = \dfrac{x}{4}$

5. $\dfrac{1}{2}x - 7 = \dfrac{2}{3}x$

6. $\dfrac{3}{2} - \dfrac{x}{3} = 5 + \dfrac{x}{6}$

7. $\dfrac{3x}{4} - \dfrac{1}{2} = \dfrac{x}{4} + \dfrac{11}{2}$

8. $\dfrac{3y}{5} + \dfrac{5}{2} = \dfrac{^-y}{5} - \dfrac{3}{2}$

8.5   Exercise Answers

1. $x = 15$

2. $x = \dfrac{^-33}{10}$

3. $y = \dfrac{13}{9}$

4. $x = 24$

5. $x = {}^-42$

6. $x = {}^-7$

7. $x = 12$

8. $y = {}^-5$

8.6   **Expressing English Phrases as Algebraic Expressions**

1 | The purpose of much of the mathematical training you receive is to permit the solving of numerical problems encountered in real life. These situations might be relatively simple, such as adjusting a recipe for a different number of people, or much more complicated, such as determining the orbit of a satellite. The ability to translate a situation into an appropriate mathematical sentence (equation) enables one to cope with many problems in everyday life, as well as the problems encountered in science and industry, in an orderly, logical manner. An important technique involves writing and solving an equation that represents the arithmetic of the situation.

Before you can solve mathematical situations, you must be able to translate English phrases into algebraic expressions. For example, letting a letter represent the number, the algebraic expressions on the right represent the English phrases on the left.

English phrase	Algebraic expression
five more than a number	$x + 5$
a number more than five	$5 + y$
three times a number	$3c$
a number minus two	$h - 2$
two minus a number	$2 - n$

Q1   Let n represent the number.* Write an algebraic expression for:

    a. seven more than the number   _____

    b. the number more than seven   _____

    c. four times the number   _____

    d. the number minus six   _____

    e. six minus the number   _____

STOP • **STOP** • **STOP** • **STOP** • **STOP** • **STOP** • **STOP** • **STOP** • **STOP**

A1   **a.** $n + 7$ **b.** $7 + n$ **c.** $4n$ **d.** $n - 6$ **e.** $6 - n$

---

2   Key words in the phrase indicate the operation involved. Some common words or phrases that indicate addition are: plus, added to, more than, sum of, and increased by. These are the clues to help decide what operation to use.

Q2   Write an algebraic expression for:

    a. the number increased by one   _____

    b. the number added to nine   _____

    c. eight plus the number   _____

    d. the sum of the number and four   _____

    e. ten more than the number   _____

STOP • **STOP** • **STOP** • **STOP** • **STOP** • **STOP** • **STOP** • **STOP** • **STOP**

A2   **a.** $n + 1$ **b.** $9 + n$ **c.** $8 + n$ **d.** $n + 4$ **e.** $n + 10$

---

3   Since addition is a commutative operation, $n + 1 = 1 + n$. However, $n + 1$ and $1 + n$ come from different English phrases. That is, "the number plus one" translates $n + 1$; "one plus the number" translates $1 + n$. It is important to translate precisely so that the meaning of the phrase does not become distorted. "The sum of two and the number" is translated $2 + n$. "The sum of the number and two" is translated $n + 2$.

Q3   Translate each phrase into an algebraic expression:

    a. the sum of the number and ten   _____

    b. the sum of ten and the number   _____

STOP • **STOP** • **STOP** • **STOP** • **STOP** • **STOP** • **STOP** • **STOP** • **STOP**

A3   **a.** $n + 10$ **b.** $10 + n$

---

4   Some common words or phrases that indicate subtraction are: difference of, difference between, take away, taken from, reduced by, less, less than, diminished by, subtracted from, decreased by, smaller than, minus, depreciate, and borrowed from.

Q4   Translate each phrase into an algebraic expression:

    a. five taken from the number   _____

    b. the number taken from five   _____

    c. the number less seven   _____

*n will be used to represent "the number" throughout this section.

     d. seven less the number   _____

     e. the number diminished by nine   _____

     f. nine diminished by the number   _____

STOP • STOP • STOP • STOP • STOP • STOP • STOP • STOP • STOP

A4   **a.** $n - 5$ **b.** $5 - n$ **c.** $n - 7$ **d.** $7 - n$ **e.** $n - 9$ **f.** $9 - n$

---

5   Since subtraction is not commutative, $n - 5 \neq 5 - n$. If you write a translation of subtraction in the wrong order, it will not only be a distortion, it will be wrong. This serves as a good example to emphasize the importance of translating precisely. "Less" and "less than" require special attention. "Five less two" translates $5 - 2$. "The number less two" translates $n - 2$. "Two less than five" translates $5 - 2$. "Two less than the number" translates $n - 2$.

Q5   Translate each phrase into an algebraic expression:

     a. the number less one   _____

     b. one less than the number   _____

     c. one less the number   _____

     d. the number less than one   _____

STOP • STOP • STOP • STOP • STOP • STOP • STOP • STOP • STOP

A5   **a.** $n - 1$ **b.** $n - 1$ **c.** $1 - n$ **d.** $1 - n$

---

6   In the phrase "the difference between" the values are listed in the order in which they are used. For example, "the difference between the number and two" is $n - 2$. "The difference between two and the number" is $2 - n$.

Q6   Translate each phrase into an algebraic expression:

     a. the difference between the number and three   _____

     b. the difference between three and the number   _____

STOP • STOP • STOP • STOP • STOP • STOP • STOP • STOP • STOP

A6   **a.** $n - 3$ **b.** $3 - n$

---

7   Some common words or phrases that indicate multiplication are: product of, times, double, twice, triple, and of.

Q7   Translate each phrase into an algebraic expression:

     a. twice the number   _____

     b. five times the number   _____

     c. the product of three and the number   _____

     d. two thirds of the number   _____

     e. triple the number   _____

STOP • STOP • STOP • STOP • STOP • STOP • STOP • STOP • STOP

A7   **a.** $2n$ **b.** $5n$ **c.** $3n$ **d.** $\frac{2}{3}n$ **e.** $3n$

---

8   It is sometimes necessary to express a percent of a number: for example, 60% of a number. Since 60% is equivalent to $\frac{3}{5}$ or 0.6, 60% of a number can be written $\frac{3}{5}n$ or $0.6n$.

Q8   Translate each phrase into an algebraic expression:

a. 20% of a number   _____

b. 15% of a number   _____

c. 50% of a number   _____

d. 75% of a number   _____

e. 100% of a number   _____

f. 125% of a number   _____

STOP • STOP • STOP • STOP • STOP • STOP • STOP • STOP • STOP

A8   a. $\frac{1}{5}n$ or $0.2n$ b. $\frac{3}{20}n$ or $0.15n$ c. $\frac{1}{2}n$ or $0.5n$ d. $\frac{3}{4}n$ or $0.75n$

e. $1n$ or n f. $1\frac{1}{4}n$ or $\frac{5}{4}n$ or $1.25n$

---

9   Some common words or phrases that indicate division are: divided into, quotient of, divided by, and divides.

Q9   Translate each phrase into an algebraic expression:

a. the number divided by six   _____

b. the number divides six   _____

c. nine divided into the number   _____

d. the quotient of the number by two   _____

e. nine divided by the number   _____

STOP • STOP • STOP • STOP • STOP • STOP • STOP • STOP • STOP

A9   a. $n \div 6$ or $\frac{n}{6}$ b. $6 \div n$ or $\frac{6}{n}$ c. $n \div 9$ or $\frac{n}{9}$ d. $n \div 2$ or $\frac{n}{2}$

e. $9 \div n$ or $\frac{9}{n}$

---

10   Skill in writing algebraic expressions that involve more than one operation is also necessary.

English phrase	Algebraic expression
three more than five times the number	$5n + 3$
twice the sum of the number and four	$2(n + 4)$

Q10   Translate each phrase into an algebraic expression:

a. the sum of the number, and five times the number   _____

b. three times as much as two more than the number   _____

c. twice the number, minus five   _____

d. the difference between three times the number and the number   _____

e. two less than the number doubled   _____

f. the number divided by three, plus eight times the number   _____

g. the number depreciated by five percent of the number   _____

h. the difference between fifteen and one-half of the number   _____

STOP • **STOP** • **STOP** • **STOP** • **STOP** • **STOP** • **STOP** • **STOP** • **STOP**

A10 **a.** $n + 5n$ **b.** $3(n + 2)$ **c.** $2n - 5$ **d.** $3n - n$

    e. $2n - 2$ **f.** $\dfrac{n}{3} + 8n$ **g.** $n - \dfrac{1}{20}n$ or $n - 0.05n$

    h. $15 - \dfrac{1}{2}n$

---

11 The word "and" plays an important role.

  Example 1: Write the sum of 2, 3, *and* 4.

  Solution

  $2 + 3 + 4$.

  Example 2: Write the difference of 4 *and* 3.

  Solution

  $4 - 3$.

  Example 3: Write the product of 2, 3, *and* 4.

  Solution

  $2(3)(4)$.

  Example 4: Write the quotient of 15 *and* 3.

  Solution

  $15 \div 3$ or $\dfrac{15}{3}$.

  In each example, "and" simply separates the last number listed from the previous numbers. The word "and" does not indicate an operation. The phrases sum, difference of, product, and quotient indicate the operation to be performed.

  Examples:

  1. The product of one more than a number *and* one less than the same number:

  $(n + 1)(n - 1)$

  2. The quotient of a number increased by six and the same number decreased by four:

  $\dfrac{n + 6}{n - 4}$

Q11   Write an algebraic expression for:
   a. the sum of a number, twice the number, and
      three times the number   _____

   b. the product of six, a number less two, and
      two less than the same number   _____

STOP • **STOP** • **STOP** • **STOP** • **STOP** • **STOP** • **STOP** • **STOP** • **STOP**

A11   **a.** $n + 2n + 3n$ **b.** $6(n - 2)(n - 2)$

This completes the instruction for this section.

8.6   Exercises

Write each phrase as an algebraic expression. Use n for the number.

1. the number added to seven
2. the sum of the number and twelve
3. three increased by the number
4. the number doubled decreased by seven
5. the number less five
6. six less than the number
7. the difference between twice the number and nine
8. three times the number, diminished by eight
9. five times the number
10. the product of three and the number, depreciated by ten
11. sixteen divided into the number
12. two-fifths of the number
13. twelve percent of the number
14. five percent of the sum of the number and fifty
15. the sum of the number and one, divided by four
16. the product of five, the number plus one, and the number minus one

8.6   Exercise Answers

1. $7 + n$ 2. $n + 12$ 3. $3 + n$
4. $2n - 7$ 5. $n - 5$ 6. $n - 6$
7. $2n - 9$ 8. $3n - 8$ 9. $5n$

10. $3n - 10$ 11. $n \div 16$ or $\dfrac{n}{16}$ 12. $\dfrac{2}{5}n$

13. $\dfrac{3}{25}n$ or $0.12n$ 14. $\dfrac{1}{20}(n + 50)$ or $0.05(n + 50)$

15. $(n + 1) \div 4$ or $\dfrac{n + 1}{4}$ 16. $5(n + 1)(n - 1)$

8.7   Solving Simple Word Problems

1	A simple word sentence such as "twice the number, minus three, is equal to seven" may be translated into an equation. Letting n be the number: "Twice the number, minus three, is equal to seven" becomes $2n - 3 = 7$. ("Is equal to" translates as "$=$.")

Q1   Translate each word sentence into an equation. Let x be the number.

    a. Seventeen subtracted from the number is equal to nine.   _____

    b. The sum of four times the number and six is equal to thirty-eight.   _____

    c. Three times the number, plus two more than the number, is equal to eighteen.   _____

    d. Three more than twice a certain number is seven less than that number (*Note:* "is" also translates as "$=$.")   _____

    e. A number increased by six is two less than three times as large as the same number.   _____

STOP • **STOP** • **STOP** • **STOP** • **STOP** • **STOP** • **STOP** • **STOP** • **STOP**

A1   **a.** $x - 17 = 9$ **b.** $4x + 6 = 38$ **c.** $3x + x + 2 = 18$
       d. $2x + 3 = x - 7$ **e.** $x + 6 = 3x - 2$

2	Any of the English sentences above could be expressed as a problem. For example, "If the sum of four times a number and six is equal to thirty-eight, what is the number?" To solve this problem and others of this type, use the procedure stated below.

    Step 1: Let some letter (variable), such as n, represent the number.

    Step 2: Translate the statement of the problem into an equation.

    Step 3: Solve the equation for the value of the letter.

    Step 4: Check the solution against the statement of the problem.

    The solution to the problem stated above would be:

    Step 1: Let $n =$ the number.

    Step 2: $4n + 6 = 38$

    Step 3: $4n = 32$
              $n = 8$

    Step 4: Since the sum of four times eight (32) and six is equal to thirty-eight, eight is the correct solution.

Q2   Complete the four steps outlined in Frame 2 to solve this problem. If seventeen subtracted from a number is nine, what is the number?

    a. Step 1: Let $x =$ _____

    b. Step 2: _____

c. Step 3:

d. Step 4: Since seventeen subtracted from _____ is nine, _____ is the correct solution.

STOP   •   STOP   •   STOP   •   STOP   •   STOP   •   STOP   •   STOP   •   STOP   •   STOP

A2   a. the number   b. $x - 17 = 9$ c. $x - 17 + 17 = 9 + 17$ (optional)
                                                              $x = 26$

      d. twenty-six (26), twenty-six (26)

Q3   Use the four-step procedure to solve each of the following word problems (do not forget the check). Use n for the number.

a. Three times the number, plus two more than the number, is equal to eighteen. What is the number?

b. If three more than twice a certain number is seven less than that number, what is the number?

STOP   •   STOP   •   STOP   •   STOP   •   STOP   •   STOP   •   STOP   •   STOP   •   STOP

A3   a. 4:   Let $n =$ the number
                  $3n + (n + 2) = 18$
                       $4n + 2 = 18$
                          $4n = 16$
                           $n = 4$

Since three times four (12) plus two more than four (6) is equal to eighteen, four (4) is the correct solution.

b. ⁻10:   Let $n =$ the number
                  $2n + 3 = n - 7$
            $2n + 3 - n = n - 7 - n$
                  $n + 3 = {}^-7$
                      $n = {}^-10$

Three more than twice negative ten ($-20 + 3$, or $^-17$) is seven less than negative ten ($^-10 - 7$, or $^-17$), so negative ten ($^-10$) is the correct solution.

---

3   The purpose of the word statements in the check is to emphasize that the solution should be checked against the statement of the problem. If an incorrect equation is written, checking the equation will only tell you that you solved the equation correctly. In this case, you still have the incorrect solution for the problem.

Q4   Solve this problem: A number increased by six is two less than three times as large as the same number. What is the number? Use x for the number.

STOP • **STOP** • **STOP** • **STOP** • **STOP** • **STOP** • **STOP** • **STOP** • **STOP**

A4   4:   Let x = the number

$$x + 6 = 3x - 2$$
$$x + 6 - x = 3x - 2 - x$$
$$6 = 2x - 2$$
$$8 = 2x$$
$$4 = x$$

Four increased by six (10) is two less than three times four (10), so the solution four (4) is correct.

Q5   Eighty percent of a number diminished by two is ten. What is the number? Use n for the number.

STOP • **STOP** • **STOP** • **STOP** • **STOP** • **STOP** • **STOP** • **STOP** • **STOP**

A5   15: $0.8n - 2 = 10$ or $\dfrac{4}{5}n - 2 = 10$

$$0.8n = 12$$

$$\frac{0.8n}{0.8} = \frac{12}{0.8}$$

$4n - 10 = 50$ (multiplying both sides by 5)

$$4n = 60$$
$$n = 15$$

$$n = \frac{120}{8} = 15$$

This completes the instruction for this section.

8.7   Exercises

Solve each of these word problems using the four-step procedure outlined in this section. Use x for the number.

1. Twice a number, minus two more than the number, is equal to fifteen. What is the number?
2. If five less than a number is divided by four, the result is seven. Find the number.
3. Five times the sum of a number and six is forty-five. What is the number?
4. If the product of five and two less than a certain number is six more than that number, what is the number?
5. If a number is added to five less than twice the number, the result is six times the original number. What is the number?
6. If one-half is subtracted from three-fifths of a number, the result is three-fourths. What is the number?
7. One-half of a number increased by five is eight less than two-thirds of the same number. What is the number?
8. A number plus twenty percent of the same number is 36. Find the number.
*9. A television repairman charges $25 for a house call plus $10 per hour for the time worked. Following a house call by a repairman, Mr. Pendergrass received a bill for $60. How long did the call last? [*Hint:* Let x = the length of time (in hours) of the house call.]
*10. The cost of a television set that was reduced by 20 percent was $164. What was the original price of the television set?

8.7   Exercise Answers

1. 17: $2x - (x + 2) = 15$

2. 33: $\dfrac{x - 5}{4} = 7$

3. 3: $5(x + 6) = 45$

4. 4: $5(x - 2) = x + 6$

5. $\dfrac{-5}{3}$: $2x - 5 + x = 6x$

6. $\dfrac{25}{12}$: $\dfrac{3}{5}x - \dfrac{1}{2} = \dfrac{3}{4}$

7. 78: $\dfrac{1}{2}x + 5 = \dfrac{2}{3}x - 8$

8. 30: $x + 0.2x = 36$ or $x + \dfrac{1}{5}x = 36$

*9. $3\dfrac{1}{2}$ hours: $25 + 10x = 60$

*10. $205: $x - 0.2x = 164$ or $x - \dfrac{1}{5}x = 164$

Chapter 8 Sample Test

At the completion of Chapter 8 it is expected that you will be able to work the following problems.

8.1 ## Equations, Open Sentences, Replacement Set, and Solution Set

1. What word best describes each phrase?
   a. a statement that two expressions are equal
   b. an equation that is neither true nor false
   c. the set of all numbers that may replace the variable in an open sentence
   d. the values from the replacement set which convert the open sentence into a true statement
2. Identify the right and left sides of each of the following equations:
   a. $3x = 7 - 2$ b. $y - 4 + 3y = 0$
3. If the replacement set for the variable x is $\{-5, 0, 1, -1, 4\}$, find the result of all possible replacements for x in the open sentence $9 - x = 8$. Label each statement as true or false.

8.2 ## Addition and Subtraction Principles of Equality

4. Label each of the following as demonstrating the addition principle of equality or the subtraction principle of equality:

   a. $$x + 7 = 12$$
   $$x + 7 - 7 = 12 - 7$$
   $$x = 5$$

   b. $$-3 = -5 + y$$
   $$-3 + 5 = -5 + y + 5$$
   $$2 = y$$

   c. $$2y = y + 4$$
   $$2y - y = y + 4 - y$$
   $$y = 4$$

   d. $$-4 = x - 4$$
   $$-4 + 4 = x - 4 + 4$$
   $$0 = x$$

5. Solve and check each of the following equations:
   a. $-3 + x = 5$ b. $-7 = y + 9$
   c. $x + 8 = 8$ d. $7 + y = 6 - 13$
   e. $15 = -25 + x$ f. $17 = y - 5$

8.3 ## Multiplication and Division Principles of Equality

6. Label each of the following as demonstrating the multiplication principle of equality or the division principle of equality:

   a. $$-5x = 30$$
   $$\frac{-5x}{-5} = \frac{30}{-5}$$
   $$x = -6$$

   b. $$-12 = \frac{3}{4}y$$
   $$\frac{4}{3}(-12) = \frac{4}{3}\left(\frac{3}{4}y\right)$$
   $$-16 = y$$

   c. $$\frac{y}{7} = \frac{5}{14}$$
   $$7\left(\frac{y}{7}\right) = 7\left(\frac{5}{14}\right)$$
   $$y = \frac{5}{2}$$

   d. $$-18 = -5x$$
   $$\frac{-18}{-5} = \frac{-5x}{-5}$$
   $$3\frac{3}{5} = x$$

7. Solve and check each of the following equations:

a. $^-5x = 25$

b. $\dfrac{5}{7}x = 35$

c. $\dfrac{-2}{3}y = \dfrac{5}{9}$

d. $^-24 = \dfrac{3}{4}x$

e. $\dfrac{y}{8} = 2$

f. $^-x = \dfrac{7}{9}$

g. $\dfrac{-4}{9} = {}^-3y$

h. $^-7y = \dfrac{14}{15}$

8.4 Solving Equations by the Use of Two or More Steps

8. Solve and check each of the following equations:

a. $2x - 5 = 55$

b. $^-5y + 1 = {}^-4$

c. $\dfrac{2}{5}y - 3 = 11$

d. $7x - 5 = 9x - 5$

e. $2 - 3y = 2y + 2$

f. $x + 11 = {}^-x + 5$

g. $6 - (2x + 3) = 9 + (x + 15)$

h. $5x - 2(x - 7) = 8(x - 2) + 10$

8.5 Equations with Fractions

9. Solve and check each of the following equations:

a. $\dfrac{3}{4}x - 3 = \dfrac{1}{2}x$

b. $\dfrac{14y}{5} = \dfrac{1}{3} + y$

8.6 Expressing English Phrases as Algebraic Expressions

10. Write each phrase as an algebraic expression. Use n for the number.
   a. the sum of fifteen and twice a number
   b. the difference between three times the number and the number
   c. five less than the number
   d. the product of nine and the number, less two
   e. the quotient of the number decreased by one and six
   f. two more than five times the quantity of the number less one
   g. twelve percent of the number added to the number
   h. two-thirds of the number subtracted from ten
   i. the opposite of the number, diminished by eight
   j. twelve, minus the quantity three minus the number

8.7 Solving Simple Word Problems

11. Solve each of these word problems using the four-step procedure outlined in this section:
   a. The sum of five times the number and three times the number, minus four, equals twenty-eight. What is the number?
   b. Two-thirds of the number taken from the number is five less than the number. Find the number.

   c. What number when increased by three and the result is multiplied by negative
   two produces ten?
   d. A number less six, subtracted from five, is four. What is the number?
   e. The sum of one-half of a number and nine is equal to four-fifths of the difference
   between the number and three. Find the number.

Chapter 8 Sample Test Answers

   1. a. equation   **b.** open sentence   **c.** replacement set   **d.** solutions
   2. a. right side: $7 - 2$; left side: $3x$ **b.** right side: 0; left side: $y - 4 + 3y$
   3. $9 - {}^-5 = 8$, false
     $9 - 0 = 8$, false
     $9 - 1 = 8$, true
     $9 - {}^-1 = 8$, false
     $9 - 4 = 8$, false
   4. a. subtraction principle of equality   **b.** addition principle of equality
     c. subtraction principle of equality   **d.** addition principle of equality
   5. a. $x = 8$ **b.** $y = {}^-16$ **c.** $x = 0$ **d.** $y = {}^-14$
     e. $x = 40$ **f.** $y = 22$
   6. a. division principle of equality   **b.** multiplication principle of equality
     c. multiplication principle of equality   **d.** division principle of equality
   7. a. $x = {}^-5$ **b.** $x = 49$ **c.** $y = \dfrac{-5}{6}$ **d.** $x = {}^-32$

   e. $y = 16$ **f.** $x = \dfrac{-7}{9}$ **g.** $y = \dfrac{4}{27}$ **h.** $y = \dfrac{-2}{15}$

   8. a. $x = 30$ **b.** $y = 1$ **c.** $y = 35$ **d.** $x = 0$
     e. $y = 0$ **f.** $x = {}^-3$ **g.** $x = {}^-7$ **h.** $x = 4$
   9. a. $x = 12$ **b.** $y = \dfrac{5}{27}$

   10. a. $15 + 2n$ **b.** $3n - n$ **c.** $n - 5$

     d. $9n - 2$ **e.** $\dfrac{n - 1}{6}$ **f.** $5(n - 1) + 2$

     g. $n + 0.12n$ or $n + \dfrac{3}{25}n$ **h.** $10 - \dfrac{2}{3}n$ **i.** ${}^-n - 8$

     j. $12 - (3 - n)$

   11. a. 4   **b.** $\dfrac{15}{2}$ or $7\dfrac{1}{2}$ **c.** ${}^-8$ **d.** 7

     e. 38

Chapter 9

Recognition and Properties of Geometric Figures

Geometric figures are found all around. Some of the properties of these geometric objects, such as the wheel, are functional and make life easier. Some objects, such as buildings and monuments, are pleasing to the eye. Other objects are intellectually interesting because of their geometric properties. To live in a geometric world a person should know some of the fundamental geometric objects and a few of their properties. This chapter will help accomplish that goal.

9.1   Points, Lines, and Planes

| 1 | The most fundamental notion in geometry is a *point*. A point is a location in space and has no thickness. Although a point is represented with a dot on paper, the dot is much too large to actually be a point. The dot helps in thinking about the point. |

Q1 Which dot is the best representation of a point?

    a. •   **b.** •   **c.** •

STOP • STOP • STOP • STOP • STOP • STOP • STOP • STOP • STOP

A1 c: it looks most like a location with no dimensions.

Q2 Is the representation in part c of Q1 actually a point? _____

STOP • STOP • STOP • STOP • STOP • STOP • STOP • STOP • STOP

A2 no: it is still too large and is actually many, many points all very close together.

| 2 | Another geometric object which is so fundamental that other objects depend on its properties is a *line*. A line is a set of points and is represented with a picture such as:

  ⟵————————————⟶

  A geometric line is always straight, has no thickness, and extends forever in both directions. The lines you draw on paper help you think about true geometric lines, which really only exist perfectly in your mind. A tightly stretched string is another representation that helps to think of a line in space, but it is still not a perfect model of a geometric line. |

Q3   List three important properties of lines. _____
        _____

STOP • STOP • STOP • STOP • STOP • STOP • STOP • STOP • STOP

A3   straightness, no thickness, extends forever in both directions

Q4   Indicate why each of the following are not lines:

        a. pencil _____

        b. equator _____

        c. telephone wire strung on poles _____

STOP • STOP • STOP • STOP • STOP • STOP • STOP • STOP • STOP

A4   **a.** A pencil has thickness, has definite length, and may not be straight.
        b. The equator is not straight but is actually curved. It also has finite length.
        c. It has thickness. It sags between poles so is not straight and has definite length.

---

3   A *plane* is usually represented by a flat surface such as a tabletop, wall, or stiff sheet of
        paper. A geometric plane, however, extends infinitely in every direction. The following
        figure represents two intersecting planes:

        The planes must be drawn in perspective and distortions occur. (For example, it may
        not look like plane *M* and plane *N* meet with square corners.) Keep in mind that all
        drawings of geometric objects, no matter how carefully they are done, are imperfect. As
        long as you use the figures to reason correctly, they are serving their purpose.

Q5   Planes *M* and *N* are shown intersecting in Frame 3. What geometric figure is formed by

        the points that are on both planes? _____

STOP • STOP • STOP • STOP • STOP • STOP • STOP • STOP • STOP

A5   a line

---

4   When more than one line is drawn, it is helpful to be able to refer to them. They are
        sometimes referred to with the notation l_1 and l_2. (This is read *l* sub one and *l* sub two.)
        The 1 and 2 are called subscripts and are used only to keep track of which *l* is being
        discussed. They are *not* exponents. The following figure contains three lines: l_1, l_2, and
        l_3:

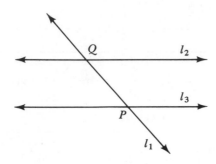

l_3 intersects l_1 at point P, but l_3 does not intersect l_2.

Two lines in one plane may intersect, or they may be parallel. If they are parallel, they have no points in common. If they intersect, they have one and only one point in common.

Intersecting lines

Parallel lines
(write $l_1 \parallel l_2$)

If all four angles formed by intersecting lines are equal, each angle is a *right* angle and the lines are said to be *perpendicular* lines:

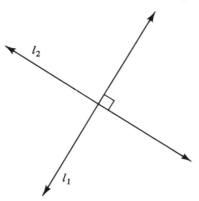

l_1 is perpendicular to l_2. Write $l_1 \perp l_2$ and place a square where they intersect.

Q6   **a.** From the figure below, write two pairs of lines that appear to be parallel (use the appropriate symbol between each pair of lines). _____

        b. Write four pairs of lines that appear to be perpendicular (use the appropriate symbol). _____

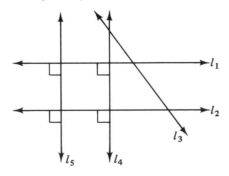

A6   **a.** $l_1 \| l_2$ and $l_4 \| l_5$
  b. $l_1 \perp l_4$, $l_1 \perp l_5$, $l_2 \perp l_4$, and $l_2 \perp l_5$

Q7   How many lines can be drawn through point A?_____

  $\cdot A$

STOP • STOP • STOP • STOP • STOP • STOP • STOP • STOP • STOP

A7   an infinite number

Q8   How many lines can be drawn that contain both point A and point B?

  _____ $\cdot A$ $\cdot B$

STOP • STOP • STOP • STOP • STOP • STOP • STOP • STOP • STOP

A8   one

5	One and only one line can be drawn through two points. This fact is sometimes stated as: *Two points determine one and only one line.*

Q9   Hold up a finger and thumb on one hand and one finger on the other. How many planes (such as a book cover) touch all three points (two fingers, one thumb)?_____

STOP • STOP • STOP • STOP • STOP • STOP • STOP • STOP • STOP

A9   one

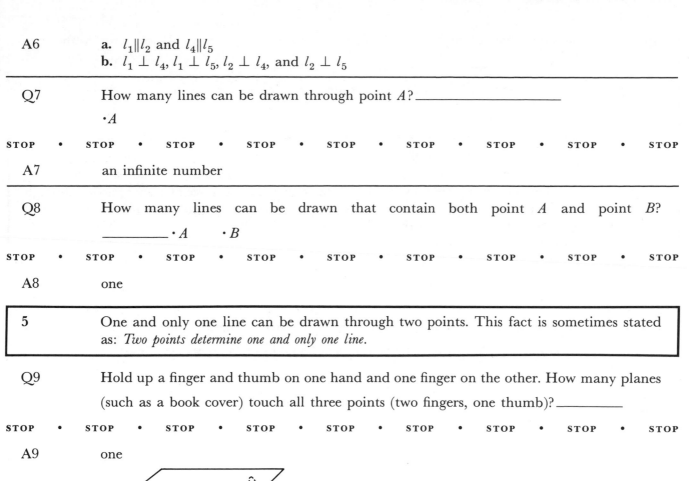

Q10   Suppose there are three points not on one line. How many planes are determined?_____

STOP • STOP • STOP • STOP • STOP • STOP • STOP • STOP • STOP

A10   one

Q11   Hold up one finger on each hand. How many ways can a flat surface touch both points?_____

STOP • STOP • STOP • STOP • STOP • STOP • STOP • STOP • STOP

A11   an infinite number

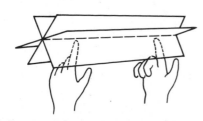

Q12   Suppose that there are two points in space. How many planes contain the two points?_____

STOP • STOP • STOP • STOP • STOP • STOP • STOP • STOP • STOP

A12   an infinite number

---

6   The answers to Q9 through Q12 are examples of two properties of planes: *Two points are contained in an infinite number of planes and three points determine one and only one plane.*

    These properties are illustrated with the plane of the surface of a door passing through two points represented by two hinges. The door can be placed in an infinite number of positions. However, when the plane must also contain a third point, such as the tip of your shoe, the door becomes stationary.

Q13   Mark three points on a sheet of paper. Hold your finger above these three points to represent a fourth point.

    a. How many ways can a flat surface touch three of the four points?_____

    b. How many lines would be determined by taking two points at a time?_____

STOP • STOP • STOP • STOP • STOP • STOP • STOP • STOP • STOP

A13   **a.** four planes: including the surface of the paper:

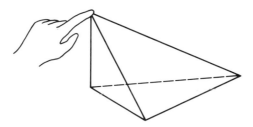

    b. six lines: three in the plane of the paper and three connecting the points on the paper with the finger.

---

7   Points are frequently labeled with capital letters. Lines are referred to by naming two points on the line and drawing a double-headed arrow above them. The following line is represented by $\overleftrightarrow{AB}$:

Q14   Label the statements true or false:

    _____ **a.** $\overleftrightarrow{AB}$ intersects $\overleftrightarrow{CD}$.

    _____ **b.** $\overleftrightarrow{AB}\|\overleftrightarrow{CD}$.

    _____ **c.** $\overleftrightarrow{CD}$ intersects $\overleftrightarrow{DB}$.

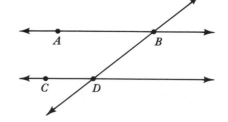

STOP • STOP • STOP • STOP • STOP • STOP • STOP • STOP • STOP

A14   **a.** false **b.** true **c.** true

8   A *line segment* (sometimes referred to as a segment) is the portion of a line between two points including the endpoints. A line segment is referred to with the symbol $\overline{AB}$, where A and B are its endpoints. Segment $\overline{CD}$ is shown as

In the next figure $\overleftrightarrow{AB}$ is drawn, but $\overline{AB}$ refers only to the points A and B and those between A and B. The segment is a part of the line.

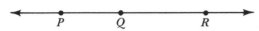

Refer to the figure below for the following examples:

1. P is on $\overleftrightarrow{QR}$.
2. P is not on $\overline{QR}$ (because P is not between Q and R).
3. Q is on $\overleftrightarrow{PR}$.
4. Q is on $\overline{PR}$.
5. R is on $\overleftrightarrow{PQ}$.
6. R is not on $\overline{PQ}$.

Q15   Label the statements true or false:

a. A is on $\overleftrightarrow{CD}$.   _____   b. A is on $\overline{CD}$.   _____

c. B is on $\overline{AC}$.   _____   d. B is on $\overleftrightarrow{AC}$.   _____

e. C is on $\overleftrightarrow{AB}$.   _____   f. C is on $\overline{AB}$.   _____

STOP • **STOP** • **STOP** • **STOP** • **STOP** • **STOP** • **STOP** • **STOP** • **STOP**

A15   a. true   b. false   c. true   d. true   e. true   f. false

Q16   a. Name the lines shown in the figure.

   _____

   b. Name the line segments shown in the

   figure. _____

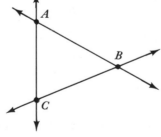

STOP • **STOP** • **STOP** • **STOP** • **STOP** • **STOP** • **STOP** • **STOP** • **STOP**

A16   a. $\overleftrightarrow{AB}$, $\overleftrightarrow{AC}$, and $\overleftrightarrow{BC}$ b. $\overline{AB}$, $\overline{AC}$, and $\overline{BC}$

9   Line segments are equal only if they have the same endpoints. Therefore, there are three different line segments $\overline{AB}$, $\overline{BC}$, and $\overline{AC}$ on the line below.

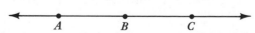

Q17   Name four segments on the figure.

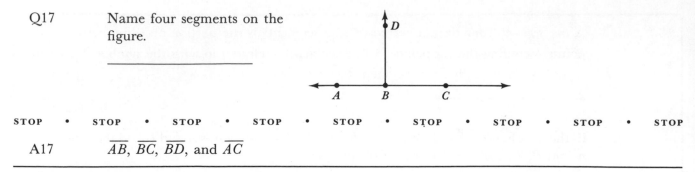

_____

STOP   •   STOP   •   STOP   •   STOP   •   STOP   •   STOP   •   STOP   •   STOP   •   STOP

A17 $\overline{AB}$, $\overline{BC}$, $\overline{BD}$, and $\overline{AC}$

This completes the instruction for this section.

9.1   Exercises

1. Name the four segments that are the four sides of the rectangle $ABCD$.

Figure 1

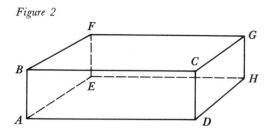

2. In Figure 1, name two line segments that intersect at B.
3. In Figure 1, name two line segments that appear to be part of parallel lines (there are two possibilities).
4. Figure 2 has the shape of a brick. The dashed lines would not be visible.

   Figure 2

   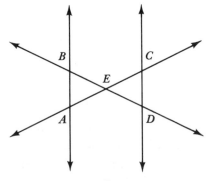

   a. Name a pair of edges that have the common point C in the plane closest to you.

   b. Name a pair of edges that appear to be part of parallel lines in the plane farthest from you (there are two possibilities).

   c. Name an edge that is not in the plane of the top or the plane of the bottom of the figure.

5. Use the figure at the right to answer the following:

   a. Name two lines that appear to be parallel.

   b. $\overleftrightarrow{AC}$ and $\overleftrightarrow{BD}$ intersect at what point?

   c. Name eight line segments on the figure.

   d. Are there any perpendicular lines in the figure?

6. Which is the best model of a point: the sun, a bowling ball, a grain of sand, or a marble?

7. a. Why is it that a table with three legs will always stand firmly on the floor, whereas a table with four legs sometimes rocks?

b. Is it always possible to adjust a wobbling table by adjusting exactly one leg?

8. Is there always a plane that contains any three-sided figure?

9. Is there always a plane that contains any four-sided figure?

9.1   Exercise Answers

1. $\overline{AB}$, $\overline{BC}$, $\overline{CD}$, and $\overline{DA}$

2. $\overline{AB}$ and $\overline{BC}$

3. $\overline{AB}\|\overline{CD}$ and $\overline{BC}\|\overline{AD}$

4. **a.** $\overline{BC}$ and $\overline{CD}$
   b. $\overline{FG}\|\overline{EH}$ and $\overline{EF}\|\overline{GH}$
   c. $\overline{AB}$, $\overline{CD}$, $\overline{GH}$, or $\overline{FE}$

5. **a.** $\overleftrightarrow{AB}\|\overleftrightarrow{CD}$
   b. E
   c. $\overline{AB}$, $\overline{AE}$, $\overline{BE}$, $\overline{EC}$, $\overline{ED}$, $\overline{CD}$, $\overline{AC}$, and $\overline{BD}$
   d. no

6. A grain of sand. However, something as large as the sun, such as a star, appears to be a point if you are far enough away.

7. **a.** The ends of three legs determine a unique plane that makes the table stable. If the fourth leg is not in the same plane, the table will rock.
   b. yes

8. yes

9. no: for example, think of raising point D in problem 1 above the plane. The figure is still four-sided but not in a plane.

9.2   **Rays and Angles**

1

The following geometric figure, which contains A and B and extends forever in each direction, is called a *line*.

A line is considered to be a set of points and, as the picture demonstrates, is assumed to be straight. A line is labeled by any two points on it. It is thus possible to speak of "line AB" or use the symbol $\overleftrightarrow{AB}$.

    The part of the line on one side of a point that includes the endpoint is called a *ray*. A ray is represented by noting its endpoint first along with another point on the ray with a single-headed arrow extending to the right above them. The name of the following rays would be $\overrightarrow{DC}$:

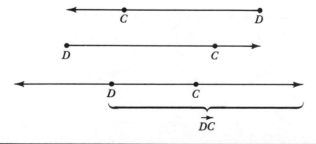

Q1   Name the rays. _____

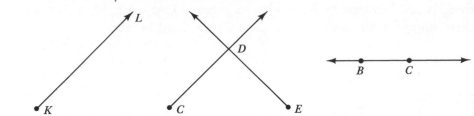

STOP • **STOP** • **STOP** • **STOP** • **STOP** • **STOP** • **STOP** • **STOP** • **STOP**

A1 $\overrightarrow{KL}$, $\overrightarrow{CD}$, $\overrightarrow{ED}$, $\overrightarrow{BC}$, and $\overrightarrow{CB}$

---

2   A good representation of a ray is a flashlight beam or searchlight beam. However, a ray has properties that these representations do not share. A ray has no thickness, extends infinitely in one direction, and has a point (with no dimensions) as its endpoint.

   The same ray can be named in several ways. For example, in the figure below $\overrightarrow{BC}$ and $\overrightarrow{BD}$ represent the same ray. We therefore say that $\overrightarrow{BC} = \overrightarrow{BD}$, because they contain the same endpoint and extend along the same line in the same direction.

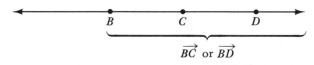

$\overrightarrow{BC}$ or $\overrightarrow{BD}$

Q2   **a.** List all the rays on the line. _____

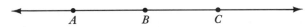

            b. Which of the rays in your answer to part a are equal? _____

STOP • **STOP** • **STOP** • **STOP** • **STOP** • **STOP** • **STOP** • **STOP** • **STOP**

A2   **a.** $\overrightarrow{AB}$, $\overrightarrow{AC}$, $\overrightarrow{BC}$, $\overrightarrow{CB}$, $\overrightarrow{CA}$, and $\overrightarrow{BA}$ (mathematicians have agreed *not* to use $\overleftarrow{BC}$ to mean the same as $\overrightarrow{CB}$)
            b. $\overrightarrow{AB} = \overrightarrow{AC}$ and $\overrightarrow{CB} = \overrightarrow{CA}$

Q3   List six different rays on the figure shown.

            _____

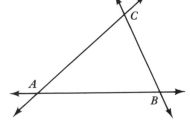

STOP • **STOP** • **STOP** • **STOP** • **STOP** • **STOP** • **STOP** • **STOP** • **STOP**

A3 $\overrightarrow{AB}$, $\overrightarrow{AC}$, $\overrightarrow{CB}$, $\overrightarrow{CA}$, $\overrightarrow{BA}$, and $\overrightarrow{BC}$

---

3   An *angle* consists of two rays with a common endpoint. The common endpoint is called the *vertex* of the angle. The rays are the *sides* of the angle. A representation of an angle is shown. The two rays are $\overrightarrow{BA}$ and $\overrightarrow{BC}$, with the vertex being the point B.

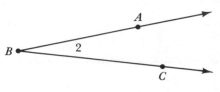

An angle can most easily be named by its vertex or by a number written at the vertex between the sides. The name of the angle is preceded by "∡." Thus, two names for the angle above are ∡B and ∡2.

However, if there is more than one angle at a vertex, a more complicated scheme is used. Another name for the angle shown is ∡ABC, where A is on one ray, B (the middle letter) is the vertex, and C is on the other ray. Another way of naming the same angle is ∡CBA. However, ∡ACB would not be correct, because C is not the vertex.

Q4   Use three letters to name the three angles in the figure.

∡1 = _____

∡2 = _____

∡3 = _____

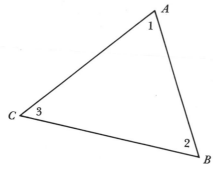

STOP • STOP • STOP • STOP • STOP • STOP • STOP • STOP • STOP

A4   ∡1 = ∡CAB or ∡BAC
∡2 = ∡ABC or ∡CBA
∡3 = ∡ACB or ∡BCA

Q5   Name the angles in the following figure by using the easiest method (fewest letters) and the points shown.

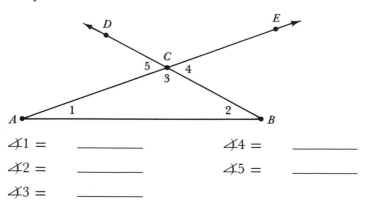

∡1 = _____   ∡4 = _____

∡2 = _____   ∡5 = _____

∡3 = _____

STOP • STOP • STOP • STOP • STOP • STOP • STOP • STOP • STOP

A5   ∡1 = ∡A ∡4 = ∡ECB
∡2 = ∡B ∡5 = ∡DCA
∡3 = ∡ACB

Of course, each of the representations could be completely reversed and it represents the same angle; that is, ∡ACB = ∡BCA.

4   The *measurement of an angle* is a number assigned to the amount of rotation required to swing one ray of the angle to the other ray. The unit of measure will be the *degree*, which is based upon the division of the rotation of 1 complete revolution into 360 parts. Each of these parts is a degree. You should become familiar with the measure of angles so you can tell the approximate size of an angle by inspection. Some frequently used angles are shown.

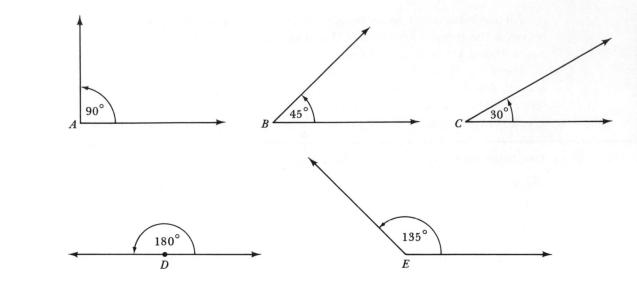

The size of an angle does not depend upon the lengths of the sides of the angle shown since the rays of an angle extend indefinitely. Therefore, angle 1 is considered to be larger than angle 2 in the following figures.

Q6   Compare the angles shown to the examples in Frame 4 to choose the approximate measure of the angle:

a.

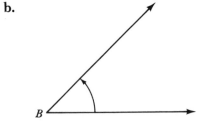

                                                          40°, 80°, or 100°   ____

b.

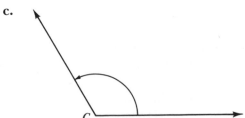

                                                          20°, 45°, or 60°   ____

c.

                                                           80°, 160°, or 120°   ____

d.

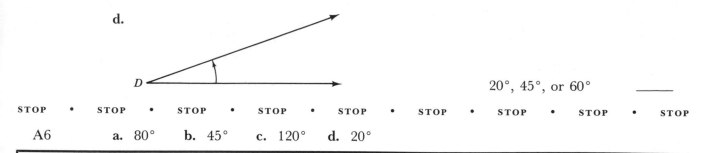

20°, 45°, or 60° _____

STOP • **STOP** • **STOP** • **STOP** • **STOP** • **STOP** • **STOP** • **STOP** • **STOP**

A6   **a.** 80°   **b.** 45°   **c.** 120°   **d.** 20°

5

The instrument used to measure angles is called a *protractor*. A protractor is needed to complete this section. The protractor is shown as it would be used to measure an angle. The base of the protractor is placed along one ray of the angle with the center mark

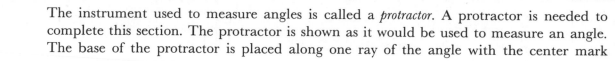

on the protractor at the vertex. The other ray of the angle will pass under the scale of the protractor and the degree measure is read from the scale. The measure of the angle shown is approximately 54° to the nearest degree. Notice that each scale mark has two numbers beside it. The numbers larger than 90 are used if an angle over 90° is being measured. Write $m\angle ABC = 54°$ (read "the measure of angle ABC is equal to 54 degrees").

Q7   Use a protractor to measure the angles (extend sides of angle if necessary):

a.   **b.**

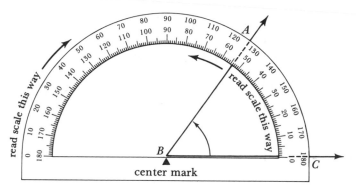

c. *K*   **d.**

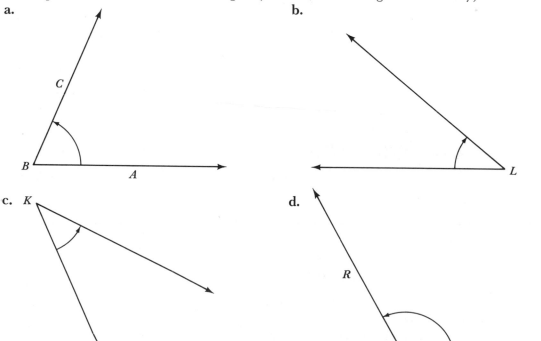

e.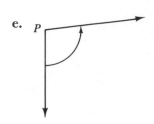

f.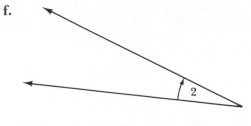

STOP • **STOP** • **STOP** • **STOP** • **STOP** • **STOP** • **STOP** • **STOP** • **STOP**

A7   **a.** $m\measuredangle ABC = 67°$ **b.** $m\measuredangle L = 40°$
     c. $m\measuredangle K = 40°$ **d.** $m\measuredangle PQR = 118°$
     e. $m\measuredangle P = 98°$ (notice that the sides must be extended)
     f. $m\measuredangle 2 = 20°$

Q8   Properties of geometric figures can be discovered by measuring angles. Add the measures of the three angles in each triangle shown. What do you discover?

a. 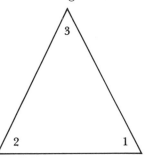   **b.**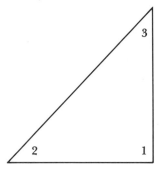

c.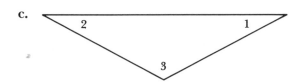

STOP • **STOP** • **STOP** • **STOP** • **STOP** • **STOP** • **STOP** • **STOP** • **STOP**

A8   **a.** $63° + 64° + 53° = 180°$
     b. $90° + 47° + 43° = 180°$
     c. $28° + 124° + 28° = 180°$
     It looks as though the sum of the measures of the angles of a triangle is $180°$. This is, in fact, true.

---

6   The results of Q8 may be summarized by saying that the sum of the measures of the angles of a triangle is $180°$.

Angles are classified according to the size of their measures.

1. Angles whose measure is less than $90°$ are called *acute* angles.
2. Angles whose measure is $90°$ are called *right* angles.
3. Angles whose measure is between $90°$ and $180°$ are called *obtuse* angles.

Q9   Write acute, right, or obtuse under each of the following angles:

a.   **b.**   **c.**

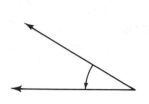   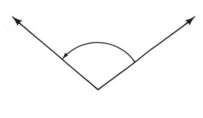   

_____   _____   _____

d.   **e.**   **f.**

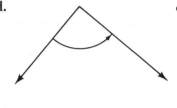   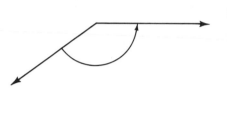   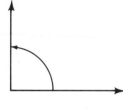

_____   _____   _____

STOP • STOP • STOP • STOP • STOP • STOP • STOP • STOP • STOP

A9   **a.** acute   **b.** obtuse   **c.** acute   **d.** right
           e. obtuse   **f.** right

---

7   Two angles whose measures added together result in a sum of 90° are said to be *complementary* angles. Each is a complement of the other. For example, if $m\angle A = 20°$ and $m\angle B = 70°$, angles A and B are complementary.

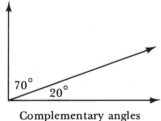   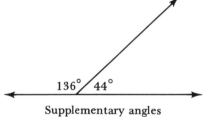

Complementary angles   Supplementary angles

Two angles whose measures added together result in a sum of 180° are said to be *supplementary* angles. Each is a supplement of the other. For example, if $m\angle C = 136°$ and $m\angle D = 44°$, angles C and D are supplementary.

Q10   **a.** The supplement of an angle of 60° would have what measure?   _____

           b. The complement of an angle of 60° would have what measure? _____

STOP • STOP • STOP • STOP • STOP • STOP • STOP • STOP • STOP

A10   **a.** 120°   **b.** 30°

Q11   **a.** Which of the given angles is the complement of the

angle to the right?_____

b. Which of the given angles is the supplement of the

angle to the right?_____

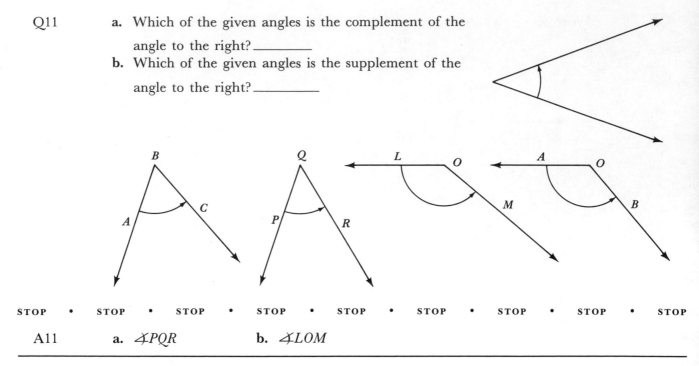

STOP   •   STOP   •   STOP   •   STOP   •   STOP   •   STOP   •   STOP   •   STOP   •   STOP

A11   **a.** ∡*PQR*   **b.** ∡*LOM*

This completes the instruction for this section.

9.2   Exercises

1. Name two rays with endpoints at *B*.

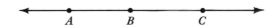

2. Give two other ways of naming the ray *AC*.

3. Name eight different rays by using the points shown on the intersecting lines.

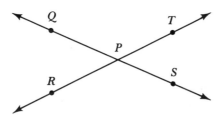

4. What is the name of the intersection of the two rays that make up an angle?
5. What is the geometric name for the sides of an angle?
6. Name the angle in three ways.

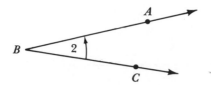

7. Name four angles in the figure.

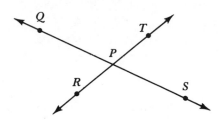

8. How many degrees are in one complete revolution?

9. Without measuring the angles, choose the most likely number of degrees in the angle:

   a.

   b.

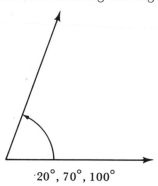

   ·20°, 70°, 100°

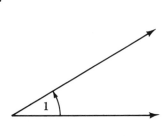

   70°, 90°, 120°

10. What is the name of the instrument used to measure angles?

11. Measure the angles below to the nearest degree:

    a.

    b.

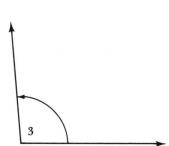

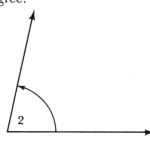

    c.

    d.

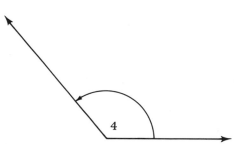

12. Consider the angles and answer the following questions:

    a.  **b.**  **c.** 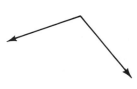 **d.**

    e.   **f.**   **g.**  **h.**

   a. Name the letters of the acute angles.
   b. Name the letters of the obtuse angles.
   c. Name the letters of the right angles.
13. What does it mean to say that two angles are complementary?
14. What does it mean to say that two angles are supplementary?
15. a. Could two acute angles be supplementary?
    b. complementary?
16. a. Could an obtuse and an acute angle be complementary?
    b. supplementary?
17. Could two obtuse angles be supplementary?

9.2   Exercise Answers

1. $\overrightarrow{BC}$ and $\overrightarrow{BA}$
2. $\overrightarrow{AB}$ and $\overrightarrow{AD}$
3. $\overrightarrow{QS}$, $\overrightarrow{PS}$, $\overrightarrow{RT}$, $\overrightarrow{PT}$, $\overrightarrow{SQ}$, $\overrightarrow{PQ}$, $\overrightarrow{TR}$, and $\overrightarrow{PR}$
4. vertex
5. rays
6. $\angle ABC$, $\angle B$, $\angle 2$, or $\angle CBA$
7. $\angle QPR$, $\angle RPS$, $\angle SPT$, and $\angle TPQ$
8. 360°
9. a. 70°   b. 120°
10. protractor
11. a. 31°   b. 78°   c. 95°   d. 130°
12. a. a, b, e, h b. d, g c. c, f
13. The sum of their measures is 90°.
14. The sum of their measures is 180°.
15. a. no   b. yes
16. a. no   b. yes
17. no

9.3   Circles and Polygons

1   A *circle* is a set of all points in a plane whose distance from a given point, P, is equal to a positive number, r. The point P is the *center* of the circle. The positive number r is the *radius* of the circle. Twice the radius is the *diameter* of the circle.

  The instrument used to draw circles is the compass. It can also be used to compare distances.

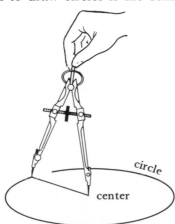

Q1   Use a compass to determine which point is the center of the circle. _____

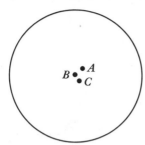

STOP • **STOP** • **STOP** • **STOP** • **STOP** • **STOP** • **STOP** • **STOP** • **STOP**

A1 B

Q2   **a.** If a circle has a radius of 3 inches, its diameter must be _____ .

      b. If a circle has a diameter of 5 inches, its radius must be _____ .

STOP • **STOP** • **STOP** • **STOP** • **STOP** • **STOP** • **STOP** • **STOP** • **STOP**

A2   **a.** 6 inches   **b.** $2\frac{1}{2}$ inches

> **2**   The word "radius" (plural form is radii) is also used to refer to a line segment with one endpoint at the center and the other on the circle. Likewise, the word "diameter" has two meanings. It is a line segment with endpoints on the circle containing the center, as well as being used to represent the length of such a line segment. You will be able to tell which meaning is intended from the context of the statement.

Q3   **a.** Identify the line segments that are radii. _____

      b. Identify the line segments that are diameters. _____

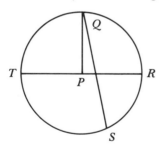

STOP • **STOP** • **STOP** • **STOP** • **STOP** • **STOP** • **STOP** • **STOP** • **STOP**

A3   **a.** $\overline{TP}$, $\overline{PR}$, and $\overline{PQ}$ **b.** $\overline{TR}$

> **3**   You may be familiar with geometric figures called polygons from your past experience. However, the geometric concept of a polygon is very carefully defined; some time will be spent just becoming familiar with the definition.
>
> A *polygon* is a set of points in a plane called *vertices* together with certain line segments called *sides* having these properties:
>
> **1.** There are a finite number of vertices (at least three) arranged in a particular order. In particular, there is a first vertex and a last vertex.
> **2.** Each vertex except the last is joined to the next vertex after it by a line segment called a *side;* and the last vertex is joined to the first vertex by a side.
> **3.** Two sides intersect only at their endpoints.

4. No two sides with a common endpoint are in the same straight line.

Some examples of polygons follow:

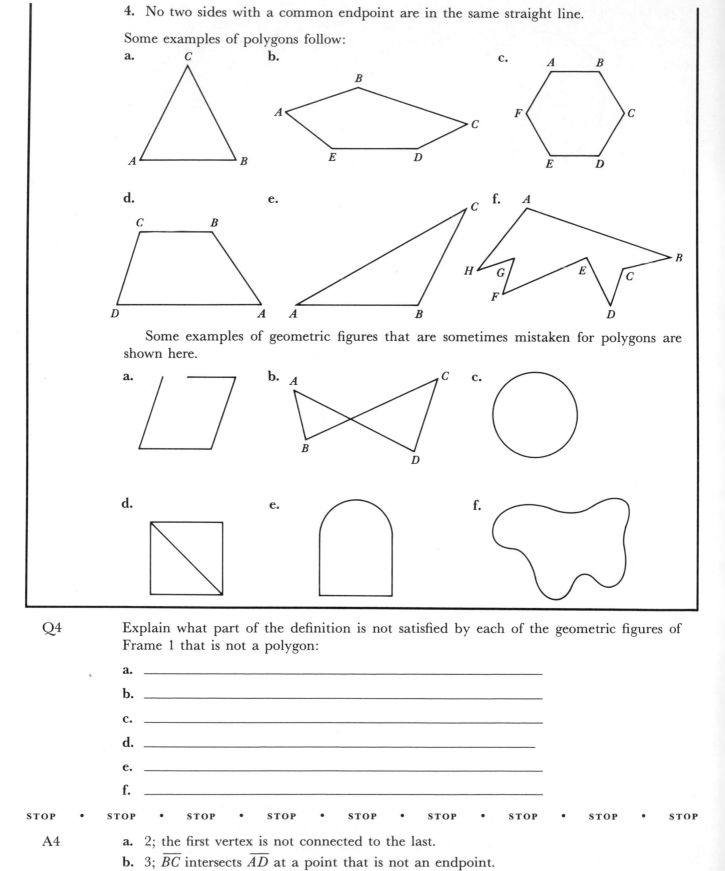

Some examples of geometric figures that are sometimes mistaken for polygons are shown here.

Q4   Explain what part of the definition is not satisfied by each of the geometric figures of Frame 1 that is not a polygon:

    a. _____

    b. _____

    c. _____

    d. _____

    e. _____

    f. _____

STOP   •   STOP   •   STOP   •   STOP   •   STOP   •   STOP   •   STOP   •   STOP   •   STOP

A4   a. 2; the first vertex is not connected to the last.
         b. 3; $\overline{BC}$ intersects $\overline{AD}$ at a point that is not an endpoint.
         c. The sides are not line segments.
         d. 2; the vertices are not joined consecutively.

    e. The sides are not all line segments.
    f. The sides are not line segments.

Q5   Indicate whether each of the figures shown is a polygon:

a.    **b.**    **c.**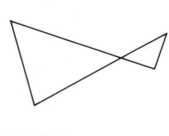

————   ————   ————

d. 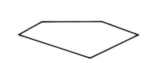   **e.**    **f.**

————   ————   ————

STOP • **STOP** • **STOP** • **STOP** • **STOP** • **STOP** • **STOP** • **STOP** • **STOP**

A5   **a.** yes   **b.** no   **c.** no   **d.** yes   **e.** no   **f.** yes

Q6   What is the least number of sides that a polygon can have?_____ Draw such a polygon.

STOP • **STOP** • **STOP** • **STOP** • **STOP** • **STOP** • **STOP** • **STOP** • **STOP**

A6   three

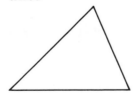

4   Polygons are named for the number of sides which they have. Some commonly used names are:

    3 sides   triangle
    4 sides   quadrilateral
    5 sides   pentagon
    6 sides   hexagon

8 sides	octagon
10 sides	decagon
12 sides	dodecagon
n sides	n-gon

Q7   Write the appropriate name under each polygon:

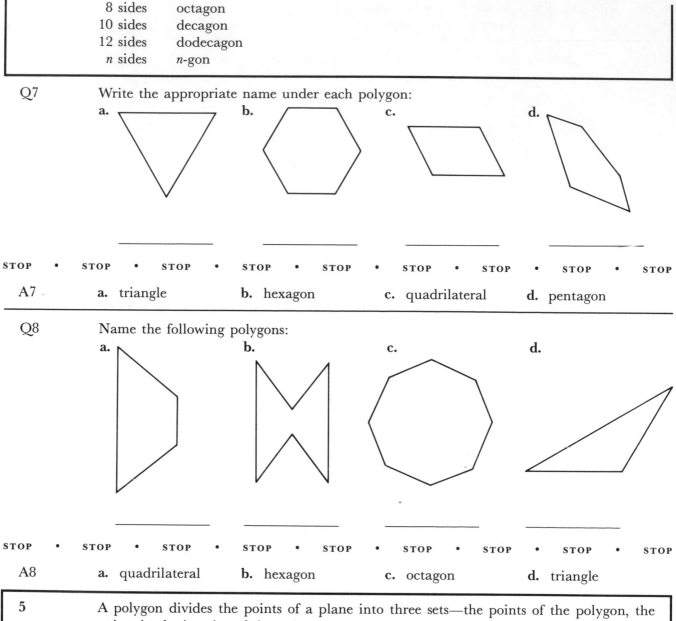

a.   **b.**   **c.**   **d.**

STOP • STOP • STOP • STOP • STOP • STOP • STOP • STOP • STOP

A7   **a.** triangle   **b.** hexagon   **c.** quadrilateral   **d.** pentagon

Q8   Name the following polygons:

a.   **b.**   **c.**   **d.**

STOP • STOP • STOP • STOP • STOP • STOP • STOP • STOP • STOP

A8   **a.** quadrilateral   **b.** hexagon   **c.** octagon   **d.** triangle

5   A polygon divides the points of a plane into three sets—the points of the polygon, the points in the interior of the polygon, and the points in the exterior of the polygon. The plane area bounded by the polygon is the *interior* of the polygon. The *exterior* of a polygon is the set of points that are not in the interior and not on the polygon.

A *triangular region* is a triangle and its interior. A *hexagonal region* is a hexagon together with its interior. Other regions can be described in a similar manner. The triangular region *ABC* is shaded in the figure.

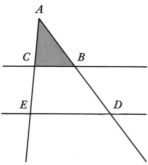

Q9   Shade the triangular region *AED* in the figure.

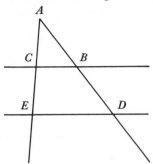

STOP   •   STOP   •   STOP   •   STOP   •   STOP   •   STOP   •   STOP   •   STOP   •   STOP

A9

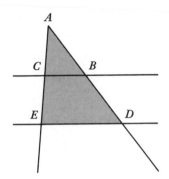

Q10   Shade a quadrilateral region in the figure (there are three possibilities.)

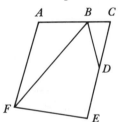

STOP   •   STOP   •   STOP   •   STOP   •   STOP   •   STOP   •   STOP   •   STOP   •   STOP

A10   Any one of the following is correct.

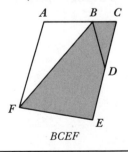   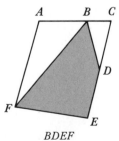   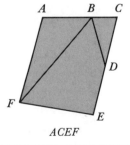

BCEF   *BDEF*   *ACEF*

Q11   Shade a pentagonal region in the figure.

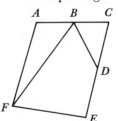

STOP   •   STOP   •   STOP   •   STOP   •   STOP   •   STOP   •   STOP   •   STOP   •   STOP

A11 *ABDEF*

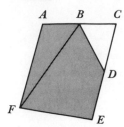

Q12 Is a point on the polygon included in the polygonal region?_____

STOP • STOP • STOP • STOP • STOP • STOP • STOP • STOP • STOP

A12 yes

Q13 Place a *P* before an object that resembles a polygon and an *R* before an object that resembles a polygonal region:

_____ **a.** picture frame

_____ **b.** picture

_____ **c.** door

_____ **d.** stop sign

_____ **e.** hexagonal sugar cookie

_____ **f.** cutting edge of a hexagonal cookie cutter

STOP • STOP • STOP • STOP • STOP • STOP • STOP • STOP • STOP

A13 **a.** *P*   **b.** *R*   **c.** *R*   **d.** *R*   **e.** *R*   **f.** *P*

---

6 Triangles are classified by properties of their sides. Some common descriptive names of triangles are given next.

An *equilateral* triangle has three equal sides.

An *isosceles* triangle has at least two equal sides.

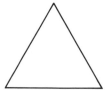

A *scalene* triangle has no pair of sides equal.

Notice that the definition of an isosceles triangle allows an equilateral triangle to also be called isosceles, because it has at least two sides equal.

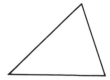

Q14   Write the descriptive name based on the lengths of the sides of each triangle:

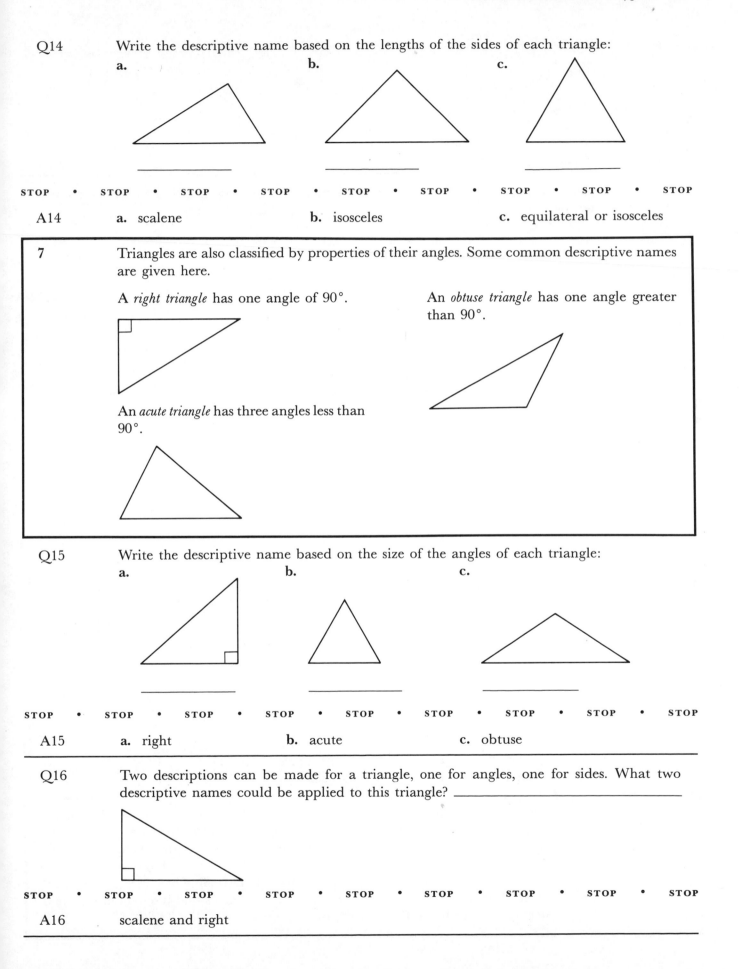

a.   **b.**   **c.**

_____   _____   _____

STOP • **STOP** • **STOP** • **STOP** • **STOP** • **STOP** • **STOP** • **STOP** • **STOP**

A14   **a.** scalene   **b.** isosceles   **c.** equilateral or isosceles

---

7   Triangles are also classified by properties of their angles. Some common descriptive names are given here.

A *right triangle* has one angle of 90°.   An *obtuse triangle* has one angle greater than 90°.

An *acute triangle* has three angles less than 90°.

Q15   Write the descriptive name based on the size of the angles of each triangle:

a.   **b.**   **c.**

_____   _____   _____

STOP • **STOP** • **STOP** • **STOP** • **STOP** • **STOP** • **STOP** • **STOP** • **STOP**

A15   **a.** right   **b.** acute   **c.** obtuse

Q16   Two descriptions can be made for a triangle, one for angles, one for sides. What two descriptive names could be applied to this triangle? _____

STOP • **STOP** • **STOP** • **STOP** • **STOP** • **STOP** • **STOP** • **STOP** • **STOP**

A16   scalene and right

Q17   What two descriptive names could be applied to this triangle?_____

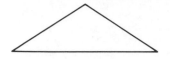

STOP • STOP • STOP • STOP • STOP • STOP • STOP • STOP • STOP

A17   obtuse and isosceles

Q18   Could there be an obtuse equilateral triangle? _____ Draw it or state why there could
           not be.

STOP • STOP • STOP • STOP • STOP • STOP • STOP • STOP • STOP

A18   no, there could not; because there could be only one obtuse angle, and the longest side is
           opposite the largest angle.

Q19   In each of the sets list the letters of the appropriate triangles:

a. 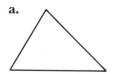   b. 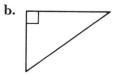   c. 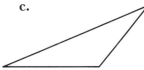   d.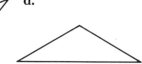

e.

           equilateral triangles = {_____}   right triangles = {_____}

           isosceles triangles   = {_____}   obtuse triangles = {_____}

           scalene triangles   = {_____}   acute triangles = {_____}

STOP • STOP • STOP • STOP • STOP • STOP • STOP • STOP • STOP

A19   equilateral triangles = {e}   right triangles = {b}
           isosceles triangles   = {d, e}   obtuse triangles = {c, d}
           scalene triangles   = {a, b, c}   acute triangles = {a, e}

---

8   Quadrilaterals are used so much that a vocabulary has developed which describes them.
           Quadrilaterals are polygons with four sides. Opposite sides of a quadrilateral need *not*
           be parallel. A few of the special types of quadrilaterals will be identified and their prop-
           erties studied.
               First, two terms that deal with quadrilaterals will be defined: adjacent sides and
           opposite sides. *Adjacent sides* of a quadrilateral are any two sides that intersect at a vertex.
           Opposite sides of a quadrilateral do not intersect.

Q20   Given the quadrilateral *ABCD:*

    a. Name the adjacent sides at *B.* _____

    b. Name the adjacent sides at *D.* _____

    c. Name all pairs of opposite sides. _____

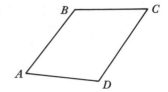

STOP • **STOP** • **STOP** • **STOP** • **STOP** • **STOP** • **STOP** • **STOP** • **STOP**

A20   **a.** $\overline{AB}$ and $\overline{BC}$ **b.** $\overline{CD}$ and $\overline{AD}$ **c.** $\overline{AB}$ and $\overline{DC}$, $\overline{AD}$ and $\overline{BC}$

---

9   Recall that parallel lines do not intersect. Two line segments are said to be parallel if the lines of which they are a part are parallel. A quadrilateral with opposite sides parallel is called a *parallelogram.* Below is a parallelogram with $\overline{AB}\|\overline{CD}$ and $\overline{BC}\|\overline{AD}$.

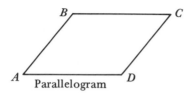

Parallelogram

Q21   The line segments $\overline{AB}$ and $\overline{CD}$ do not intersect. Why are they not considered parallel?

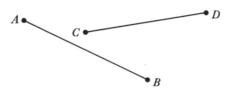

_____

STOP • **STOP** • **STOP** • **STOP** • **STOP** • **STOP** • **STOP** • **STOP** • **STOP**

A21   They are not parallel, because the lines of which they are a part do intersect.

Q22   Which of the following figures appear to be parallelograms? _____

    a.   **b.**   **c.**   **d.**

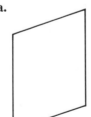 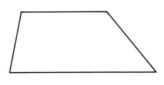 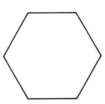

STOP • **STOP** • **STOP** • **STOP** • **STOP** • **STOP** • **STOP** • **STOP** • **STOP**

A22   a and b: notice that the opposite sides of figure d appear to be parallel. However, a parallelogram must first be a quadrilateral.

---

10   If the sides of a parallelogram meet at right angles, it is a *rectangle.* If the parallelogram's sides meet at right angles and are of equal length, it is a *square.*

The following quadrilateral is a rectangle as well as being a parallelogram:

Rectangle

The following quadrilateral is a square as well as being a rectangle and a parallelogram:

Square

Q23   In each set, list the letters of the appropriate geometric figures:

a.   b.   c.   d.   e.

polygons   = { _____ }

quadrilaterals = { _____ }

parallelograms = { _____ }

rectangles   = { _____ }

squares   = { _____ }

STOP • STOP • STOP • STOP • STOP • STOP • STOP • STOP • STOP

A23   polygons   = {a, b, c, d, e}
       quadrilaterals = {b, c, d, e}
       parallelograms = {b, c, e}
       rectangles   = {b, e}
       squares   = {e}

11   If one pair of opposite sides of a quadrilateral are parallel but the other pair are not, the quadrilateral is a *trapezoid*.
       Following are two trapezoids:

Q24   Were any of the polygons in Q23 trapezoids? _____ Which? _____

STOP • STOP • STOP • STOP • STOP • STOP • STOP • STOP • STOP

A24   yes, d

---

12   The *midpoint* is the point on a line segment that divides the line segment into two segments of equal length. The midpoint of line segment $\overline{AB}$ below is C, since the length of $\overline{AC}$ equals the length of $\overline{CB}$.

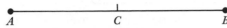

Q25   Connect the midpoints of the quadrilaterals consecutively.

    a. What does the figure obtained appear to be?   _____

    b. Do you think that always happens?   _____

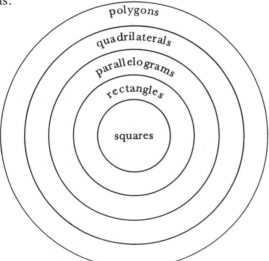

Try drawing some odd-shaped quadrilaterals of your own and test your guess.

STOP • **STOP** • **STOP** • **STOP** • **STOP** • **STOP** • **STOP** • **STOP** • **STOP**

A25   **a.** a parallelogram   **b.** yes: they will always be parallelograms.

---

13   A diagram is sometimes used to classify information. One that is useful for quadrilaterals is:

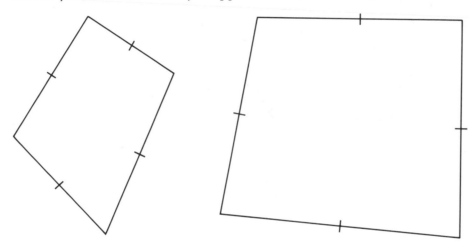

The diagram can be used to show that all rectangles are parallelograms but not all parallelograms are rectangles.

Q26   Place the letter of each of the following in the appropriate space in the diagram of Frame 13:

a. trapezoid   b. triangle   c.   d.

STOP • **STOP** • **STOP** • **STOP** • **STOP** • **STOP** • **STOP** • **STOP** • **STOP**

A26

Q27   Use the diagram of Frame 13 to fill in the following five statements:

a. All _____ are quadrilaterals but not all quadrilaterals are _____.

b. All _____ are quadrilaterals but not all quadrilaterals are _____.

c. All _____ are quadrilaterals but not all quadrilaterals are _____.

d. All _____ are parallelograms but not all parallelograms are _____.

e. All _____ are rectangles but not all rectangles are _____.

STOP • **STOP** • **STOP** • **STOP** • **STOP** • **STOP** • **STOP** • **STOP** • **STOP**

A27
a. parallelograms, parallelograms ⎫
b. rectangles, rectangles   ⎬   a, b, and c may be interchanged
c. squares, squares   ⎭
d. squares, squares (or rectangles, but that is the same statement as in Frame 13)
e. squares, squares

This completes the instruction for this section.

9.3   Exercises

1. Use the figure to identify the following by name:

   a. P is the _____.

   b. $\overline{PB}$ is a(n) _____.

   c. $\overline{AC}$ is a(n) _____.

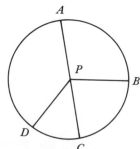

2. How long is the diameter of a circle with a radius of $5\frac{1}{2}$ centimeters?

3. What is the radius of a circle with a diameter of 7 inches?
4. Indicate which of the following are polygons:

   a.    **b.**    **c.** 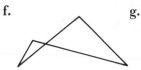   **d.**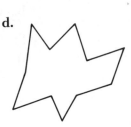

   e.    **f.**    **g.** 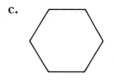   **h.**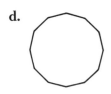

5. Indicate the name of each of the polygons:

   a. 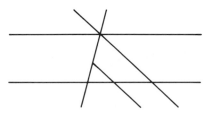   **b.** 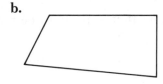   **c.**    **d.**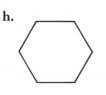

6. **a.** Shade a quadrilateral region in the figure.

   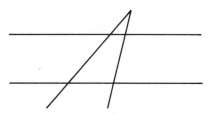

   b. Shade a triangular region in the figure (two possible).

7. In each of the following sets, list the letters of the appropriate triangles:

   a.   **b.** 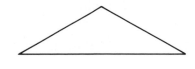   **c.**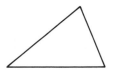

   d. 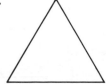   **e.** 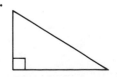   **f.**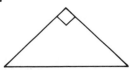

   a. equilateral triangles = {_____}

   b. isosceles triangles = {_____}

   c. scalene triangles   = {_____}

   d. right triangles   = {_____}

   e. obtuse triangles   = {_____}

   f. acute triangles   = {_____}

8. Consider quadrilateral *ABCD:*

   a. Name two sides adjacent to $\overline{DC}$.

   b. Name the side opposite $\overline{AD}$.

9. Shade a polygonal region that appears to be bounded by a parallelogram.

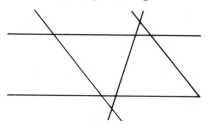

10. How many pairs of parallel sides are contained in a parallelogram?
11. Is a square a rectangle?
12. Is a rectangle a parallelogram?
13. Is a square a parallelogram?
14. Is a square a polygon?
15. Is a rectangle a square?
16. Is a parallelogram a rectangle?
17. Is a parallelogram a trapezoid?
18. Draw a trapezoid.
19. A square is a rectangle with what further restriction?
20. A rectangle is a parallelogram with what further restriction?

9.3   Exercise Answers

1. a. center   **b.** radius   **c.** diameter
2. 11 centimeters

3. $3\frac{1}{2}$ inches

4. a, d, g, and h
5. a. triangle   **b.** quadrilateral   **c.** hexagon   **d.** dodecagon
6. a.   **b.**   or

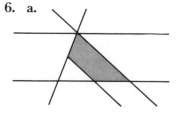
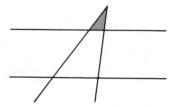
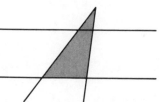

7. a. {d}   **b.** {a, b, d, f}   **c.** {c, e}   **d.** {e, f}
   e. {b}   **f.** {a, c, d}

8. a. $\overline{AD}$ and $\overline{BC}$ **b.** $\overline{BC}$

9.

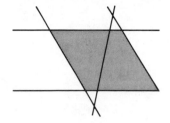

10. 2
11. yes
12. yes
13. yes
14. yes
15. no, not necessarily
16. no, not necessarily
17. no
18.

19. The sides are of equal measure.
20. The angles at the vertices have measure 90°.

9.4   Polyhedra, Prisms, and Pyramids

1   There is a geometric term for solid shapes with flat surfaces. They are called polyhedra. A *polyhedron* is defined as a geometric solid whose surfaces are polygonal regions. Some examples of polyhedrons are shown.

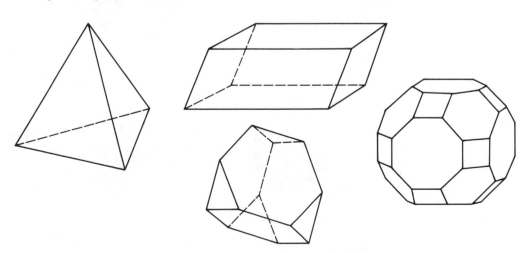

   The polygonal regions which form the surface of the polyhedron are called *faces*. The sides of the faces are now called *edges* of the polyhedron. The vertices of the faces are the *vertices* of the polyhedron.

Q1   One common polyhedron is the cube. Examine the cube shown below. You see seven of its eight vertices. Suppose the one you do not see is labeled *E*. Use your imagination to name all the members in the sets.

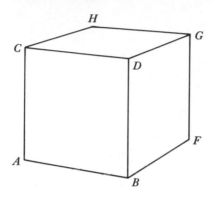

vertices = { _____ }

edges = { _____ }

STOP • STOP • STOP • STOP • STOP • STOP • STOP • STOP • STOP

A1   vertices = {*A, B, C, D, E, F, G, H*}

edges = {$\overline{AB}$, $\overline{BD}$, $\overline{DC}$, $\overline{CA}$, $\overline{BF}$, $\overline{DG}$, $\overline{CH}$, $\overline{AE}$, $\overline{EF}$, $\overline{FG}$, $\overline{GH}$, $\overline{HE}$}

Q2   How many faces does the cube have in Q1? _____

STOP • STOP • STOP • STOP • STOP • STOP • STOP • STOP • STOP

A2   6

Q3   Below is an array of solid figures. Some are polyhedra and some are not. List the letters of all figures that are polyhedra. _____

a.
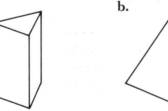

b.

c.

d.

e.

f.

g.

h.
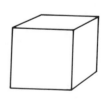

STOP • STOP • STOP • STOP • STOP • STOP • STOP • STOP • STOP

A3   a, b, e, and h

---

2   Polyhedra are classified by the number of faces they contain. A *tetrahedron* has four faces, a *octahedron* eight, and a *dodecahedron* twelve. Except for the tetrahedron, polyhedrons with a certain number of faces are named with the same prefix as a polygon with that number

of sides. For example, a polyhedron with five faces would be called a pentahedron, because a polygon with five sides is called a pentagon. The following polyhedron is an octahedron:

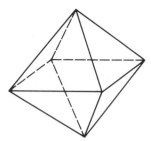

Notice that the figure looks transparent. However, line segments that would actually be hidden if the figure were solid are dashed. This helps a person "view" all the faces.

Q4   Indicate (1) the number of faces and (2) the name of each polyhedron:

a.

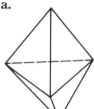

b.

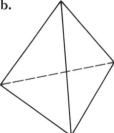

c.

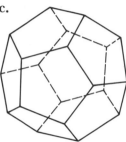

(1) ——————   (1) ——————   (1) ——————

(2) ——————   (2) ——————   (2) ——————

STOP • **STOP** • **STOP** • **STOP** • **STOP** • **STOP** • **STOP** • **STOP** • **STOP**

A4   **a.** (1) 6 faces   (2) hexahedron
      b. (1) 4 faces   (2) tetrahedron
      c. (1) 12 faces   (2) dodecahedron

3   If a polygon has sides and angles of equal measure, it is called a *regular polygon*. A *regular polyhedron* has faces that are regular polygons. Since there are an infinite number of regular polygons, it may be surprising that there are only five regular polyhedra. They are the regular tetrahedron, the cube (actually a regular hexahedron), the regular octahedron, the regular dodecahedron, and the regular icosahedron (20 faces).

Q5   Write the appropriate name under each regular polyhedron:

a.   b.   c.   d.   e.

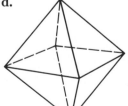

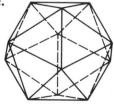

a. ——————————   b. ——————————

c. ——————————   d. ——————————

e. ——————————

STOP • **STOP** • **STOP** • **STOP** • **STOP** • **STOP** • **STOP** • **STOP** • **STOP**

A5   **a.** regular tetrahedron   **b.** cube
     c. regular dodecahedron   **d.** regular octahedron
     e. regular icosahedron

Q6   It is possible to cut a model of a regular solid on some of its edges and "flatten" it out. Examine the flattened solids here and match them with the regular solids in Q5.

a.

matches _____

b.

matches _____

c.

matches _____

d.

matches _____

e.

matches _____

STOP • **STOP** • **STOP** • **STOP** • **STOP** • **STOP** • **STOP** • **STOP** • **STOP**

A6   **a.** cube (b)   **b.** octahedron (d)   **c.** tetrahedron (a)
     d. dodecahedron (c)   **e.** icosahedron (e)

4   A prism is a special kind of polyhedron. A *prism* has two faces which have the same size and shape and which lie in parallel planes; the remaining faces are parallelograms. The two faces that are parallel to each other are called *bases*. The remaining faces are called *lateral* faces. Some examples:

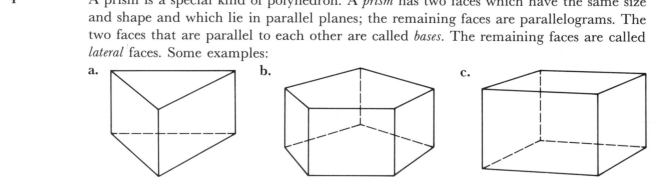

a.   **b.**   **c.**

     The base determines the name of the prism. A prism whose bases are triangular regions is called a *triangular prism;* a prism whose base is a hexagonal region is a *hexangular prism;* and so on.

Q7   What would the three prisms in Frame 4 be called?

     a. _____

     b. _____

     c. _____

STOP • **STOP** • **STOP** • **STOP** • **STOP** • **STOP** • **STOP** • **STOP** • **STOP**

A7 **a.** triangular prism **b.** pentangular prism **c.** rectangular prism

Q8 Give the descriptive name of each prism; otherwise, indicate "not a prism":

a. **b.** **c.** **d.** **e.** **f.**

a. _____ b. _____

c. _____ d. _____

e. _____ f. _____

STOP • **STOP** • **STOP** • **STOP** • **STOP** • **STOP** • **STOP** • **STOP** • **STOP**

A8 **a.** rectangular prism (any of the rectangular faces could be considered the base)
b. triangular prism
c. quadrangular prism
d. not a prism
e. not a prism
f. hexangular prism (the prism bases must be polygons of the same size and shape and lie in parallel planes)

5 Most people can draw prisms with a small amount of instruction and some practice. To begin with, two views of a common rectangular prism called the cube will be examined. Below is a picture of the cube and a geometric diagram of the cube. The edges that appear on both figures are drawn with solid lines. The edges that are not visible in the picture are called *hidden* edges and are drawn with dashed lines in the diagram.

Q9 Consider the cube in Frame 5:

a. What is the one hidden vertex? _____

b. Name the three hidden edges. _____

c. One of the hidden faces is *ABFE*. Name two others. _____

STOP • **STOP** • **STOP** • **STOP** • **STOP** • **STOP** • **STOP** • **STOP** • **STOP**

A9 **a.** *E* **b.** $\overline{AE}$, $\overline{EF}$, and $\overline{EH}$
c. *AEHD* and *EFGH* (The letters should be arranged in order starting with any one of them; that is, *AEHD = EHDA*.)

Q10 Notice that faces on opposite sides of the prism (in Frame 5) have exactly the same shape:

a. Name the front and back lateral faces. _____

    b. Name the left and right lateral faces.   _____

    c. Name the top base and then the bottom base.   _____

STOP • STOP • STOP • STOP • STOP • STOP • STOP • STOP • STOP

A10   **a.** *ABCD* and *EFGH*   **b.** *AEHD* and *BFGC*   **c.** *DCGH* and *ABFE*

---

6   Now look at the same cube from a different angle. Suppose that the cube of Frame 5 had face *ABFE* resting on a transparent table. Move your eye down below the table and look up at the cube. Different edges become hidden.

     Compare this view of the cube with the view in Frame 5 until you "see" in your mind how the angle of sight has changed.

Q11   **a.** Is the cube in Frame 5 being viewed from above the cube or from below?   _____

     b. Is the cube in Frame 6 being viewed from above the cube or from below?   _____

STOP • STOP • STOP • STOP • STOP • STOP • STOP • STOP • STOP

A11   **a.** above   **b.** below

Q12   Consider the cube of Frame 6:

     a. Name the hidden vertex.   _____

     b. Name the hidden edges.   _____, _____, _____

     c. Were any of the hidden edges of Frame 6 also hidden in Frame 5?   _____

     Which ones?   _____

     d. Name the hidden faces.   _____, _____, _____

     e. Were any of the hidden faces of Frame 6 also hidden in Frame 5?   _____

     Which ones?   _____

STOP • STOP • STOP • STOP • STOP • STOP • STOP • STOP • STOP

A12   **a.** *G*   **b.** $\overline{HG}$, $\overline{FG}$, and $\overline{CG}$
         c. no, none   **d.** *EFGH*, *BCGF*, and *CDHG*
         e. yes, *EFGH* (the face on the back of the cube)

Q13   Which sketches are drawn correctly?   _____

     a.   **b.**   **c.**   **d.**

STOP • STOP • STOP • STOP • STOP • STOP • STOP • STOP • STOP

A13   d

---

7   To sketch a cube some of its faces must be distorted in order to "look" right. However, the bottom hidden base is always the same shape as the visible top base. Several views of a cube are shown here. Notice that some of the hidden edges have been sketched. Use these as a guide to complete Q14.

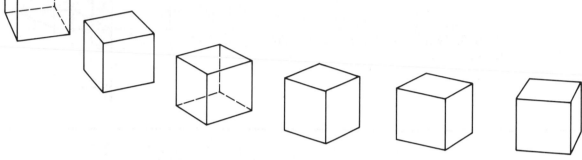

Q14   Sketch the hidden edges in each of these cubes.

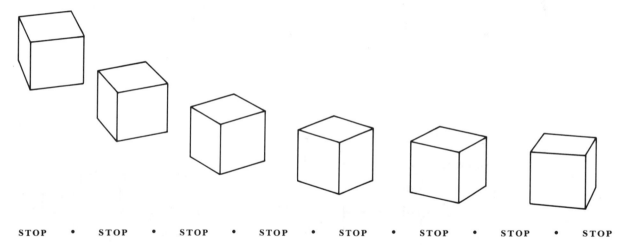

STOP • **STOP** • **STOP** • **STOP** • **STOP** • **STOP** • **STOP** • **STOP** • **STOP**

A14

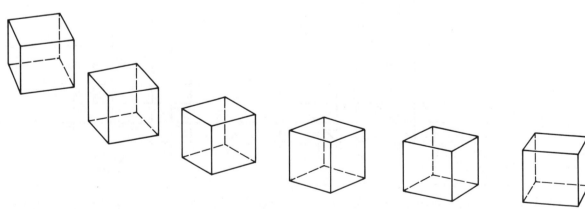

---

8   You have had a chance to interpret sketches of cubes. You now will get an opportunity to make your own sketches. It is helpful to have a background of graph paper on which to draw, although the same procedures may be followed without it. To sketch a cube follow these steps:

Step 1: Draw the top face. Make opposite edges parallel and the same length.

Step 2: Draw another edge from each vertex as shown. Make them parallel and the same length.

Step 3: Draw the bottom face. The top and bottom faces should have the same shape and size.

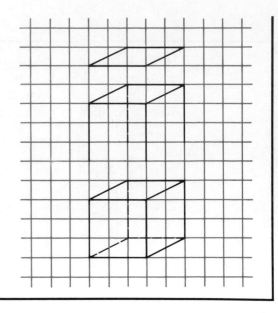

Q15   Here is another example. Copy it on the graph provided. Count the squares carefully.

Step 1:

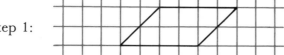

Step 2:

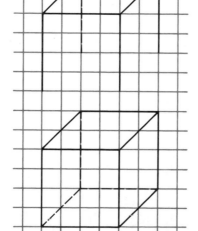

Step 3:

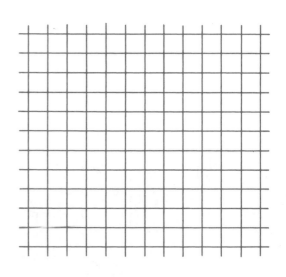

STOP • STOP • STOP • STOP • STOP • STOP • STOP • STOP • STOP

A15   Compare your sketch with the finished cube.

Q16   Copy each of the following sketches. Count squares to make your drawings the same size and shape.

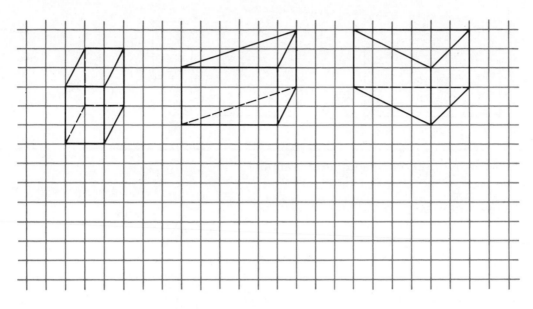

STOP   •   STOP   •   STOP   •   STOP   •   STOP   •   STOP   •   STOP   •   STOP   •   STOP

A16   Compare your sketch with the original prism.

Q17   Which of the following sketches has all the hidden edges drawn correctly?   _____

a. 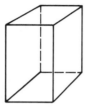   **b.** 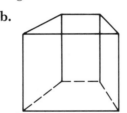   **c.** 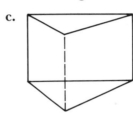   **d.**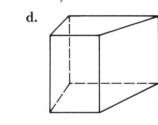

STOP   •   STOP   •   STOP   •   STOP   •   STOP   •   STOP   •   STOP   •   STOP   •   STOP

A17   d

Q18   Sketch the following prisms without using a background of graph paper. Include all hidden edges in your sketch.

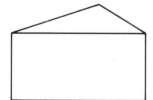   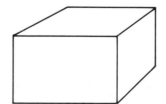   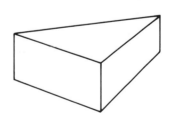

A18

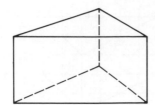 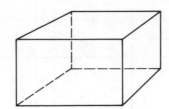 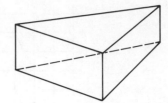

9   The following polyhedra are pyramids:

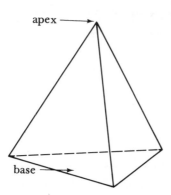 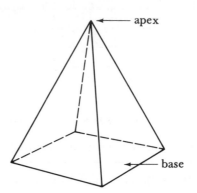

A *pyramid* is formed by a polygonal region called the *base* and a point outside the plane containing the base called the *apex*. A pyramid is the solid formed by all the points in the polygonal region of the base and all triangular regions formed by segments between the vertices of the base and the apex together with all interior points of the figure.

The *altitude* of a pyramid is a segment from the apex perpendicular to the base.

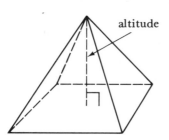

Q19   Choose the sketches with the hidden lines shown correctly: _____

a.   b.   c.   d.

 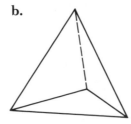 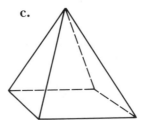 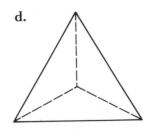

STOP • STOP • STOP • STOP • STOP • STOP • STOP • STOP • STOP

A19   c and d

Q20   Use the polygonal regions shown below for the base and the point A for the apex and sketch the pyramids.

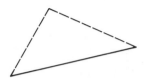

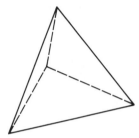

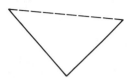

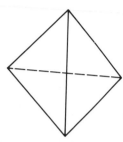

STOP • STOP • STOP • STOP • STOP • STOP • STOP • STOP • STOP

A20

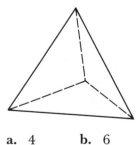

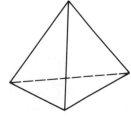

---

10   Pyramids are classified and named according to the shape of their base. One with a triangular base is a triangular pyramid; one with a rectangular base is a rectangular pyramid; and so on.

Q21   Make a sketch of a triangular pyramid.

a. How many vertices does it have?   _____

b. How many edges does it have?   _____

c. How many faces does it have?   _____

STOP • STOP • STOP • STOP • STOP • STOP • STOP • STOP • STOP

A21   or

a. 4   **b.** 6   **c.** 4

Q22   If all the faces, including the base, are the same size and shape, a triangular pyramid could also be called a regular _____.

STOP • STOP • STOP • STOP • STOP • STOP • STOP • STOP • STOP

A22   tetrahedron

Q23   If the center of a cube is connected with each of its vertices, how many pyramids are formed? _____

STOP • **STOP** • **STOP** • **STOP** • **STOP** • **STOP** • **STOP** • **STOP** • **STOP**

A23   6

This completes the instruction for this section.

9.4   Exercises

1. Consider the polyhedron and answer the questions:

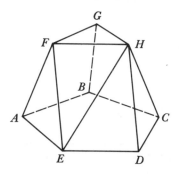

     a. Name the "hidden" vertices.
     b. Name the "hidden" edges.
     c. Name two quadrilateral faces.
     d. Name a pentagonal face.
     e. Name five triangular faces.
     f. What is a more specific name for the polyhedron?

2. How many regular polyhedra are there?

3. Name the regular polyhedra and give the number of faces of each.

4. Use the figure here as an upper base and sketch a cube.

5.   **a.** Use the triangle here as the upper base and sketch a triangular prism.

     b. How many faces does a triangular prism have?
     c. How many edges does it have?
     d. How many vertices does it have?

6. Consider the pyramid and answer the following:
     a. Name the three hidden edges.
     b. Name the triangles that bound the lateral faces.
     c. Name the three edges that intersect at B.
     d. Name the three angles formed at B.
     e. Are any two of the angles formed at B in the same plane?

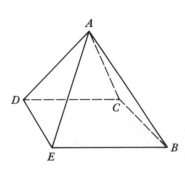

7. Think of a point 1 inch in the air above the polygon shown. Think of the polyhedron formed by connecting each vertex of the polygon with the point above the polygon.

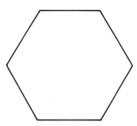

    a. What is the name of this polyhedron?
    b. How many vertices does it have?
    c. How many edges does it have?
    d. How many faces does it have?

8. What is the least number of faces that a prism can have?

9. **a.** Are all prisms polyhedra?
    b. Are all polyhedra prisms?

10. Sketch the hidden lines in the following figures:
    a.   **b.**

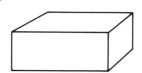
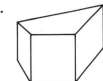

11. Use the polygonal region below for the base and the point *A* for the apex and sketch a pyramid. Sketch the altitude of the pyramid.

          • *A*

9.4   Exercise Answers

1. **a.** *B*   **b.** $\overline{AB}$, $\overline{BC}$, and $\overline{BG}$ **c.** *ABGF* and *BCHG*
    d. *ABCDE*   **e.** *AEF*, *EHF*, *EDH*, *DCH*, and *FGH*
    f. octahedron

2. five

3. regular tetrahedron, 4; cube, 6; regular octahedron, 8; regular dodecahedron, 12; regular icosahedron, 20

4.

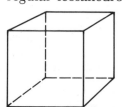

5. a. 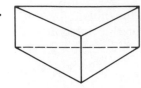   **b.** 5 faces   **c.** 9 edges

    d. 6 vertices

6. a. $\overline{BC}$, $\overline{CD}$, and $\overline{AC}$
    b. ABE, AED, ADC, and ACB
    c. $\overline{BE}$, $\overline{BC}$, and $\overline{BA}$
    d. $\angle ABE$, $\angle ABC$, and $\angle CBE$
    e. no

7. a. hexagonal pyramid   **b.** 7 vertices   **c.** 12 edges
    d. 7 faces

8. 5 faces

9. a. yes   **b.** no

10. a. 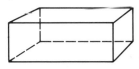   **b.**

11.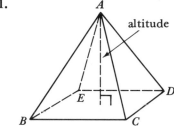

9.5   Cylinders, Cones, and Spheres

1   Circular cylinders are not difficult to recognize. Some examples are shown here.

 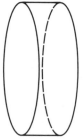 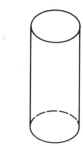

A right-circular *cylinder* consists of two circular regions with the same radius, called *bases*, connected by line segments perpendicular to the planes of the two circles. Because only right-circular cylinders are discussed in this text, we shall refer to them simply as "cylinders."

Q1 Which of the following have a cylindrical shape? _____

    a. round pencil   **b.** wastebasket

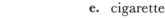

    c. section of a water pipe   **d.** beer can

    e. cigarette   **f.** cigar

STOP • STOP • STOP • STOP • STOP • STOP • STOP • STOP • STOP

A1 a, c, d, and e: b is not because the circles do not have the same radius, f is not because of the semipointed end.

Q2 Complete the sketch of the cylinders.

STOP • STOP • STOP • STOP • STOP • STOP • STOP • STOP • STOP

A2

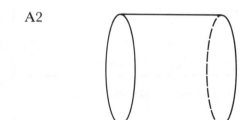

---

2   The curved surface connecting the bases is called the *lateral surface*. The *axis* of a cylinder is the line segment that connects the centers of the bases of the cylinder. The length of the axis is the *altitude* of a right-circular cylinder.

Q3 Sketch the cylinders and draw in the axis.

STOP • STOP • STOP • STOP • STOP • STOP • STOP • STOP • STOP

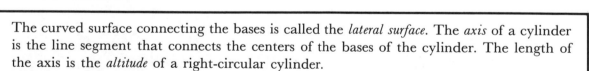

A3

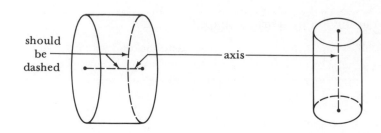

Q4   **a.** How many centimeters long is the radius of the cylinder? _____

    b. How many centimeters is the altitude of the cylinder? _____

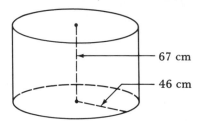

— 67 cm

— 46 cm

STOP • STOP • STOP • STOP • STOP • STOP • STOP • STOP • STOP

A4   **a.** 46 cm   **b.** 67 cm

---

3   A right-circular *cone* is a circular region and the surface made up of line segments connecting the circle with a point located on a line through the center of the circle and perpendicular to the plane of the circle.

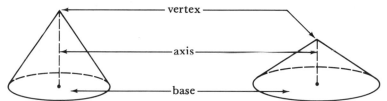

vertex

axis

base

The circular region is called the *base*. The point to which the base is connected is the *vertex*. The line segment connecting the vertex to the center of the base is the *axis*.

Q5   Make sketches of cones with the base and vertex shown. Include a sketch of the axes.

    a.   •   **b.**

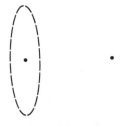

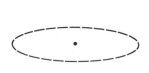

STOP • STOP • STOP • STOP • STOP • STOP • STOP • STOP • STOP

A5   **a.**   **b.** Two possible sketches:

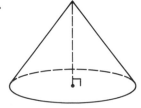

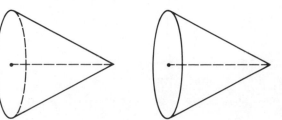

> **4** The *altitude* of a right-circular cone is the length of its axis. Since the altitude of a cone is a length, it is a number rather than a geometric object. The *slant height* of a right-circular cone is the length of a line segment from the vertex to a point on the circle of the base.

Q6 Use the sketch to complete the sentences below:

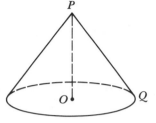

a. $\overline{OP}$ is the _____ of the right-circular cone.

b. The length of $\overline{OP}$ is the _____.

c. The length of $\overline{PQ}$ is the _____ of the cone.

STOP • STOP • STOP • STOP • STOP • STOP • STOP • STOP • STOP

A6 **a.** axis **b.** altitude **c.** slant height

> **5** A *sphere* is a set of points in space at some fixed distance from a given point. The given point is the *center* of the sphere. A *radius* of a sphere is a line segment from the center to the sphere. The radius also refers to the measure of such a line segment. You will be able to tell which meaning is intended by the context in which the word "radius" is used. A *diameter* of the sphere is a line segment containing the center with endpoints on the sphere. A sketch of a sphere is shown. Notice how the dashed lines are used to give the sketch depth.
>
>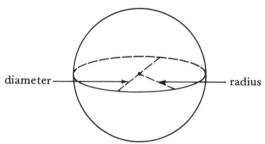
>
> To draw a sphere use a compass to sketch a circle of the proper radius. Then sketch the elliptical shape in the interior of the circle. Be sure to make the hidden lines dashed.

Q7 Sketch a sphere and draw a radius of the sphere.

STOP • STOP • STOP • STOP • STOP • STOP • STOP • STOP • STOP

A7

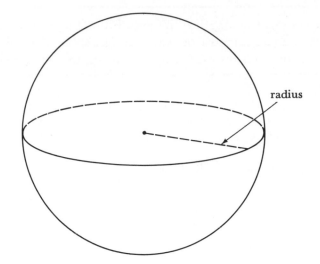

radius

---

6   A circle formed by the intersection of the sphere and a plane containing the center of the sphere is a *great circle*.

    The sphere in Figure 1 has two great circles shown, *AEBF* and *CEDF*. The circle *ABCD* in Figure 2 is not a great circle because it does not contain the center of the sphere.

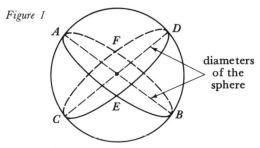

Figure 1

Figure 2

diameters of the sphere

not a diameter of the sphere

Q8   Use the figure to answer the questions:

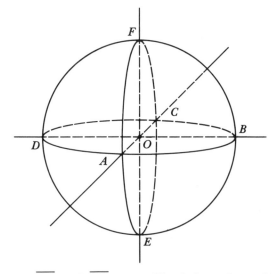

a. $\overline{OC}$ and $\overline{OF}$ are radii of the sphere. Name four more radii.   _____

b. $\overline{EF}$ is a diameter of the sphere. Name two other diameters.   _____

c. *DEBF* is a great circle. Name two others.   _____

A8   **a.** $\overline{OB}$, $\overline{OA}$, $\overline{OD}$, and $\overline{OE}$ **b.** $\overline{AC}$ and $\overline{DB}$
           c. *AECF* and *ABCD*

Q9   The lines of longitude are shown on a diagram of the earth's surface in Figure 1. Lines
           of latitude are shown in Figure 2.

Figure 1   *Figure 2*

   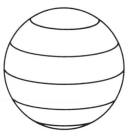

           The lines of _____ are all great circles. Only one of the lines of _____

is a great circle. This great circle is called the _____ .

STOP • **STOP** • **STOP** • **STOP** • **STOP** • **STOP** • **STOP** • **STOP** • **STOP**

A9   longitude, latitude, equator

This completes the instruction for this section.

9.5   **Exercises**

   1. Name the geometric figures:
       a.   **b.**   **c.**

   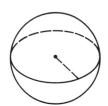   

   2. Name the term that each of the following describes:
       a. the line connecting the centers of the ends of a cylinder
       b. the point outside the plane of the base but connected to the base of a cone with
           line segments
       c. the perpendicular distance between bases of a cylinder
       d. the distance from the center of a sphere to the sphere
       e. the line connecting the vertex of a cone to the center of the base
       f. the geometric figure formed by a cross section of a sphere through its center
       g. the length of a line segment from the vertex to the base of a right-circular cone.
   3. Name the parts of the cylinder to which the arrows point:

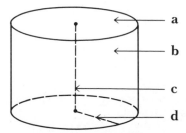

4. Name the parts of the cone to which the arrows point:

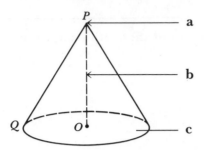

5. Name the line segment whose length is the slant height in problem 4.
6. Name the parts of the sphere to which the arrows point:

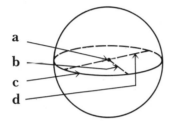

7. Complete the sketch of a cylinder. Sketch the axis.

8. Complete the sketch of a cone. Sketch the axis.

9. Complete the sketch of a sphere. Sketch a radius and two great circles.

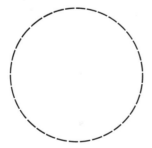

9.5   Exercise Answers

1. **a.** cylinder   **b.** sphere   **c.** cone
2. **a.** axis   **b.** vertex   **c.** altitude   **d.** radius
    e. axis   **f.** great circle **g.** altitude

3. **a.** base   **b.** lateral surface   **c.** axis   **d.** radius
4. **a.** vertex   **b.** axis   **c.** base
5. $\overline{PQ}$
6. **a.** center   **b.** radius   **c.** great circle   **d.** diameter
7.   8.   9.

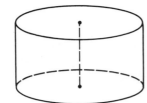
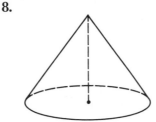
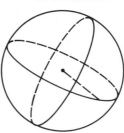

Chapter 9 Sample Test

At the completion of Chapter 9 it is expected that you will be able to work the following problems.

9.1   Points, Lines, and Planes

1. The following suggest one of the three geometric objects: a point, a line, or a plane. Indicate which.
   a. intersection of two walls in a room
   b. tabletop
   c. stretched rubber band
   d. grain of sand
   e. stiff piece of cardboard
2. Answer true or false to the following statements. All lines referred to are on the figure.

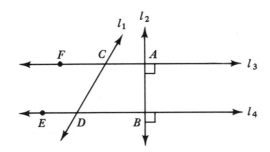

   a. A point has no dimensions.
   b. $l_3 \| l_4$.
   c. $l_2 \perp l_3$.
   d. $l_1 \| l_2$.
   e. $l_2 \| l_4$.
   f. Pictures are imperfect representations of points, lines, and planes.
   g. l_2 and l_4 intersect in another point besides point B.
   h. Many different planes contain all three of the points A, B, and C.
   i. Other lines besides the one drawn pass through C and D.
   j. Many other line segments different from $\overline{FA}$ contain point C.

   k. ∡*CAB* is a right angle.
   l. ∡*EBA* = ∡*DBA*.
   m. ∡*ACD* is a right angle.
   n. *DB* represents a line segment.
   o. $\overleftrightarrow{CA}$ is a line.
   p. *E* is on $\overline{DB}$.
   q. *F* is on $\overleftrightarrow{CA}$.

9.2 Rays and Angles

3. Match the number of the geometric figures with their names:
   a. obtuse angle
   b. ray
   c. acute angle
   d. complementary angles
   e. right angle
   f. supplementary angles

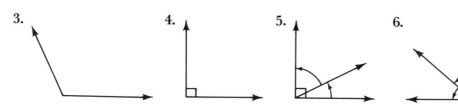

4. Answer true or false to the following statements. All angles and rays referred to are on the figure shown.
   a. *F* is a point on ∡*ABC*.
   b. The vertex of ∡*ACB* is *B*.
   c. $\overrightarrow{EC} = \overrightarrow{CE}$.
   d. $\overrightarrow{EB} = \overrightarrow{EC}$.
   e. $\overrightarrow{BF}$ is a side of ∡*ABC*.
   f. $\overrightarrow{BC}$ is a side of ∡*ABC*.
   g. ∡*EBD* has vertex *B*.
   h. ∡*EBF* is obtuse.
   i. ∡*ACB* is acute.
   j. The sum of the measures of the angles of triangle *ABC* is 360°.
   k. An angle is two line segments with a common endpoint.
   l. A protractor measures the size of an angle.

5. Match the angles with their approximate measures:

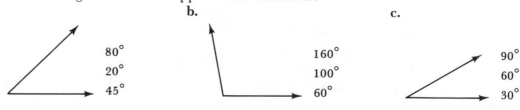

   a.   80°   **b.**   160°   **c.**   90°
            20°   100°   60°
            45°   60°   30°

9.3   **Circles and Polygons**

6. Name an example of each geometric term from the figure:
   a. diameter
   b. radius
   c. center

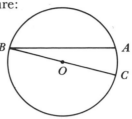

7. Complete the following sentences:
   a. The instrument used to draw circles is a _____.
   b. The length of a line segment from the center to a point on the circle is the

      _____.

   c. A line segment with endpoints on the circle containing the center is a

      _____.

8. List the letters of the figures that are polygons:

   a.    b.    c.    d.

   e.   f. 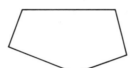   g.   h.

9. Match the number of the appropriate polygon with its name:
   a. triangle
   b. octagon
   c. pentagon
   d. hexagon
   e. quadrilateral

   1.    2.    3.

   4.   5.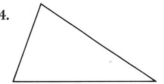

10. Explain how a triangular region differs from a triangle.

11. Each descriptive word applies to several of the triangles at the right. List *all* the numbers of triangles that can be described by each term. The number in parentheses indicates how many there are of each.
    a. equilateral (1)
    b. right (2)
    c. scalene (2)
    d. isosceles (4)
    e. obtuse (2)
    f. acute (2)

    1.    2.

    5.    6.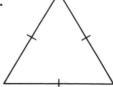

    3.    4.

12. Each term applies to several of the following figures. List *all* the numbers of figures that can be described by each term. The number in parentheses indicates how many there are of each.
   a. parallelogram (4)
   b. quadrilateral (7)
   c. square (1)
   d. rectangle (2)
   e. trapezoid (2)

1. 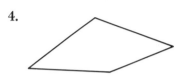 2.   3.

4.  5. 6. 7.

8.

9.4 Polyhedra, Prisms, and Pyramids

13. Each term applies to several of the following figures. List *all* the numbers of figures that can be described by each term. The number in parentheses indicates how many there are of each.
   a. polyhedra (7)
   b. prism (4)
   c. pyramid (2)
   d. regular polyhedra (3)
   e. cube (1)

1. 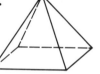 2.   3.

4.  5. 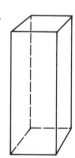 6.   7. 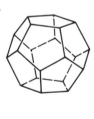 8.

14. Match the name of the geometric term with the number of the arrow pointing to the object:
   a. face
   b. edge
   c. vertex
   d. apex
   e. base
   f. altitude

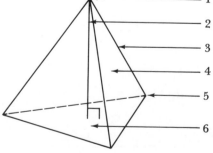

15. Indicate the letters of the figures that have the hidden lines drawn correctly.

a.   **b.** 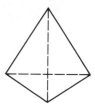   **c.** 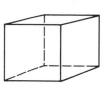   **d.**

e.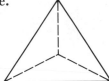

16. Sketch the hidden lines of the prisms:

a. 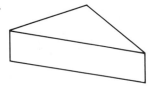   **b.**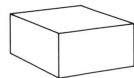

9.5   Cylinders, Cones, and Spheres

17. Match the name of the geometric term with the number of the arrow pointing to the object:

Cylinder
a. axis
b. base
c. lateral surface

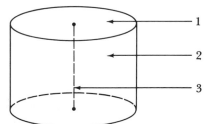

Cone
d. base
e. vertex
f. axis

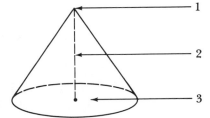

Sphere
g. center
h. radius
i. diameter
j. great circle

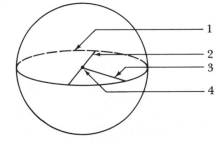

18. Complete the sketches of the figures, including the hidden lines.

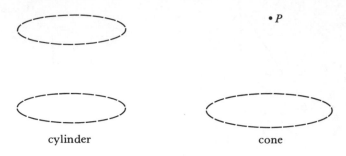

cylinder   cone

sphere

Chapter 9 Sample Test Answers

1. **a.** line   **b.** plane   **c.** line   **d.** point
   e. plane

2. **a.** true   **b.** true   **c.** true   **d.** false
   e. false   **f.** true   **g.** false   **h.** false
   i. false   **j.** true   **k.** true   **l.** true
   m. false   **n.** false   **o.** true   **p.** false
   q. true

3. **a.** 3   **b.** 2   **c.** 1   **d.** 5
   e. 4   **f.** 6

4. **a.** true   **b.** false   **c.** false   **d.** true
   e. true   **f.** true   **g.** true   **h.** true
   i. true   **j.** false   **k.** false   **l.** true

5. **a.** 45°   **b.** 100°   **c.** 30°

6. **a.** $\overline{BC}$ **b.** $\overline{OB}$ or $\overline{OC}$ **c.** 0

7. **a.** compass   **b.** radius   **c.** diameter

8. a, b, and f

9. **a.** 4   **b.** 3   **c.** 1   **d.** 5
   e. 2

10. A triangular region includes the points inside the triangle.

11. **a.** 6   **b.** 2 and 4   **c.** 3 and 4   **d.** 1, 2, 5, and 6
    e. 1 and 3   **f.** 5 and 6

12. **a.** 2, 3, 6, and 7   **b.** 2, 3, 4, 5, 6, 7, and 8   **c.** 3
    d. 3 and 6   **e.** 5 and 8

13. **a.** 1, 2, 4, 5, 6, 7 and 8   **b.** 2, 4, 5, and 6   **c.** 1 and 7
    d. 2, 7, and 8   **e.** 2

14. **a.** 4 (or 6)   **b.** 3   **c.** 5 (or 1)   **d.** 1
    e. 6   **f.** 2

15. c and e

16. **a.**   **b.**

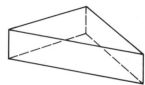   

17. **a.** 3 **b.** 1 **c.** 2 **d.** 3
e. 1 **f.** 2 **g.** 4 **h.** 3
i. 2 **j.** 1

18. **a.** 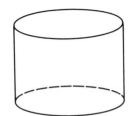 **b.**  **c.**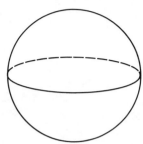

Chapter 10

Measurement and Geometry

In Chapter 9 certain geometric objects were defined. Relationships between these objects were stated as properties. At that time measurement of line segments was not required. In this chapter measurement is introduced and properties of geometry that make use of the idea of measurement are studied. This geometry is called *metric* geometry. Some of the basic notions of measurement will be discussed first. *An inch ruler and a centimeter ruler are required to study Chapter 10.* Such a ruler is provided on page 405.

10.1   Measurement of Line Segments

| 1 | How many people are in the room in which you are located? How many rooms are there in the building? The answers to these questions can be obtained by counting and an *exact* result obtained. |

Q1   How many words are there in the statement of this question?_____

STOP　•　STOP　•　STOP　•　STOP　•　STOP　•　STOP　•　STOP　•　STOP　•　STOP

A1   11

| 2 | Measurement is the comparison of an observed quantity with a standard unit. The result of the measurement is a number times the standard unit. When counting a set of objects, the standard unit of measurement is one object. |

Q2   **a.** How many letters are in the set?

{a, b, c, d, e, . . . , z} _____

b. What was the standard unit of measurement in part a?_____

STOP　•　STOP　•　STOP　•　STOP　•　STOP　•　STOP　•　STOP　•　STOP　•　STOP

A2   **a.** 26　　　　　　　　**b.** one letter

| 3 | A standard unit for measuring length is an inch. One inch is a length equal to the following line segment:

————————

To measure a line segment in inches, a ruler with subdivisions showing inches is held next to the line segment, and the number of inches is read as carefully as possible. For greater accuracy, look down on the end of the segment from directly above the ruler. Although counting results in a measurement that is exact, measuring with a ruler gives an approximate result. |

Q3   Decide whether the measurement would be approximate or exact:

         a. length of your index finger   _____

         b. number of letters in the word "bump"   _____

         c. circumference of your head   _____

STOP • STOP • STOP • STOP • STOP • STOP • STOP • STOP • STOP

A3   **a.** approximate   **b.** exact   **c.** approximate

---

4   The *precision* of a measurement is dependent upon the instrument used to measure. The precision of the instrument is determined by the smallest subdivision of the instrument. A ruler marked only at the inch marks is precise to a whole number of inches. A ruler marked at the quarter-inch positions is precise to a quarter of an inch. A ruler marked at the one-sixteenth-inch positions is precise to sixteenths of an inch. The precision of a measurement is the same as the precision of the instrument used to obtain the measurement.

     Study each point on the ruler to which the arrow is directed until you agree with the measurement given.

A scale marked in $\frac{1}{4}$-inch units.

A scale marked in $\frac{1}{8}$-inch units.

A scale marked in $\frac{1}{16}$-inch units.

Q4   **a.** What is the precision of the measurements given

         in the first example in Frame 4?   _____

     b. What is the precision of the measurements given

         in the second example in Frame 4?   _____

     c. What is the precision of the measurements given

         in the third example in Frame 4?   _____

STOP • STOP • STOP • STOP • STOP • STOP • STOP • STOP • STOP

A4   **a.** $\frac{1}{4}$ inch   **b.** $\frac{1}{8}$ inch   **c.** $\frac{1}{16}$ inch

---

5   When measuring a line segment one must first decide how precise the resulting measurement needs to be. If only a measurement to the nearest inch is necessary, a ruler with only the inch marks is required. The bar below would have a measurement of three inches if it was measured to the *nearest* inch.

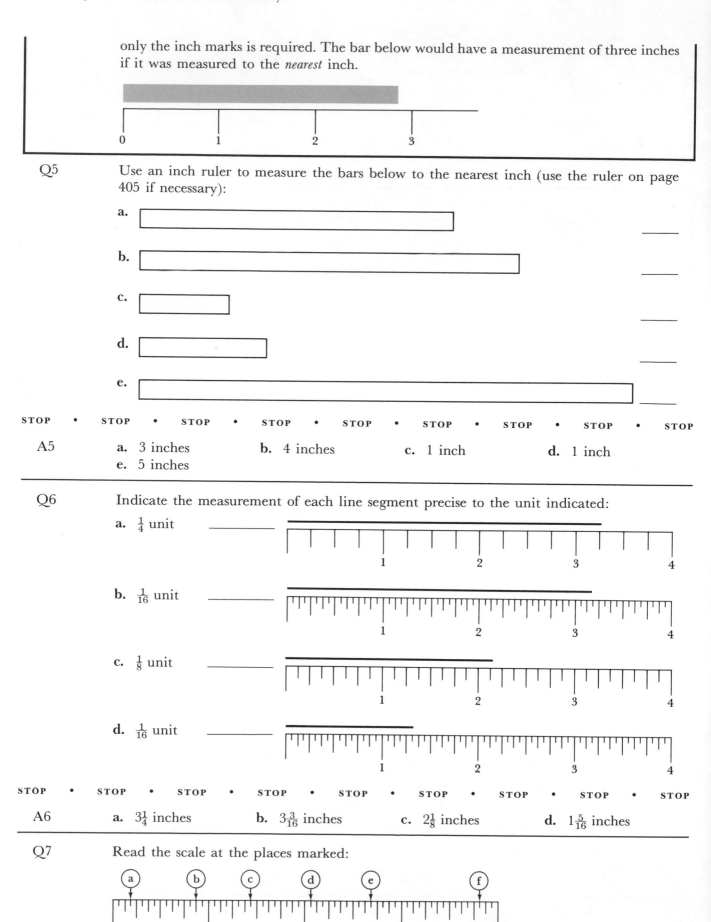

Q5   Use an inch ruler to measure the bars below to the nearest inch (use the ruler on page 405 if necessary):

a.

b.

c.

d.

e.

STOP • STOP • STOP • STOP • STOP • STOP • STOP • STOP • STOP

A5   **a.** 3 inches   **b.** 4 inches   **c.** 1 inch   **d.** 1 inch
      e. 5 inches

Q6   Indicate the measurement of each line segment precise to the unit indicated:

a. $\frac{1}{4}$ unit

b. $\frac{1}{16}$ unit

c. $\frac{1}{8}$ unit

d. $\frac{1}{16}$ unit

STOP • STOP • STOP • STOP • STOP • STOP • STOP • STOP • STOP

A6   **a.** $3\frac{1}{4}$ inches   **b.** $3\frac{3}{16}$ inches   **c.** $2\frac{1}{8}$ inches   **d.** $1\frac{5}{16}$ inches

Q7   Read the scale at the places marked:

a. _____   b. _____   c. _____   d. _____

e. _____   f. _____

STOP • **STOP** • **STOP** • **STOP** • **STOP** • **STOP** • **STOP** • **STOP** • **STOP**

A7   a. $\frac{3}{16}$ inches   b. $\frac{7}{8}$ inches   c. $1\frac{7}{16}$ inches   d. $2\frac{1}{16}$ inches

e. $2\frac{11}{16}$ inches   f. $3\frac{13}{16}$ inches

---

6   All measurements of line segments are approximate. The degree of precision of your reported measurement must be decided before you measure. The line segments below are all the same length. The measurement of their length is approximated to different degrees of precision.

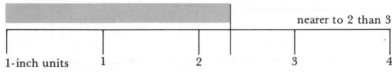

nearer to 2 than 3

1-inch units   1   2   3   4

The length is 2 inches to the nearest inch.

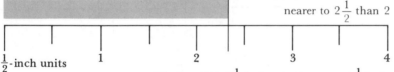

nearer to $2\frac{1}{2}$ than 2

$\frac{1}{2}$-inch units   1   2   3   4

The length is $2\frac{1}{2}$ inches to the nearest $\frac{1}{2}$ inch.

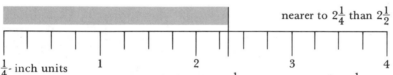

nearer to $2\frac{1}{4}$ than $2\frac{1}{2}$

$\frac{1}{4}$-inch units   1   2   3   4

The length is $2\frac{1}{4}$ inches to the nearest $\frac{1}{4}$ inch.

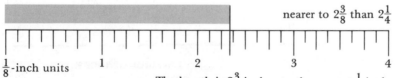

nearer to $2\frac{3}{8}$ than $2\frac{1}{4}$

$\frac{1}{8}$-inch units   1   2   3   4

The length is $2\frac{3}{8}$ inches to the nearest $\frac{1}{8}$ inch.

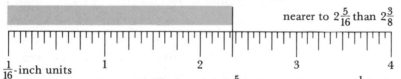

nearer to $2\frac{5}{16}$ than $2\frac{3}{8}$

$\frac{1}{16}$-inch units   1   2   3   4

The length is $2\frac{5}{16}$ inches to the nearest $\frac{1}{16}$ inch.

Notice that the bars are the same length, but the measurements all differ from each other. Each is a correct measurement to the precision indicated for that measurement.

Q8   Approximate the length of the line segments to the degree of precision indicated:

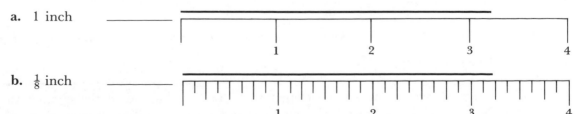

a.   1 inch   _____

1   2   3   4

b. $\frac{1}{8}$ inch   _____

1   2   3   4

c. $\frac{1}{16}$ inch

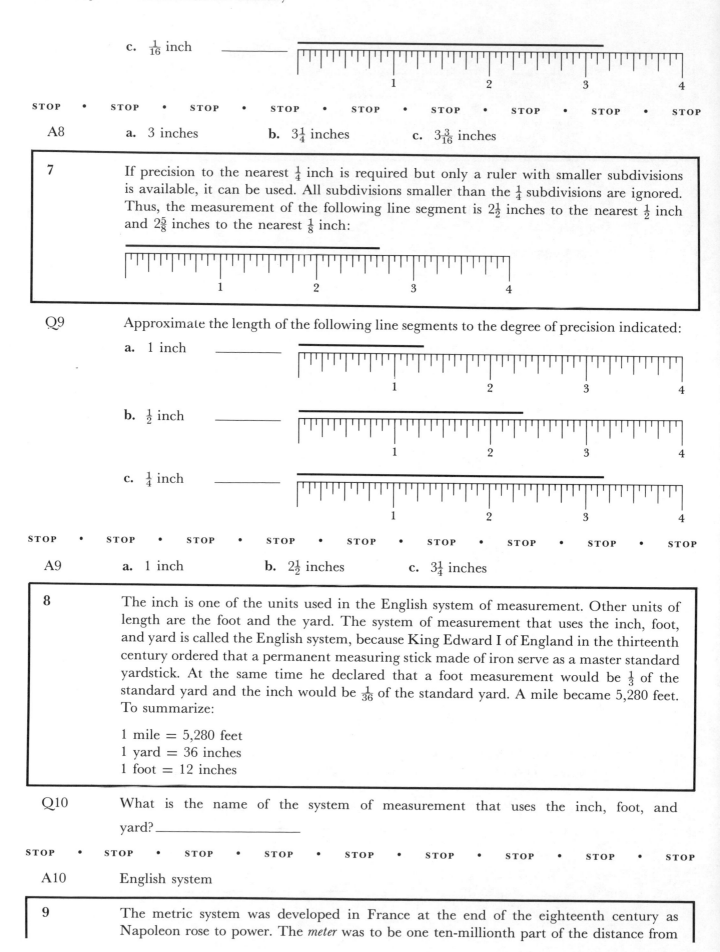

STOP • STOP • STOP • STOP • STOP • STOP • STOP • STOP • STOP

A8   **a.** 3 inches   **b.** $3\frac{1}{4}$ inches   **c.** $3\frac{3}{16}$ inches

---

7

If precision to the nearest $\frac{1}{4}$ inch is required but only a ruler with smaller subdivisions is available, it can be used. All subdivisions smaller than the $\frac{1}{4}$ subdivisions are ignored. Thus, the measurement of the following line segment is $2\frac{1}{2}$ inches to the nearest $\frac{1}{2}$ inch and $2\frac{5}{8}$ inches to the nearest $\frac{1}{8}$ inch:

Q9   Approximate the length of the following line segments to the degree of precision indicated:

a. 1 inch

b. $\frac{1}{2}$ inch

c. $\frac{1}{4}$ inch

STOP • STOP • STOP • STOP • STOP • STOP • STOP • STOP • STOP

A9   **a.** 1 inch   **b.** $2\frac{1}{2}$ inches   **c.** $3\frac{1}{4}$ inches

---

8

The inch is one of the units used in the English system of measurement. Other units of length are the foot and the yard. The system of measurement that uses the inch, foot, and yard is called the English system, because King Edward I of England in the thirteenth century ordered that a permanent measuring stick made of iron serve as a master standard yardstick. At the same time he declared that a foot measurement would be $\frac{1}{3}$ of the standard yard and the inch would be $\frac{1}{36}$ of the standard yard. A mile became 5,280 feet. To summarize:

1 mile = 5,280 feet
1 yard = 36 inches
1 foot = 12 inches

Q10   What is the name of the system of measurement that uses the inch, foot, and yard? _____

STOP • STOP • STOP • STOP • STOP • STOP • STOP • STOP • STOP

A10   English system

9

The metric system was developed in France at the end of the eighteenth century as Napoleon rose to power. The *meter* was to be one ten-millionth part of the distance from the north pole to the equator on a great circle through Paris. The meter is slightly longer than the yard. The metric system is presently used in most countries of the world.

If the bar shown here were attached at the end of a standard yardstick, the result would be a meterstick (1 meter long):

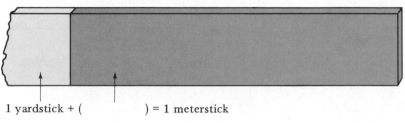

1 yardstick + (   ) = 1 meterstick

Q11   What system of measurement is used in most countries of the world? _____

STOP • STOP • STOP • STOP • STOP • STOP • STOP • STOP • STOP

A11   metric

Q12   Arrange the following units of measurement in order from shortest to longest: foot, meter,

inch, mile, yard. _____, _____, _____, _____,

_____

STOP • STOP • STOP • STOP • STOP • STOP • STOP • STOP • STOP

A12   inch, foot, yard, meter, mile

Q13   Would the 100-meter dash be a longer or shorter race than the 100-yard dash? _____

STOP • STOP • STOP • STOP • STOP • STOP • STOP • STOP • STOP

A13   longer

10   The meter that was the base unit of the metric system was multiplied and divided by powers of 10 to obtain other linear units of measurement. Greek prefixes to the term *meter* were used for multiples of the unit, and Latin prefixes were used for subdivisions. The results were as follows:

1 kilometer = 1000 meters   1 decimeter = 0.1 meter = $\frac{1}{10}$ meter

1 hectometer = 100 meters   1 centimeter = 0.01 meter = $\frac{1}{100}$ meter

1 dekameter = 10 meters   1 millimeter = 0.001 meter = $\frac{1}{1000}$ meter

1 meter = 1 meter   1 micrometer (micron) = 0.000 001 meter = $\frac{1}{1\,000\,000}$ meter

In the metric system when there are more than four digits to the right or left of the decimal point, the numerals are separated by a space into groups of three digits, starting at the decimal point. No commas are inserted.

Q14   Between each of the pairs of metric units place a > (greater than) or < (less than) symbol to make a true statement:

a. 1 kilometer _____ 1 meter

b. 1 centimeter _____ 1 meter

c. 1 dekameter _____ 1 decimeter

d. 1 millimeter _____ 1 centimeter

STOP • STOP • STOP • STOP • STOP • STOP • STOP • STOP • STOP

A14   a. >   b. <   c. >   d. <

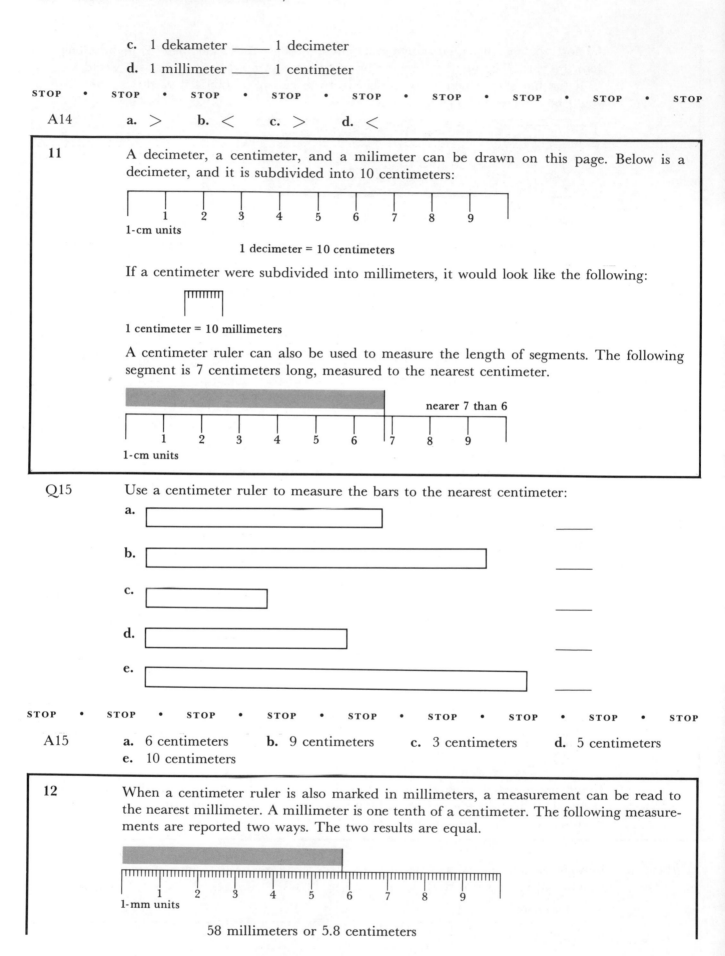

11   A decimeter, a centimeter, and a milimeter can be drawn on this page. Below is a decimeter, and it is subdivided into 10 centimeters:

1-cm units

1 decimeter = 10 centimeters

If a centimeter were subdivided into millimeters, it would look like the following:

1 centimeter = 10 millimeters

A centimeter ruler can also be used to measure the length of segments. The following segment is 7 centimeters long, measured to the nearest centimeter.

nearer 7 than 6

1-cm units

Q15   Use a centimeter ruler to measure the bars to the nearest centimeter:

a. _____

b. _____

c. _____

d. _____

e. _____

STOP • STOP • STOP • STOP • STOP • STOP • STOP • STOP • STOP

A15   a. 6 centimeters   b. 9 centimeters   c. 3 centimeters   d. 5 centimeters
           e. 10 centimeters

12   When a centimeter ruler is also marked in millimeters, a measurement can be read to the nearest millimeter. A millimeter is one tenth of a centimeter. The following measurements are reported two ways. The two results are equal.

1-mm units

58 millimeters or 5.8 centimeters

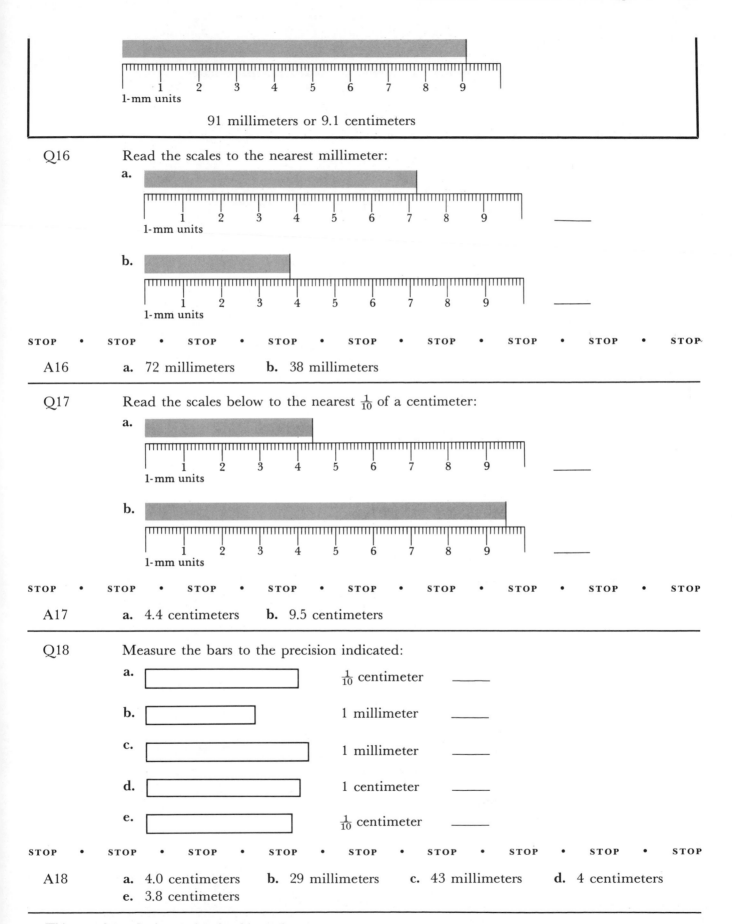

91 millimeters or 9.1 centimeters

Q16   Read the scales to the nearest millimeter:

a.

1-mm units

b.

1-mm units

STOP • STOP • STOP • STOP • STOP • STOP • STOP • STOP • STOP

A16   a. 72 millimeters   b. 38 millimeters

Q17   Read the scales below to the nearest $\frac{1}{10}$ of a centimeter:

a.

1-mm units

b.

1-mm units

STOP • STOP • STOP • STOP • STOP • STOP • STOP • STOP • STOP

A17   a. 4.4 centimeters   b. 9.5 centimeters

Q18   Measure the bars to the precision indicated:

a. $\frac{1}{10}$ centimeter   _____

b.   1 millimeter   _____

c.   1 millimeter   _____

d.   1 centimeter   _____

e. $\frac{1}{10}$ centimeter   _____

STOP • STOP • STOP • STOP • STOP • STOP • STOP • STOP • STOP

A18   a. 4.0 centimeters   b. 29 millimeters   c. 43 millimeters   d. 4 centimeters
e. 3.8 centimeters

This completes the instruction for this section.

10.1   Exercises

1. Which of the following measurements would be approximate?
    a. time for a person to run the 100-yard dash
    b. number of floors in the Empire State Building
    c. length of a room
    d. height of a person

2. Read the inch ruler at each of the arrows:

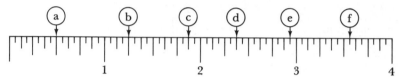

3. Approximate the length of the line segments to the precision indicated:

    a. $\frac{1}{4}$ inch

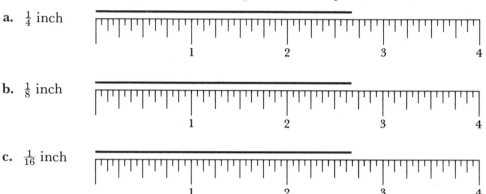

    b. $\frac{1}{8}$ inch

    c. $\frac{1}{16}$ inch

4. Arrange the following units from shortest to longest: foot, meter, yard, inch, centimeter, mile, millimeter.

5. Arrange the following units from shortest to longest: meter, millimeter, kilometer, decimeter, centimeter.

6. Approximate the length of the bars to the precision indicated:

    a. 1 centimeter

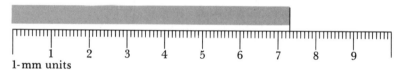

    b. $\frac{1}{10}$ centimeter

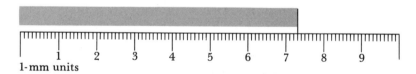

    c. 1 millimeter

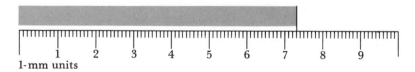

10.1   Exercise Answers

1. a, c, and d
2. a. $\frac{1}{2}$ inch   b. $1\frac{1}{4}$ inches   c. $1\frac{7}{8}$ inches
   d. $2\frac{3}{8}$ inches   e. $2\frac{15}{16}$ inches   f. $3\frac{9}{16}$ inches
3. a. $2\frac{3}{4}$ inches   b. $2\frac{5}{8}$ inches   c. $2\frac{11}{16}$ inches
4. millimeter, centimeter, inch, foot, yard, meter, mile
5. millimeter, centimeter, decimeter, meter, kilometer
6. a. 7 centimeters   b. 7.3 centimeters   c. 73 millimeters

The ruler at the right may be cut from the text and used whenever a measurement is required. The right-hand portion has two scales, one marked off each $\frac{1}{16}$ inch and the other each $\frac{1}{10}$ inch. The left-hand part is graduated in millimeters. On the left-hand scale, the numbers indicate centimeters.

10.2   Introduction to Measurement of Distance, Area, and Volume (English and Metric)

1	If an object is to be measured, the first task is to decide the unit of measurement to be used. One has a choice of units in some cases. However, the choice is not entirely arbitrary (one would not measure height in bushels, for example). When measuring distances on a line, *linear* units are used. Distance is said to be a one-dimensional measure. Examples include: the height of a person, the width of a page, the distance to the moon, and the length of a kite string. Notice that the idea of distance on a line is extended to the length of a curve (kite string), which is not straight. The length of the curve is the same as the length would be if the string were laid out in a straight line.

Q1   Choose the examples that illustrate need for a linear measurement: _____
    a. height of a flagpole
    b. weight of a person
    c. flying distance between New York City and Detroit
    d. distance around your waist

STOP • STOP • STOP • STOP • STOP • STOP • STOP • STOP • STOP

A1   a, c, and d

2	You have a choice of systems to use when measuring distance in linear units. Within each system there is also a choice of units. In the English system of measurement there are inches, feet, yards, rods, and miles. In the metric system, micrometers, millimeters, centimeters, decimeters, meters, and kilometers are all linear units. Inches were used as one of the units of measurement in Section 10.1. If twelve of these units are placed end to end, the result is a line segment 1 foot long. This is the length of the standard foot ruler. Three of the foot rulers placed end to end form a yard, the length of a standard yardstick. These relationships are summarized below with the abbreviations for each unit: 1 foot (ft) = 12 inches (in) 1 rod (rd) = $16\frac{1}{2}$ feet (ft) 1 yard (yd) = 3 feet (ft) 1 mile (mi) = 1,760 yards (yd) Another symbol for 12 inches is 12″; 3 feet is 3′.

Q2   Choose the linear units from the following: _____
    a. centimeter   **b.** pound   **c.** degree   **d.** yard

STOP • STOP • STOP • STOP • STOP • STOP • STOP • STOP • STOP

A2   a and d: a pound measures weight and a degree measures an angle or a temperature.

Q3   If a distance is measured in feet, could it also be measured in centimeters? _____

STOP • STOP • STOP • STOP • STOP • STOP • STOP • STOP • STOP

A3   yes

Q4   If a distance is measured in inches, could it also be measured in pounds?_____

STOP • **STOP** • **STOP** • **STOP** • **STOP** • **STOP** • **STOP** • **STOP** • **STOP**

A4   no: pounds do not represent a linear measurement.

---

3   Metric units are easier to work with mathematically, compared with English units, because each is a multiple of 10 times another. A centimeter is shown here. A decimeter, which is 10 times as long as a centimeter, is also shown.

1-cm unit

1 centimeter

1-cm units

1 decimeter = 10 centimeters

Unfortunately, the size of the page does not allow a line segment 1 meter long. It is slightly longer than 1 yard. One thousand meters makes a kilometer that is over $\frac{1}{2}$ mile.

If a centimeter is divided into 10 parts, each is 1 millimeter long.

1 centimeter = 10 millimeters

A micrometer (micron) is the length of one of 1,000 equal parts of a millimeter. A micrometer cannot be shown on this page because of limitations on the accuracy of printing machines. A human red blood cell is about 7 micrometers across. Special measuring devices are used to measure to such accuracy. (One device is called a micrometer.) The diameter of a hair from your head is probably about 50 micrometers. These relationships are summarized below with symbols for each unit:

1 millimeter (mm) = 1000 micrometers (microns) (μm)

1 centimeter (cm) = 10 millimeters (mm)

1 decimeter (dm) = 10 centimeters (cm)

1 meter (m) = 10 decimeters (dm)

1 dekameter (dam) = 10 meters (m)

1 hectometer (hm) = 10 dekameters (dam)

1 kilometer (km) = 10 hectometers (hm)

Q5   Choose the most likely measurement for each of the following:

a. width of a finger_____
(1) 1 centimeter   (2) 1 meter   (3) 1 micrometer

b. thickness of a fingernail_____
(1) 300 micrometers   (2) 300 decimeters   (3) 300 centimeters

c. diameter of the wood part of a wooden pencil_____
(1) $\frac{3}{4}$ kilometer   (2) $\frac{3}{4}$ centimeter   (3) $\frac{3}{4}$ millimeter

    d. diameter of a fine pencil lead for a mechanical pencil_____

       (1) 1 meter   (2) 1 millimeter   (3) 1 centimeter

STOP • STOP • STOP • STOP • STOP • STOP • STOP • STOP • STOP

A5   **a.** (1)   **b.** (1)   **c.** (2)   **d.** (2)

Q6   Choose the most appropriate answer:

    a. length of a horse racetrack_____

       (1) 1 meter   (2) 1 decimeter   (3) 1 kilometer

    b. one car length_____

       (1) 5 meters   (2) 5 centimeters   (3) 5 kilometers

    c. the height of a telephone pole_____

       (1) 10 kilometers   (2) 10 centimeters   (3) 10 meters

    d. diameter of the wheel of a 10-speed bicycle_____

       (1) $\frac{3}{4}$ centimeter   (2) $\frac{3}{4}$ meter   (3) $\frac{3}{4}$ kilometer

STOP • STOP • STOP • STOP • STOP • STOP • STOP • STOP • STOP

A6   **a.** (3)   **b.** (1)   **c.** (3)   **d.** (2)

Q7   Match the symbol for a metric unit with the unit:

    a. cm   _____   (1) hectometer

    b. μm   _____   (2) decimeter

    c. mm   _____   (3) meter

    d. m   _____   (4) micrometer

    e. km   _____   (5) dekameter

    f. dm   _____   (6) kilometer

    g. hm   _____   (7) millimeter

    h. dam   _____   (8) centimeter

STOP • STOP • STOP • STOP • STOP • STOP • STOP • STOP • STOP

A7   **a.** (8)   **b.** (4)   **c.** (7)   **d.** (3)   **e.** (6)   **f.** (2)   **g.** (1)

    h. (5)

---

4   *Area* is a measure of the region within a closed curve (including polygons) in a plane. Geometric figures that do not enclose a finite region do not have an area. The angle of Figure 1 and the object of Figure 2 do not have area.

    Figure 1   *Figure 2*

Irregular shapes that enclose a finite region have area, but the measurement of that area is difficult to obtain and usually is only roughly approximated. The following figure is of this nature:

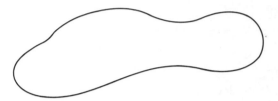

The areas of certain geometric regions, such as circular regions, triangular regions, and many others, are obtained from standard formulas that will be studied in this chapter.

Q8　　Choose the geometric regions that have an area._____

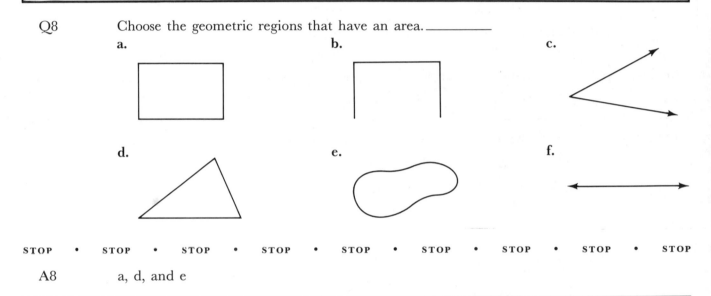

a.　　　　　　　　b.　　　　　　　　c.

d.　　　　　　　　e.　　　　　　　　f.

STOP • **STOP** • **STOP** • **STOP** • **STOP** • **STOP** • **STOP** • **STOP** • **STOP**

A8　　　　a, d, and e

5　　　A square region with all four sides 1 inch long is said to have an area of 1 square inch. This area is considered to be a unit, and areas of other closed figures are given as some number of square inches. "Square inch" will be symbolized as in^2. Other square units will be symbolized in a similar manner, that is, square feet as ft^2, square centimeters as cm^2, and so on. The figures below show a square inch and also a rectangular region with an area of 6 square inches. This area may be obtained by counting.

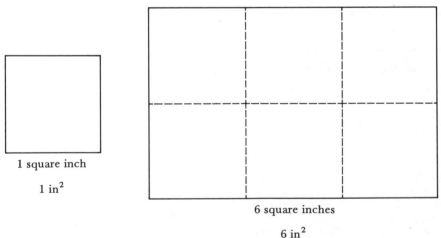

1 square inch

1 in^2

6 square inches

6 in^2

In the metric system a square region with all four of its sides 1 centimeter long has an area of 1 square centimeter. Areas of other geometric figures are given as some number of square centimeters. A square centimeter and a rectangle of 12 square centimeters are shown.

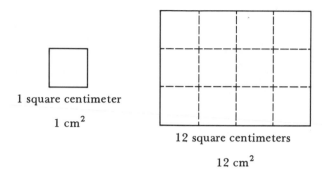

1 square centimeter

1 cm^2

12 square centimeters

12 cm^2

Area is a measure of the polygonal region rather than the polygon, because the region includes all the points inside. It is the space inside that is being measured. (Other texts may use the less-precise phrase "area of a square." If you read such a phrase, interpret it as "area of a square region").

Q9   Use the 1-centimeter grid to count the number of square centimeters in the following polygonal regions:

a.

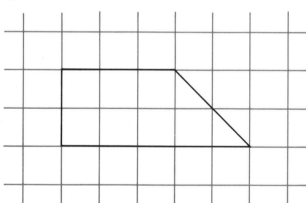

b.

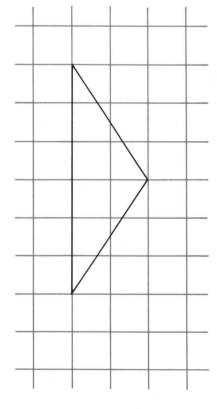

a. _____ cm^2

b. _____ cm^2

A9   **a.** 8 cm^2
          b. 6 cm^2: you may obtain this by mentally piecing triangles together to make whole
          squares, as follows:

a.

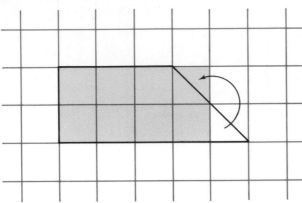

b.

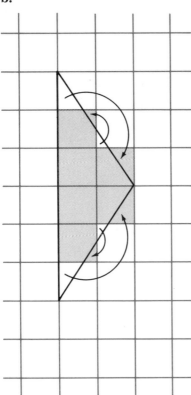

---

6   Other units that can be used in the metric system to measure area besides the square
             centimeter are: square millimeter, square decimeter, square meter, and square kilometer.
             Another unit used to measure area in the metric system is the *hectare*. It equals 10 000
             square meters.
                 The relationship in size between a square centimeter and a square decimeter is possible
             to show with a drawing. It illustrates the problem for all conversions of square units. A
             decimeter contains 10 centimeters, but a square decimeter contains 100 square centimeters,
             as follows:

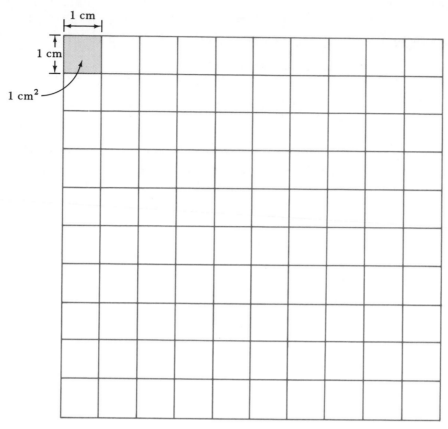

1 square decimeter = 100 square centimeters

Similar figures could be drawn to show the relationships contained in this chart:

1 square centimeter = 100 square millimeters

1 square decimeter = 100 square centimeters

1 square meter = 100 square decimeters = 10 000 square centimeters

1 square kilometer = 1 000 000 square meters

1 hectare = 10 000 square meters

Q10   Use the diagrams (drawn to scale) below to answer the questions:

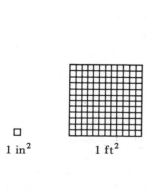

1 in^2 1 ft^2

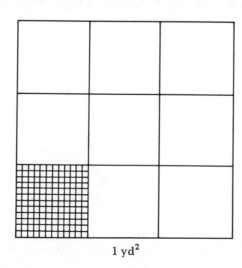

1 yd^2

a.   1 square foot = _____ square inches

b.   1 square yard = _____ square feet

c.   1 square yard = _____ square inches

STOP   •   STOP   •   STOP   •   STOP   •   STOP   •   STOP   •   STOP   •   STOP   •   STOP

A10   **a.** 144   **b.** 9   **c.** 1,296

---

7   A unit of square measure usually used for land measurement in the English system is
             the acre. One acre contains 43,560 square feet. One hectare in the metric system contains
             approximately 2.471 acres. The symbol for "approximately equals" is "$\doteq$."
                Summarizing the answers to Q10 and the above remarks:

             1 square foot = 144 square inches

             1 square yard = 9 square feet

                  1 acre = 43,560 square feet

                  1 hectare $\doteq$ 2.471 acres

             The number of the smaller square units in a larger square unit may be obtained by
             multiplying. For example, since there are 3 feet in a yard, there are 3×3, or 9, square
             feet in a square yard. Also, since there are 12 inches in a foot, there are 12×12, or 144,
             square inches in a square foot.

Q11   **a.** Since there are 36 inches in a yard, there are _____ $\times$ _____, or _____,
                    square inches in a square yard.

           b. Since there are 1,760 yards in a mile, there are _____ $\times$ _____, or _____,
                    square yards in a square mile.

           c. Since there are 1000 micrometers in a millimeter, there are _____ $\times$ _____,

                    or _____, square micrometers in a square millimeter.

STOP   •   STOP   •   STOP   •   STOP   •   STOP   •   STOP   •   STOP   •   STOP   •   STOP

A11   **a.** 36, 36, 1,296
           b. 1,760, 1,760, 3,097,600
           c. 1000, 1000, 1 000 000

---

8   *Volume* is a measure of the space within a closed solid figure. Irregular-shaped solid figures
             have volume, but the measurement of these volumes is difficult to obtain. The volumes
             of certain solid figures, such as prisms, cones, and spheres, are obtained by using standard
             formulas that will be studied in this chapter.
                One basic unit of volume is the cubic inch. This is a cube with each edge measuring
             1 inch. To find the volume of a rectangular prism in cubic inches, think of the number
             of cubes, 1 inch on each edge, that could be stacked inside the prism.
                The prism shown below would have 15 cubes on the bottom level. This is the same
             number as the area of the rectangular base. Above *each* of these bottom cubes would be
             three more, making a stack 4 cubes high. Therefore, the total volume would be 60 cubic
             inches. "Cubic inch" will be symbolized by "in^3." Other cubic units will be symbolized
             in a similar manner, that is, cubic centimeters as cm^3, cubic yards as yd^3, and so on.

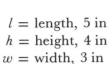

length, 1 in
height, 1 in
width, 1 in

l = length, 5 in
h = height, 4 in
w = width, 3 in

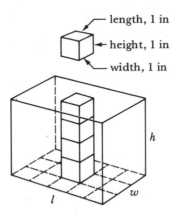

h

l w

Q12   Use the method of Frame 8 to determine the volume of the prism at the right. _____

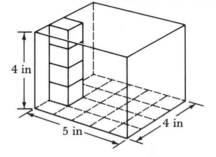

4 in

5 in   4 in

STOP • STOP • STOP • STOP • STOP • STOP • STOP • STOP • STOP

A12   80 in^3

9   The volume of a prism can be obtained from the dimensions: the length (l), width (w), and height (h). Just as the volume of the rectangular prism in Frame 8 could be found by multiplying the length times the width to get the area of the base and then multiplying that times the height, the volume of any rectangular prism may be found the same way. The formula would be $V = lwh$.

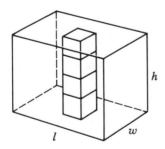

h

l w

To find the number of cubic inches in a cubic foot, the same formula may be used. The dimensions would be $l = 12$ in, $w = 12$ in, $h = 12$ in.

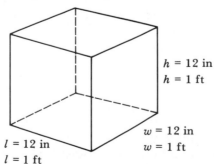

$V = lwh$
$V = 12 \times 12 \times 12$
1 ft^3 = $1,728$ in^3

$h = 12$ in
$h = 1$ ft

$w = 12$ in
$w = 1$ ft

$l = 12$ in
$l = 1$ ft

Q13   Find the number of cubic inches in a rectangular solid with dimensions: $l = 8.0$ in, $w = 10.0$ in, $h = 4.5$ in.

STOP • STOP • STOP • STOP • STOP • STOP • STOP • STOP • STOP

A13   360 in³: $V = lwh$
$V = (8.0)(10.0)(4.5)$
$V = 360$

Q14   Use the formula for the volume of a prism to find the number of cubic feet in 1 cubic yard.

STOP • STOP • STOP • STOP • STOP • STOP • STOP • STOP • STOP

A14   27 ft³: $V = lwh$
$V = 3(3)(3)$
$V = 27$
$1 \text{ yd}^3 = 27 \text{ ft}^3.$

Q15   Find the number of cubic centimeters in 1 cubic decimeter.

STOP • STOP • STOP • STOP • STOP • STOP • STOP • STOP • STOP

A15   1000 cm³: $V = lwh$
$V = 10(10)(10)$
$V = 1000$
$1 \text{ dm}^3 = 1000 \text{ cm}^3$

---

10   Summarizing the relationships between different units of volume:

1 cubic foot = 1,728 cubic inches
1 cubic yard = 27 cubic feet
1 cubic decimeter = 1000 cubic centimeters

Other units of liquid volume are the pint, quart, and gallon in the English system and the liter in the metric system. The relationships that exist between them and relate them to the cubic measurements follow:

1 quart (qt) = 2 pints (pt)
1 gallon (gal) = 4 quarts (qt) = 8 pints (pt)

> 1 quart (qt) = 57.75 cubic inches (in^3)
> 1 liter (l) $\doteq$ 1.05 quarts (qt)
> 1 liter (l) = 1000 cubic centimeters (cm^3)

Q16   Indicate whether each unit measures distance, area, or volume:

    a. ft^2 _____   **b.** quart _____

    c. cm^2 _____   **d.** cm^3 _____

    e. liter _____   **f.** millimeter _____

    g. in^3 _____   **h.** pint _____

STOP • STOP • STOP • STOP • STOP • STOP • STOP • STOP • STOP

A16   **a.** area   **b.** volume   **c.** area
         d. volume   **e.** volume   **f.** distance
         g. volume   **h.** volume

Q17   Use the statement 1 liter $\doteq$ 1.05 quarts to tell which is larger, a liter or a quart.

    _____

STOP • STOP • STOP • STOP • STOP • STOP • STOP • STOP • STOP

A17   liter: because it takes more than 1 quart to equal 1 liter

Q18   Indicate all units of measurement that could be used to report the measurement required (there can be more than one correct):

    a. length of a room_____
       (1) in^3 (2) liters   (3) feet   (4) decimeters

    b. volume of a barrel_____
       (1) gallons   (2) ft^2 (3) cm^3 (4) micrometers

    c. area of a bulletin board_____
       (1) quarts   (2) in^2 (3) feet   (4) cm^2

    d. surface area of a pyrimid_____
       (1) in^2 (2) quart   (3) feet   (4) yd^2

STOP • STOP • STOP • STOP • STOP • STOP • STOP • STOP • STOP

A18   **a.** (3) and (4)   **b.** (1) and (3)   **c.** (2) and (4)   **d.** (1) and (4)

---

11   It helps to visualize some of the units presented in this section by comparing them with familiar objects which are approximately that size.

English

cubic inch	cube the size of a marshmallow
1 quart	quart milk carton
1 gallon	gallon milk carton or a gallon ice cream container
1 cubic foot	cube with each edge the length of a foot ruler such as a box of 12 in $\times$ 12 in floor tile 12 in high

	Metric
millimeter	diameter of a paper-clip wire
centimeter	width of a large paper clip
cubic centimeter	sugar cube
liter	slightly larger than 1 quart

Q19   Arrange the following units in the following set from smallest to largest: liter, cubic foot,

cubic inch, quart. _____

STOP • **STOP** • **STOP** • **STOP** • **STOP** • **STOP** • **STOP** • **STOP** • **STOP**

A19   cubic inch, quart, liter, cubic foot

This completes the instruction for this section.

10.2   Exercises

1. Label each of the following as (1) a distance, (2) an area, (3) a volume, or (4) none of these:
   a. length of a pencil
   b. floor space in a room
   c. amount you weigh
   d. amount of water in a bucket
   e. circumference of your head
   f. amount of window space in a house
   g. amount of gravel on a driveway
   h. amount of blood in your body
   i. depth of water in a lake

2. What are the names of the two systems of measurement used in the world today?

3. What is the quantity that is measured in linear units?

4. Choose the linear units from the following: centimeters, square miles, gallons, inches, meters, cubic yards, rods.

5. If a distance is measured in meters, could it also be measured in centimeters?

6. If an area is measured in square feet, could it also be measured in decimeters?

7. Choose the most likely object to have the given measurement:
   a. 7 in: (1) length of an adult arm, (2) length of a pencil, (3) diameter of a light bulb
   b. $2\frac{1}{2}$ cm: (1) diameter of the cap of a pop bottle, (2) width of a tooth, (3) width of an adult hand
   c. 1 mm: (1) thickness of a human hair, (2) thickness of a fine pencil lead, (3) length of an adult's big toe
   d. 1 km: (1) length of an automobile, (2) length of a football field, (3) length of six city blocks

8. Choose the units that could be used to measure area from the following: centimeters, square feet, square meters, gallons, cubic inches, yards, hectares, acres.

9. Use the fact that there are 100 centimeters in a meter to find the number of square centimeters in a square meter.

10. Choose the most likely object to have the given measurement:
    a. 1 in^2: (1) fingernail on an adult's little finger (2) cross section of a wooden pencil, (3) postage stamp
    b. 1 ft^2: (1) surface of a brick, (2) surface of an asbestos floor tile, (3) surface of an adult's hand

c. 20 ft²: (1) surface of a door to a house, (2) area of a living room, (3) area of the windshield on a car

d. 1 cm²:

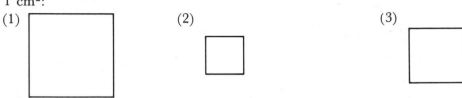

(1)   (2)   (3)

11. Choose the units of measure that could be used to measure volume: cubic inch, acre, gallon, pint, cubic centimeter, decimeter, square foot, liter, cubic yard, hectare.

12. Use the fact that there are 100 centimeters in a meter to find the number of cubic centimeters in a cubic meter.

13. Choose all units of measurement that could be used to report the measurement required:

   a. Space in the trunk of an automobile: (1) square feet, (2) cubic meters, (3) feet, (4) cubic feet

   b. Size of a field: (1) cubic yards, (2) hectares, (3) acres, (4) square yards

   c. Milk in a bottle: (1) pints, (2) cubic centimeters, (3) centimeters, (4) gallons

   d. Capacity of the lungs: (1) cubic inches, (2) square centimeters, (3) cubic centimeters, (4) hectares

10.2   Exercise Answers

1. **a.** (1)   **b.** (2)   **c.** (4)   **d.** (3)   **e.** (1)   **f.** (2)   **g.** (3)
   h. (3)   **i.** (1)
2. English and metric
3. distance
4. centimeters, inches, meters, rods
5. yes
6. no
7. **a.** (2)   **b.** (1)   **c.** (2)   **d.** (3)
8. square feet, square meters, hectares, acres
9. 10 000
10. **a.** (3)   **b.** (2)   **c.** (1)   **d.** (2)
11. cubic inch, gallon, pint, cubic centimeter, liter, cubic yard
12. 1 000 000
13. **a.** (2) and (4)   **b.** (2), (3), and (4)   **c.** (1), (2), and (4)
    d. (1) and (3)

10.3   Perimeters

1   The *perimeter* of a polygon is the total length of all its sides. For any geometric figure in a plane, the perimeter is the length of the boundary of the geometric figure. Notice that perimeter is a measure of length (distance).

The perimeter of a rectangle can be found by adding the lengths of the sides:

The perimeter is the total of 12 cm.

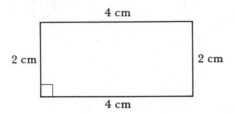

The perimeter of the rectangle above would be

$p = 2 + 4 + 2 + 4$
$p = 12$ Therefore, the perimeter is 12 cm.

Q1   Which would have a larger perimeter, a triangle with each side 2 inches or a square with each side 2 inches? _____

STOP • STOP • STOP • STOP • STOP • STOP • STOP • STOP • STOP

A1   square

Q2   Find the perimeter of the figures:

a.

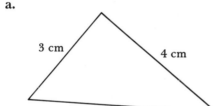

b.

```
        1.8 in
┌─────────────────┐
│                 │  1.0 in
│ ┌─              │
│ │               │
└─────────────────┘
```

STOP • STOP • STOP • STOP • STOP • STOP • STOP • STOP • STOP

A2   **a.** 12 cm: $p = 3 + 4 + 5$ **b.** 5.6 in: $p = 1.8 + 1.0 + 1.8 + 1.0$
                        $p = 12$ $p = 5.6$

---

2   Since perimeter is a length, each time a perimeter is found, the unit of measurement must be reported. Many units of length have been introduced. For example, in the English system there are the inch, foot, yard, rod, mile. In the metric system there are the micrometer, millimeter, centimeter, decimeter, meter, and kilometer. The perimeter of a polygon could be measured in any of these units or in other units of linear measure.

Q3   Indicate yes if the unit could be used to measure perimeter. Otherwise indicate no.

        a. inch   _____   **b.** square foot   _____

        c. mile   _____   **d.** acre   _____

        e. centimeter   _____   **f.** yard   _____

        g. cubic feet   _____   **h.** rod   _____

STOP • STOP • STOP • STOP • STOP • STOP • STOP • STOP • STOP

A3   **a.** yes   **b.** no   **c.** yes   **d.** no   **e.** yes   **f.** yes   **g.** no
         h. yes

---

3   For irregular-shaped polygons, there is no other way to find perimeters but to add the sides. Formulas for finding the perimeters of these polygons are very similar to each other, but most people do not memorize the formula. Rather, they remember the idea of what perimeter means. Some examples are as follows:

triangle $p = a + b + c$

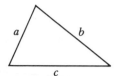

quadrilateral $p = a + b + c + d$

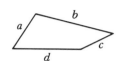

pentagon $p = a + b + c + d + e$

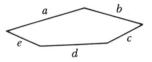

Q4   Find the perimeter of the polygons:

a.

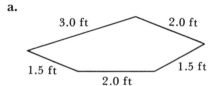

3.0 ft   2.0 ft
1.5 ft   1.5 ft
2.0 ft

b.

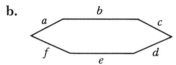

$a = 9$ yd $b = 18$ yd $c = 9$ yd
$d = 11$ yd $e = 16$ yd $f = 11$ yd

STOP • **STOP** • **STOP** • **STOP** • **STOP** • **STOP** • **STOP** • **STOP** • **STOP**

A4   **a.**   10.0 ft: $p = a + b + c + d + e$
                $p = 3.0 + 2.0 + 1.5 + 2.0 + 1.5$
                $p = 10.0$
    b.   74 yd: $p = a + b + c + d + e + f$
                $p = 9 + 18 + 9 + 11 + 16 + 11$
                $p = 74$

---

4   For rectangles and squares, formulas can be written which shorten the work somewhat. If the length is l and the width is w,

$p = l + w + l + w$
$p = l + l + w + w$
$p = 2l + 2w$

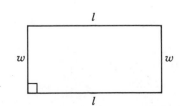

This gives a formula for the perimeter of a rectangle:

$p = 2l + 2w$

A formula for the perimeter of a square may be obtained from the formula for the perimeter of a rectangle $l = s$ and $w = s$:

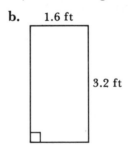

$p = 2l + 2w$
$p = 2s + 2s$
$p = 4s$

Therefore, $p = 4s$.

Find the perimeter of a rectangle with width 7.0 in and length 8.4 in.

$p = 2l + 2w$
$p = 2(8.4) + 2(7.0)$
$p = 16.8 + 14.0$
$p = 30.8$

Therefore, the perimeter is 30.8 in.

Q5   Find the perimeters of the following rectangles by substituting in the appropriate formula:

a.
8.0 mi
1.4 mi

b.
1.6 ft
3.2 ft

c.
22.0 cm
22.0 cm

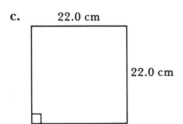

d.
$1\frac{3}{4}$ in
$1\frac{3}{4}$ in

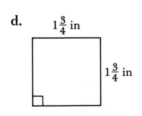

STOP • **STOP** • **STOP** • **STOP** • **STOP** • **STOP** • **STOP** • **STOP** • **STOP**

A5   **a.**   18.8 mi: $p = 2l + 2w$
                $p = 2(8.0) + 2(1.4)$
                $p = 16.0 + 2.8$
                $p = 18.8$

b.   9.6 ft: $p = 2l + 2w$
                $p = 2(3.2) + 2(1.6)$
                $p = 6.4 + 3.2$
                $p = 9.6$

c.   88.0 cm: $p = 4s$
$p = 4(22.0)$
$p = 88.0$

d.   7 in: $p = 4s$
$p = 4(1\frac{3}{4})$
$p = 4(\frac{7}{4})$
$p = 7$

---

5   The perimeter of a circle is called its *circumference*. The circumference of a circle is found with another formula. To obtain the circumference it is necessary to know the diameter of the circle and the value of the number π (pi). The number π is the ratio of the circumference to the diameter of a circle. The value of π is a constant; that is, it never changes. However, it is impossible to write a decimal or fraction that is exactly equal to π. Since all rational numbers can be written as fractions, we conclude that π is not a rational number. One is therefore forced to use approximations of π rounded off to various numbers of decimal places. Since it is not possible to say that π *equals* another number, the symbol "$\doteq$" is used to mean approximately equal. The symbol "$\doteq$" says "is approximately equal to." Various approximations follow:

1. $\pi \doteq 3.14159265$, precise to 8 decimal places.
2. $\pi \doteq 3.14$, precise to 2 decimal places. This approximation is acceptable for calculations in this book.
3. $\pi \doteq 3\frac{1}{7}$, also precise to 2 decimal places. This value is commonly used when calculating with fractions.
4. $\pi \doteq \frac{355}{113}$, precise to 6 decimal places. This is an easy approximation to remember (think 1, 1, 3, 3, 5, 5) for use with an electronic calculator.

Q6   Is π a variable or a constant? _____

STOP   •   STOP   •   STOP   •   STOP   •   STOP   •   STOP   •   STOP   •   STOP   •   STOP

A6   constant

Q7   **a.**   Give two approximations of π that are precise to two decimal places. _____ and

_____

b.   Give two approximations of π that are precise to six decimal places. _____ and

_____

STOP   •   STOP   •   STOP   •   STOP   •   STOP   •   STOP   •   STOP   •   STOP   •   STOP

A7   **a.**   3.14 and $3\frac{1}{7}$ **b.**   3.141593 and $\frac{355}{113}$

Q8   Is π a rational number? _____

STOP   •   STOP   •   STOP   •   STOP   •   STOP   •   STOP   •   STOP   •   STOP   •   STOP

A8   no: rational numbers can be written as a fraction and π is not equal to any fraction.

---

6   The circumference of a circle is found by using the formula $c = \pi d$, where d is the diameter. Since the diameter is twice the radius ($d = 2r$), a second formula for the circumference is $c = 2\pi r$, where r is the radius:

$c = 2\pi r = \pi d$

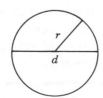

If the diameter of a circle is 2.5 in, what is its circumference?

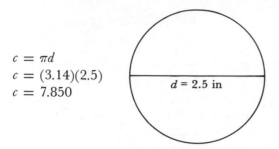

$c = \pi d$
$c = (3.14)(2.5)$
$c = 7.850$

$d = 2.5$ in

Therefore, the circumference is 7.850 in.

Use $\pi \doteq \frac{22}{7}$ when measurements are given as common fractions. Use $\pi \doteq 3.14$ when measurements are given as decimal fractions.

Q9   Find the circumference of a circle with diameter 8.5 cm (use $\pi \doteq 3.14$).

STOP • **STOP** • **STOP** • **STOP** • **STOP** • **STOP** • **STOP** • **STOP** • **STOP**

A9   26.690 cm: $c = \pi d$
                             $c = (3.14)(8.5)$
                             $c = 26.690$

Q10   What is the circumference of a circle whose *radius* is $4\frac{3}{4}$ ft? (Use $c = 2\pi r$ and $\pi \doteq 3\frac{1}{7}$.)

STOP • **STOP** • **STOP** • **STOP** • **STOP** • **STOP** • **STOP** • **STOP** • **STOP**

A10 $29\frac{6}{7}$ ft: $c = 2\pi r$
                              $c = 2(3\frac{1}{7})(4\frac{3}{4})$
                              $c = \frac{2}{1}(\frac{22}{7})(\frac{19}{4})$
                              $c = 29\frac{6}{7}$

7   Problems have been given in which the dimensions of the polygons or circles have been supplied. One should also be able to measure the appropriate dimensions and compute the perimeter of a figure. The following problems will require that you measure with an inch or centimeter ruler and calculate the perimeters.

Q11   Compute the perimeter in inches (measure to the nearest $\frac{1}{8}$ inch):

a.

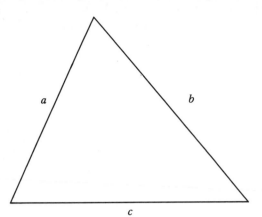

b.

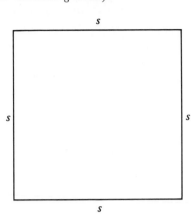

STOP • STOP • STOP • STOP • STOP • STOP • STOP • STOP • STOP

A11   **a.** $7\frac{1}{8}$ in: $p = a + b + c$

$\qquad p = 2\frac{1}{8} + 2\frac{1}{2} + 2\frac{1}{2}$

$\qquad p = 7\frac{1}{8}$

b. 7 in: $p = 4s$

$\qquad p = 4(1\frac{3}{4})$

$\qquad p = 7$

Q12   Compute the perimeter in centimeters (measure to the nearest $\frac{1}{10}$ centimeter):

a.

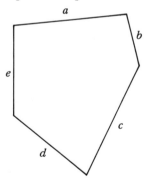

b.

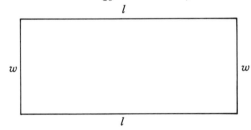

STOP • STOP • STOP • STOP • STOP • STOP • STOP • STOP • STOP

A12   **a.** 12.5 cm: $p = a + b + c + d + e$

$\qquad p = 3.0 + 1.4 + 3.2 + 2.5 + 2.4$

$\qquad p = 12.5$

b. 16.4 cm: $p = 2l + 2w$

$\qquad p = 2(5.7) + 2(2.5)$

$\qquad p = 11.4 + 5.0$

$\qquad p = 16.4$

Q13   Compute the circumference of each circle:
  a. Measure the diameter of circle 1 to the nearest $\frac{1}{4}$ inch.
  b. Measure the diameter of circle 2 to the nearest $\frac{1}{2}$ centimeter.

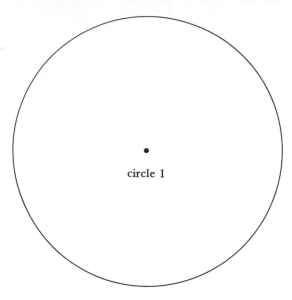

circle 1

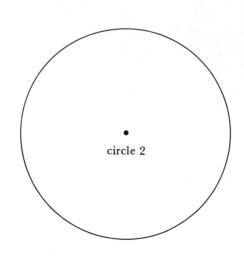

circle 2

STOP • STOP • STOP • STOP • STOP • STOP • STOP • STOP • STOP

A13   **a.** $8\frac{9}{14}$ in: $c = \pi d$

$$c = 3\frac{1}{7}\left(2\frac{3}{4}\right)$$

$$c = \frac{\overset{11}{\cancel{22}}}{7} \cdot \frac{11}{\underset{2}{\cancel{4}}}$$

$$c = \frac{121}{14}$$

$$c = 8\frac{9}{14}$$

b. 17.27 cm: $c = \pi d$
$c = (3.14)(5.5)$
$c = 17.27$

or

$17\frac{2}{7}$ cm: $c = \pi d$

$$c = \frac{22}{7}\left(5\frac{1}{2}\right)$$

$$c = \frac{22}{7} \cdot \frac{11}{2}$$

$$c = \frac{242}{14}$$

$$c = 17\frac{2}{7}$$

---

8   A summary of formulas for perimeters is provided. Some have been considered in the previous material. Others may be useful in further work.

Geometric figure	Formula for perimeter
Triangle: sides a, b, c	$p = a + b + c$
Quadrilateral: sides a, b, c, d	$p = a + b + c + d$

Pentagon: sides a, b, c, d, e $p = a + b + c + d + e$
Rectangle: length l, width w $p = 2l + 2w$
Square: side s $p = 4s$
Circle: radius r $c = 2\pi r$ or $c = \pi d$
Equilateral triangle: side s $p = 3s$
Regular pentagon: side s $p = 5s$
Parallelogram: sides a and b $p = 2a + 2b$

This completes the instruction for this section.

10.3   Exercises

1. The total length of the boundary of a polygon is called its _____.
2. The perimeter of a circle is called its _____.
3. Find the perimeter of the polygons below:

   a.
   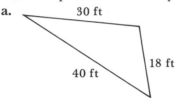
   30 ft   18 ft   40 ft

   b.
   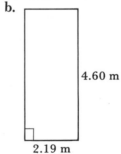
   20 mm   15 mm   22 mm   15 mm   10 mm

4. Find the perimeter of a pentagon with sides to the nearest $\frac{1}{8}$ inch of the following lengths: $1\frac{3}{4}$, $2\frac{1}{4}$, 3, $1\frac{7}{8}$, and $2\frac{1}{2}$ inches.
5. Use the formula $p = 2l + 2w$ to compute the perimeter of the rectangles:

   a.
   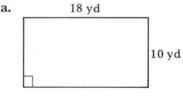
   18 yd   10 yd

   b.
   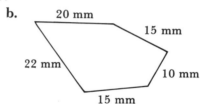
   4.60 m   2.19 m

6. Use the formula $p = 4s$ to compute the perimeter of the following:
   a. square with a side of 4.7 centimeters
   b. square with a side of $1\frac{7}{8}$ inches
7. Use the formula $c = 2\pi r$ or $c = \pi d$ to compute the circumference of the circles:

   a.
   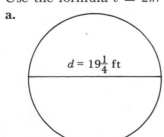
   $d = 19\frac{1}{4}$ ft

   b.
   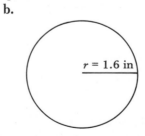
   $r = 1.6$ in

8. Use a centimeter rule to measure the figures to the nearest 0.1 centimeter and compute their perimeters in centimeters:

a.

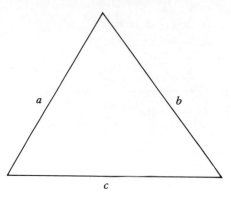

b.

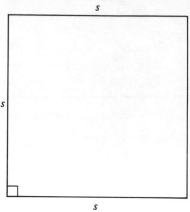

c.

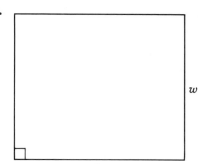

d.

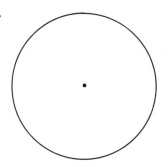

9. Use an inch ruler to measure the figures to the nearest $\frac{1}{8}$ inch, and compute their perimeters:

a.

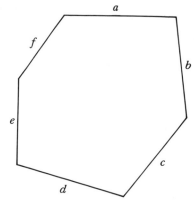

b.

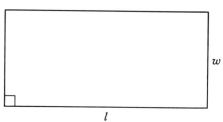

c.

d.

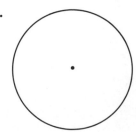

10. To order wallpaper it is necessary to know the perimeter of a room. If the dimensions of the room to the nearest inch were 10 ft 6 in × 15 ft, what was its perimeter?
11. Match the formula with the geometric figure:

   a. $p = 2l + 2w$ (1) regular pentagon
   b. $c = \pi d$ (2) equilateral triangle
   c. $p = 4s$ (3) circle
   d. $p = 5s$ (4) rectangle
   e. $c = 2\pi r$ (5) any triangle
   f. $p = a + b + c$ (6) quadrilateral
   g. $p = 3s$ (7) square
   h. $p = a + b + c + d$

10.3   Exercise Answers

1. perimeter
2. circumference
3. **a.** 88 ft   **b.** 82 mm
4. $11\frac{3}{8}$ in
5. **a.** 56 yd   **b.** 13.58 m
6. **a.** 18.8 cm   **b.** $7\frac{1}{2}$ in
7. **a.** $60\frac{1}{2}$ ft   **b.** 10.048 in
8. **a.** 16.1 cm   **b.** 19.2 cm   **c.** 16.6 cm   **d.** 11.932 cm
9. **a.** $5\frac{7}{8}$ in   **b.** $6\frac{1}{4}$ in   **c.** $3\frac{1}{2}$ in   **d.** $3\frac{13}{14}$ in
10. 51 ft
11. **a.** 4   **b.** 3   **c.** 7   **d.** 1
    e. 3   **f.** 5   **g.** 2   **h.** 6

10.4   Areas

1   The units that are used to report area within geometric figures were discussed in Section 10.2. Some examples are square inches, square feet, square centimeters, and square meters. The numerical value of the area is usually determined by a formula. For example, the area of a rectangular region is obtained with the formula $A = lw$, where l is the length and w is the width. The length and width are linear measurements. However, when inches are multiplied by inches, the result is a number with the unit square inches. Before the numbers are substituted into formulas, all measurements must be in the same units. For example,

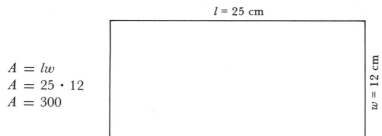

$$A = lw$$
$$A = 25 \cdot 12$$
$$A = 300$$

Therefore, the area is 300 cm².

Q1   Find the area of a floor with one dimension 10.5 ft and the other dimension 12 ft.

STOP • STOP • STOP • STOP • STOP • STOP • STOP • STOP • STOP

A1   126 ft^2: $A = lw$
            $A = 12 \cdot 10.5$
            $A = 126$

Q2   Find the area in square yards of a lot with dimensions 30 yd 2 ft by 53 yd 1 ft.

STOP • STOP • STOP • STOP • STOP • STOP • STOP • STOP • STOP

A2 $1,635\frac{5}{9}$ yd^2: 30 yd 2 ft $= 30\frac{2}{3}$ yd $A = lw$
                53 yd 1 ft $= 53\frac{1}{3}$ yd $A = 53\frac{1}{3} \cdot 30\frac{2}{3}$
                                $A = 1,635\frac{5}{9}$

---

2   The formula for the area of a square region is a special case of the formula for the area of a rectangular region. If $l = w$ in the formula $A = lw$, then it becomes $A = ww = w^2$. Rather than using w for the width of a square, the variable s is usually used. Therefore, the area of a square region is computed with the formula

$$A = s^2$$

Example: Find the area of a square card table that is 30 inches on a side.

Solution

$A = s^2$
$A = (30)^2$
$A = 900$

Therefore, the area is 900 in^2.

Q3   Find the area of a field that is 45 yards square.

STOP • STOP • STOP • STOP • STOP • STOP • STOP • STOP • STOP

A3   2,025 yd^2: $A = s^2$
                $A = 45^2$
                $A = 2,025$

3   The following figure is a parallelogram. The opposite sides are parallel.

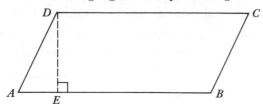

The bases of a parallelogram are any two opposite sides (usually the longest side is considered a base). The altitude of a parallelogram is the perpendicular distance between the bases.

Example 1: Name the bases of the parallelogram.

Solution

$\overline{AB}$ and $\overline{DC}$ ($\overline{AD}$ and $\overline{BC}$ could be considered bases, but it would not be as convenient to find the altitude)

Example 2: The altitude of parallelogram *ABCD* is the length of what line segment?

Solution

$\overline{DE}$

The area of a parallelogram is found by multiplying the length of a base, *b*, times the altitude (height), *h*. The formula is

$A = bh$

Example 3: Find the area of a parallelogram with base 22 cm and altitude 14 cm.

Solution

$A = bh$
$A = 22(14)$
$A = 308$ Therefore, the area is 308 cm².

Q4   **a.** Find the area of a parallelogram with base 202 m and altitude 26 m.

b. Find the area of the parallelogram.

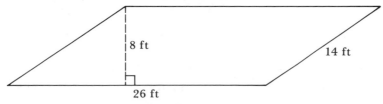

STOP • STOP • STOP • STOP • STOP • STOP • STOP • STOP • STOP

A4   **a.** 5252 m²: $A = bh$ **b.** 208 ft²: $A = bh$
                              $A = 202(26)$ $A = 26(8)$
                              $A = 5252$ $A = 208$

4 To derive a formula for the area of a region within a parallelogram with a base b and altitude (height) h, a perpendicular is drawn from D to $\overline{AB}$, and from C to $\overleftrightarrow{AB}$. $\overline{AE}$ and $\overline{BF}$ have the same length, and $\overline{DE}$ and $\overline{CF}$ have the same length (h), so the triangular regions AED and BFC have the same area. Since the area of the triangular region AED

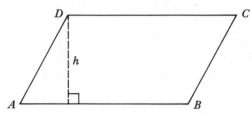 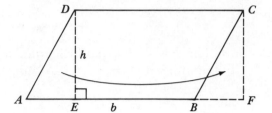

is replaced with the area of the triangular region BFC, the parallelogram $ABCD$ and the rectangle $EFCD$ enclose the same area. To find the area of region $EFCD$, the length and width are needed. The length is h and the width is b, because $\overline{EF}$ and $\overline{AB}$ have the same length.

area of region $ABCD$ = area of region $EFCD$ = bh

Therefore, the area within a parallelogram with base b and altitude h is given by the formula

$A = bh$

Example: Find the area of a parallelogram with altitude 4.06 inches and base 8.05 inches.

Solution

$A = bh$
$A = (8.05)(4.06)$
$A = 32.6830$

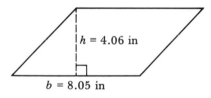

Therefore, the area is 32.6830 in².

Q5 Find the area of the region within a parallelogram with base 12.5 inches and altitude 6.0 inches.

STOP • STOP • STOP • STOP • STOP • STOP • STOP • STOP • STOP

A5 75 in²: $A = bh$
$A = (12.5)(6.0)$
$A = 75.00$

Q6   Find the total area of the space enclosed in the following floor plan:

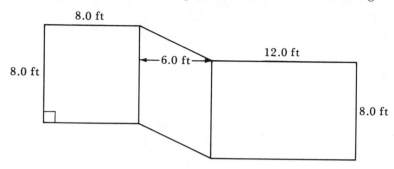

STOP • **STOP** • **STOP** • **STOP** • **STOP** • **STOP** • **STOP** • **STOP** • **STOP**

A6   208 ft²: $A = s^2$ $A = bh$ $A = lw$
$A = 8.0^2$ $A = (8.0)(6.0)$ $A = (12.0)(8.0)$
$A = 64$ $A = 48$ $A = 96$

---

5   The formula for the area of the triangular region ABC with base b and altitude h may be found by drawing another triangle with the same size and shape upside down above it.

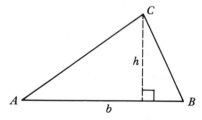   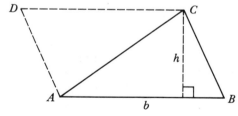

Since the triangular regions ABC and ACD have the same area, the area of the triangular region ABC is equal to one half the area within parallelogram $ABCD$.

area of region $ABC = \dfrac{1}{2}$ area of region $ABCD = \dfrac{1}{2}bh$

The formula for the area of a triangle with base b and altitude h is $A = \frac{1}{2}bh$.

Example: Find the area of a triangular region with base 10.5 m and altitude 5.6 m.

Solution

$A = \dfrac{1}{2}bh$

$A = \dfrac{1}{2}(10.5)(5.6)$

$A = 29.40$

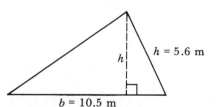

$h = 5.6$ m   Therefore, the area is 29.4 m².

Q7 Find the area of a triangular region with base 4.6 cm and altitude 3.7 cm.

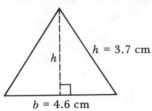

$b = 4.6$ cm

STOP • **STOP** • **STOP** • **STOP** • **STOP** • **STOP** • **STOP** • **STOP** • **STOP**

A7 8.51 cm^2: $A = \frac{1}{2}bh$
$A = \frac{1}{2}(4.6)(3.7)$
$A = 8.51$

Q8 Find the area of a triangular region with base 4.2 ft and altitude 4 yd.

STOP • **STOP** • **STOP** • **STOP** • **STOP** • **STOP** • **STOP** • **STOP** • **STOP**

A8 25.2 ft^2: $A = \frac{1}{2}bh$
$A = \frac{1}{2}(4.2)(12.0)$

6 The formula for the area of a trapezoidal region with altitude h and bases b_1 (read "bee sub-one") and b_2 (read "bee sub-two") can be found by drawing another trapezoid upside down beside it which is equal in size and shape. (In b_1 and b_2, the 1 and 2 are called *subscripts* and are used to distinguish between the two bases.)

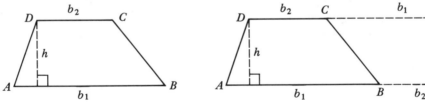

A parallelogram $AEFD$ is formed which includes twice the area of the trapezoidal region $ABCD$. The base of the new parallelogram is $b_1 + b_2$ and its altitude is h. Therefore, the area of the trapezoid region $ABCD$ is $\frac{1}{2}$ the area of the region within parallelogram $AEFD$.

area of region $ABCD = \frac{1}{2}$ area of region $AEFD = \frac{1}{2}h(b_1 + b_2)$

Therefore, the area of a trapezoidal region with altitude h and bases b_1 and b_2 is given by the formula

$$A = \frac{1}{2}h(b_1 + b_2)$$

Example: Find the area of a trapezoidal region with altitude 7.2 cm and bases 5.3 cm and 4.7 cm.

Solution

$$A = \frac{1}{2}h(b_1 + b_2)$$

$$A = \frac{1}{2}(7.2)(5.3 + 4.7)$$

$$A = (3.6)(10.0)$$

$$A = 36.00$$

Therefore, the area is 36 cm^2.

Q9　　**a.** Find the area of the trapezoidal region shown.

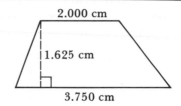

2.000 cm

1.625 cm

3.750 cm

　　b. Find the area of a trapezoidal region if $h = 5$ m, $b_1 = 7.8$ m, and $b_2 = 4.3$ m.

STOP　•　STOP　•　STOP　•　STOP　•　STOP　•　STOP　•　STOP　•　STOP　•　STOP

A9　　**a.** 4.671 875 cm^2:　$A = \frac{1}{2}h(b_1 + b_2)$
$A = \frac{1}{2}(1.625)(2.000 + 3.750)$
$A = 4.671\ 875$

　　b. 30.25 m^2

Q10　　Find the area of a trapezoidal region with altitude 20.0 meters and bases 19.7 and 14.3 meters.

STOP　•　STOP　•　STOP　•　STOP　•　STOP　•　STOP　•　STOP　•　STOP　•　STOP

A10　　340 m^2:　$A = \frac{1}{2}h(b_1 + b_2)$
$A = \frac{1}{2}(20.0)(19.7 + 14.3)$
$A = 340.0$

7　　The area of a circular region is found with a formula that contains the constant π. The values of 3.14 or $3\frac{1}{7}$ may be used for the problems of this section.

　　The area of a circular region with radius r is given by

$$A = \pi r^2$$

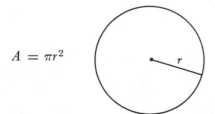

Example: If the radius of a circle is 3.5 cm, find the area of the region within it.

Solution

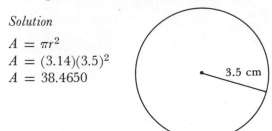

$A = \pi r^2$

$A = (3.14)(3.5)^2$

$A = 38.4650$

Therefore, the area is 38.4650 cm^2.

Q11 Find the area of a circular region with radius 2.5 feet (use $\pi \doteq 3.14$ when you are given a measurement in decimal form).

STOP • STOP • STOP • STOP • STOP • STOP • STOP • STOP • STOP

A11 19.6250 ft^2: $A = \pi r^2$

$A = (3.14)(2.5)(2.5)$

$A = 19.6250$

Q12 Find the area of the following region. (*Hint:* To find the area of a semicircular region, take $\frac{1}{2}$ the area of a circular region of the same radius.) Round off to a whole number of square feet.

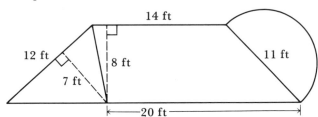

STOP • STOP • STOP • STOP • STOP • STOP • STOP • STOP • STOP

A12 225 ft^2: $A = \frac{1}{2}bh$ $A = \frac{1}{2}h(b_1 + b_2)$ $A = \frac{1}{2}\pi r^2$

$A = \frac{1}{2}(12)(7.0)$ $A = \frac{1}{2}(8.0)(20 + 14)$ $A = \frac{1}{2}(3.14)(5.5)^2$

$A = 42$ $A = 136$ $A = 47.4925$

$A = 47$

total area $= 42 + 136 + 47 = 225$ ft^2

Q13   Measure the following geometric figures and determine their area in in². Measure to the nearest $\frac{1}{8}$ inch.

a.

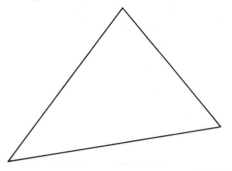

b.

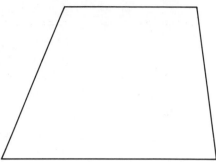

c.

d.

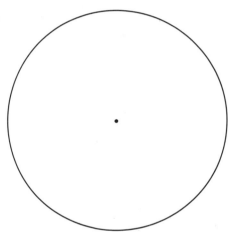

STOP • STOP • STOP • STOP • STOP • STOP • STOP • STOP • STOP

A13   **a.** $1\frac{35}{64}$ in²: $A = \frac{1}{2}bh$
$A = \frac{1}{2}(2\frac{1}{4})(1\frac{3}{8})$
$A = 1\frac{35}{64}$

b. $2\frac{121}{128}$ in²: $A = \frac{1}{2}h(b_1 + b_2)$
$A = \frac{1}{2}(1\frac{5}{8})(2\frac{1}{4} + 1\frac{3}{8})$
$A = 2\frac{121}{128}$

c. $2\frac{1}{2}$ in²: $A = lw$
$A = (2\frac{1}{2})(1)$
$A = 2\frac{1}{2}$

d. $3\frac{219}{224}$ in²: $A = \pi r^2$
$A = 3\frac{1}{7}(1\frac{1}{8})^2$
$A = 3\frac{219}{224}$

Q14   Measure the geometric figures in Q13 with a centimeter ruler to the nearest 0.1 cm and compute the area.

STOP • STOP • STOP • STOP • STOP • STOP • STOP • STOP • STOP

A14   **a.** 9.975 cm^2: $A = \frac{1}{2}bh$ **b.** 18.4 cm^2: $A = \frac{1}{2}h(b_1 + b_2)$
$A = \frac{1}{2}(5.7)(3.5)$ $A = \frac{1}{2}(4.0)(5.7 + 3.5)$
$A = 9.975$ $A = 18.4$

c. 15.75 cm^2: $A = lw$ **d.** 26.4074 cm^2: $A = \pi r^2$
$A = (6.3)(2.5)$ $A = (3.14)(2.9)^2$
$A = 15.75$ $A = 26.4074$

---

8   This section will be concluded with a summary of the formulas that are used to compute area.

Geometric regions	Formula for area
Triangle	$A = \frac{1}{2}bh$
Square	$A = s^2$
Rectangle	$A = lw$
Parallelogram	$A = bh$
Trapezoid	$A = \frac{1}{2}h(b_1 + b_2)$
Circle	$A = \pi r^2$

This completes the instruction for this section.

10.4   Exercises

1. Find the area of a rectangular region with length 25.7 feet and width 9.6 feet.
2. Find the area of a square region with side 14.8 cm.
3. Find the area of the region within a parallelogram with base 120 feet and altitude 82 feet.
4. Find the area of a triangular region with base 79 meters and altitude 84 meters.
5. Find the area of a trapezoidal region with altitude 1.42 cm and bases 0.36 cm and 0.48 cm.
6. Find the area of a circular region with radius 1.8 cm.
7. Find the areas of the following regions:

a.

13 ft

27.7 ft

b.

6.2 in

6.2 in

c.

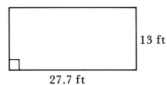

8.6 m

12 m

d.

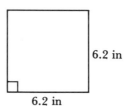

10.6 cm

14.9 cm

e.

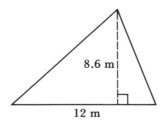

18.6 in

10.2 in

20.1 in

f.

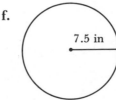

7.5 in

8. Find the area of the region. Round off to the nearest whole number.

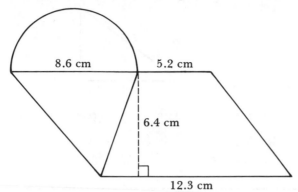

8.6 cm   5.2 cm

6.4 cm

12.3 cm

9. Match the geometric figure with the formula for the area of that figure:

a. Square $\qquad$ (1) $A = bh$
b. Rectangle $\qquad$ (2) $A = \frac{1}{2}bh$
c. Trapezoid $\qquad$ (3) $A = s^2$
d. Circle $\qquad$ (4) $A = \frac{1}{2}h(b_1 + b_2)$
e. Triangle $\qquad$ (5) $A = lw$
f. Parallelogram $\qquad$ (6) $A = \pi r^2$

10.4 Exercise Answers

1. 246.72 ft^2
2. 219.04 cm^2
3. $9{,}840 \text{ ft}^2$
4. 3318 m^2
5. 0.5964 cm^2
6. 10.1736 cm^2
7. **a.** 360.1 ft^2 **b.** 38.44 in^2 **c.** 51.6 m^2 **d.** 157.94 cm^2
   e. 197.37 in^2 **f.** 176.625 in^2
8. 113 cm^2: 29 cm^2 in the semicircular region, 28 cm^2 in the triangular region, and 56 cm^2 in the region within the trapezoid.
9. **a.** 3   **b.** 5   **c.** 4   **d.** 6   **e.** 2   **f.** 1

10.5 The Pythagorean Theorem

1

A relationship among the three sides of any right triangle was known to man far back into history. The relationship was used to lay out land after the floods in Egypt and continues to be useful for many purposes. The relationship is named for the ancient Greek mathematician Pythagoras.

First consider some vocabulary for right triangles.

A right triangle is a triangle with one right ($90°$) angle.

In the right triangle shown here, side c is opposite $\angle C$.
Side b is opposite $\angle B$. Side a is opposite $\angle A$. Side c,
the side opposite the right angle, is the *hypotenuse*.
The other two are called the *legs* of the right triangle.

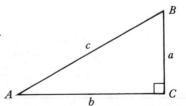

Q1 Given the right triangle shown here, identify:

  a. the legs _____

  b. the hypotenuse _____

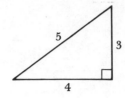

STOP • **STOP** • **STOP** • **STOP** • **STOP** • **STOP** • **STOP** • **STOP** • **STOP**

A1 **a.** 3 and 4 (either order) **b.** 5

2 Consider the right triangle with legs of 3 cm and 4 cm and a hypotenuse of 5 cm. A square is constructed on each of the three sides of the right triangle. If the number of square centimeters in the square regions constructed on the legs are added, they exactly equal the number of square centimeters in the square region on the hypotenuse.

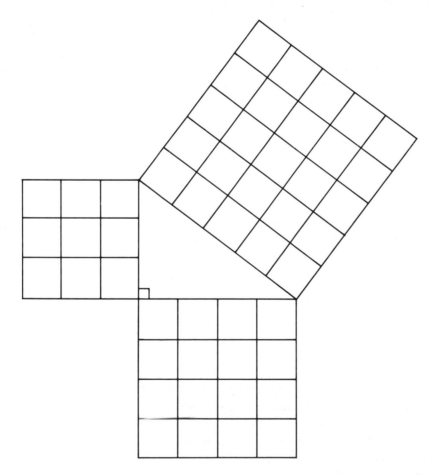

The relationship can be stated mathematically as $3^2 + 4^2 = 5^2$. This triangle is an example of the *Pythagorean theorem*, which could be stated: *The sum of the squares of the legs of a right triangle is equal to the square of the hypotenuse.*

Q2   Given the right triangle shown, verify that the Pythagorean relationship holds.

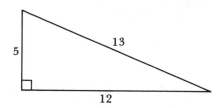

STOP • STOP • STOP • STOP • STOP • STOP • STOP • STOP • STOP

A2 $5^2 + 12^2 = 13^2$
      $25 + 144 = 169$

---

3   The Pythagorean theorem is true for any right triangle. In general, given any right triangle with legs a and b and hypotenuse c,

$a^2 + b^2 = c^2$

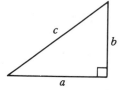

Q3   Given the triangle at the right, complete the following:

$r^2 = $ _____

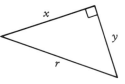

STOP • STOP • STOP • STOP • STOP • STOP • STOP • STOP • STOP

A3 $x^2 + y^2$ or $y^2 + x^2$

---

4   When working with the Pythagorean theorem, an equation such as $x^2 = 25$ often arises. One can examine the meaning of the equation and see that a number that could be squared to get 25 would be either 5 or $^-5$. Five and $^-5$ are called the *square roots* of 25. Notice that there are two square roots.

Q4   Write the square roots of the following:

a. 81 _____ and _____

b. 9 _____ and _____

c. 100 _____ and _____

d. $\frac{1}{4}$ _____ and _____

STOP • STOP • STOP • STOP • STOP • STOP • STOP • STOP • STOP

A4   **a.** 9 and $^-9$ **b.** 3 and $^-3$ **c.** 10 and $^-10$ **d.** $\frac{1}{2}$ and $^-\frac{1}{2}$

---

5   When measuring the length of a side of a triangle only a positive number would be used. The positive square root of 25 is written $\sqrt{25}$.

Q5   Answer the following:

     a. The positive square root of 49 is written _____.

     b. The positive square root of 53 is written _____.

     c. $\sqrt{17}$ means _____.

STOP • STOP • STOP • STOP • STOP • STOP • STOP • STOP • STOP

A5   **a.** $\sqrt{49}$ **b.** $\sqrt{53}$ **c.** the positive square root of 17

---

6   Determine the value of the hypotenuse of the right triangle shown.

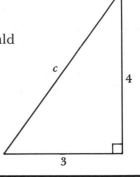

$$a^2 + b^2 = c^2$$
$$9 + 16 = c^2$$
$$25 = c^2$$
$$\sqrt{25} = c \quad \text{(only the positive root is given; negative root would}$$
$$5 = c \quad \text{have no meaning)}$$

Q6   Determine the value of the hypotenuse of the right triangle shown.

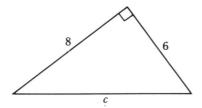

STOP • STOP • STOP • STOP • STOP • STOP • STOP • STOP • STOP

A6   10: $8^2 + 6^2 = c^2$
$$64 + 36 = c^2$$
$$100 = c^2$$
$$\sqrt{100} = c$$
$$10 = c$$

---

7   The symbol $\sqrt{}$ is called a *radical sign.* When the number under the radical sign is so large that the square root is not recognized, a table can be used to find the square root. A table is provided in the Appendix. The numbers in the column with heading "Square" are called *perfect squares.* They are obtained by squaring the natural numbers in the column with heading "Number." The first few perfect squares are 1, 4, 9, 16, 25, ...

Example: $\sqrt{1,764} = ?$

Solution

Find a number in the table which when squared gives 1,764. The number is 42. Therefore, $\sqrt{1,764} = 42$.

Q7   Use the table to find the following positive square roots:

   a. $\sqrt{361} =$ _____   **b.** $\sqrt{1,225} =$ _____   **c.** $\sqrt{3,969} =$ _____   **d.** $\sqrt{121} =$ _____

STOP   •   STOP   •   STOP   •   STOP   •   STOP   •   STOP   •   STOP   •   STOP   •   STOP

A7   **a.** 19   **b.** 35   **c.** 63   **d.** 11

Q8   Determine the value of the hypotenuse of the right triangle shown.

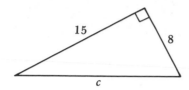

STOP   •   STOP   •   STOP   •   STOP   •   STOP   •   STOP   •   STOP   •   STOP   •   STOP

A8   17: $8^2 + 15^2 = c^2$

$$64 + 225 = c^2$$
$$289 = c^2$$
$$\sqrt{289} = c$$
$$17 = c$$

---

8   The following two examples illustrate the use of the Pythagorean theorem when solving for the value of a leg given the other leg and the hypotenuse.

Example 1: Find a.

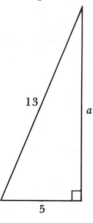

Example 2: Find b.

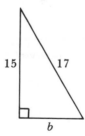

Solution

$$15^2 + b^2 = 17^2$$
$$225 + b^2 = 289$$
$$b^2 = 64$$
$$b = 8$$

Solution

$$a^2 + 5^2 = 13^2$$
$$a^2 + 25 = 169$$
$$a^2 = 144$$
$$a = \sqrt{144} \text{ *}$$
$$a = 12$$

*This step is usually done mentally when the number is a perfect square.

Q9   Use the technique of Frame 8 to determine a in the right triangle shown.

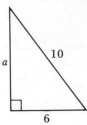

STOP • STOP • STOP • STOP • STOP • STOP • STOP • STOP • STOP

A9   8: $a^2 + 6^2 = 10^2$
              $a^2 + 36 = 100$
                   $a^2 = 64$
                    $a = 8$

---

9   If there is no whole number which, when squared, is equal to the number under the radical sign, the table can still be used to approximate the square root.

Example: $\sqrt{24} = ?$

Solution

Since no whole number squared is 24, look for 24 in the "Number" column and find a decimal approximation of $\sqrt{24}$ in the "Square Root" column. Therefore, $\sqrt{24} \doteq 4.899$.
          The positive square root of 24 is approximately equal to 4.899 rounded off to three decimal places.

Q10   Find the following positive square roots to three decimal places:

          a. $\sqrt{5} \doteq$ _____   **b.** $\sqrt{68} \doteq$ _____   **c.** $\sqrt{64} =$ _____   **d.** $\sqrt{14} \doteq$ _____

STOP • STOP • STOP • STOP • STOP • STOP • STOP • STOP • STOP

A10   **a.** 2.236   **b.** 8.246   **c.** 8.000   **d.** 3.742

Q11   Find c in the figure shown.

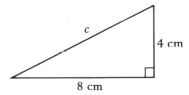

STOP • STOP • STOP • STOP • STOP • STOP • STOP • STOP • STOP

A11   8.944 cm: $8^2 + 4^2 = c^2$
$$64 + 16 = c^2$$
$$80 = c^2$$
$$\sqrt{80} = c$$
$$8.944 \doteq c$$

Q12   Find a in the figure shown.

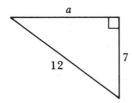

STOP • STOP • STOP • STOP • STOP • STOP • STOP • STOP • STOP

A12   9.747: $a^2 + 7^2 = 12^2$
$$a^2 + 49 = 144$$
$$a^2 = 95$$
$$a = \sqrt{95}$$
$$a \doteq 9.747$$

Q13   Find the area of the triangle shown. (*Hint:* First find the altitude h.)

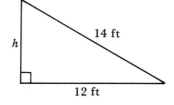

STOP • STOP • STOP • STOP • STOP • STOP • STOP • STOP • STOP

A13   43.266 ft^2: $h^2 + 12^2 = 14^2$ $A = \frac{1}{2}hb$
$$h^2 + 144 = 196 \qquad A = \frac{1}{2}(7.211)(12)$$
$$h^2 = 52 \qquad A = 43.266$$
$$h = \sqrt{52}$$
$$h \doteq 7.211$$

Q14 Find the area of the triangle shown. (*Hint:* First find altitude *h*.)

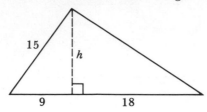

STOP • **STOP** • **STOP** • **STOP** • **STOP** • **STOP** • **STOP** • **STOP** • **STOP**

A14 162: $h^2 + 9^2 = 15^2$ $A = \frac{1}{2}bh$

$h^2 + 81 = 225$ $A = \frac{1}{2}(27)(12)$

$h^2 = 144$ $A = 162$

$h = \sqrt{144}$

$h = 12$

This completes the instruction for this section.

10.5   Exercises

Find the unknown side on each of the following triangles:

1.

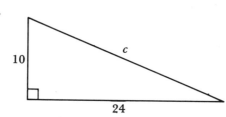

2.

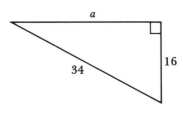

3.

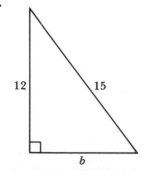

4.

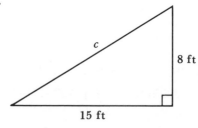

5.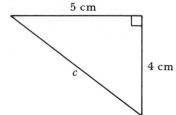

6.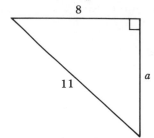

7. Find the area of the triangle:

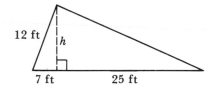

10.5   Exercise Answers

1. 26 **2.** 30 **3.** 9 **4.** 17 ft

5. 6.403 cm **6.** 7.55 **7.** 155.952 ft^2

10.6   Volumes

1

The *volume* of a solid object can be thought of as the amount that the object "holds." It is a measure of the enclosed space within the object. As was discussed in Section 10.2, the units of measurement for volume are cubic units. Examples are: cubic inches, cubic feet, cubic yards, cubic centimeters, and cubic meters.

Volumes are computed by using a formula to combine various linear measurements of the solid. For example, the volume of a rectangular prism is found with the formula

$V = lwh$

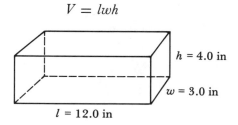

The volume of the rectangular prism is

$V = (12.0)(3.0)(4.0)$
$V = 144.000$

Therefore, the volume is 144 cubic inches.

Q1   Find the volumes of the following prisms:

a.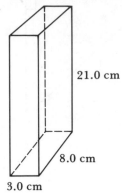

21.0 cm

8.0 cm

3.0 cm

b.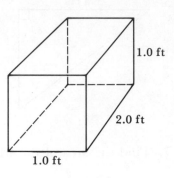

1.0 ft

2.0 ft

1.0 ft

STOP • STOP • STOP • STOP • STOP • STOP • STOP • STOP • STOP

A1   a. 504 cm³: $V = lwh$ b. 2 ft³: $V = lwh$
$V = (3.0)(8.0)(21.0)$ $V = (1.0)(2.0)(1.0)$
$V = 504.000$ $V = 2.000$

2   The formula for the volume of a rectangular prism fits a general form that can also be used for all prisms and cylinders. The volume of each is found by multiplying the area of the base by the altitude (height perpendicular to the base):

$V = Bh$

More specific formulas follow:

Rectangular prism   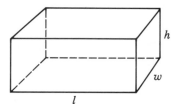   Area of the base
$V = Bh$ $B = lw$
$V = lwh$

Cube   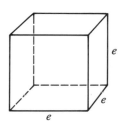   Area of the base
$V = Be$ $B = e^2$
$V = e^2 e$
$V = e^3$

Cylinder   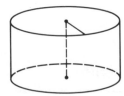   Area of the base
$V = Bh$ $B = \pi r^2$
$V = \pi r^2 h$

Q2   Find the volume of each of the prisms (round answer off to nearest tens):

a.

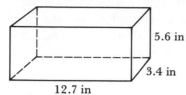

b.

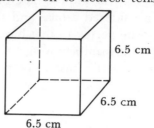

STOP • STOP • STOP • STOP • STOP • STOP • STOP • STOP • STOP

A2   **a.** 240 in³: $V = lwh$
              $V = (12.7)(3.4)(5.6)$
              $V = 241.808$

   b. 270 cm³: $V = e^3$
              $V = 6.5^3$
              $V = 274.625$

Q3   Find the volume of the following cylinders (round off to whole numbers):

a.

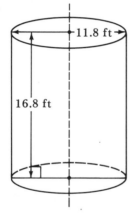

b.

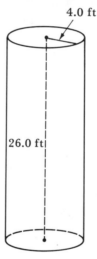

STOP • STOP • STOP • STOP • STOP • STOP • STOP • STOP • STOP

A3   **a.** 1,836 ft³: $V = \pi r^2 h = (3.14)(5.9)(5.9)(16.8)$
                 $V = 1,836.29712$

   b. 1,306 ft³: $V = \pi r^2 h = (3.14)(4.0)(4.0)(26.0)$
                 $V = 1,306.24$

3   In a triangular prism the parallel faces of the prism are called bases. If the prism rests on one of these parallel faces, it is called *the* base. In the figure shown here the base of the prism is the triangular region *ABC*.

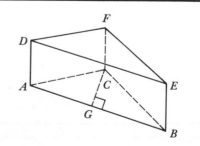

In a triangle the base of the triangle is arbitrary. Any of the sides of the triangle can be considered a base and the altitude to that base is a perpendicular segment from the opposite vertex to that base. The following three triangles have the same area. However, different numbers would be used to calculate that area, depending upon which side was chosen as the base to be placed in the formula $A = \frac{1}{2}bh$.

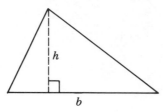 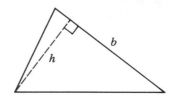 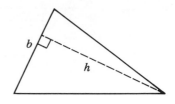

Q4   Use the triangular prism shown here to answer the following:

a. Name the bases of the prism: _____ and _____.

b. Name *the* base of the prism _____.

c. Name the sides of the base triangle that could be used as the base of the triangle. _____

d. Name three segments whose length is the altitude of the prism. _____

e. Name a segment that could be used as an altitude of the base triangle. _____

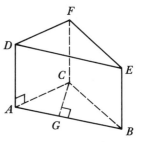

STOP • **STOP** • **STOP** • **STOP** • **STOP** • **STOP** • **STOP** • **STOP** • **STOP**

A4   a. *ABC* and *DEF*   b. *ABC*   c. $\overline{AB}$, $\overline{BC}$, or $\overline{AC}$

d. $\overline{AD}$, $\overline{BE}$, or $\overline{CF}$ e. $\overline{CG}$

---

4   The volume of a triangular prism is found by multiplying the area of the base times the height of the prism. Notice that the area of the base is found by multiplying the base of the triangular base by the altitude of the triangular base.

Triangular prism
$$V = Bh$$

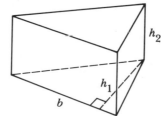

Area of the base
$$B = \frac{1}{2}bh_1$$

So,

$$V = \frac{1}{2}bh_1h_2$$

Q5   Compute the volume of the triangular prisms:
      a. Round off to tenths.   **b.** Round off to tens.

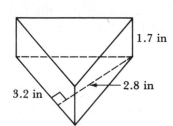

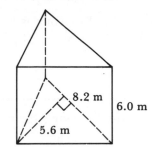

STOP • STOP • STOP • STOP • STOP • STOP • STOP • STOP • STOP

A5   **a.** 7.6 in^3: $V = \frac{1}{2}bh_1h_2$
                       $V = \frac{1}{2}(3.2)(2.8)(1.7)$
                       $V = 7.616$

      b. 140 m^3: $V = \frac{1}{2}bh_1 \cdot h_2$
                       $V = \frac{1}{2}(8.2)(5.6)(6.0)$
                       $V = 137.760$

Q6   Find the volume of the prism shown. (Notice that the prism is not resting on its base.)
      Round off to a whole number.

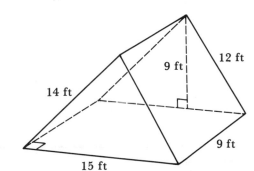

STOP • STOP • STOP • STOP • STOP • STOP • STOP • STOP • STOP

A6 608 ft^3: $V = \frac{1}{2}bh_1h_2$
                   $V = \frac{1}{2}(15)(9)(9)$
                   $V = 607.5$

5   The volume of pyramids and cones are related to the volume of the prism or cylinder
      with the same base and altitude. Consider the following figures:

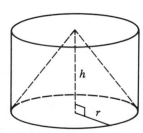

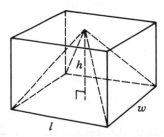

Volume of cylinder: $V = \pi r^2 h$ Volume of prism: $V = lwh$

Volume of cone: $V = \frac{1}{3}\pi r^2 h$ Volume of pyramid: $V = \frac{1}{3}lwh$

The volume of a pyramid or cone is always one-third of the area of the base times the altitude. This applies to all pyramids, no matter what the shape of the base.

$$V = \frac{1}{3}Bh$$

Q7   Find the volume of the following figures (round off to tens):

a.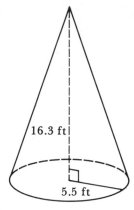

b.

28.6 ft

110 ft   81 ft

16.3 ft

5.5 ft

STOP • **STOP** • **STOP** • **STOP** • **STOP** • **STOP** • **STOP** • **STOP** • **STOP**

A7   **a.** 520 ft³: $V = \frac{1}{3}\pi r^2 h$
$V = \frac{1}{3}(3.14)(5.5)(5.5)(16.3)$
$V = 516.08517$

b. 84,940 ft³: $V = \frac{1}{3}lwh$
$V = \frac{1}{3}(110)(81)(28.6)$
$V = 84,942$

6   The volume of a sphere is found with the formula

$$V = \frac{4}{3}\pi r^3$$

$r = 2.4$ ft

Example: Find the volume of the sphere above with $r = 2.4$ ft rounded off to the nearest whole number.

Solution

$$V = \frac{4}{3}\pi r^3$$

$$V = \frac{4}{3}(3.14)(2.4)(2.4)(2.4)$$

$V = 57.87648$ Therefore the volume is 58 ft³.

Q8   Find the volume of a sphere with radius 8.4 in. Round off to hundreds.

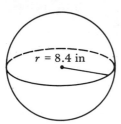

STOP • STOP • STOP • STOP • STOP • STOP • STOP • STOP • STOP

A8 $2,500 \text{ in}^3$: $V = \frac{4}{3}\pi r^3$

$V = \frac{4}{3}(3.14)(8.4)(8.4)(8.4)$

$V = 2,481.4540$

---

7   This section will close with a summary of the formulas for volume.

Geometric figure	Formula for volume
Prism or cylinder	$V = Bh$, where B is the area of the base
Cube	$V = e^3$
Rectangular prism	$V = lwh$
Cylinder	$V = \pi r^2 h$
Triangular prism	$V = \frac{1}{2}bh_1h_2$
Cone	$V = \frac{1}{3}\pi r^2 h$
Pyramid	$V = \frac{1}{3}Bh$, where B is the area of the base
Sphere	$V = \frac{4}{3}\pi r^3$

This completes the instruction for this section.

10.6   Exercises

Find the volumes of the following figures:

1.

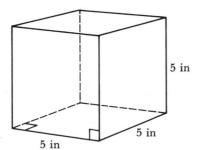

5 in

5 in

5 in

2. Round off to tenths.

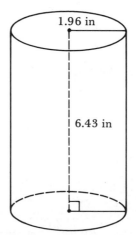

1.96 in

6.43 in

3.

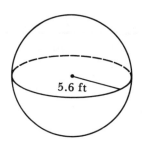

12 cm   7 cm   3 cm

4. Round off to tens.

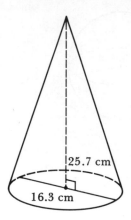

25.7 cm

16.3 cm

5. Round off to tens.

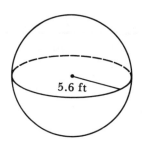

5.6 ft

6. Round off to tens.

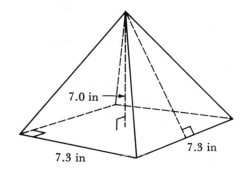

7.0 in

7.3 in   7.3 in

7.

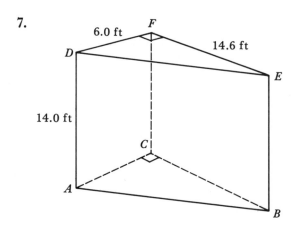

6.0 ft   F   14.6 ft

D   E

14.0 ft

C

A   B

8. Find the total volume of air space in the building. Round off to thousands.

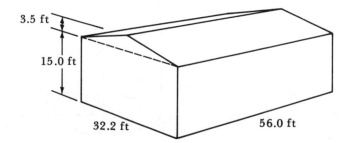

3.5 ft

15.0 ft

32.2 ft   56.0 ft

9. Find the volume of the pentagonal pyramid at right. (Find the area of the rectangular and triangular regions and add them to determine the area of the base.)

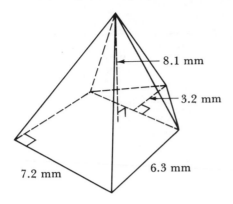

8.1 mm

3.2 mm

7.2 mm

6.3 mm

10.6   Exercise Answers

1. 125 in³   **2.** 77.6 in³   **3.** 252 cm³
4. 1790 cm³   **5.** 740 ft³   **6.** 120 in³
7. 613.2 ft³   **8.** 30,000 ft³   **9.** 153.576 m³

10.7   Solving Word Problems Involving Formulas

1

In previous sections you have worked with many formulas to find missing information. For example, determine the perimeter of a rectangle given the length and width to be 17.0 and 2.1 centimeters, respectively.

The formula, $p = 2l + 2w$, for the perimeter of a rectangle would be used.

$p = 2l + 2w$
$p = 2(17.0) + 2(2.1)$
$p = 34.0 + 4.2$
$p = 38.2$

Therefore, the perimeter of the desired rectangle is 38.2 centimeters.

Q1

Given $A = \frac{1}{2}h(b_1 + b_2)$, determine the area of a trapezoidal region if the parallel sides are 12 inches and $7\frac{1}{2}$ inches and the altitude (height) is $3\frac{3}{4}$ inches. (Do not forget to record the final answer in square inches.)

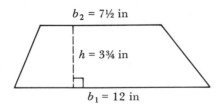

$b_2 = 7\frac{1}{2}$ in

$h = 3\frac{3}{4}$ in

$b_1 = 12$ in

A1 $36\frac{9}{16}$ square inches or 36.5625 square inches:

$A = \frac{1}{2}h(b_1 + b_2)$

$A = \frac{1}{2}(3\frac{3}{4})(12 + 7)$ or $(0.5)(3.75)(12 + 7.5)$

---

2   In the formula $p = 2l + 2w$, p is called the *subject* of the formula. In this formula, the perimeter, p, is expressed in terms of the length, l, and the width, w. The subject of any formula is expressed in terms of the other variables present.

Q2   What is the subject of each of the following formulas?

a. $A = \frac{1}{2}bh$ _____   **b.** $c = 2\pi r$ _____

c. $V = \frac{1}{3}\pi r^2 h$ _____   **d.** $a + b + c = p$ _____

STOP • STOP • STOP • STOP • STOP • STOP • STOP • STOP • STOP

A2   **a.** A **b.** c **c.** V **d.** p

---

3   Often it is necessary to find the value of a variable that is not the subject of the formula. This can be done as long as values for all other variables in the formula are known.

Example: Find the length of a rectangle if the perimeter of the rectangle is 52 meters and the width of the rectangle is 12 meters.

Solution

Step 1: Determine the appropriate formula. The formula for the perimeter of a rectangle is $p = 2l + 2w$, where

p = perimeter
l = length
w = width

Step 2: Identify the given and unknown information from the problem.

p = 52 meters
l = unknown
w = 12 meters

Step 3: Substitute the values of the known information into the formula.

$p = 2l + 2w$
$52 = 2l + 2(12)$

Step 4: Solve the resulting equation.

$p = 2l + 2w$
$52 = 2l + 2(12)$
$52 = 2l + 24$
$28 = 2l$
$14 = l$

Step 5: Check the solution.

$2(14) + 2(12) \overset{?}{=} 52$
$28 + 24 \overset{?}{=} 52$
$52 = 52$

Therefore, the required length of the rectangle is 14 meters.

Q3   Find the height of a rectangular prism if its volume is 315 cubic inches, the length of its base is 25 inches, and the width of its base is 3 inches. The formula for the volume of a rectangular prism is $V = lwh$, where V = volume, l = length of the base, w = width of the base, and h = height of the prism.

a. Identify the given and unknown information.

$V = \underline{\hspace{1.5cm}}$ $l = \underline{\hspace{1.5cm}}$

$w = \underline{\hspace{1.5cm}}$ $h = \underline{\hspace{1.5cm}}$

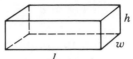

b. Substitute the values of the known information into the formula $V = lwh$.

c. Solve the resulting equation.

d. Check the solution.

e. Therefore, the required height is $\underline{\hspace{1.5cm}}$.

STOP • STOP • STOP • STOP • STOP • STOP • STOP • STOP • STOP

A3   **a.** $V = 315 \text{ in}^3$
$l = 25$ inches
$w = 3$ inches
$h = $ unknown

d. $315 \overset{?}{=} 25(3)(4.2)$
$315 = 315$

b. $V = lwh$
$315 = 25(3)h$

e. 4.2 inches

c. $315 = 75h$
$4.2 = h$

Q4   Determine the required height for a cylindrical tank if the volume is to be 660 cubic feet and the diameter of the tank is to be 7 feet. $\left(Hint:\ V = \pi r^2 h,\ \text{where}\ r = \dfrac{d}{2}.\right.$

Use $\pi \doteq \dfrac{22}{7}.$ $^{*}\Big)$

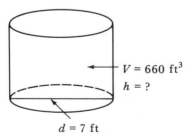

$V = 660 \text{ ft}^3$
$h = ?$

$d = 7 \text{ ft}$

*Recall that π is approximately equal to $\frac{22}{7}$. $\frac{22}{7}$ equals 3.14 only to the nearest hundredth. $\pi = 3.14159\cdots.$

STOP • STOP • STOP • STOP • STOP • STOP • STOP • STOP • STOP

A4 $\qquad$ $17\frac{1}{7}$ feet:
$$V = \pi r^2 h$$
$$660 = \frac{22}{7}(\tfrac{7}{2})(\tfrac{7}{2})h$$
$$660 = \frac{77}{2}h$$
$$\tfrac{2}{77}(660) = \tfrac{2}{77}(\tfrac{77}{2}h)$$
$$17\tfrac{1}{7} = h$$

Therefore, the required height is $17\frac{1}{7}$ feet (17 ft $1\frac{5}{7}$ in).

Q5 $\qquad$ Find the other parallel side of a trapezoid if the area is 24 square meters, the altitude is 3 meters, and one of the parallel sides is 2 meters. Use $A = \frac{1}{2}h(b_1 + b_2)$.

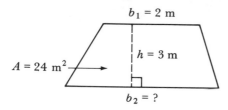

$b_1 = 2$ m

$h = 3$ m

$A = 24$ m²

$b_2 = ?$

STOP • STOP • STOP • STOP • STOP • STOP • STOP • STOP • STOP

A5 $\qquad$ 14 meters:
$$A = \tfrac{1}{2}h(b_1 + b_2)$$
$$24 = \tfrac{1}{2}(3)(2 + b_2)$$
$$24 = \tfrac{3}{2}(2 + b_2)$$
$$\tfrac{2}{3}(24) = \tfrac{2}{3}\cdot\tfrac{3}{2}(2 + b_2)$$
$$16 = 2 + b_2$$
$$14 = b_2$$

---

4 $\qquad$ Whenever a formula can be used to determine missing information, the technique of substituting the known values and solving the resulting equation is often convenient.

This completes the instruction for this section.

10.7 Exercises

1. Given $p = a + b + c$, find a when $p = 32$, $b = 9$, and $c = 5$.
2. Given $p = 2l + 2w$, find w when $p = 29$ and $l = 4.5$.
3. Given $A = \tfrac{1}{2}bh$, find b when $A = 54$ and $h = 9$.
4. Given $c = 2\pi r$, find r when $c = 37.68$. Use $\pi \doteq 3.14$.
5. Given $V = \tfrac{1}{3}Bh$, find B when $V = 2,500$ and $h = 75$.
6. Given $p = 2l + 2w$, determine the length of a rectangle if the perimeter is 35 centimeters and the width is 5.5 centimeters.
7. Given $A = bh$, determine the base of a parallelogram whose area is 51.8 yd² and whose height (altitude) is 14 yards.

8. Given $V = \pi r^2 h$, determine the height of a cylinder whose volume is 1,884 cm³ when the radius of the cylinder is 10 centimeters. Use $\pi \doteq 3.14$.

9. Given $A = \frac{1}{2}h(b_1 + b_2)$, find the height of a trapezoid if the area of the trapezoidal region is 168 m² when the two parallel bases are 15 meters and 17 meters.

10. Given $A = \pi r^2$, find the radius necessary to form a circle whose area is 154 in². Use $\pi \doteq \frac{22}{7}$.

*11. Find the length of a rectangle if the rectangle's diagonal is 13 feet and its width is 5 feet. (*Hint:* Use the Pythagorean theorem: $c^2 = a^2 + b^2$, where a = width of the rectangle, b = length of the rectangle, and c = diagonal of the rectangle.)

10.7   Exercise Answers

1. 18	**2.** 10	**3.** 12	**4.** 6
5. 100	**6.** 12 centimeters	**7.** 3.7 yards	**8.** 6 centimeters
9. 10.5 meters	**10.** 7 inches	***11.** 12 feet	

Chapter 10 Sample Test

At the completion of Chapter 10 it is expected that you will be able to complete the following problems.

10.1   Measurement of Line Segments

1. Which of the following measurements are approximate rather than exact?
   a. height of the Empire State Building
   b. number of brothers and sisters of George Washington
   c. capacity of a soda bottle

2. Indicate the precision of each measurement:

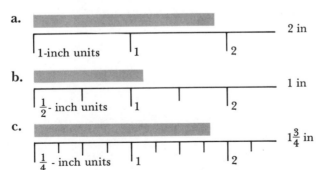

3. Measure each segment to the precision indicated with an inch or centimeter ruler:
   a. $\frac{1}{4}$ in   _____
   b. 1 in   _____
   c. $\frac{1}{8}$ in   _____
   d. $\frac{1}{10}$ cm   _____
   e. 1 cm   _____
   f. 1 mm   _____

10.2 Introduction to Measurement of Distance, Area, and Volume

4. Label each as being a distance, an area, or a volume:
   a. amount of wallpaper needed in a room
   b. length of a brick
   c. amount of water in a swimming pool
   d. capacity of a human lung
5. Label each unit as a unit of distance, area, or volume:
   a. cubic inch
   b. square centimeter
   c. decimeter
   d. kilometer
   e. square yard
6. Choose the most likely measurement for each object:
   a. diameter of a 60-watt light bulb:
      (1) $2\frac{1}{2}$ in   (2) $2\frac{1}{2}$ cm   (3) $2\frac{1}{2}$ m
   b. area of the surface of a $1 bill:
      (1) $15\,cm^2$ (2) $15\,yd^2$ (3) $15\,in^2$

10.3 Perimeters

(Formulas for perimeters may be found immediately preceding the sample test answers.)

7. Find the perimeter of each of the following figures:
   a.

   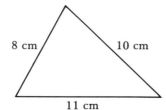

   8 cm   10 cm

   11 cm

   b.

   5.9 in

   12.3 in

   c.

   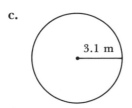

   3.1 m

8. Measure the following figures and compute the perimeter:

    a. Nearest $\frac{1}{8}$ in

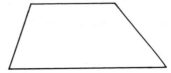

    b. Nearest 0.1 cm

10.4   Areas

(A summary of formulas is included with this test, immediately preceding the sample test answers.)

9. Find the area of each of the following figures:

    a.

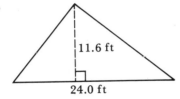

11.6 ft

24.0 ft

    b.

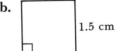

1.5 cm

    c.

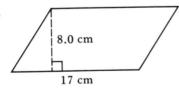

8.0 cm

17 cm

10. Measure the following figures and compute the area:

    a. Nearest $\frac{1}{8}$ in

    b. Nearest 0.1 cm

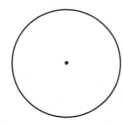

10.5 The Pythagorean Theorem

(Use the square-root table in the Appendix.)

11. Find the missing side on each of the following triangles:

a.

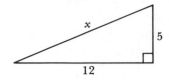

b.

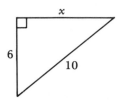

c.

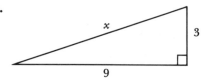

10.6 Volumes

(A summary of formulas is included with this test, immediately preceding the sample test answers.)

12. Find the volumes of each of the following figures:

a.

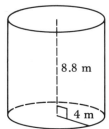

b.

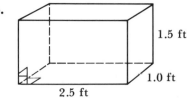

c.

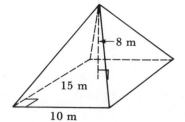

10.7   Solving Word Problems Involving Formulas

13. Solve these problems by substituting the given information into the appropriate formula and solving the resulting equation for the desired value:

   a. Given $p = 2l + 2w$, find the width of a rectangle if the perimeter is 41 centimeters and the length is 13 centimeters.

   b. Given $A = \frac{1}{2}bh$, find the length of the base of a triangle whose area is 90 in^2 and whose altitude (height) is 12 inches.

   c. Given $c = 2\pi r$, find the radius of a circle with a circumference of 47.1 feet. Use $\pi \doteq 3.14$.

   d. Given $V = \frac{1}{3}\pi r^2 h$, find the radius of a cone 4.2 millimeters high which has a volume of 215.6 mm^3. Use $\pi \doteq \frac{22}{7}$.

   e. Given $T = 2\pi r^2 + 2\pi rh$, where

   $\quad\quad T = $ total surface of a cylinder

   $\quad\quad r = $ radius of the circular base

   $\quad\quad h = $ height (altitude) of the cylinder

   determine the height of the cylinder if its total surface is 748 m^2 and the radius of the base is 7 meters. Use $\pi \doteq \frac{22}{7}$.

Summary of Formulas

Geometric figure	Perimeter	Area
Triangle	$p = a + b + c$	$A = \frac{1}{2}bh$
Quadrilateral	$p = a + b + c + d$	—
Rectangle	$p = 2l + 2w$	$A = lw$
Square	$p = 4s$	$A = s^2$
Circle	$c = 2\pi r = \pi d$	$A = \pi r^2$
Parallelogram	$p = 2a + 2b$	$A = bh$
Trapezoid	—	$A = \frac{1}{2}h(b_1 + b_2)$

Geometric figure	Volume
Prism or cylinder	$V = Bh$
Cube	$V = e^3$
Rectangular prism	$V = lwh$
Triangular prism	$V = \frac{1}{2}bh_1 h_2$
Cylinder	$V = \pi r^2 h$
Cone	$V = \frac{1}{3}\pi r^2 h$
Pyramid	$V = \frac{1}{3}Bh$
Sphere	$V = \frac{4}{3}\pi r^3$

Chapter 10 Sample Test Answers

1. a and c
2. **a.** 1 in   **b.** $\frac{1}{2}$ in   **c.** $\frac{1}{4}$ in
3. **a.** 2 in   **b.** 1 in   **c.** $1\frac{3}{8}$ in   **d.** 2.7 cm
   e. 2 cm   **f.** 44 mm
4. **a.** area   **b.** distance   **c.** volume   **d.** volume
5. **a.** volume   **b.** area   **c.** distance   **d.** distance
   e. area

6. **a.** 1   **b.** 3
7. **a.** 29 cm   **b.** 36.4 in   **c.** 19.468 m
8. **a.** $4\frac{1}{8}$ in   **b.** 10.2 cm
9. **a.** 139.2 ft^2 **b.** 2.25 cm^2 **c.** 136 cm^2
10. **a.** $\frac{49}{64}$ in^2 **b.** 6.1544 cm^2
11. **a.** 13   **b.** 8   **c.** 9.487
12. **a.** 442.112 m^3 **b.** 3.75 ft^3 **c.** 400 m^3
13. **a.** $7\frac{1}{2}$ or 7.5 cm **b.** 15 in   **c.** 7.5 ft   **d.** 7 mm
    e. 10 m

Chapter 11

Comparison of Metric and English Systems of Measurement

11.1   English System of Measurement

---

1   There are four basic types of measure that are important for an understanding of the English system of measurement: length (linear units), area, weight, and volume (capacity). Some of the more common linear units are the inch, foot, yard, and mile. The relationships between these units are summarized as follows:

12 inches (in) = 1 foot (ft)
3 feet (ft) = 1 yard (yd)
36 in = 1 yd
1,760 yd = 1 mile (mi)
5,280 ft = 1 mi

Q1   Complete the following equations from memory, wherever possible. Use the relationships of Frame 1 to complete the remaining equations:

a. _____ in = 1 ft   b. 3 ft = _____ yd

c. 1 yd = _____ ft   d. _____ ft = 1 mi

e. 1 mi = _____ yd

STOP • STOP • STOP • STOP • STOP • STOP • STOP • STOP • STOP

A1   a. 12   b. 1   c. 3   d. 5,280   e. 1,760

---

2   To convert one unit to another unit, it is important to notice that each of the equations of Frame 1 can be used to derive a second equation. For example, consider the result of dividing both sides of the equation 12 in = 1 ft by 12.

12 in = 1 ft

$$\frac{12 \text{ in}}{12} = \frac{1 \text{ ft}}{12}$$

$$1 \text{ in} = \frac{1}{12} \text{ ft}$$

The derived equation 1 in = $\frac{1}{12}$ ft still expresses a relationship between inches and feet, but in a different way.

Q2   Divide both sides of the equation 3 ft = 1 yd by 3 to derive a second relationship between feet and yards.

STOP • STOP • STOP • STOP • STOP • STOP • STOP • STOP • STOP

A2   1 ft = $\frac{1}{3}$ yd: 3 ft = 1 yd
                         $\frac{3}{3}$ ft = $\frac{1}{3}$ yd
                         1 ft = $\frac{1}{3}$ yd

---

3   The equation 36 in = 1 yd expresses the equivalent of 1 *yard* in inches. To find the equivalent of 1 *inch* in yards, divide by 36 as follows:

$$36 \text{ in} = 1 \text{ yd}$$

$$\frac{36 \text{ in}}{36} = \frac{1}{36} \text{ yd}$$

$$1 \text{ in} = \frac{1}{36} \text{ yd}$$

Q3   Consider the equation 1,760 yd = 1 mi:

a. To complete the equation 1 yd = _____ mi it is necessary to divide both sides by

_____ .

b. Complete the equation 1 yd = _____ mi by performing the division.

STOP • STOP • STOP • STOP • STOP • STOP • STOP • STOP • STOP

A3   a. 1,760   b. 1 yd = $\frac{1}{1,760}$ mi

Q4   Use the equation 5,280 ft = 1 mi to complete the equation 1 ft = _____ mi.

STOP • STOP • STOP • STOP • STOP • STOP • STOP • STOP • STOP

A4   1 ft = $\frac{1}{5,280}$ mi

---

4   A summary of the linear unit relationships is as follows:

12 in = 1 ft   3 ft = 1 yd   36 in = 1 yd

1 in = $\frac{1}{12}$ ft   1 ft = $\frac{1}{3}$ yd   1 in = $\frac{1}{36}$ yd

1,760 yd = 1 mi   5,280 ft = 1 mi

1 yd = $\frac{1}{1,760}$ mi   1 ft = $\frac{1}{5,280}$ mi

Using these relationships it is possible to perform conversion between the various units. For example, to complete the equation 5 ft = _____ in, it is only necessary to choose the equation that expresses the equivalent of 1 ft in inches. That is, 1 ft = 12 in. Multiplying both sides of 1 ft = 12 in by 5 will give the desired equivalent of 5 ft in inches.

5 ft = _____ in

1 ft = 12 in   (multiply by 5)

5(1 ft) = 5(12 in)

5 ft = 60 in

Thus, the answer is 60.

Q5   Complete the equation 3.2 ft = _____ in.

STOP   •   STOP   •   STOP   •   STOP   •   STOP   •   STOP   •   STOP   •   STOP   •   STOP

A5   38.4:   3.2 ft = _____ in
1 ft = 12 in
3.2(1 ft) = 3.2(12 in)
3.2 ft = 38.4 in

Q6   Complete the equation $2\frac{3}{4}$ ft = _____ in.

STOP   •   STOP   •   STOP   •   STOP   •   STOP   •   STOP   •   STOP   •   STOP   •   STOP

A6   33: $2\frac{3}{4}$ ft = _____ in
1 ft = 12 in
$2\frac{3}{4}$(1 ft) = $2\frac{3}{4}$(12 in)
$2\frac{3}{4}$ ft = $\frac{11}{4}(\frac{12}{1})$ in
$2\frac{3}{4}$ ft = 33 in

5   To complete the equation 27 in = _____ yd, choose the equation that expresses 1 in in terms of yards and multiply both sides by 27.

27 in = _____ yd

$1 \text{ in} = \frac{1}{36} \text{ yd}$

$27(1 \text{ in}) = 27\left(\frac{1}{36} \text{ yd}\right)$

$27 \text{ in} = \frac{3}{4} \text{ yd}$

Thus, the answer is $\frac{3}{4}$.

Q7   Complete the equation 32 in = _____ yd.

STOP • STOP • STOP • STOP • STOP • STOP • STOP • STOP • STOP

A7 $\frac{8}{9}$:   32 in = _____ yd
                1 in = $\frac{1}{36}$ yd
            32(1 in) = 32($\frac{1}{36}$ yd)
                32 in = $\frac{8}{9}$ yd

Q8   a. Which of the equations from Frame 4 would you choose to complete the equation

            39 ft = _____ yd? _____

          b. Use the equation of part a to complete the equation 39 ft = _____ yd.

STOP • STOP • STOP • STOP • STOP • STOP • STOP • STOP • STOP

A8   a. 1 ft = $\frac{1}{3}$ yd   b. 13:   39 ft = _____ yd
                                                1 ft = $\frac{1}{3}$ yd
                                            39(1 ft) = 39($\frac{1}{3}$ yd)
                                                39 ft = 13 yd

Q9   a. Which of the relationships in Frame 4 is appropriate to complete the equation

            660 yd = _____ mi? _____

          b. Complete the equation 660 yd = _____ mi.

STOP • STOP • STOP • STOP • STOP • STOP • STOP • STOP • STOP

A9   a. 1 yd = $\frac{1}{1,760}$ mi   b. $\frac{3}{8}$:   660 yd = _____ mi
                                                    1 yd = $\frac{1}{1,760}$ mi
                                                660(1 yd) = 660($\frac{1}{1,760}$ mi)
                                                    660 yd = $\frac{660}{1,760}$ mi
                                                    660 yd = $\frac{3}{8}$ mi

---

6   To complete the equation 4.3 mi = _____ ft, proceed as follows:

            4.3 mi = _____ ft
            1 mi = 5,280 ft
        4.3(1 mi) = 4.3(5,280 ft)
            4.3 mi = 22,704 ft

        Therefore, the answer is 22,704.

Q10   Complete the equation 1,320 yd = _____ mi.

STOP • **STOP** • **STOP** • **STOP** • **STOP** • **STOP** • **STOP** • **STOP** • **STOP**

A10 $\frac{3}{4}$:   1,320 yd = _____ mi
                      1 yd = $\frac{1}{1,760}$ mi
              1,320(1 yd) = 1,320($\frac{1}{1,760}$ mi)
                  1,320 yd = $\frac{3}{4}$ mi

---

7   Common units of measure for area in the English system are square inch, square foot, square yard, square mile, and acre. Some of the relationships between these units are summarized below:

$$9 \text{ ft}^2 = 1 \text{ yd}^2$$
$$144 \text{ in}^2 = 1 \text{ ft}^2$$
$$1,296 \text{ in}^2 = 1 \text{ yd}^2$$
$$4,840 \text{ yd}^2 = 1 \text{ acre}$$
$$640 \text{ acres} = 1 \text{ mi}^2$$

Q11   Complete the equation 24 yd^2 = _____ ft^2.

STOP • **STOP** • **STOP** • **STOP** • **STOP** • **STOP** • **STOP** • **STOP** • **STOP**

A11   216:   24 yd^2 = _____ ft^2
                     1 yd^2 = 9 ft^2
              24(1 yd^2) = 24(9 ft^2)
                  24 yd^2 = 216 ft^2

Q12   Complete the equation 5.2 mi^2 = _____ acres.

STOP • **STOP** • **STOP** • **STOP** • **STOP** • **STOP** • **STOP** • **STOP** • **STOP**

A12   3,328:   5.2 mi^2 = _____ acres
                        1 mi^2 = 640 acres
               5.2(1 mi^2) = 5.2(640 acres)
                   5.2 mi^2 = 3,328 acres

---

8   To complete the equation 324 in^2 = _____ ft^2, it is necessary to find the equivalent of 1 in^2 in terms of ft^2. This may be done by dividing both sides of the equation 144 in^2 = 1 ft^2 by 144:

$$\frac{144 \text{ in}^2}{144} = \frac{1 \text{ ft}^2}{144}$$

$$1 \text{ in}^2 = \frac{1}{144} \text{ ft}^2$$

The equation can now be completed as follows:

$$324 \text{ in}^2 = \underline{\hspace{2cm}} \text{ ft}^2$$

$$1 \text{ in}^2 = \frac{1}{144} \text{ ft}^2$$

$$324(1 \text{ in}^2) = 324\left(\frac{1}{144} \text{ ft}^2\right)$$

$$324 \text{ in}^2 = 2\frac{1}{4} \text{ (or 2.25)} \text{ft}^2$$

Q13   Complete the equation $324 \text{ in}^2 = \underline{\hspace{2cm}} \text{ yd}^2$.

STOP • **STOP** • **STOP** • **STOP** • **STOP** • **STOP** • **STOP** • **STOP** • **STOP**

A13 $\frac{1}{4}$ or 0.25: $324 \text{ in}^2 = \underline{\hspace{2cm}} \text{ yd}^2$
$$1 \text{ in}^2 = \frac{1}{1,296} \text{ yd}^2$$
$$324(1 \text{ in}^2) = 324(\frac{1}{1,296} \text{ yd}^2)$$
$$324 \text{ in}^2 = \frac{1}{4} \text{ yd}^2$$

Q14   Complete each of the following equations:

a. $144 \text{ ft}^2 = \underline{\hspace{2cm}} \text{ yd}^2$ b. $2.7 \text{ mi}^2 = \underline{\hspace{2cm}}$ acres

c. $\underline{\hspace{2cm}} \text{ mi}^2 = 128$ acres   d. 1.5 acres $= \underline{\hspace{2cm}} \text{ yd}^2$

e. $5\frac{3}{8} \text{ yd}^2 = \underline{\hspace{2cm}} \text{ in}^2$ f. $1,815 \text{ yd}^2 = \underline{\hspace{2cm}}$ acres

STOP • **STOP** • **STOP** • **STOP** • **STOP** • **STOP** • **STOP** • **STOP** • **STOP**

A14   **a.** 16   **b.** 1,728   **c.** $\frac{1}{5}$ **d.** 7,260   **e.** 6,966   **f.** $\frac{3}{8}$

9	Common measures of weight in the English system of measurement are the ounce, pound, and ton. The relationships between these units is given by the following equations:

16 ounces (oz) = 1 pound (lb)
2,000 lb = 1 ton

Q15   Complete the equation 3.5 lb = _____ oz.

STOP • **STOP** • **STOP** • **STOP** • **STOP** • **STOP** • **STOP** • **STOP** • **STOP**

A15   56:   3.5 lb = _____ oz
1 lb = 16 oz
3.5(1 lb) = 3.5(16 oz)
3.5 lb = 56 oz

Q16   Complete the equation $1\frac{3}{8}$ tons = _____ lb.

STOP • **STOP** • **STOP** • **STOP** • **STOP** • **STOP** • **STOP** • **STOP** • **STOP**

A16   2,750: $1\frac{3}{8}$ tons = _____ lb
1 ton = 2,000 lb
$1\frac{3}{8}$(1 ton) = $1\frac{3}{8}$(2,000 lb)
$1\frac{3}{8}$ tons = $\frac{11}{8}$(2,000 lb)
$1\frac{3}{8}$ tons = 2,750 lb

10	To complete the equation 850 lb = _____ ton, it is necessary to use the relationship that expresses 1 pound in terms of tons. This can be derived from the equation 2,000 lb = 1 ton by dividing both sides by 2,000:

2,000 lb = 1 ton

$$\frac{2,000 \text{ lb}}{2,000} = \frac{1 \text{ ton}}{2,000}$$

$$1 \text{ lb} = \frac{1}{2,000} \text{ ton}$$

The equation 850 lb = _____ ton can now be completed as follows:

$$850 \text{ lb} = \underline{\hspace{1cm}} \text{ ton}$$

$$1 \text{ lb} = \frac{1}{2,000} \text{ ton}$$

$$850(1 \text{ lb}) = 850\left(\frac{1}{2,000} \text{ ton}\right)$$

$$850 \text{ lb} = \frac{17}{40} \text{ ton}$$

Thus, the answer is $\frac{17}{40}$.

Q17 Complete the equation 1,750 lb = _____ ton.

STOP • STOP • STOP • STOP • STOP • STOP • STOP • STOP • STOP

A17 $\frac{7}{8}$: 1,750 lb = _____ ton
$$1 \text{ lb} = \frac{1}{2,000} \text{ ton}$$
$$1,750(1 \text{ lb}) = 1,750(\frac{1}{2,000} \text{ ton})$$
$$1,750 \text{ lb} = \frac{7}{8} \text{ ton}$$

Q18 **a.** Use the equation 16 oz = 1 lb to derive an equation that expresses 1 oz in terms of pounds.

b. Use the result of part a to complete the equation 86 oz = _____ lb.

STOP • STOP • STOP • STOP • STOP • STOP • STOP • STOP • STOP

A18 **a.** $1 \text{ oz} = \frac{1}{16} \text{ lb}$ **b.** $5\frac{3}{8}$: 86 oz = _____ lb
$$1 \text{ oz} = \frac{1}{16} \text{ lb}$$
$$86(1 \text{ oz}) = 86(\frac{1}{16} \text{ lb})$$
$$86 \text{ oz} = 5\frac{3}{8} \text{ lb}$$

Q19 Complete each of the following equations:

a. 24 oz = _____ lb   **b.** 5.2 tons = _____ lb

c. 4,800 lb = _____ tons   **d.** 2.2 lb = _____ oz

STOP • STOP • STOP • STOP • STOP • STOP • STOP • STOP • STOP

A19   **a.** $1\frac{1}{2}$ **b.** 10,400 **c.** $2\frac{2}{5}$ **d.** 35.2

---

11   To complete the equation _____ ton = 1,500 lb, notice that the equation can also be
written 1,500 lb = _____ ton. The procedure used to complete the equation is now the
same as before:

$$1{,}500 \text{ lb} = \underline{\hspace{2cm}} \text{ ton}$$

$$1 \text{ lb} = \frac{1}{2{,}000} \text{ ton}$$

$$1{,}500(1 \text{ lb}) = 1{,}500\left(\frac{1}{2{,}000} \text{ ton}\right)$$

$$1{,}500 \text{ lb} = \frac{3}{4} \text{ ton}$$

Q20   Complete the equation _____ lb = 3.4 tons.

STOP • STOP • STOP • STOP • STOP • STOP • STOP • STOP • STOP

A20   6,800: _____ lb = 3.4 tons
                       3.4 tons = _____ lb
                       1 ton = 2,000 lb
                       3.4(1 ton) = 3.4(2,000 lb)
                       3.4 tons = 6,800 lb

Q21   Complete the equation _____ yd = $14\frac{3}{5}$ ft.

STOP • STOP • STOP • STOP • STOP • STOP • STOP • STOP • STOP

A21　　　　$4\frac{13}{15}$:　　_____ yd $= 14\frac{3}{5}$ ft

$14\frac{3}{5}$ ft $=$ _____ yd

1 ft $= \frac{1}{3}$ yd

$14\frac{3}{5}(1 \text{ ft}) = \frac{73}{5}(\frac{1}{3}\text{ yd})$

$14\frac{3}{5}$ ft $= \frac{73}{15}$ yd

$14\frac{3}{5}$ ft $= 4\frac{13}{15}$ yd

12　　　　The volume or capacity of an object can be thought of as the amount that the object "holds." Some of the more common measures of liquid volume in the English system of measurement are the ounce, pint, quart, and gallon. The relationships between these units are summarized as follows:

16 ounces (oz) $=$ 1 pint (pt)

32 oz $=$ 1 quart (qt)

2 pt $=$ 1 qt

4 qt $=$ 1 gallon (gal)

8 pt $=$ 1 gal

128 oz $=$ 1 gal

Q22　　　　Complete the equation 15 gal $=$ _____ qt.

STOP • STOP • STOP • STOP • STOP • STOP • STOP • STOP • STOP

A22　　　　60:　　15 gal $=$ _____ qt

1 gal $=$ 4 qt

15(1 gal) $=$ 15(4 qt)

15 gal $=$ 60 qt

Q23　　　　Complete the equation $3\frac{1}{4}$ qt $=$ _____ oz.

STOP • STOP • STOP • STOP • STOP • STOP • STOP • STOP • STOP

A23　　　　104:　　$3\frac{1}{4}$ qt $=$ _____ oz

1 qt $=$ 32 oz

$3\frac{1}{4}(1 \text{ qt}) = 3\frac{1}{4}(32)$ oz

$3\frac{1}{4}$ qt $= \frac{13}{4}(32)$ oz

$3\frac{1}{4}$ qt $=$ 104 oz

13　　　　To complete the equation 12 oz $=$ _____ qt, it is necessary to use the relationship that expresses 1 ounce in terms of quarts. This can be derived from the equation 32 oz $=$ 1 qt by dividing both sides by 32:

$$32 \text{ oz} = 1 \text{ qt}$$

$$\frac{32}{32} \text{ oz} = \frac{1}{32} \text{ qt}$$

$$1 \text{ oz} = \frac{1}{32} \text{ qt}$$

This equation can now be used to complete 12 oz = _____ qt as follows:

$$12 \text{ oz} = \underline{\hspace{1cm}} \text{ qt}$$

$$1 \text{ oz} = \frac{1}{32} \text{ qt}$$

$$12(1 \text{ oz}) = 12\left(\frac{1}{32} \text{ qt}\right)$$

$$12 \text{ oz} = \frac{3}{8} \text{ qt}$$

Q24 **a.** Use the equations of Frame 12 to derive the equation needed to complete 104 oz = _____ qt.

b. Complete the equation 104 oz = _____ qt.

STOP • **STOP** • **STOP** • **STOP** • **STOP** • **STOP** • **STOP** • **STOP** • **STOP**

A24 **a.** $1 \text{ oz} = \frac{1}{32} \text{ qt}$: $32 \text{ oz} = 1 \text{ qt}$
$$\frac{32}{32} \text{ oz} = \frac{1}{32} \text{ qt}$$
$$1 \text{ oz} = \frac{1}{32} \text{ qt}$$

b. $3\frac{1}{4}$: $104 \text{ oz} = 1 \text{ qt}$
$$1 \text{ oz} = \frac{1}{32} \text{ qt}$$
$$104(1 \text{ oz}) = 104(\frac{1}{32} \text{ qt})$$
$$104 \text{ oz} = 3\frac{1}{4} \text{ qt}$$

Q25 Use the relationships of Frame 12 to derive each of the following equations:

a. 1 pt = _____ gal **b.** 1 oz = _____ qt

    c. 1 qt = _____ gal   **d.** 1 oz = _____ gal

    e. 1 oz = _____ pt   **f.** 1 pt = _____ qt

STOP • **STOP** • **STOP** • **STOP** • **STOP** • **STOP** • **STOP** • **STOP** • **STOP**

A25   **a.** $\frac{1}{8}$ **b.** $\frac{1}{32}$ **c.** $\frac{1}{4}$ **d.** $\frac{1}{128}$ **e.** $\frac{1}{16}$ **f.** $\frac{1}{2}$

Q26   Complete the following equations:

    a. 12 oz = _____ pt   **b.** _____ gal = 80 oz

    c. 4 qt = _____ gal   **d.** _____ oz = 7.2 gal

STOP • **STOP** • **STOP** • **STOP** • **STOP** • **STOP** • **STOP** • **STOP** • **STOP**

A26   **a.** $\frac{3}{4}$ **b.** $\frac{5}{8}$ **c.** 1   **d.** 921.6

14	Volume in the English system may also be measured in cubic units such as cubic inches, cubic feet, and cubic yards. The relationships among these units are:

$$1 \text{ ft}^3 = 1{,}728 \text{ in}^3$$
$$1 \text{ yd}^3 = 27 \text{ ft}^3$$

Q27 $4.5 \text{ ft}^3 = $ _____ in^3

STOP • **STOP** • **STOP** • **STOP** • **STOP** • **STOP** • **STOP** • **STOP** • **STOP**

A27   7,776: $4.5 \text{ ft}^3 = $ _____ in^3
$$1 \text{ ft}^3 = 1{,}728 \text{ in}^3$$
$$4.5(1 \text{ ft}^3) = 4.5(1{,}728 \text{ in}^3)$$
$$4.5 \text{ ft}^3 = 7{,}776 \text{ in}^3$$

Q28   15 yd³ = _____ ft³

STOP • STOP • STOP • STOP • STOP • STOP • STOP • STOP • STOP

A28   405:   15 yd³ = _____ ft³
                     1 yd³ = 27 ft³
              15(1 yd³) = 15(27 ft³)
                  15 yd³ = 405 ft³

This completes the instruction for this section.

11.1   Exercises

1. Complete each of the following equations:
   a. 12 in = _____ ft   b. _____ yd = 3 ft
   c. _____ yd = 1 mi   d. _____ lb = 1 ton
   e. _____ pt = 1 qt   f. 1 qt = _____ oz
   g. _____ mi² = 640 acres   h. _____ ft² = 1 yd²
2. Complete each of the following equations:
   a. 660 ft = _____ mi   b. _____ in = 5.7 yd
   c. _____ qt = 48 oz   d. $5\frac{1}{2}$ gal = _____ pt
   e. _____ oz = $4\frac{1}{8}$ lb f. 120 acres = _____ mi²
   g. 176 oz = _____ gal   h. 20 in = _____ yd
   i. 1.07 tons = _____ lb   j. _____ yd² = 73 ft²

11.1   Exercise Answers

1. a. 1   b. 1   c. 1,760   d. 2,000
   e. 2   f. 32   g. 1   h. 9
2. a. $\frac{1}{8}$ b. 205.2   c. $1\frac{1}{2}$ d. 44
   e. 66   f. $\frac{3}{16}$ g. $1\frac{3}{8}$ h. $\frac{5}{9}$
   i. 2,140   j. $8\frac{1}{9}$

11.2   Metric System of Measurement

1 | The metric system of measurement is used today throughout the world. Because of its widespread use, it is important to learn the various units of the metric system. There are three main units of measure in the metric system: the *meter,* the *kilogram,* and the *liter.* The meter is a measure of length, the kilogram a measure of weight, and the liter a measure of volume or capacity. The *meter* is used in the same way as feet and yards to measure

distance. The *kilogram* is used as one would use ounces and pounds. The *liter* is used in the same way as pints, quarts, and gallons. The original base unit for weight was the gram but was later changed to the kilogram as the standard unit, owing to the difficulty involved in working with something as small as the gram.

Q1   Write the appropriate unit, meter, kilogram, or liter, for measuring each of the following:

    a. carton of milk   _____

    b. your height   _____

    c. brick of cheese   _____

    d. tank of gas   _____

    e. your weight   _____

    f. length of a football field   _____

STOP • STOP • STOP • STOP • STOP • STOP • STOP • STOP • STOP

A1   **a.** liter   **b.** meter   **c.** kilogram   **d.** liter

        e. kilogram   **f.** meter

---

2   The base unit of length in the metric system is the *meter*. Originally defined as one ten-millionth of the distance from the North Pole to the Equator, the meter is today defined for greater accuracy in terms of the orange-red wavelength of radiating krypton gas (1 650 763.73 wavelengths in vacuum of the orange-red line of the spectrum of krypton 86). As shown in the following scale drawing of a meter stick and a yardstick, the length of a meter is slightly longer than 1 yard. More precisely, 1 meter is approximately equal to 1.1 yd. That is,

$$1 \text{ meter (m)} \doteq 1.1 \text{ yd}$$
$$1 \text{ m} \doteq 39.37 \text{ in}$$

The symbol "$\doteq$" is used to mean "approximately equal."

Q2   Answer true or false:

    a. The meter is the base unit of length in the metric system.   _____

    b. A yard is longer than a meter.   _____

    c. A meter is longer than a yard.   _____

STOP • STOP • STOP • STOP • STOP • STOP • STOP • STOP • STOP

A2   **a.** true   **b.** false   **c.** true

---

3   Metric measurement is based on a decimal system, which means that its units are related by multiples of 10 (10, 100, 1000, . . .) or divisions of ten ($\frac{1}{10}$, $\frac{1}{100}$, $\frac{1}{1000}$, . . .). The multiples of 10 most commonly used are denoted by the Greek-derived prefixes deka (10), hecto (100), and kilo (1000). The divisions of 10 most commonly used are denoted by the Latin-derived prefixes deci ($\frac{1}{10}$ or 0.1), centi ($\frac{1}{100}$ or 0.01), and milli ($\frac{1}{1000}$ or 0.001). Thus, a dekameter has a measure of 10 meters, while a millimeter is the equivalent of $\frac{1}{1000}$, or 0.001, of a meter.

Q3   Use the prefixes of Frame 3 to give the equivalents of each of the following in terms of a meter:

a. 1 hectometer = _____   **b.** 1 decimeter = _____

c. 1 centimeter = _____   **d.** 1 kilometer = _____

e. 1 millimeter = _____   **f.** 1 dekameter = _____

STOP • **STOP** • **STOP** • **STOP** • **STOP** • **STOP** • **STOP** • **STOP** • **STOP**

A3   **a.** 100 meters   **b.** $\frac{1}{10}$ or 0.1 meter
            c. $\frac{1}{100}$ or 0.01 meter   **d.** 1000 meters
            e. $\frac{1}{1000}$ or 0.001 meter   **f.** 10 meters

---

4   A summary of the equations resulting from the prefixes of Frame 3 are as follows:

$$1 \text{ kilometer} = 1000 \text{ meters}$$
$$1 \text{ hectometer} = 100 \text{ meters}$$
$$1 \text{ dekameter} = 10 \text{ meters}$$
$$1 \text{ meter} = 1 \text{ meter}$$

$$1 \text{ decimeter} = \frac{1}{10} \text{ meter or } 0.1 \text{ meter}$$

$$1 \text{ centimeter} = \frac{1}{100} \text{ meter or } 0.01 \text{ meter}$$

$$1 \text{ millimeter} = \frac{1}{1000} \text{ meter or } 0.001 \text{ meter}$$

Notice that the units on the left are arranged in order from largest (kilometer) to smallest (millimeter), and that any two consecutive units differ either by a multiple or a division of 10.

Q4   Place a $>$ (greater than) or $<$ (less than) between each of the pairs of metric units to make a true statement:

a. 1 kilometer _____ 1 hectometer

b. 1 meter _____ 1 centimeter

c. 1 millimeter _____ 1 centimeter

d. 1 meter _____ 1 kilometer

e. 1 dekameter _____ 1 meter

STOP • **STOP** • **STOP** • **STOP** • **STOP** • **STOP** • **STOP** • **STOP** • **STOP**

A4   **a.** $>$ **b.** $>$ **c.** $<$ **d.** $<$ **e.** $>$

Q5   Fill in a multiple or division of 10 to make a true statement:

a. _____ meter = 1 millimeter   **b.** 1 kilometer = _____ meters

c. 1 decimeter = _____ meter   **d.** 1 dekameter = _____ meters

e. 1 centimeter = _____ meter   **f.** _____ meters = 1 hectometer

STOP • **STOP** • **STOP** • **STOP** • **STOP** • **STOP** • **STOP** • **STOP** • **STOP**

A5   **a.** $\frac{1}{1000}$ or 0.001   **b.** 1000   **c.** $\frac{1}{10}$ or 0.1   **d.** 10
            e. $\frac{1}{100}$ or 0.01   **f.** 100

5   The symbols for units of length in the metric system are listed below.

kilometer	km
hectometer	hm
dekameter	dam
meter	m
decimeter	dm
centimeter	cm
millimeter	mm

The units are listed from largest (km) to smallest (mm).

Q6   Write symbols for each of the following:

     a. decimeter _____   **b.** dekameter _____

     c. millimeter _____   **d.** hectometer _____

     e. centimeter _____   **f.** kilometer _____

STOP • STOP • STOP • STOP • STOP • STOP • STOP • STOP • STOP

A6   **a.** dm   **b.** dam   **c.** mm
        d. hm   **e.** cm   **f.** km

Q7   Place $>$ or $<$ between each of the following pairs to form a true statement:

     a. dm _____ dam   **b.** mm _____ cm   **c.** m _____ km

     d. m _____ dm   **e.** km _____ hm   **f.** dam _____ hm

STOP • STOP • STOP • STOP • STOP • STOP • STOP • STOP • STOP

A7   **a.** $<$ **b.** $<$ **c.** $<$
        d. $>$ **e.** $>$ **f.** $<$

6   Computation is simplified in the metric system, if one understands the movement of the decimal point when multiplying or dividing a number by a multiple of 10 (10, 100, 1000). Recall that to multiply a number by a multiple of 10, the decimal point is moved to the *right* the same number of places as there are zeros in the multiple of 10. That is,

Product	Movement of decimal point
$(12.34)1\underline{0} = 123.4$	1 place to the right
$(12.34)1\underline{00} = 1234.$	2 places to the right
$(12.34)1\underline{000} = 12\ 340.$	3 places to the right

Since division is the opposite operation of multiplication, the movement of the decimal point in dividing by a multiple of 10 is a movement to the *left* the same number of places as there are zeros in the divisor. That is,

Quotient	Movement of the decimal point
$\dfrac{12.34}{1\underline{0}} = 1.234$	1 place to the left
$\dfrac{12.34}{1\underline{00}} = 0.1234$	2 places to the left
$\dfrac{12.34}{1\underline{000}} = 0.012\ 34$	3 places to the left

Q8   Find the product or quotient as indicated for each of the following:

a. $5.2(10) =$ _____

b. $\dfrac{5.2}{10} =$ _____

c. $100(0.423) =$ _____

d. $\dfrac{0.423}{100} =$ _____

e. $1.7(1000) =$ _____

f. $\dfrac{1.7}{1000} =$ _____

STOP • STOP • STOP • STOP • STOP • STOP • STOP • STOP • STOP

A8   a. 52   b. 0.52   c. 42.3
     d. 0.004 23 e. 1700   f. 0.0017

---

7   To determine the movement of the decimal point when multiplying by 0.1, 0.01, or 0.001, recall that multiplying by $\frac{1}{10}$ or 0.1, $\frac{1}{100}$ or 0.01, $\frac{1}{1000}$ or 0.001 is the same as dividing by 10, 100, or 1000.

Examples:

$$(12.34)0.1 = (12.34)\frac{1}{10}$$

$$= \frac{12.34}{10}$$

$$(12.34)0.1 = 1.234$$

Thus, the result of multiplying by 0.1 was a movement of the decimal point one place to the left.

$$(12.34)0.01 = (12.34)\frac{1}{100}$$

$$= \frac{12.34}{100}$$

$$(12.34)0.01 = 0.1234$$

Thus, the result of multiplying by 0.01 was a movement of the decimal point two places to the left.

Q9   Find the following products:

a. $\frac{12.34}{1000} =$ _____

b. $(12.34)0.001 =$ _____

c. $\frac{253}{100} =$ _____

d. $(253)0.01 =$ _____

e. $(0.43)0.1 =$ _____

f. $(946)0.001 =$ _____

STOP • STOP • STOP • STOP • STOP • STOP • STOP • STOP • STOP

A9   a. 0.012 34   b. 0.012 34   c. 2.53
     d. 2.53   e. 0.043   f. 0.946

---

8   The equations of Frame 4 can be used to perform conversions between units using the procedures of Section 11.1. For example, to complete the equation 45 dm = _____ m, it is necessary to choose the equation from Frame 4 that expresses the equivalent of 1 dm in terms of m. The desired equation is 1 dm = 0.1 m. The conversion is completed as follows:

$$45 \text{ dm} = \underline{\hspace{2cm}} \text{ m}$$
$$1 \text{ dm} = 0.1 \text{ m} \quad \text{(multiply both sides by 45)}$$
$$45(1 \text{ dm}) = 45(0.1 \text{ m})$$
$$45 \text{ dm} = 4.5 \text{ m}$$

Some additional examples are as follows:

$$125 \text{ cm} = \underline{\hspace{2cm}} \text{ m}$$
$$1 \text{ cm} = 0.01 \text{ m} \quad \text{(multiply both sides by 125)}$$
$$125(1 \text{ cm}) = 125(0.01 \text{ m}) \quad \text{(to multiply by 0.01, move the decimal point two places to the left)}$$

$$125 \text{ cm} = 1.25 \text{ cm}$$

$$4.73 \text{ km} = \underline{\hspace{2cm}} \text{ m}$$
$$1 \text{ km} = 1000 \text{ m} \quad \text{(multiply both sides by 4.73)}$$
$$4.73(1 \text{ km}) = 4.73(1000 \text{ m}) \quad \text{(to multiply by 1000, move the decimal three places to the right)}$$

$$4.73 \text{ km} = 4730 \text{ m}$$

Q10   Complete the equation $1.3 \text{ hm} = \underline{\hspace{2cm}} \text{ m}$.

STOP • **STOP** • **STOP** • **STOP** • **STOP** • **STOP** • **STOP** • **STOP** • **STOP**

A10   130: $1.3 \text{ hm} = \underline{\hspace{2cm}} \text{ m}$
$$1 \text{ hm} = 100 \text{ m}$$
$$1.3(1 \text{ hm}) = 1.3(100 \text{ m}) \quad \text{(to multiply by 100, move the decimal point two places to the right)}$$
$$1.3 \text{ hm} = 130 \text{ m}$$

Q11   Complete the equation $528 \text{ mm} = \underline{\hspace{2cm}} \text{ m}$.

STOP • **STOP** • **STOP** • **STOP** • **STOP** • **STOP** • **STOP** • **STOP** • **STOP**

A11   0.528: $528 \text{ mm} = \underline{\hspace{2cm}} \text{ m}$
$$1 \text{ mm} = 0.001 \text{ m}$$
$$528(1 \text{ mm}) = 528(0.001 \text{ m}) \quad \text{(to multiply by 0.001, move the decimal point three places to the left)}$$

$$528 \text{ mm} = 0.528 \text{ m}$$

Q12   Complete each of the following:

a. $150 \text{ m} = \underline{\hspace{2cm}} \text{ hm}$ **b.** $0.05 \text{ km} = \underline{\hspace{2cm}} \text{ m}$

STOP • **STOP** • **STOP** • **STOP** • **STOP** • **STOP** • **STOP** • **STOP** • **STOP**

A12   **a.** 1.5   **b.** 50

---

9 | Each of the equations of Frame 4 can be used to derive a second equation which expresses a relationship between the same two units but in a different way. For example, the equation 1 kilometer = 1000 meters expresses the equivalent of *1 kilometer* in terms of meters. If both sides of this equation are divided by 1000, the result is an equation that expresses the equivalent of *1 meter* in terms of kilometers.

$$\boxed{1 \text{ km} = 1000 \text{ m}}$$

$$\frac{1 \text{ km}}{1000} = \frac{1000 \text{ m}}{1000} \qquad \text{(divide both sides by 1000)}$$

$$\boxed{\frac{1}{1000} \text{ km} = 1 \text{ m}}$$

Thus, the equivalent of 1 meter is $\frac{1}{1000}$ of a kilometer.

Q13   Use the equation 1 hm = 100 m to express the equivalent of 1 meter in terms of hecto-meters.

STOP • STOP • STOP • STOP • STOP • STOP • STOP • STOP • STOP

A13 $\frac{1}{100}$ hm = 1 m: 1 hm = 100 m   (divide both sides by 100)

$$\frac{1 \text{ hm}}{100} = \frac{100 \text{ m}}{100}$$

$$\frac{1}{100} \text{ hm} = 1 \text{ m}$$

---

10 | The equation 1 centimeter = 0.01 meter expresses the equivalent of *1 centimeter* in terms of meters. If both sides of this equation are multiplied by 100, the result is an equation that expresses the equivalent of *1 meter* in terms of centimeters.

$$\boxed{1 \text{ cm} = 0.01 \text{ m}} \qquad \text{(multiply both sides by 100)}$$

$$100(1 \text{ cm}) = 100(0.01 \text{ m})$$

$$\boxed{100 \text{ cm} = 1 \text{ m}}$$

Thus, the equivalent of 1 meter is 100 centimeters.

Q14   Use the equation 1 dm = 0.1 m to express the equivalent of 1 meter in terms of decimeters.

STOP • STOP • STOP • STOP • STOP • STOP • STOP • STOP • STOP

A14   10 dm = 1 m: 1 dm = 0.1 m
                                   10(1 dm) = 10(0.1 m)
                                      10 dm = 1 m

Q15   Find an equivalent equation for each of the following equations from Frame 4:

Example: 1 km = 1000 m   0.001   km = 1 m

a. 1 hm = 100 m   _____ hm = 1 m

b. 1 dam = 10 m   _____ dam = 1 m

c. 1 dm = $\frac{1}{10}$ m   _____ dm = 1 m

d. 1 mm = 0.001 m   _____ mm = 1 m

STOP • **STOP** • **STOP** • **STOP** • **STOP** • **STOP** • **STOP** • **STOP** • **STOP**

A15   **a.** 0.01   **b.** 0.1   **c.** 10   **d.** 1000

Q16   Use the results of Q15 to complete each of the following:

a. 5.23 m = _____ cm   b. 428 m = _____ km

c. 0.04 m = _____ dm   d. 51.7 km = _____ m

e. 12 m = _____ dam   f. 0.017 m = _____ mm

STOP • **STOP** • **STOP** • **STOP** • **STOP** • **STOP** • **STOP** • **STOP** • **STOP**

A16   **a.** 523:   5.23 m = _____ cm   **b.** 0.428:   428 m = _____ km
                      1 m = 100 cm   1 m = 0.001 km
              5.23(1 m) = 5.23(100 cm)   428(1 m) = 428(0.001 km)
                 5.23 m = 523 cm   428 m = 0.428 km
              (to *multiply* by 100, move the   (to *multiply* by 0.001, move
              decimal point two places to   the decimal point three places
              the *right*)   to the *left*)

c. 0.4   **d.** 51 700   **e.** 1.2   **f.** 17

---

11   Recall from the list in Frame 4 that any two consecutive units differ by a multiple or
             division of 10. That this is true for the units mm, cm, and dm is shown by a metric rule:

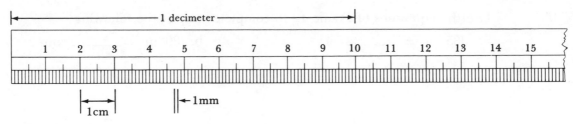

The relationships demonstrated by the metric rule are:

1 dm = 10 cm   or   0.1 dm = 1 cm
1 cm = 10 mm   or   0.1 cm = 1 mm

Q17   Use the equations of Frame 11 to complete each of the following:

a. 52 dm = _____ cm   b. 37 mm = _____ cm

STOP   •   STOP   •   STOP   •   STOP   •   STOP   •   STOP   •   STOP   •   STOP   •   STOP

A17   a. 520:   52 dm = _____ cm   b. 3.7:   37 mm = _____ cm
                    1 dm = 10 cm   1 mm = 0.1 cm
              52(1 dm) = 52(10 cm)   37(1 mm) = 37(0.1 cm)
                    52 dm = 520 cm   37 mm = 3.7 cm

12   Just as mm, cm, and dm are related by multiples or divisions of 10, so also are km, hm, and dam. For example, consider the following equations from Frame 4:

1 km = 1000 m

$\bigg\uparrow \times 10 \quad \bigg\downarrow \times \dfrac{1}{10}$

1 hm = 100 m

$\bigg\uparrow \times 10 \quad \bigg\downarrow \times \dfrac{1}{10}$

1 dam = 10 m

$\bigg\uparrow \times 10 \quad \bigg\downarrow \times \dfrac{1}{10}$

1 m = 1 m

Notice that as one moves from meter to kilometer, each unit is 10 times the preceding unit. As one moves from kilometer to meter, each unit is $\frac{1}{10}$ times the preceding unit. These relationships are summarized as follows:

1 hm = 10 dam   or   0.1 hm = 1 dam
1 km = 10 hm   or   0.1 km = 1 hm

Q18 Use the equations of Frame 12 to complete each of the following:

a. 385 hm = _____ dam

b. 80 hm = _____ km

STOP • STOP • STOP • STOP • STOP • STOP • STOP • STOP • STOP

A18 a. 3850:   385 hm = _____ dam   b. 8:   80 hm = _____ km

$$1 \text{ hm} = 10 \text{ dam}$$
$$385(1 \text{ hm}) = 385(10 \text{ dam})$$
$$385 \text{ hm} = 3850 \text{ dam}$$

$$1 \text{ hm} = 0.1 \text{ km}$$
$$80(1 \text{ hm}) = 80(0.1 \text{ km})$$
$$80 \text{ hm} = 8 \text{ km}$$

---

13 The relationship between the units hm and dam can be diagrammed as follows:

$$\text{hm} \; \overset{\overset{\textstyle 1}{\searrow}{}^{10}}{\underset{0.1\underset{\textstyle 1}{\nwarrow}}{\rule{1.5em}{0.4pt}}} \; \text{dam}$$

If we begin reading with the 1's (above and below the units) and read in the direction of the arrow, the above diagram gives two equations. They are:

1 hm = 10 dam   and   1 dam = 0.1 hm

Q19 Write two equations from the diagram:

$$\text{km} \; \overset{\overset{\textstyle 1}{\searrow}{}^{10}}{\underset{0.1\underset{\textstyle 1}{\nwarrow}}{\rule{1.5em}{0.4pt}}} \; \text{hm}$$

STOP • STOP • STOP • STOP • STOP • STOP • STOP • STOP • STOP

A19 1 km = 10 hm, 1 hm = 0.1 km

Q20 Write two equations from the diagram:

$$\text{cm} \; \overset{\overset{\textstyle 1}{\searrow}{}^{10}}{\underset{0.1\underset{\textstyle 1}{\nwarrow}}{\rule{1.5em}{0.4pt}}} \; \text{mm}$$

STOP • STOP • STOP • STOP • STOP • STOP • STOP • STOP • STOP

A20 1 cm = 10 mm, 1 mm = 0.1 cm

---

14 The relations of Frames 11 and 12 can be combined and expressed in the following diagram:

larger   smaller

km $\overset{10}{\underset{0.1}{—}}$ hm $\overset{10}{\underset{0.1}{—}}$ dam $\overset{10}{\underset{0.1}{—}}$ m $\overset{10}{\underset{0.1}{—}}$ dm $\overset{10}{\underset{0.1}{—}}$ cm $\overset{10}{\underset{0.1}{—}}$ mm

Notice the following things about the diagram:

1. The units are arranged from larger (km) to smaller (mm).
2. The multiple relationships are shown above the units. For example, 1 km = 10 hm, 1 hm = 10 dam, 1 dam = 10 m, and so on.
3. The division relationships are shown below the units. For example, 1 mm = 0.1 cm, 1 cm = 0.1 dm, 1 dm = 0.1 m, and so on.
4. For any two consecutive units, two equations result. For example, consider the consecutive units dm and cm. Looking only at these two units on the diagram, we see

$$\overset{1}{\underset{\underset{\displaystyle 1}{0.1}}{\text{dm} \overset{10}{\underline{\qquad}} \text{cm}}}$$
. Reading with the arrows, the two equations are 1 dm = 10 cm and 1 cm =

0.1 dm.

Q21   **a.** Place > or < in the blank to make a true statement. mm _____ cm
      b. Write the two equations that relate mm and cm.

_____ , _____

STOP • **STOP** • **STOP** • **STOP** • **STOP** • **STOP** • **STOP** • **STOP** • **STOP**

A21   **a.** <   **b.** 1 mm = 0.1 cm, 1 cm = 10 mm

---

15   Consider the completion of the equation 42.5 m = _____ dam.

$$42.5 \text{ m} = \underline{\qquad} \text{ dam}$$
$$1 \text{ m} = 0.1 \text{ dam}$$
$$42.5(1 \text{ m}) = 42.5(0.1 \text{ dam})$$
$$42.5 \text{ m} = 4.25 \text{ dam}$$

Notice that the result of multiplying by 0.1 was a movement of the decimal point one place to the *left*. Notice, also, that the direction of the arrow on the diagram of Frame 14 is to the *left*.

$$\text{dam} \underset{0.1}{\underline{\qquad}} \text{m}$$
$$\overset{\displaystyle 1}{\uparrow}$$
arrow points *left* and multiplication by 0.1 moves the decimal point to the *left*

A second example is the following:

$$0.56 \text{ cm} = \underline{\qquad} \text{ mm}$$
$$1 \text{ cm} = 10 \text{ mm}$$
$$0.56(1 \text{ cm}) = 0.56(10 \text{ mm})$$
$$0.56 \text{ cm} = 5.6 \text{ mm}$$

The result of multiplying by 10 was a movement of the decimal point one place to the *right*. The direction of the arrow on the diagram of Frame 14 is to the right.

$$\overset{1}{\underset{10}{\text{cm}}} \underline{\qquad} \text{mm}$$
arrow points right and multiplication by 10 moves the decimal point to the *right*

Thus, another important fact about the diagram of Frame 14 is that *the arrows indicate the way the decimal point moves when completing conversions between any two units.*

Q22   Use the diagram of Frame 14 to complete the following equations:

    a. 10.5 cm = _____ dm   **b.** 0.6 km = _____ hm

    c. 55 dm = _____ m   **d.** 150 m = _____ dam

STOP • **STOP** • **STOP** • **STOP** • **STOP** • **STOP** • **STOP** • **STOP** • **STOP**

A22   **a.** 1.05:   10.5 cm = _____ dm   **b.** 6:   0.6 km = _____ hm
                       1 cm = 0.1 dm   1 km = 10 hm
               10.5(1 cm) = 10.5(0.1 dm)   0.6(1 km) = 0.6(10 hm)
                  10.5 cm = 1.05 dm   0.6 km = 6 hm

    c. 5.5   **d.** 15

---

16   Notice that the diagram of Frame 14 can also be used to give equations relating units that are not consecutive. For example, to find an equation that relates cm and dam, look only at that portion of the diagram:

$$\text{dam} \xrightarrow[\underset{1}{0.1}]{10} \text{m} \xrightarrow[\underset{1}{0.1}]{10} \text{dm} \xrightarrow[\underset{1}{0.1}]{10} \text{cm}$$

The relation is found by finding the product of the factors present between the units. Thus, since $10 \cdot 10 \cdot 10 = 1000$, 1 dam = 1000 cm. Similarly, since $(0.1)(0.1)0.1 = 0.001$, 1 cm = 0.001 dam.

Q23   Write two equations relating each of the pairs of units below (use the diagram, only if you do not already know the relations):

    a. cm, mm   _____   _____

    b. km, dam   _____   _____

    c. dm, dam   _____   _____

STOP • **STOP** • **STOP** • **STOP** • **STOP** • **STOP** • **STOP** • **STOP** • **STOP**

A23   **a.** 1 cm = 10 mm, 1 mm = 0.1 cm
        b. 1 km = 100 dam, 1 dam = 0.01 km
        c. 1 dm = 0.01 dam, 1 dam = 100 dm

Q24   Complete the equation:

    a. 3.3 cm = _____ mm   **b.** 457 dm = _____ dam

STOP • STOP • STOP • STOP • STOP • STOP • STOP • STOP • STOP

A24   **a.** 33   **b.** 4.57

---

17   In the metric system, as in the English system of measurement, area is measured using square units. A common metric unit of area is the *square meter* (m^2) or the area covered by a square with sides equal to 1 meter. Other common metric units of area are the square decimeter (dm^2), square centimeter (cm^2), square millimeter (mm^2), are (a), and hectare (ha). The relationship between m^2 and dm^2 can be seen using the area formula and the relationship 1 m = 10 dm.

$A = l \cdot w$
$A = 1\,m \cdot 1\,m = 10\,dm \cdot 10\,dm$
$A = 1\,m^2 = 100\,dm^2$

Thus, $1\,m^2 = 100\,dm^2$ or $0.01\,m^2 = 1\,dm^2$. The relationships among dm^2, cm^2, and mm^2 can be seen in the following actual-size drawing of $1\,dm^2$:

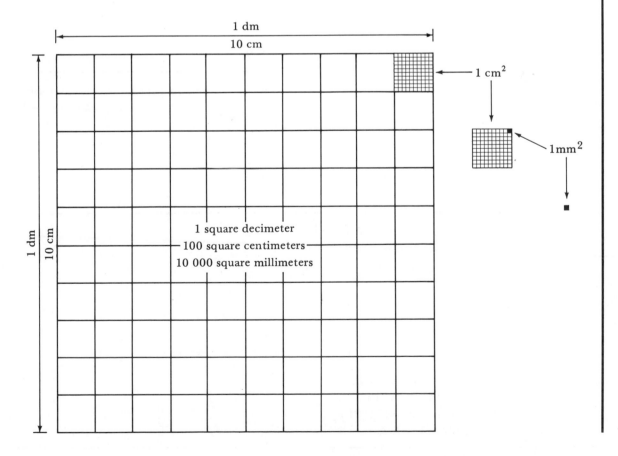

The relationships demonstrated by the drawing are:

$$1 \text{ dm}^2 = 100 \text{ cm}^2 \quad \text{or} \quad 0.01 \text{ dm}^2 = 1 \text{ cm}^2$$
$$1 \text{ cm}^2 = 100 \text{ mm}^2 \quad \text{or} \quad 0.01 \text{ cm}^2 = 1 \text{ mm}^2$$

Notice that consecutive units of area from mm^2 to m^2 are related by multiples or divisions of 100.

Q25   Complete each of the following using the information of Frame 17:

a. $1 \text{ m}^2 = $ _____ dm^2 **b.** $1 \text{ dm}^2 = $ _____ cm^2

c. $1 \text{ mm}^2 = $ _____ cm^2 **d.** $1 \text{ cm}^2 = $ _____ dm^2

STOP • STOP • STOP • STOP • STOP • STOP • STOP • STOP • STOP

A25   **a.** 100   **b.** 100
        c. 0.01 or $\frac{1}{100}$ **d.** 0.01 or $\frac{1}{100}$

---

18   The metric units of area *are* and *hectare* also differ by a multiple or division of 100 and are defined as follows:

$$1 \text{ are (a)} = 100 \text{ m}^2$$
$$1 \text{ hectare (ha)} = 100 \text{ a } (10\ 000 \text{ m}^2)$$

Q26   Complete each of the following equations:

a. $1 \text{ m}^2 = $ _____ a   **b.** $100 \text{ a} = $ _____ hectare

c. $1 \text{ a} = $ _____ hectare   **d.** $1 \text{ a} = $ _____ m^2

STOP • STOP • STOP • STOP • STOP • STOP • STOP • STOP • STOP

A26   **a.** 0.01 or $\frac{1}{100}$ **b.** 1
        c. 0.01 or $\frac{1}{100}$ **d.** 100

---

19   The relations between units of area can be shown in diagram form as follows:

larger   smaller

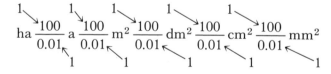

Q27   Complete the equation $800 \text{ m}^2 = $ _____ a.

STOP • STOP • STOP • STOP • STOP • STOP • STOP • STOP • STOP

A27   8: $800 \text{ m}^2 = $ _____ a
               $1 \text{ m}^2 = 0.01 \text{ a}$
        $800(1 \text{ m}^2) = 800(0.01 \text{ a})$
               $800 \text{ m}^2 = 8 \text{ a}$

Q28   Complete each of the following:

a. $3460 \text{ cm}^2 = $ _____ dm^2 **b.** $48 \text{ a} = $ _____ m^2

c. $7.5 \text{ ha} = $ _____ m^2

STOP   •   STOP   •   STOP   •   STOP   •   STOP   •   STOP   •   STOP   •   STOP   •   STOP

A28   **a.** 34.6   **b.** 4 800
           c. 75 000: $7.5 \text{ ha} = $ _____ m^2
                                 $1 \text{ ha} = 10\,000 \text{ m}^2$
                         $7.5(1 \text{ ha}) = 7.5(10\,000 \text{ m}^2)$
                              $7.5 \text{ ha} = 75\,000 \text{ m}^2$

---

20   The base unit of weight in the metric system is the *kilogram*. The only base unit still defined by a man-made object, the kilogram, is defined as the weight of a cylinder of platinum–iridium alloy kept by the International Bureau of Weights and Measures in Sèvres, France. A duplicate of this cylinder is present in the National Bureau of Standards in the United States. When compared with a unit of the English system, the weight of 1 kilogram is approximately 2.2 pounds.

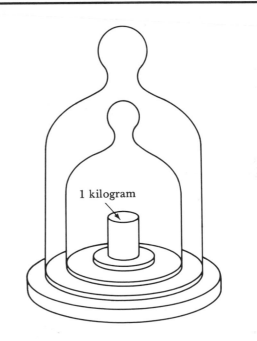

1 kilogram

The original base unit for weight was the gram, defined as the weight of a cubic centimeter of water at the temperature of melting ice. The change was later made to the kilogram as the base unit, owing to the difficulty involved in working with something as small as the gram. The units for weight are multiples and divisions of 10, using the same prefixes as with the units of length. The units of weight, their symbols, and equivalents are:

          1 kilogram (kg) = 1000 grams
          1 hectogram (hg) = 100 grams
          1 dekagram (dag) = 10 grams
                1 gram (g) = 1 gram
          1 decigram (dg) = 0.1 gram
          1 centigram (cg) = 0.01 gram
          1 milligram (mg) = 0.001 gram

Of the units listed, the most commonly used are the kilogram, the milligram by pharmacists, and the gram.

          A unit that is not listed above is the metric ton. It is defined as follows:

          1 metric ton (t) = 1000 kg

Q29   Complete each equation using the information of Frame 20:

a. $1 \text{ cg} = $ _____ g   **b.** _____ $\text{g} = 1 \text{ kg}$

c. _____ $\text{mg} = 1 \text{ g}$ **d.** $10 \text{ g} = $ _____ dag

e. _____ $\text{kg} = 1 \text{ t}$

STOP   •   STOP   •   STOP   •   STOP   •   STOP   •   STOP   •   STOP   •   STOP   •   STOP

A29   **a.** 0.01 or $\frac{1}{100}$ **b.** 1000   **c.** 1000
            d. 1   **e.** 1000

Q30   Place $<$ or $>$ between each of the following pairs to form a true statement:

            a. g ——— mg   **b.** mg ——— cg

            c. g ——— kg   **d.** kg ——— hg

STOP • STOP • STOP • STOP • STOP • STOP • STOP • STOP • STOP

A30   **a.** $>$ **b.** $<$ **c.** $<$ **d.** $>$

---

21   The relations between units of weight can be shown in a diagram similar to that of Frame 14:

            larger   smaller

            Notice again that the units are ordered from larger (kg) to smaller (mg) and that for any two units there are two equations relating them which result.
                For example, to relate mg and cg, look at only that portion of the diagram:

            Reading with the arrows, the two equations that result are 1 cg = 10 mg and 1 mg = 0.1 cg. To find the relation between kg and dag, look at the following portion of the diagram:

            Since two factors appear between the units kg and dag, the *product* of the units is used. Thus, the two equations are   1 kg = 100 dag (10 · 10 = 100)   and   1 dag = 0.01 kg (0.1 · 0.1 = 0.01).

Q31   Use the diagram of Frame 21 (if necessary) to write two equations that relate each of the following pairs of units:

            a. g, mg   ———   ———

            b. g, cg   ———   ———

            c. cg, dg   ———   ———

            d. hg, kg   ———   ———

            e. kg, dag   ———   ———

STOP • STOP • STOP • STOP • STOP • STOP • STOP • STOP • STOP

A31   **a.** 1 g = 1000 mg, 1 mg = 0.001 g
        b. 1 g = 100 cg, 1 cg = 0.01 g
        c. 1 cg = 0.1 dg, 1 dg = 10 cg
        d. 1 hg = 0.1 kg, 1 kg = 10 hg
        e. 1 kg = 100 dag, 1 dag = 0.01 kg

Q32   Complete the equation 57.5 g = _____ cg.

STOP • **STOP** • **STOP** • **STOP** • **STOP** • **STOP** • **STOP** • **STOP** • **STOP**

A32   5750:   57.5 g = _____ cg
                    1 g = 100 cg
           57.5(1 g) = 57.5(100 cg)   (to *multiply* by 100, move the decimal
                                         point two places to the *right*)

                   57.5 g = 5750 cg

Q33   Complete the equation _____ g = 750 mg.

STOP • **STOP** • **STOP** • **STOP** • **STOP** • **STOP** • **STOP** • **STOP** • **STOP**

A33   0.75:   _____ g = 750 mg
                  750 mg = _____ g
                    1 mg = 0.001 g
           750(1 mg) = 750(0.001 g)   (to *multiply* by 0.001, move the decimal point
                                         three places to the *left*)

                  750 mg = 0.75 g

Q34   Complete each of the following:

        a. 50 mg = _____ cg   **b.** 0.6 kg = _____ g

        c. _____ g = 240 cg   **d.** 4297 g = _____ hg

e.   2.3 t = _____ kg   **f.**   250 kg = _____ t

STOP   •   **STOP**   •   **STOP**   •   **STOP**   •   **STOP**   •   **STOP**   •   **STOP**   •   **STOP**   •   **STOP**

A34   **a.** 5   **b.** 600   **c.** 2.4   **d.** 42.97   **e.** 2300   **f.** 0.25

22   The volume or capacity of an object can be thought of as the amount that the object holds. Volume can be expressed in units of liquid measure or in cubic units. A cube that has edges of length 1 cm is said to have a volume of 1 cubic centimeter (1 cm³ or 1 cc). A cubic centimeter is shown here, its size being 1 cubic centimeter (cm³), approxi-

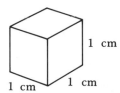

1 cm

1 cm   1 cm

mately that of a sugar cube. If a cube with edges of length 10 cm (1 dm) is formed, the result is 1000 cubic centimeters (1 cubic decimeter).

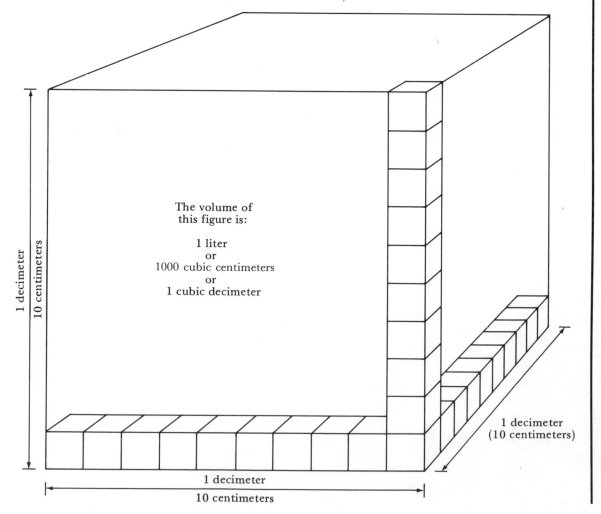

The volume of
this figure is:

1 liter
or
1000 cubic centimeters
or
1 cubic decimeter

1 decimeter
10 centimeters

1 decimeter
(10 centimeters)

1 decimeter
10 centimeters

The unit that is generally used as the metric unit of liquid volume or capacity is the *liter* (l). One liter is defined as 1000 cubic centimeters; the large cube shown has a volume of 1 liter. Compared with the English system, the liter is slightly larger than a quart (1 l $\doteq$ 1.057 qt).

The metric units of liquid volume are defined as follows:

$$1 \text{ kiloliter (kl)} = 1000 \text{ liters (l)}$$
$$1 \text{ hectoliter (hl)} = 100 \text{ liters}$$
$$1 \text{ dekaliter (dal)} = 10 \text{ liters}$$
$$1 \text{ liter (l)} = 1 \text{ liter}$$
$$1 \text{ deciliter (dl)} = 0.1 \text{ liter}$$
$$1 \text{ centiliter (cl)} = 0.01 \text{ liter}$$
$$1 \text{ milliliter (ml)} = 0.001 \text{ liter}$$

Q35   Answer true or false:

a. A liter is larger than a quart. _____

b. 1 l $= 1000 \text{ cm}^3$ _____

c. 1 l $= 1 \text{ dm}^3$ _____

STOP • STOP • STOP • STOP • STOP • STOP • STOP • STOP • STOP

A35   **a.** true   **b.** true   **c.** true

Q36   Complete each equation using the information of Frame 22:

a. 1 ml $=$ _____ l   **b.** 1000 cm^3 $=$ _____ l

STOP • STOP • STOP • STOP • STOP • STOP • STOP • STOP • STOP

A36   **a.** 0.001 or $\frac{1}{1000}$ **b.** 1

23   In Frame 22 a liter was defined as 1 l $= 1000 \text{ cm}^3$. Dividing both sides by 1000, the equation obtained is $\frac{1}{1000}$ l $= 1 \text{ cm}^3$. From Frame 22 it was also given that $\frac{1}{1000}$ l $= 1$ ml. Since both 1 cm^3 and 1 ml are equal to $\frac{1}{1000}$ l, the two units must be equal to each other. That is,

$$1 \text{ cm}^3 \text{ (cc)} = 1 \text{ ml}$$

Q37   Complete the equation 1 cm^3 $=$ _____ ml.

STOP • STOP • STOP • STOP • STOP • STOP • STOP • STOP • STOP

A37   1

Q38   Write two ways to name a cube that has edges of length 1 cm. _____

STOP • STOP • STOP • STOP • STOP • STOP • STOP • STOP • STOP

A38   1 cm^3 (1 cc), 1 ml

24   The relations between units of liquid volume can be shown in diagram form as follows:

larger   smaller

Equations involving volume units are completed, using the diagram, in the same way as with length and weight units.

Examples:

1.   750 ml = _____ l
       1 ml = 0.001 l
      750(1 ml) = 750(0.001 l)   (to *multiply* by 0.001, move the decimal point three places to the *left*)

      750 ml = 0.75 l

(Notice that the arrows on the diagram connecting ml and l point *left*.)

2. 5.3 ml = _____ cm^3
       1 ml = 1 cm^3
      5.3 ml = 5.3 cm^3

3.   0.03 dal = _____ l
       1 dal = 10 l
      0.03(1 dal) = 0.03(10 l)   (to *multiply* by 10, move the decimal point one place to the *right*)

      0.03 dal = 0.3 l

Q39   Complete the equation 50 kl = _____ l.

STOP • STOP • STOP • STOP • STOP • STOP • STOP • STOP • STOP

A39   50 000:   50 kl = _____ l
                        1 kl = 1000 l
                     50(1 kl) = 50(1000 l)
                     50 kl = 50 000 l

Q40   Complete each of the following:

a. 25 l = _____ ml   b. _____ l = 0.1 kl

c. 45 ml = _____ cm^3 d. 7.5 cl = _____ ml

   e. $1000 \text{ cm}^3 = $ _____ l   **f.** $1 \text{ dm}^3 = $ _____ l

STOP • STOP • STOP • STOP • STOP • STOP • STOP • STOP • STOP

A40   **a.** 25 000   **b.** 100   **c.** 45   **d.** 75
          e. 1   **f.** 1

---

25   For volume measured in cubic units in the metric system, a common unit is the cubic meter (m^3). One cubic meter is the volume of a cube with edges equal to 1 meter. Other common units of volume are the cubic centimeter (cm^3) and the cubic decimeter (dm^3).

   The relationship between m^3 and dm^3 can be seen using the volume formula and the relationship $1 \text{ m} = 10 \text{ dm}$.

$V = lwh$
$V = 1 \text{ m} \cdot 1 \text{ m} \cdot 1 \text{ m} = 10 \text{ dm} \cdot 10 \text{ dm} \cdot 10 \text{ dm}$
$V = 1 \text{ m}^3 \qquad\qquad = 1000 \text{ dm}^3$

   Thus, $1 \text{ m}^3 = 1000 \text{ dm}^3$. $\frac{1}{1000}$ or $0.001 \text{ m}^3 = 1 \text{ dm}^3$. The relationship between dm^3 and cm^3 is demonstrated in Frame 22. It is $1 \text{ dm}^3 = 1000 \text{ cm}^3$. Also, $\frac{1}{1000}$ or $0.001 \text{ dm}^3 = 1 \text{ cm}^3$.

Q41   **a.** For volume measured in cubic units, a common unit is the _____ .

          b. For liquid volume, a common unit is the _____ .

STOP • STOP • STOP • STOP • STOP • STOP • STOP • STOP • STOP

A41   **a.** m^3 **b.** liter

Q42   Complete each equation using the information of Frame 25:

          a. $1 \text{ m}^3 = $ _____ dm^3 **b.** $1 \text{ cm}^3 = $ _____ dm^3

STOP • STOP • STOP • STOP • STOP • STOP • STOP • STOP • STOP

A42   **a.** 1000   **b.** 0.001 or $\frac{1}{1000}$

---

26   Notice that any two units of volume from the list m^3, dm^3, cm^3 differ by a multiple or division of 1000. This fact is helpful in performing conversions between cubic units and is summarized in the following diagram:

   larger   smaller

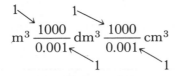

Q43   Use the diagram of Frame 26 to write two equations relating each of the pairs of units.

          a. cm^3, dm^3 _____   _____

          b. m^3, cm^3 _____   _____

STOP • STOP • STOP • STOP • STOP • STOP • STOP • STOP • STOP

A43   **a.** $1 \text{ cm}^3 = 0.001 \text{ dm}^3$, $1 \text{ dm}^3 = 1000 \text{ cm}^3$
           b. $1 \text{ m}^3 = 1\,000\,000 \text{ cm}^3$, $0.000\,001 \text{ m}^3 = 1 \text{ cm}^3$

Q44   Complete the equation $500 \text{ dm}^3 =$ _____ m^3.

STOP • STOP • STOP • STOP • STOP • STOP • STOP • STOP • STOP

A44   0.5: $500 \text{ dm}^3 =$ _____ m^3
                     $1 \text{ dm}^3 = 0.001 \text{ m}^3$
              $500(1 \text{ dm})^3 = 500(0.001 \text{ m}^3)$
                  $500 \text{ dm}^3 = 0.5 \text{ m}^3$

Q45   Complete the equation $8.5 \text{ m}^3 =$ _____ dm^3.

STOP • STOP • STOP • STOP • STOP • STOP • STOP • STOP • STOP

A45   8500

---

27   Volume can be expressed in units of liquid measure or in cubic units. The relations
           between the two volume measures are as follows:

           $1 \text{ kl} = 1 \text{ m}^3$
           $1 \text{ l} = 1 \text{ dm}^3$
           $1 \text{ ml} = 1 \text{ cm}^3 \, (1 \text{ cc})$

           Thus, a volume of 500 m^3 is the same as a volume of 500 kl. Similarly, $45 \text{ ml} = 45 \text{ cm}^3$
           and $0.93 \text{ dm}^3 = 0.93 \text{ l}$.

Q46   Complete each of the following:

           a. $7.5 \text{ l} =$ _____ dm^3 **b.** $3{,}500 \text{ kl} =$ _____ m^3

           c. _____ $\text{cm}^3 = 100 \text{ ml}$

STOP • STOP • STOP • STOP • STOP • STOP • STOP • STOP • STOP

A46   **a.** 7.5   **b.** 3500   **c.** 100

---

28   In countries using the metric system, two scales are used to measure temperature. The
           Kelvin scale is used for scientific work, and the *Celsius* (centigrade) scale is used when
           temperature is being measured for other purposes (measurement of air temperature, for
           example).
               The Kelvin scale, named after the British physicist Lord Kelvin, has its starting point
           at absolute zero (the temperature at which all molecular activity ceases) and a fixed point

of 273.16 K at the triple point of water (the temperature at which water exists in all three states—vapor, liquid, and solid). The triple point of water is slightly above what we call the freezing point.

The Celsius scale is named after the Swedish astronomer Anders Celsius (1701–1744), who invented it. Water freezes at 0 °C and boils at 100 °C. A comfortable room temperature is 20 °C.

The Kelvin and Celsius temperature scales are compared:

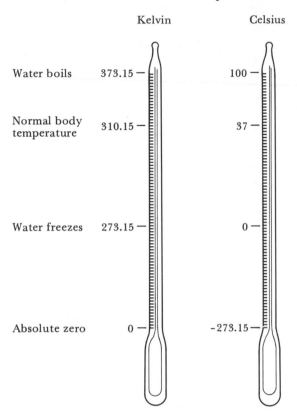

Notice that the difference between the freezing and boiling points is 100 on both the Celsius and Kelvin scales. In fact, the only difference between the two scales is that each Kelvin temperature is 273.15 units more than the corresponding Celsius temperature. That is, Celsius (C) and Kelvin (K) temperatures are related by the equation

K = C + 273.15

Q47   Complete each of the following:

a. The freezing point of water is _____ °C and _____ K.

b. The boiling point of water is _____ °C and _____ K.

c. Normal body temperature is _____ °C.

d. What is the Kelvin temperature for 50 °C? _____

STOP   •   STOP   •   STOP   •   STOP   •   STOP   •   STOP   •   STOP   •   STOP   •   STOP

A47   **a.** 0, 273.15   **b.** 100, 373.15   **c.** 37

d. 323.15:   K = 50 + 273.15
                     K = 323.15

This completes the instruction for this section.

11.2 Exercises

1. Write a common unit of measure in the metric system for each of the following:
   a. length _____
   b. weight _____
   c. liquid volume _____
   d. volume (cubic unit) _____
   e. area _____

2. Place $>$, $<$, or $=$ between each pair to make a true statement:
   a. meter _____ yard
   b. quart _____ liter
   c. g _____ mg
   d. m^2 _____ a
   e. ml _____ cm^3
   f. $1000\ cm^3$ _____ l

3. Complete each of the following equations using 1 or a multiple of 10 (10, 100, 1000) or a division of 10 (0.1, 0.01, 0.001):
   a. _____ m $= 1000$ mm
   b. _____ g $= 1000$ mg
   c. 1 dm $=$ _____ m
   d. 1 cg $=$ _____ g
   e. 1 ml $=$ _____ cm^3
   f. 1 cm $=$ _____ dm
   g. 10 mg $=$ _____ cg
   h. $1\ cm^2 =$ _____ dm^2
   i. 1 ha $=$ _____ a
   j. $1\ m^2 =$ _____ dm^2

4. Complete each of the following equations:
   a. 0.5 cm $=$ _____ mm
   b. 85 dam $=$ _____ hm
   c. 76.3 g $=$ _____ cg
   d. _____ t $= 1000$ kg
   e. 1 l $=$ _____ cm^3
   f. 450 mm $=$ _____ km
   g. 12 dg $=$ _____ mg
   h. _____ $cm^3 = 40$ ml
   i. 3.9 g $=$ _____ mg
   j. 495 l $=$ _____ hl

5. Any two metric units of *area* differ by a multiple or division of _____.

6. Complete each of the following:
   a. $1\ m^2 =$ _____ dm^2
   b. $1\ mm^2 =$ _____ cm^2
   c. 2.7 a $=$ _____ m^2
   d. 45 ha $=$ _____ m^2
   e. $750\ m^2 =$ _____ a
   f. $5\ cm^2 =$ _____ dm^2

7. Any two cubic units of *volume* in the metric system differ by a multiple or division of _____.

8. Complete each of the following:
   a. $548\ cm^3 =$ _____ dm^3
   b. $2\ m^3 =$ _____ dm^3
   c. _____ $cm^3 = 1\ dm^3$
   d. $1\ dm^3 =$ _____ m^3

9. Complete the following table:

	Kelvin temperature	Celsius temperature
a. Absolute zero	_____	_____
b. Water freezes	_____	_____
c. Water boils	_____	_____
d. Normal body temperature	_____	_____

10. Complete the following equations:
    a. 15 °C $=$ _____ K
    b. 40.37 °C $=$ _____ K
    c. 293.15 K $=$ _____ °C
    d. 273.15 K $=$ _____ °C

11.2 Exercise Answers

1. a. m   b. kg   c. l   d. m^3 e. m^2
2. a. $>$ b. $<$ c. $>$ d. $<$ e. $=$ f. $=$
3. a. 1   b. 1   c. 0.1   d. 0.01   e. 1   f. 0.1   g. 1
   h. 0.01   i. 100   j. 100

4. **a.** 5   **b.** 8.5   **c.** 7630   **d.** 1
   e. 1000   **f.** 0.000 45   **g.** 1200   **h.** 40
   i. 3900   **j.** 4.95
5. 100
6. **a.** 100   **b.** 0.01   **c.** 270   **d.** 450 000
   e. 7.5   **f.** 0.05
7. 1000
8. **a.** 0.548   **b.** 2000   **c.** 1000   **d.** 0.001
9. **a.** 0 K, ⁻273.15 °C **b.** 273.15 K, 0 °C **c.** 373.15 K, 100 °C **d.** 310.15 K, 37 °C
10. **a.** 288.15   **b.** 313.52   **c.** 20   **d.** 0

11.3   Conversions Between Metric and English Systems of Measurement

1 | During the period of gradual change from an English system of measurement to a metric system there will be initially, for some, confusion as to what a given measurement in one system means in terms of the other system. Although, ideally, everyone will quickly learn to "think metric," most likely this will not happen overnight. The purpose of this section, therefore, is to develop the ability to perform conversions between units of the two systems.
   Recall from Section 11.2 the scale drawing comparing 1 meter and 1 yard.

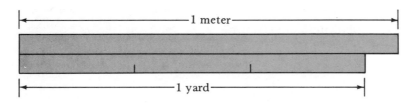

The approximate relationship between the meter and yard is:

1 m ≐ 1.1 yd   or   1 yd ≐ 0.9 m

(*Note:* ≐ means "is approximately equal to.") The approximate relationship between a meter and inches is:

1 m ≐ 39.37 in

Q1   **a.** Which is longer, a meter or a yard? _____

b. If a person is 2 m tall, is he taller or shorter than 6 ft (2 yd)? _____

c. Write the following in order from smallest to largest: yard, inch, meter.

_____

STOP   •   STOP   •   STOP   •   STOP   •   STOP   •   STOP   •   STOP   •   STOP   •   STOP

A1   **a.** meter   **b.** taller   **c.** inch, yard, meter

2   Conversions between English and metric units can be performed using the procedures of the previous two sections.

Examples:

$$4.0 \text{ m} = \underline{\hspace{1.5cm}} \text{ yd}$$
$$1 \text{ m} = 1.1 \text{ yd}$$
$$4.0(1 \text{ m}) = 4.0(1.1 \text{ yd})$$
$$4.0 \text{ m} = 4.4 \text{ yd}$$

(*Note:* Although the "$=$" sign is used above and in the remainder of this section, the "$\doteq$" sign is more accurate, because approximate relationships are being used.)

$$5.5 \text{ yd} = \underline{\hspace{1.5cm}} \text{ m} \qquad\qquad 2.5 \text{ m} = \underline{\hspace{1.5cm}} \text{ in}$$
$$1 \text{ yd} = 0.9 \text{ m} \qquad\qquad 1 \text{ m} = 39.37 \text{ in}$$
$$5.5(1 \text{ yd}) = 5.5(0.9 \text{ m}) \qquad 2.5(1 \text{ m}) = 2.5(39.37 \text{ in})$$
$$5.5 \text{ yd} = 4.95 \text{ m} \qquad\qquad 2.5 \text{ m} = 98.425 \text{ in}$$

Q2 $9 \text{ yd} = \underline{\hspace{1.5cm}} \text{ m}$

STOP • STOP • STOP • STOP • STOP • STOP • STOP • STOP • STOP

A2   8.1: $9 \text{ yd} = \underline{\hspace{1.5cm}} \text{ m}$
$$1 \text{ yd} = 0.9 \text{ m}$$
$$9(1 \text{ yd}) = 9(0.9 \text{ m})$$
$$9 \text{ yd} = 8.1 \text{ m}$$

Q3   Perform each of the following conversions:

a. $7.5 \text{ m} = \underline{\hspace{1.5cm}} \text{ yd}$ **b.** $1.3 \text{ m} = \underline{\hspace{1.5cm}} \text{ in}$

STOP • STOP • STOP • STOP • STOP • STOP • STOP • STOP • STOP

A3   **a.** 8.25: $7.5 \text{ m} = \underline{\hspace{1.5cm}} \text{ yd}$ **b.** 51.18: $1.3 \text{ m} = \underline{\hspace{1.5cm}} \text{ in}$
$$1 \text{ m} = 1.1 \text{ yd} \qquad\qquad\qquad 1 \text{ m} = 39.37 \text{ in}$$
$$7.5(1 \text{ m}) = 7.5(1.1 \text{ yd}) \qquad 1.3(1 \text{ m}) = 1.3(39.37 \text{ in})$$
$$7.5 \text{ m} = 8.25 \text{ yd} \qquad\qquad 1.3 \text{ m} = 51.18 \text{ in}$$

3   One yard is approximately equal to 1 meter. It is, therefore, also true that $\frac{1}{3}$ of a yard (1 ft) is approximately equal to $\frac{1}{3}$ meter. The relationship between ft and m is:

$$1 \text{ m} = 3.3 \text{ ft} \qquad \text{or} \qquad 1 \text{ ft} = 0.3 \text{ m}$$

Q4   **a.** Which is larger, a foot or a meter? _____
      b. Write the following in order from smallest to largest: foot, inch, meter, yard.

      _____

STOP • STOP • STOP • STOP • STOP • STOP • STOP • STOP • STOP

A4   **a.** meter   **b.** inch, foot, yard, meter

---

4   The relationship among the inch, centimeter, and millimeter can be seen in a comparison of the metric and English rulers:

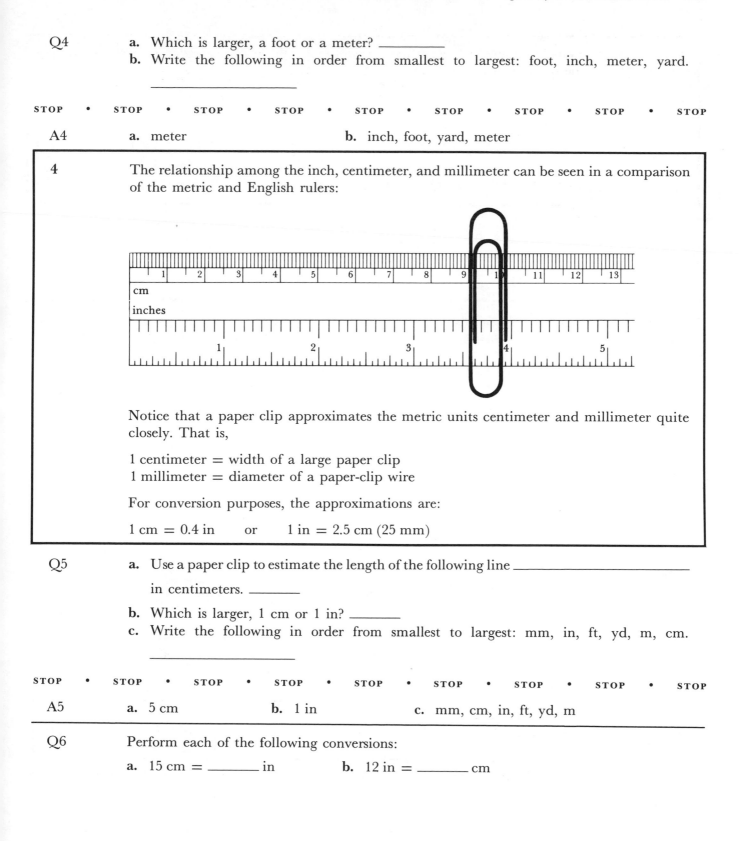

Notice that a paper clip approximates the metric units centimeter and millimeter quite closely. That is,

1 centimeter = width of a large paper clip
1 millimeter = diameter of a paper-clip wire

For conversion purposes, the approximations are:

1 cm = 0.4 in   or   1 in = 2.5 cm (25 mm)

Q5   **a.** Use a paper clip to estimate the length of the following line _____

      in centimeters. _____

      b. Which is larger, 1 cm or 1 in? _____
      c. Write the following in order from smallest to largest: mm, in, ft, yd, m, cm.

      _____

STOP • STOP • STOP • STOP • STOP • STOP • STOP • STOP • STOP

A5   **a.** 5 cm   **b.** 1 in   **c.** mm, cm, in, ft, yd, m

Q6   Perform each of the following conversions:

      a. 15 cm = _____ in   **b.** 12 in = _____ cm

   c. $3.2 \, \text{cm} = \underline{\hspace{2cm}} \text{in}$ **d.** $1.375 \, \text{in} = \underline{\hspace{2cm}} \text{mm}$

STOP   •   STOP   •   STOP   •   STOP   •   STOP   •   STOP   •   STOP   •   STOP   •   STOP

A6   **a.** 6   **b.** 30   **c.** 1.28   **d.** 34.375

---

5   Recall that 5,280 ft = 1 mi. Since 1 meter is approximately 3 ft, 1 kilometer is 1000 meters or 1,000 (3 ft) = 3,000 ft. Thus, 1 km is a little larger than a $\frac{1}{2}$ mile. More precisely, the relationship between mile and kilometer is:

   $1 \, \text{km} = 0.6 \, \text{mi}$ or $1 \, \text{mi} = 1.6 \, \text{km}$

Q7   **a.**   Which is larger, a kilometer or a mile? $\underline{\hspace{2cm}}$
        b.   Write the following in order from smallest to largest: in, cm, mm, yd, ft, m, km, mi.

        $\underline{\hspace{6cm}}$

STOP   •   STOP   •   STOP   •   STOP   •   STOP   •   STOP   •   STOP   •   STOP   •   STOP

A7   **a.** mile   **b.** mm, cm, in, ft, yd, m, km, mi

Q8   Perform each of the following conversions:

        a. $72 \, \text{km} = \underline{\hspace{2cm}} \text{mi}$ **b.** $50 \, \text{mi} = \underline{\hspace{2cm}} \text{km}$

STOP   •   STOP   •   STOP   •   STOP   •   STOP   •   STOP   •   STOP   •   STOP   •   STOP

A8   **a.** 43.2   **b.** 80

---

6   A summary of the approximate relationships between metric and English units of length is:

        $1 \, \text{m} = 39.37 \, \text{in}$

        $1 \, \text{cm} = 0.4 \, \text{in}$ or $1 \, \text{in} = 2.5 \, \text{cm} \, (25 \, \text{mm})$
        $1 \, \text{m} = 3.3 \, \text{ft}$ $1 \, \text{ft} = 0.3 \, \text{m}$
        $1 \, \text{m} = 1.1 \, \text{yd}$ $1 \, \text{yd} = 0.9 \, \text{m}$
        $1 \, \text{km} = 0.6 \, \text{mi}$ $1 \, \text{mi} = 1.6 \, \text{km}$

Q9   Perform each of the following conversions:

        a. $32 \, \text{in} = \underline{\hspace{2cm}} \text{cm}$ **b.** $4.1 \, \text{m} = \underline{\hspace{2cm}} \text{in}$

    c. 10 mi = _____ km   **d.** 45 km = _____ mi

    e. 440 yd = _____ m   **f.** 1000 m = _____ yd

STOP • STOP • STOP • STOP • STOP • STOP • STOP • STOP • STOP

A9   **a.** 80   **b.** 161.417   **c.** 16
    d. 27   **e.** 396   **f.** 1,100

7   The following is a scale that gives approximate relationships between kilometers and miles. It can be used to check answers to conversion problems or to quickly estimate one reading in terms of the other.

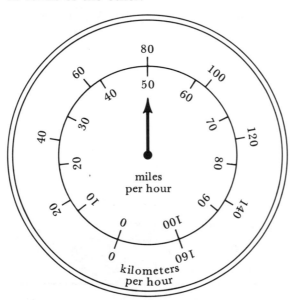

Q10   What is the approximate speed limit, in miles per hour, indicated by each of the following metric speed-limit signs (kilometers per hour):

a. speed limit 100 km/h _____ mph

b. speed limit 40 km/h _____ mph

c. speed limit 80 km/h _____ mph

STOP • STOP • STOP • STOP • STOP • STOP • STOP • STOP • STOP

A10   **a.** 63   **b.** 25   **c.** 50

Q11   Approximate the distance to each of the following cities:

       a. Detroit, 70 kilometers   **a.** _____

       b. Lansing, 105 kilometers   **b.** _____

       c. Mackinac Bridge, 450 kilometers   **c.** _____

STOP • **STOP** • **STOP** • **STOP** • **STOP** • **STOP** • **STOP** • **STOP** • **STOP**

A11   **a.** 42 mi   **b.** 63 mi   **c.** 270 mi

---

8   Units of area in the metric and English systems are related in the same way as their corresponding linear (length) units. Thus, since 1 meter is larger than 1 yard, 1 square meter is larger than 1 square yard. Similarly, just as 1 inch is larger than 1 cm, 1 square inch is larger than 1 square centimeter.

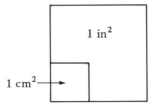

Some of the approximate relationships for units of area are:

$$1 \text{ cm}^2 = 0.16 \text{ in}^2 \quad \text{or} \quad 1 \text{ in}^2 = 6.5 \text{ cm}^2$$
$$1 \text{ m}^2 = 11 \text{ ft}^2 \qquad\qquad\quad 1 \text{ ft}^2 = 0.09 \text{ m}^2$$
$$1 \text{ m}^2 = 1.2 \text{ yd}^2 \qquad\qquad 1 \text{ yd}^2 = 0.8 \text{ m}^2$$

Q12   A 9 × 12 yd rug covers an area of 108 yd². What is the area measure in square meters?

       $108 \text{ yd}^2 = $ _____ m^2

STOP • **STOP** • **STOP** • **STOP** • **STOP** • **STOP** • **STOP** • **STOP** • **STOP**

A12   86.4: $108 \text{ yd}^2 = $ _____ m^2
                 $1 \text{ yd}^2 = 0.8 \text{ m}^2$
        $108(1 \text{ yd}^2) = 108(0.8 \text{ m}^2)$
           $108 \text{ yd}^2 = 86.4 \text{ m}^2$

Q13   The area of a rectangle is 300 cm². Find the equivalent area in square inches.

$$300 \text{ cm}^2 = \underline{\hspace{1.5cm}} \text{ in}^2$$

STOP • STOP • STOP • STOP • STOP • STOP • STOP • STOP • STOP

A13   48: $300 \text{ cm}^2 = \underline{\hspace{1.5cm}} \text{ in}^2$
$$1 \text{ cm}^2 = 0.16 \text{ in}^2$$
$$300(1 \text{ cm}^2) = 300(0.16 \text{ in}^2)$$
$$300 \text{ cm}^2 = 48 \text{ in}^2$$

Q14   Complete each of the following:

  a. $4.7 \text{ m}^2 = \underline{\hspace{1.5cm}} \text{ ft}^2$ **b.** $\underline{\hspace{1.5cm}} \text{ cm}^2 = 150 \text{ in}^2$

STOP • STOP • STOP • STOP • STOP • STOP • STOP • STOP • STOP

A14   **a.** 51.7   **b.** 975

---

9   Large areas are measured in hectares or square kilometers in the metric system of measurement and in acres or square miles in the English system of measurement. These units are related as follows:

  1 hectare $(10\,000 \text{ m}^2) = 2.5$ acres   or   1 acre $= 0.4$ hectare
  $1 \text{ km}^2 = 0.39 \text{ mi}^2$ $1 \text{ mi}^2 = 2.6 \text{ km}^2$

Q15   The area of a field is 40 acres. What is its measure in hectares? 40 acres $= \underline{\hspace{1.5cm}}$ hectare

STOP • STOP • STOP • STOP • STOP • STOP • STOP • STOP • STOP

A15   16

Q16   Complete each of the following:

  a. $10 \text{ mi}^2 = \underline{\hspace{1.5cm}} \text{ km}^2$ **b.** $5000 \text{ km}^2 = \underline{\hspace{1.5cm}} \text{ mi}^2$

STOP • STOP • STOP • STOP • STOP • STOP • STOP • STOP • STOP

A16   **a.** 26   **b.** 1,950

10   Recall from Section 11.2 that the metric units gram and kilogram were used in the same way as the English units ounce and pound. The approximate relationships between these units are:

  1 g = 0.035 oz   or   1 oz = 28.3 g
  1 kg = 2.2 lb   or   1 lb = 454 g (0.454 kg)

To help yourself think metric, notice that a kilogram is more than twice as large as a pound, while a gram is only approximately $\frac{1}{28}$ of an ounce. Some common comparisons with the gram are the weight of a paper clip (1 g), the weight of a nickel (5 g), and the weight of a pound of butter (454 g).

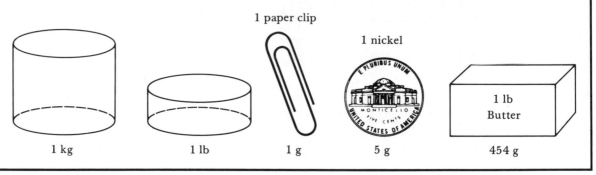

1 kg	1 lb	1 g	5 g	454 g

Q17   Write the following in order from smallest to largest:

g, lb, kg, oz, mg _____

STOP • **STOP** • **STOP** • **STOP** • **STOP** • **STOP** • **STOP** • **STOP** • **STOP**

A17   mg, g, oz, lb, kg

Q18   Robert weighs 70 kg and John's weight is 150 lb. Who is heavier? _____

STOP • **STOP** • **STOP** • **STOP** • **STOP** • **STOP** • **STOP** • **STOP** • **STOP**

A18   Robert: 70 kg = 154 lb

Q19   350 g = _____ oz

STOP • **STOP** • **STOP** • **STOP** • **STOP** • **STOP** • **STOP** • **STOP** • **STOP**

A19   12.25:   350 g = _____ oz
                1 g = 0.035 oz
        350(1 g) = 350(0.035 oz)
          350 g = 12.25 oz

Q20   _____ g = 8 oz

A20   226.4: _____ g = 8 oz

$$8 \text{ oz} = \underline{\hspace{1cm}} \text{ g}$$
$$1 \text{ oz} = 28.3 \text{ g}$$
$$8(1 \text{ oz}) = 8(28.3 \text{ g})$$
$$8 \text{ oz} = 226.4 \text{ g}$$

---

11   Notice that since 8 oz = $\frac{1}{2}$ lb, Q20 could have been completed differently. Since 1 lb = 454 g, the equivalent of $\frac{1}{2}$ lb is just $\frac{1}{2}$ (454 g) or 227 g. The fact that this answer is different from that of A20 (226.4 g) brings up an important point about performing conversions with *approximate* relationships between units. When using approximate relations between units, answers may vary slightly depending on the approximation that is used. Although the answers vary slightly, both are considered correct.

Q21   Perform the conversion 16 oz = _____ g in the two ways indicated:

    a. 16 oz = _____ g (use 1 oz = 28.3 g)

    b. 16 oz (1 lb) = _____ g (use 1 lb = 454 g)

    c. Do the answers to parts a and b differ slightly?   _____

    d. Which answer is considered correct?   _____

A21   **a.** 452.8   **b.** 454   **c.** yes   **d.** both

Q22   Perform each of the following conversions:

    a. 2.5 oz = _____ g   **b.** 10 g = _____ oz

       c. $\frac{3}{4}$ lb = _____ g   d. 175 lb = _____ kg

STOP • STOP • STOP • STOP • STOP • STOP • STOP • STOP • STOP

A22   a. 70.75   b. 0.35   c. 340.5   d. 79.45

---

12   Common units of liquid volume in the metric and English systems are the liter and quart, respectively. Recall from Section 11.2 that a liter is slightly larger than a quart. More specifically,

1 l = 1.05 qt or 1 qt = 0.95 l

1 liter   1 quart

Q23   a. Which is larger, a quart of milk or a liter of milk?   _____

         b. Which is larger, $\frac{1}{2}$ gal of milk (2 qt) or 2 l of milk?   _____

         c. Which is larger, 1 pt ($\frac{1}{2}$ qt) of cream or $\frac{1}{2}$ l?   _____

         d. Which is larger, 1 cup ($\frac{1}{4}$ qt) of tea or $\frac{1}{4}$ l?   _____

STOP • STOP • STOP • STOP • STOP • STOP • STOP • STOP • STOP

A23   a. liter   b. 2 l   c. $\frac{1}{2}$ l   d. $\frac{1}{4}$ l

---

Q24   a. 4 qt = _____ l   b. 8 l = _____ qt

STOP • STOP • STOP • STOP • STOP • STOP • STOP • STOP • STOP

A24   a. 3.8   b. 8.4

---

13   The approximate relationships among cup, pint, and liter are as follows:

     1 l = 2.1 pt   or   1 pt = 0.47 l

     1 l = 4.2 cups   or   1 cup = 0.24 l

Q25   Perform each of the following conversions:

a. 2 pt = _____ l   b. 0.5 l = _____ pt

c. 4 l = _____ cups   d. 4 cups = _____ l

STOP • STOP • STOP • STOP • STOP • STOP • STOP • STOP • STOP

A25   **a.** 0.94   **b.** 1.05   **c.** 16.8   **d.** 0.96

---

14   Small amounts of liquid volume are measured in milliliters or ounces.

1 ml = 0.03 oz   or   1 oz = 30 ml

The approximate size of a milliliter can be imagined from the fact that it takes 5 milliliters to equal 1 teaspoonful.

Q26   **a.** Which is larger, 1 ml or 1 oz? _____
b. Write the following in order from smallest to largest: qt, l, pt, ml, oz

_____

STOP • STOP • STOP • STOP • STOP • STOP • STOP • STOP • STOP

A26   **a.** 1 oz   **b.** ml, oz, pt, qt, l

Q27   **a.** 195 ml = _____ oz   **b.** 32 oz = _____ ml

STOP • STOP • STOP • STOP • STOP • STOP • STOP • STOP • STOP

A27   **a.** 5.85   **b.** 960

---

15   Volume measures in cubic units are related in the metric and English systems by the following:

$1 \text{ cm}^3 = 0.06 \text{ in}^3$ or $1 \text{ in}^3 = 16 \text{ cm}^3$
$1 \text{ m}^3 = 35 \text{ ft}^3$ $1 \text{ ft}^3 = 0.03 \text{ m}^3$
$1 \text{ m}^3 = 1.3 \text{ yd}^3$ $1 \text{ yd}^3 = 0.8 \text{ m}^3$

Q28   Perform each of the following conversions:

     a. $12 \, m^3 =$ _____ yd^3 **b.** $2.5 \, ft^3 =$ _____ m^3

     c. $1.2 \, in^3 =$ _____ cm^3 **d.** $0.5 \, yd^3 =$ _____ m^2

STOP • STOP • STOP • STOP • STOP • STOP • STOP • STOP • STOP

A28   **a.** 15.6   **b.** 0.075   **c.** 19.2   **d.** 0.4

---

16   Temperature in the metric system is measured on the Kelvin and the Celsius scales. The English system uses the Fahrenheit scale for temperature measurement. The three scales are compared:

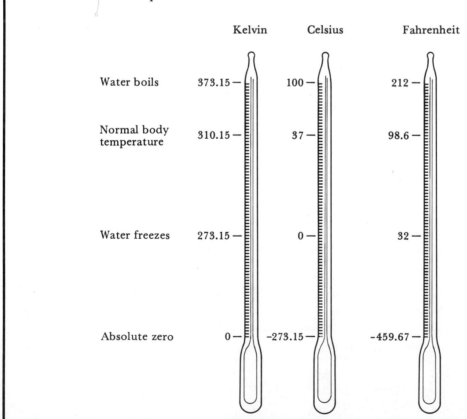

	Kelvin	Celsius	Fahrenheit
Water boils	373.15	100	212
Normal body temperature	310.15	37	98.6
Water freezes	273.15	0	32
Absolute zero	0	−273.15	−459.67

Q29   Complete the table:

	Normal body temperature	Freezing point	Boiling point
a. Celsius	_____	_____	_____
b. Fahrenheit	_____	_____	_____

STOP • **STOP** • **STOP** • **STOP** • **STOP** • **STOP** • **STOP** • **STOP** • **STOP**

A29   **a.** 37 °C, 0 °C, 100 °C   **b.** 98.6 °F, 32 °F, 212 °F

---

17   The conversion formulas for Celsius (C) and Fahrenheit (F) temperatures are as follows:

$$C = \frac{F - 32}{1.8} \qquad F = 1.8C + 32$$

To change 75 °F to its corresponding Celsius temperature, we use the formula $C = \frac{F - 32}{1.8}$.

$$C = \frac{F - 32}{1.8}$$

$$C = \frac{75 - 32}{1.8}$$

$$C = \frac{43}{1.8}$$

$C = 23.9$ Hence, 75 °F = 23.9 °C.

All temperatures will be rounded off to the nearest tenth.

Q30   50 °F = _____ °C

STOP • **STOP** • **STOP** • **STOP** • **STOP** • **STOP** • **STOP** • **STOP** • **STOP**

A30   10: $C = \dfrac{F - 32}{1.8}$

$$C = \frac{50 - 32}{1.8}$$

$$C = \frac{18}{1.8}$$

$C = 10$

---

18   To change 25 °C to its corresponding Fahrenheit temperature, we use the formula F = 1.8C + 32 as follows:

F = 1.8C + 32
F = 1.8(25) + 32
F = 45 + 32
F = 77   Hence, 25 °C = 77 °F

Q31   80 °C = _____ °F

STOP • STOP • STOP • STOP • STOP • STOP • STOP • STOP • STOP

A31   176: F = 1.8C + 32
                   F = 1.8(80) + 32
                   F = 144 + 32
                   F = 176

Q32   Complete each of the following:

a. 95 °F = _____ °C   b. 0 °C = _____ °F

c. 0 °F = _____ °C   d. 10 °C = _____ °F

STOP • STOP • STOP • STOP • STOP • STOP • STOP • STOP • STOP

A32   a. 35   b. 32   c. ⁻17.8   d. 50

| 19 | Notice in the formula F = 1.8C + 32 that a Fahrenheit temperature is approximately two (1.8) times the Celsius temperature plus 32. For example, if C = 20, F is approximately 2(20) + 32 = 72. To approximate the Fahrenheit equivalent of a Celsius temperature, therefore, double the Celsius temperature and add 32. |

Q33   Use the procedure of Frame 19 to match the Celsius temperatures with the appropriate
             description:

a. 35 °C   _____   (1) freezing point of water

b. 90 °C   _____   (2) spring or summer day

c. 0 °C   _____   (3) heat-wave conditions

d. 25 °C   _____   (4) a winter day

e. 5 °C   _____   (5) boiling point of water

STOP • STOP • STOP • STOP • STOP • STOP • STOP • STOP • STOP

A33   a. (3)   b. (5)   c. (1)   d. (2)   e. (4)

| 20 | The following chart is a comparison of the Celsius and Fahrenheit scales. It can be used for quick approximations between the two scales. |

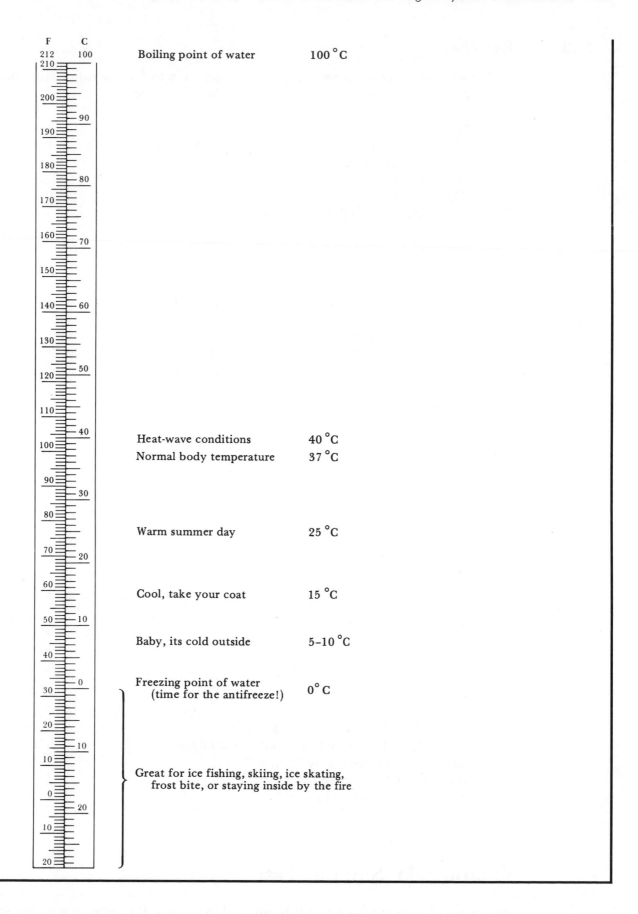

Boiling point of water   100 °C

Heat-wave conditions   40 °C
Normal body temperature   37 °C

Warm summer day   25 °C

Cool, take your coat   15 °C

Baby, its cold outside   5–10 °C

Freezing point of water
   (time for the antifreeze!)   0° C

Great for ice fishing, skiing, ice skating,
   frost bite, or staying inside by the fire

This completes the instruction for this section.

11.3   Exercises

1. Place $>$ or $<$ between each of the following pairs to form a true statement:
    a. m _____ yd   b. g _____ oz
    c. l _____ qt   d. km _____ mi
    e. kg _____ lb
2. Place each of the following sets of units in order from smaller to larger:
    a. m, cm, in, yd, mm, ft, km, mi
    b. oz, g, lb, kg, mg
    c. ml, oz, qt, l, gal
3. Perform each of the following conversions:
    a. 10 m = _____ yd   b. 4.5 m = _____ ft
    c. 0.5 m = _____ in   d. 2.5 cm = _____ in
    e. 7 lb = _____ kg   f. 1.5 kg = _____ lb
    g. 250 g = _____ oz   h. 4 l = _____ qt
    i. 1 pt = _____ l   j. 1 gal = _____ l
4. Complete each of the following:
    a. 50 cm^3 = _____ in^3 b. 144 yd^3 = _____ m^3
    c. 16 m^3 = _____ ft^3 d. 1.5 m^2 = _____ ft^2
    e. 81 yd^2 = _____ m^2 f. 85 acres = _____ hectare
5. Perform the following conversions between the Celsius and Fahrenheit temperature scales:
    a. 212 °F = _____ °C   b. 45 °C = _____ °F
    c. 0 °F = _____ °C   d. 0 °C = _____ °F
6. Do the following make sense?
    a. gas purchase of 40 l   _____
    b. man's height of 6 m   _____
    c. speed limit of 120 km/h   _____
    d. waist size of 70 cm   _____

11.3   Exercise Answers

1. a. $>$ b. $<$ c. $>$ d. $<$ e. $>$
2. a. mm, cm, in, ft, yd, m, km, mi
    b. mg, g, oz, lb, kg
    c. ml, oz, qt, l, gal
3. a. 11   b. 14.85   c. 19.685 d. 1   e. 3.178 f. 3.3   g. 8.75
    h. 4.2   i. 0.47   j. 3.8
4. a. 3   b. 115.2   c. 560   d. 16.5   e. 64.8   f. 34
5. a. 100   b. 113   c. $^-$17.8 d. 32
6. a. yes: 40 l = 38 qt, which is about 9$\frac{1}{2}$ gal of gas.
    b. no: 6 m is approximately 18 ft.
    c. yes: 120 km/h is approximately 75 mi/h.
    d. yes: 70 cm is approximately 28 in.

Chapter 11 Sample Test

At the completion of Chapter 11 it is expected that you will be able to work the following problems.

11.1   English System of Measurement

1. Complete each of the following relationships between linear units to make a true statement:

   a. _____ in = 1 ft
   b. _____ ft = 1 yd
   c. 1 yd = _____ in
   d. _____ yd = 1 mi
   e. 1 mi = _____ ft

2. Complete the following equations:

   a. 4.5 ft = _____ in
   b. 1,100 yd = _____ mi
   c. _____ yd = 50 ft
   d. 726 ft = _____ mi

3. Complete each of the following relationships between area units to make a true statement:

   a. 1 yd^2 = _____ ft^2
   b. 1 ft^2 = _____ in^2
   c. 1 acre = _____ yd^2
   d. 1 mi^2 = _____ acres

4. Complete each of the following equations:

   a. 3.3 mi^2 = _____ acres
   b. 12 yd^2 = _____ ft^2
   c. _____ acres = 968 yd^2
   d. 804 in^2 = _____ ft^2

5. Complete each of the following relationships between weight units or volume units to form a true statement:

   a. _____ oz = 1 lb
   b. _____ lb = 1 ton
   c. _____ oz = 1 pt
   d. _____ oz = 1 qt
   e. _____ pt = 1 qt
   f. _____ oz = 1 gal

6. Complete each of the following equations:

   a. 9 pt = _____ qt
   b. 2 qt = _____ gal
   c. 1,500 lb = _____ ton
   d. 56 oz = _____ lb

11.2   Metric System of Measurement

7. List a common unit of measure in the metric system for each of the following:
   a. weight
   b. volume
   c. length

8. Place > or < between each of the following to form a true statement:

   a. 1 qt _____ 1 l
   b. 1 yd _____ 1 m
   c. 1 kg _____ 1 lb
   d. 1 m _____ 1 cm
   e. 1 dam _____ 1 hm
   f. 1 cm _____ 1 mm

9. Find the product or quotient:

   a. $\frac{52.7}{100}$
   b. (20.5)(1000)
   c. 234(0.01)
   d. (0.0072)(10)

10. Complete each of the following using a multiple of 10 (10, 100, 1000) or a division of 10 (0.1, 0.01, 0.001):

    a. 1 cm = _____ mm
    b. _____ kg = 1 g
    c. _____ ml = 1 l
    d. 1 dam = _____ hm
    e. _____ m = 1 cm
    f. _____ cg = 1 g

11. Complete each of the following equations:

    a. 7.35 km = _____ m
    b. 70 g = _____ cg

    c. _____ l = 350 ml   **d.** 0.45 cm = _____ mm

    e. 1.75 hg = _____ dag   **f.** _____ dg = 0.05 mg

12. A common unit of area in the metric system is the _____.

13. Consecutive units of area from mm² to m² are related by multiples or divisions of _____.

14. Complete each of the following equations:

    a. 400 ha = _____ a   **b.** 25.5 mm² = _____ cm²

    c. _____ a = 50 m²   **d.** 100 cm² = _____ dm²

15. Volume is measured in liquid units or cubic units in the metric system of measurement.

    a. A common metric unit of liquid volume is the _____.

    b. A common metric unit of volume measured in cubic units is the _____.

    c. Consecutive units of volume measured in cubic units are related by a multiple or division of _____.

16. Complete each of the following equations:

    a. 250 ml = _____ l   **b.** 2000 cm³ = _____ m³

    c. 4.7 cm³ = _____ ml   **d.** 0.5 m³ = _____ cm³

    e. _____ ml = 2.7 l   **f.** _____ kl = 25 000 l

17. Temperature in the metric system is measured in the _____ and _____ scales.

18. Complete the table:

	Celsius	Kelvin
a. Water boils	_____	_____
b. Water freezes	_____	_____
c. Normal body temperature	_____	_____
d. Absolute zero	_____	_____

19. Complete each of the following equations:

    a. 25 °C = _____ K   **b.** ⁻273.15 °C = _____ K

    c. 73 K = _____ °C   **d.** 300 K = _____ °C

11.3   Conversions Between Metric and English Systems of Measurement

20. Place $>$ or $<$ between each of the following to form a true statement:

    a. cm _____ in   **b.** m _____ ft

    c. yd _____ m   **d.** km _____ mi

    e. l _____ qt   **f.** oz _____ g

    g. kg _____ lb   **h.** mm _____ in

21. Place each of the following sets in order from _larger_ to _smaller:_

    a. g, oz, mg, lb, kg

    b. m, mi, cm, km, in, yd, ft, mm

    c. gal, ml, l, oz, qt

22. Perform each of the following conversions:

    a. 4.2 yd = _____ m   **b.** 35 km = _____ mi

    c. _____ in = 25 cm   **d.** 3 m = _____ yd

    e. 52 g = _____ oz   **f.** _____ g = 2.5 lb

    g. 6.6 lb = _____ kg   **h.** 4 qt = _____ l

    i. 2 pt = _____ l   **j.** 1 l = _____ cups

    k. 1 oz = _____ ml   **l.** _____ oz = 100 ml

23. Complete each of the following:

    a. 3.5 m² = _____ yd²   **b.** 0.5 in² = _____ cm²

    c. 70 acres = _____ ha   **d.** 1000 m³ = _____ ft³

    e. 3 in³ = _____ cm³

24. Complete the table:

	Celsius	Fahrenheit
a. Normal body temperature	_____	_____
b. Water freezes	_____	_____
c. Water boils	_____	_____

25. Perform the following conversions:
    a. 5 °C = _____ °F   **b.** 68 °F = _____ °C
    c. 80 °C = _____ °F   **d.** 50 °F = _____ °C

26. Do the following make sense?
    a. 3 × 4 m rug   **b.** 1 kg of meat
    c. hat size of 17.5 cm   **d.** snow at 32 °C
    e. woman's weight of 140 kg

Chapter 11 Sample Test Answers

1. a. 12   **b.** 3   **c.** 36   **d.** 1,760
  e. 5,280
2. a. 54   **b.** $\frac{5}{8}$ **c.** $16\frac{2}{3}$ **d.** $\frac{11}{80}$ or 0.1375
3. a. 9   **b.** 144   **c.** 4,840   **d.** 640
4. a. 2,112   **b.** 108   **c.** $\frac{1}{5}$ **d.** $5\frac{7}{12}$
5. a. 16   **b.** 2,000   **c.** 16   **d.** 32
  e. 2   **f.** 128
6. a. $4\frac{1}{2}$ **b.** $\frac{1}{2}$ **c.** $\frac{3}{4}$ **d.** $3\frac{1}{2}$
7. a. kilogram   **b.** liter   **c.** meter
8. a. <   **b.** <   **c.** >   **d.** >
  e. <   **f.** >
9. a. 0.527   **b.** 20 500   **c.** 2.34   **d.** 0.072
10. a. 10   **b.** 0.001   **c.** 1000   **d.** 0.1
  e. 0.01   **f.** 100
11. a. 7350   **b.** 7000   **c.** 0.35   **d.** 4.5
  e. 17.5   **f.** 0.0005
12. m²
13. 100
14. a. 40 000   **b.** 0.255   **c.** 0.5   **d.** 1
15. a. liter   **b.** m³   **c.** 1000
16. a. 0.25   **b.** 0.002   **c.** 4.7   **d.** 500 000
  e. 2700   **f.** 25
17. Kelvin, Celsius
18. a. 100 °C, 373.15 K   **b.** 0 °C, 273.15 K   **c.** 37 °C, 310.15 K
  d. ⁻273.15 °C, 0 K
19. a. 298.15   **b.** 0   **c.** ⁻200.15   **d.** 26.85
20. a. <   **b.** >   **c.** <   **d.** <
  e. >   **f.** >   **g.** >   **h.** <
21. a. kg, lb, oz, g, mg   **b.** mi, km, m, yd, ft, in, cm, mm
  c. gal, l, qt, oz, ml
22. a. 3.78   **b.** 21   **c.** 10   **d.** 3.3
  e. 1.82   **f.** 1135   **g.** 2.9964   **h.** 3.8
  i. 0.94   **j.** 4.2   **k.** 30   **l.** 3
23. a. 4.2   **b.** 3.25   **c.** 28   **d.** 35,000
  e. 48

24. **a.** 37 °C, 98.6 °F　**b.** 0 °C, 32 °F　　**c.** 100 °C, 212 °F
25. **a.** 41　　　　　　**b.** 20　　　　　**c.** 176　　　　　　**d.** 10
26. **a.** yes: 3 × 4 m is approximately 9 × 12 ft.
　　b. yes: 1 kg ≐ 2.2 lb.
　　c. yes: 17.5 cm ≐ 7 in.
　　d. no: 32 °C ≐ 89.6 °F.
　　e. no: 140 kg ≐ 308 lb.

Chapter 12

Ratio and Proportion

12.1   Introduction to Ratio and Proportion

1	A fraction can be used to compare two quantities or numbers. Suppose that Sally has $13 and Linda has $39. Sally then has $\frac{13}{39}$ or $\frac{1}{3}$ as much money as Linda. When a fraction is used in this way to compare two numbers, the fraction is called a ratio. A *ratio* is the quotient of two quantities or numbers. The ratio $\frac{1}{3}$ is read "one to three" and can be written using a colon as $1:3$.

Q1   **a.** 4 of 7 square regions are shaded. Write the ratio of the shaded regions to the total region. _____

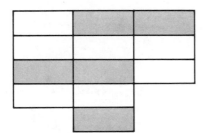

b. Of the 12 rectangular regions, 5 are shaded. Write the ratio of shaded regions to the total region. _____

c. 6 of the 11 hexagonal regions are shaded. Write the indicated ratio. _____

d. 5 of the 11 hexagonal regions are *not* shaded. What ratio represents this statement? _____

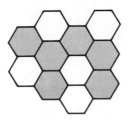

e. _____ of 6 circular regions are shaded.

f. Write the ratio. _____

g. 4 of _____ squares are circled.

h. Write the ratio. _____

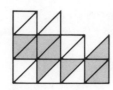

i. _____ of _____ triangular regions are shaded.

j. Write the indicated ratio. _____

STOP • STOP • STOP • STOP • STOP • STOP • STOP • STOP • STOP

A1   **a.** $\frac{4}{7}$ or $4:7$ **b.** $\frac{5}{12}$ or $5:12$ **c.** $\frac{6}{11}$ or $6:11$ **d.** $\frac{5}{11}$ or $5:11$

e. 5   **f.** $\frac{5}{6}$ or $5:6$ **g.** 9   **h.** $\frac{4}{9}$ or $4:9$

i. 9, 18   **j.** $\frac{9}{18}$ or $9:18\left(\frac{1}{2}$ or $1:2\right)$

Q2   The comparison of one quantity to another by division is called a _____.

STOP • STOP • STOP • STOP • STOP • STOP • STOP • STOP • STOP

A2   ratio

Q3   **a.** What is the ratio of 24 years to 9 years?   _____

b. What is the ratio of 9 years to 24 years?   _____

STOP • STOP • STOP • STOP • STOP • STOP • STOP • STOP • STOP

A3   **a.** $\frac{8}{3}:\frac{24}{9}=\frac{8}{3}$ **b.** $\frac{3}{8}$

Q4   Find the following ratios as fractions reduced to the lowest terms:

a. 25 to 125   **b.** 25 to 10   **c.** 25 to 5   **d.** 25 to $\frac{1}{5}$

e. $\frac{1}{4}$ to $\frac{1}{2}$ **f.** $\frac{1}{4}$ to 2

STOP • STOP • STOP • STOP • STOP • STOP • STOP • STOP • STOP

A4   **a.** $\frac{1}{5}$ **b.** $\frac{5}{2}$ **c.** $\frac{5}{1}$ **d.** $\frac{125}{1}$

e. $\frac{1}{2}$ **f.** $\frac{1}{8}$

Q5   The ratio $\frac{1}{2}$ is read "_____."

STOP • STOP • STOP • STOP • STOP • STOP • STOP • STOP • STOP

A5   one to two

---

2   Writing the ratio of 2 feet to 15 inches as $\frac{2}{15}$ would be incorrect, because 2 feet is not

$\frac{2}{15}$ of 15 inches. Two feet is actually 24 inches, so the ratio is really $\frac{24}{15}$ or $\frac{8}{5}$.

> When a comparison between two quantities is desired, the quantities should be expressed in the same unit of measurement. However, the ratio is an abstract number (written without a unit of measurement).

Q6   Determine the ratio of $5 to 40 cents.

STOP • **STOP** • **STOP** • **STOP** • **STOP** • **STOP** • **STOP** • **STOP** • **STOP**

A6 $\frac{25}{2}$: $5 = 500 cents, hence, $\frac{500}{40}$

Q7   Express as a ratio:
    a. 15 inches to 2 feet.   **b.** 10 ounces to 1 pound.

    c. 45 minutes to 1 hour.   **d.** a nickel to a dime.

STOP • **STOP** • **STOP** • **STOP** • **STOP** • **STOP** • **STOP** • **STOP** • **STOP**

A7   **a.** $\frac{5}{8}$: 2 feet = 24 inches   **b.** $\frac{5}{8}$: 1 pound = 16 ounces

    c. $\frac{3}{4}$: 1 hour = 60 minutes   **d.** $\frac{1}{2}$

---

3   A statement of equality between two ratios (or fractions) is called a *proportion*. A proportion is actually a special type of equation. The statement $\frac{1}{3} = \frac{13}{39}$ is a proportion and could be written

1:3 = 13:39

    A proportion consists of four terms labeled as follows:

1st term:2nd term = 3rd term:4th term

In the proportion $\frac{1}{3} = \frac{13}{39}$, the first term is 1, the second term is 3, the third term is 13, and the fourth term is 39.

Q8   **a.** A statement of equality between two ratios is called a _____.

    b. The four parts of a proportion are called _____.

    c. In the proportion $\frac{2}{3} = \frac{10}{15}$, what is the third term? _____

    d. In the proportion 1:8 = 2:16, what is the fourth term? _____

STOP • **STOP** • **STOP** • **STOP** • **STOP** • **STOP** • **STOP** • **STOP** • **STOP**

A8   **a.** proportion   **b.** terms   **c.** 10   **d.** 16

Q9 Use colons to write the proportion $\dfrac{7}{2} = \dfrac{21}{6}$.

STOP • STOP • STOP • STOP • STOP • STOP • STOP • STOP • STOP

A9 $7:2 = 21:6$

Q10 Write the proportion $25:5 = 5:1$ in fraction **form**.

STOP • STOP • STOP • STOP • STOP • STOP • STOP • STOP • STOP

A10 $\dfrac{25}{5} = \dfrac{5}{1}$

---

4 In a proportion, the first and fourth terms are called the *extremes* and the second and third terms are called the *means*. In the proportion

$$\frac{2}{5} = \frac{14}{35}$$

2 and 35 are the extremes and 5 and 14 are the means.

$2:5 = 14:35$

An important property of proportions is that *the product of the means is equal to the product of the extremes.* In the proportion $\dfrac{2}{5} = \dfrac{14}{35}$ the product of the means is $5 \cdot 14 = 70$ and the product of the extremes is $2 \cdot 35 = 70$.

Q11 **a.** In a proportion the first and fourth terms are called the _____.

  b. In a proportion the product of the _____ is equal to the product of the

  _____.

  c. Name the means of $\dfrac{7}{21} = \dfrac{1}{3}$. _____

  d. What is the product of the extremes in $\dfrac{2}{5} = \dfrac{8}{20}$? _____

STOP • STOP • STOP • STOP • STOP • STOP • STOP • STOP • STOP

A11 **a.** extremes **b.** means, extremes **c.** 21 and 1 **d.** 40: $2 \cdot 20$

---

5 The truth of a proportion can be determined by using the property that the product of the means is equal to the product of the extremes. The statement of equality between the ratios will be true only if the above property is satisfied. The proportion $\dfrac{7}{2} = \dfrac{14}{4}$ is true because

$$\underbrace{2 \cdot 14}_{} = \underbrace{7 \cdot 4}_{}$$

product of means   product of extremes

However, the proportion $\dfrac{6}{7} = \dfrac{5}{8}$ is false because

$7 \cdot 5 \neq 6 \cdot 8$

Q12   **a.** In the proportion $\dfrac{7}{8} = \dfrac{8}{9}$ the product of the means is _____ and

b. the product of the extremes is _____.

c. Is the proportion true or false? _____.

STOP • **STOP** • **STOP** • **STOP** • **STOP** • **STOP** • **STOP** • **STOP** • **STOP**

A12   **a.** $8 \cdot 8 = 64$ **b.** $7 \cdot 9 = 63$ **c.** false

Q13   Express $\dfrac{7}{3} = \dfrac{21}{9}$ as a product of its means and extremes.

STOP • **STOP** • **STOP** • **STOP** • **STOP** • **STOP** • **STOP** • **STOP** • **STOP**

A13 $3 \cdot 21 = 7 \cdot 9$

6   The property that the product of the means is equal to the product of the extremes, in a proportion, can be developed and generalized using the principles of equality. That is, for the proportion

$$\frac{a}{b} = \frac{c}{d}$$

the means are b and c and the extremes are a and d. The proportion can be rewritten as follows:

$$\frac{a}{b} = \frac{c}{d}$$

$$bd\left(\frac{a}{b}\right) = bd\left(\frac{c}{d}\right)$$ (multiplying both sides by the LCD, bd)

Hence,

$ad = bc$

which states that the product of the extremes is equal to the product of the means. The last equation could also be written

$bc = ad$

Q14   Is $\dfrac{7}{3} = \dfrac{21}{9}$ true or false? _____

STOP • **STOP** • **STOP** • **STOP** • **STOP** • **STOP** • **STOP** • **STOP** • **STOP**

A14   true: $3 \cdot 21 = 7 \cdot 9$
       $63 = 63$

Q15   Is each of the following a true proportion?

a. $\dfrac{7}{14} = \dfrac{21}{42}$ _____

b. $\dfrac{4}{6} = \dfrac{20}{24}$ _____

c. $3:8 = 15:40$ _____

d. $6:9 = 48:72$ _____

e. $\dfrac{15}{20} = \dfrac{5}{4}$ _____

f. $\dfrac{9}{15} = \dfrac{72}{120}$ _____

STOP • **STOP** • **STOP** • **STOP** • **STOP** • **STOP** • **STOP** • **STOP** • **STOP**

A15   a. yes: $14 \cdot 21 = 7 \cdot 42$ b. no: $6 \cdot 20 \ne 4 \cdot 24$
       c. yes: $8 \cdot 15 = 3 \cdot 40$ d. yes: $9 \cdot 48 = 6 \cdot 72$
       e no: $20 \cdot 5 \ne 15 \cdot 4$ f. yes: $15 \cdot 72 = 9 \cdot 120$

Q16   Write the proportion $\dfrac{m}{n} = \dfrac{x}{y}$ so that the product of the means is equal to the product
       of the extremes.

STOP • **STOP** • **STOP** • **STOP** • **STOP** • **STOP** • **STOP** • **STOP** • **STOP**

A16 $nx = my$

This completes the instruction for this section.

12.1   Exercises

1. The comparison of one quantity to another by division is called a _____.
2. Find the following ratios in fraction form:

a. 16 to 8   b. $\dfrac{9}{27}$ c. 25 to 10   d. 6 to 3

e. $\dfrac{1}{4}$ to $\dfrac{1}{12}$ f. 0.2 to 2   g. 16 to 64   h. 9 to $\dfrac{1}{3}$

i. 25 to 5   j. 6 to 2   k. $\dfrac{1}{4}$ to 2   l. 0.2 to $\dfrac{1}{10}$

3. Find the following ratios in fraction form:

a. a quart to a pint   b. a quart to a gallon
c. a yard to two inches   d. an inch to a foot
e. five minutes to an hour f. a day to an hour

4. The statement of equality between two ratios is called a _____ .

5. Name the means in the proportion $\dfrac{6}{13} = \dfrac{12}{26}$.

6. Name the extremes in the proportion $1:5 = 7:35$.

7. In a proportion the product of the _____ is equal to the product of the _____ .

8. Which of the following represent true proportions?

   a. $\dfrac{7}{2} = \dfrac{35}{10}$ 	b. $\dfrac{2}{7} = \dfrac{3}{11}$

   c. $6:2 = 23:8$ 	d. $9:11 = 63:77$

   e. $\dfrac{7}{15} = \dfrac{8}{14}$ 	f. $\dfrac{3}{18} = \dfrac{1}{6}$

12.1 Exercise Answers

1. ratio

2. a. $\dfrac{2}{1}$ b. $\dfrac{1}{3}$ c. $\dfrac{5}{2}$ d. $\dfrac{2}{1}$ e. $\dfrac{3}{1}$ f. $\dfrac{1}{10}$

   g. $\dfrac{1}{4}$ h. $\dfrac{27}{1}$ i. $\dfrac{5}{1}$ j. $\dfrac{3}{1}$ k. $\dfrac{1}{8}$ l. $\dfrac{2}{1}$

3. a. $\dfrac{2}{1}$ b. $\dfrac{1}{4}$ c. $\dfrac{18}{1}$ d. $\dfrac{1}{12}$ e. $\dfrac{1}{12}$ f. $\dfrac{24}{1}$

4. proportion
5. 13 and 12
6. 1 and 35
7. means, extremes
8. a, d, and f

12.2 Solution of a Proportion

1   Since a proportion is a special type of equation, it can be solved using the principles established for solving equations. For example, the proportion

$$\frac{n}{3} = \frac{4}{21}$$

may be solved as follows:

$$\frac{n}{3} = \frac{4}{21}$$

$$3\left(\frac{n}{3}\right) = 3\left(\frac{4}{21}\right) \qquad \text{(multiplying both sides by 3)}$$

$$n = \frac{12}{21} = \frac{4}{7}$$

Q1   Solve the proportion $\dfrac{n}{14} = \dfrac{51}{7}$.

STOP • STOP • STOP • STOP • STOP • STOP • STOP • STOP • STOP

A1   102: $14\left(\dfrac{n}{14}\right) = 14\left(\dfrac{51}{7}\right)$

$n = 2 \cdot 51$

Q2   Solve the following proportions:

    a. $\dfrac{4}{7} = \dfrac{n}{28}$ **b.** $\dfrac{5}{14} = \dfrac{n}{42}$

STOP • STOP • STOP • STOP • STOP • STOP • STOP • STOP • STOP

A2   **a.** 16: $28\left(\dfrac{4}{7}\right) = 28\left(\dfrac{n}{28}\right)$ **b.** 15

Q3   Solve the following proportions:

    a. $\dfrac{4}{17} = \dfrac{n}{34}$ **b.** $\dfrac{n}{30} = \dfrac{70}{175}$

STOP • STOP • STOP • STOP • STOP • STOP • STOP • STOP • STOP

A3   **a.** 8   **b.** 12

---

2   The missing term in a proportion can be any of the four terms of the proportion. For example, in the proportion

$$\frac{15}{540} = \frac{2}{n}$$

the missing term is the fourth term. This proportion can be solved using principles of equality; however, the solution is simplified by using the property that the product of the means is equal to the product of the extremes. That is,

$$540 \cdot 2 = 15 \cdot n$$

$$\frac{540 \cdot 2}{15} = n$$

$$72 = n$$

Q4   Solve the proportion $\dfrac{12}{n} = \dfrac{72}{30}$.

STOP • STOP • STOP • STOP • STOP • STOP • STOP • STOP • STOP

A4   5: $n \cdot 72 = 12 \cdot 30$

$$n = \frac{12 \cdot 30}{72}$$

Q5   Solve the proportion $\dfrac{2}{n} = \dfrac{6}{15}$.

STOP • STOP • STOP • STOP • STOP • STOP • STOP • STOP • STOP

A5   5: $n \cdot 6 = 2 \cdot 15$

$$n = \frac{2 \cdot 15}{6}$$

Q6   Solve the following proportions:

    a. $\dfrac{18}{63} = \dfrac{440}{n}$ **b.** $\dfrac{108}{27} = \dfrac{52}{n}$

STOP • STOP • STOP • STOP • STOP • STOP • STOP • STOP • STOP

A6   **a.** 1,540: $63 \cdot 440 = 18 \cdot n$ **b.** 13

$$\frac{63 \cdot 440}{18} = n$$

Q7   Solve the following proportions:

    a. $\dfrac{n}{12} = \dfrac{2.5}{5}$ **b.** $\dfrac{0.75}{n} = \dfrac{19}{114}$

STOP • STOP • STOP • STOP • STOP • STOP • STOP • STOP • STOP

A7   **a.** 6   **b.** 4.5

3   Various types of problems can be solved by writing the conditions of the problem as a proportion and solving for the missing term. Two general statements will aid in properly placing the conditions of a problem in the proportion.

1. Each ratio should be a comparison of similar things.
2. In every proportion both ratios must be written in the same order of value.

Example: If 6 pencils cost 25 cents, how much will 12 pencils cost?

Solution

Let c represent the cost of 12 pencils and substitute the conditions of the problem in the following proportion:

$$\frac{\text{small number of pencils}}{\text{large number of pencils}} = \frac{\text{small cost}}{\text{large cost}}$$

Each ratio is a comparison of similar things. Each ratio is written in the same order of value. Hence,

$$\frac{6}{12} = \frac{25}{c}$$

$$12 \cdot 25 = 6 \cdot c$$

$$50 = c$$

Therefore, 12 pencils will cost 50 cents. The quantities in the above proportion are *directly* related because an increase in the number of pencils causes a corresponding increase in the cost.

Q8   If 3 cans of cola cost 25 cents, how much will 15 cans cost?

$$\frac{\text{large number}}{\text{small number}} = \frac{\text{large cost}}{\text{small cost}}$$

(Note that the order of the quantities in the ratios has been reversed.)

STOP • STOP • STOP • STOP • STOP • STOP • STOP • STOP • STOP

A8 $1.25: $\dfrac{15}{3} = \dfrac{c}{0.25}$ ⟵ large cost because 15 cans will cost more

Q9   In a proportion, if an increase in one quantity causes a corresponding increase in another quantity, the quantities are _____ related.

STOP • STOP • STOP • STOP • STOP • STOP • STOP • STOP • STOP

A9   directly

Q10   An increase in the height of an object will cause a corresponding _____ in the length of its shadow.
                                                      increase/decrease

STOP • STOP • STOP • STOP • STOP • STOP • STOP • STOP • STOP

A10   increase

Q11   If an 18-foot pole casts a 15-foot shadow, what will be the length of the shadow cast by a 50-foot pole?

STOP • **STOP** • **STOP** • **STOP** • **STOP** • **STOP** • **STOP** • **STOP** • **STOP**

A11 $41\frac{2}{3}$ feet: $\dfrac{18}{50} = \dfrac{15}{s}$ (s is the unknown length and will be larger than 15 feet)

Q12   A car can travel 47 miles on 3 gallons of gas. How far can it go on 18 gallons?

STOP • **STOP** • **STOP** • **STOP** • **STOP** • **STOP** • **STOP** • **STOP** • **STOP**

A12   282 miles: $\dfrac{18}{3} = \dfrac{N}{47}$ or $\dfrac{3}{18} = \dfrac{47}{N}$

Q13   If $1,000 worth of insurance cost $20.80, what will $15,000 of insurance cost?

STOP • **STOP** • **STOP** • **STOP** • **STOP** • **STOP** • **STOP** • **STOP** • **STOP**

A13 $312: $\dfrac{1}{15} = \dfrac{20.80}{N}$

Q14   If valve stems cost 3 for 10 cents, what is the cost of 11 valve stems?

STOP • **STOP** • **STOP** • **STOP** • **STOP** • **STOP** • **STOP** • **STOP** • **STOP**

A14   37 cents, rounded off to the nearest cent: $\dfrac{3}{11} = \dfrac{10}{c}$

Q15   If 3.5 tons of top soil cost $20, what will 12.5 tons of top soil cost?

STOP • **STOP** • **STOP** • **STOP** • **STOP** • **STOP** • **STOP** • **STOP** • **STOP**

A15 $71.43, rounded off to the nearest cent

4   If an increase in one quantity causes a corresponding decrease in another quantity or a decrease in one quantity causes a corresponding increase in the other, the quantities are *indirectly* or inversely related. For example, if the number of workers on a job is increased, the time required to do the work is decreased. Problems involving indirect relationships can be solved using the principles established earlier.

Example: If 2 workers can build a garage in 5 days, how long will it take 6 workers, assuming that they all work at the same rate?

Solution

Since the number of days required to build the garage becomes smaller as the number of workers becomes larger, the quantities vary indirectly. Thus,

$$\frac{2 \text{ (smaller no. workers)}}{6 \text{ (larger no. workers)}} = \frac{N \text{ (smaller no. days)}}{5 \text{ (larger no. days)}}$$

$$6N = 10$$

$$N = 1\frac{2}{3} \text{ days}$$

Q16   Three workers can build a house in 10 days. Will it take 8 workers working at the same rate more or less time to build the house? _____

STOP • STOP • STOP • STOP • STOP • STOP • STOP • STOP • STOP

A16   less

Q17   A plane flies from Detroit to Philadelphia in 1 hour and 40 minutes at 440 miles per hour. Will it take more or less time to make the trip flying at 380 miles per hour? _____

STOP • STOP • STOP • STOP • STOP • STOP • STOP • STOP • STOP

A17   more

Q18   Eight students sell 3,800 tickets to a football game. Will it take more or fewer students to sell 5,200 tickets selling at the same rate? _____

STOP • STOP • STOP • STOP • STOP • STOP • STOP • STOP • STOP

A18   more

Q19   If 4 men can clear the snow near a school in 6 hours, how long will it take 12 men working at the same rate to do the job?

$$\frac{4}{12} = \frac{(\ )}{(\ )}$$

(Let N equal the unknown time and ask yourself, "Does the time get smaller or larger?" The smaller time must be placed in the numerator and the larger time in the denominator.)

STOP • STOP • STOP • STOP • STOP • STOP • STOP • STOP • STOP

A19   2:   smaller number of men $\longrightarrow \dfrac{4}{12} = \dfrac{N}{6} \longleftarrow$ smaller time

Q20   A plane takes 3 hours at a speed of 320 miles per hour to go from Chicago to New York. How fast must the plane fly to make the trip in 2.5 hours?

STOP • STOP • STOP • STOP • STOP • STOP • STOP • STOP • STOP

A20   384: greatest time $\longrightarrow \dfrac{3}{2.5} = \dfrac{N}{320} \longleftarrow$ greatest speed

(The problem could have been solved by placing the smaller numbers on top; that is,

$\dfrac{2.5}{3} = \dfrac{320}{N}$.)

Q21   When two pulleys are belted together, the revolutions per minute (rpm) vary inversely as the size of the pulleys. A 20-inch pulley running at 180 rpm drives an 8-inch pulley. Find the rpm of the 8-inch pulley.

STOP • STOP • STOP • STOP • STOP • STOP • STOP • STOP • STOP

A21   450 rpm: $\dfrac{20}{8} = \dfrac{N}{180}$

This completes the instruction for this section.

12.2   Exercises

1. If an increase in one quantity causes a decrease in another quantity, the quantities are _____ related.
2. If a decrease in one quantity causes a decrease in another quantity, the quantities are _____ related.
3. Solve the following proportions:

   a. $\dfrac{5}{7} = \dfrac{N}{49}$ **b.** $\dfrac{49}{98} = \dfrac{7}{N}$ **c.** $\dfrac{6}{18} = \dfrac{N}{21}$ **d.** $\dfrac{5}{2} = \dfrac{15}{N}$

   e. $\dfrac{4}{3} = \dfrac{N}{21}$ **f.** $\dfrac{25}{N} = \dfrac{15}{24}$ **g.** $\dfrac{2}{N} = \dfrac{4}{8}$ **h.** $\dfrac{3}{N} = \dfrac{18}{27}$

   i. $\dfrac{9}{2} = \dfrac{5}{x}$ **j.** $\dfrac{\frac{2}{3}}{9} = \dfrac{6}{x}$ **k.** $\dfrac{\frac{1}{2}}{\frac{1}{4}} = \dfrac{x}{\frac{3}{4}}$ **l.** $\dfrac{0.3}{0.8} = \dfrac{x}{\frac{1}{8}}$

4. If 30 men do a job in 12 days, how long will it take 20 men to do the same work?
5. If the wing area of a model airplane is 3 square feet and it can lift 10 pounds, how much can a wing whose area is 149 square feet lift?
6. A swimming pool large enough for 6 people must hold 5,000 gallons of water. How much water must a pool large enough for 15 people hold?
7. If a car goes 52 miles on 3 gallons of gas, how far will it go on 11 gallons?
8. If lobster tails are 3 boxes for $3.60, how much will 14 boxes cost?
9. A train takes 24 hours at a rate of 17 miles per hour to travel a certain distance. How fast would a plane have to fly in order to cover the same distance in 3 hours?
10. If electricity costs 10 cents for 4 kilowatt-hours, how many kilowatt-hours could you use for $7.20?

12.2   Exercise Answers

1. indirectly
2. directly
3. **a.** 35   **b.** 14   **c.** 7   **d.** 6

   e. 28   **f.** 40   **g.** 4   **h.** $\frac{9}{2}$ or $4\frac{1}{2}$

   i. $\frac{10}{9}$ or $1\frac{1}{9}$ **j.** 81   **k.** $\frac{3}{2}$ **l.** 0.046875

4. 18 days
5. $496\frac{2}{3}$ pounds
6. 12,500 gallons
7. $190\frac{2}{3}$ miles
8. $16.80
9. 136 miles per hour
10. 288 kilowatt-hours

Chapter 12 Sample Test

At the completion of Chapter 12 it is expected that you will be able to work the following problems.

12.1   Introduction to Ratio and Proportion

1. Determine the ratio of (write in fraction form):
   a. 15 years to 10 years   **b.** $5.00 to 60 cents

   c. 1 foot to 4 yards   **d.** $2\frac{1}{2}$ to $3\frac{1}{4}$

2. **a.** Name the means in the proportion $\frac{3}{8} = \frac{12}{32}$.

   b. What is the product of the extremes in the proportion $7:12 = 14:24$?

   c. Is $\frac{2}{3} = \frac{4}{5}$ a true proportion?

   d. Why?

12.2 **Solution of a Proportion**

3. Solve the following proportions:

   a. $\dfrac{16}{52} = \dfrac{M}{36}$ b. $\dfrac{5}{12} = \dfrac{M}{36}$ c. $\dfrac{M}{12} = \dfrac{2.5}{5}$ d. $\dfrac{4}{5} = \dfrac{7}{M}$

4. Solve the following problems:
   a. Four bricklayers can brick a building in 15 hours. How long would it take with 6 men working at the same rate?
   b. If rain is falling at the rate of 0.5 inch per hour, how many inches will fall in 15 minutes?
   c. A well produces 4,800 gallons of water in 120 minutes. How long will it take to produce 5,700 gallons of water?
   d. An airplane travels from one city to another in 160 minutes at the rate of 270 miles per hour. How long will it take to make the return trip, traveling at 240 miles per hour?

Chapter 12 Sample Test Answers

1. a. $\dfrac{3}{2}$ b. $\dfrac{25}{3}$ c. $\dfrac{1}{12}$ d. $\dfrac{10}{13}$

2. a. 8 and 12 b. 168 c. no d. $3 \times 4 \neq 2 \times 5$

3. a. $\dfrac{144}{13}$ or $11\dfrac{1}{13}$ b. 15 c. 6 d. $\dfrac{35}{4}$ or $8\dfrac{3}{4}$

4. a. 10 hours b. $\dfrac{1}{8}$ inch

   c. 142 minutes 30 seconds, or $142\dfrac{1}{2}$ minutes

   d. 180 minutes

Chapter 13

Interpretation and Construction of Circle, Bar, and Line Graphs

13.1   Choosing the Type of Graph

1   Frequently people are asked to read and understand material that contains many numbers. The method of presentation of those numbers determines the ease with which the number relationships are understood. Written paragraphs containing numbers are difficult to understand unless there are very few numbers. Tables in which numbers are located in a uniform pattern are more easily read, and graphs will often show relationships much more forcefully and possibly help the reader recognize and remember relationships.

A graph is used to present information artistically and accurately. A few mathematical principles are used which help to ensure that the graphs convey the truth. If these principles are ignored, the relationships between the numbers are distorted and the reader is left with an inaccurate impression of the facts. A person should be able to recognize distortions of graphs that he reads and avoid introducing distortions in graphs that he constructs.

Three types of graphs are used so frequently that everyone should be able to interpret and construct them. They are the *circle graph*, the *bar graph*, and the *line graph*. The name of the type of graph describes its character. Numbers are represented by sectors* of a circle on a circle graph, by bars in a bar graph, and by a broken line on a line graph.

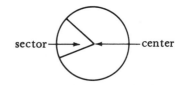

*A sector is a section of a circle shaped like a piece of pie.

Q1   Examine the differences in the following graphs and identify each type of graph by writing its name on the line under the graph.

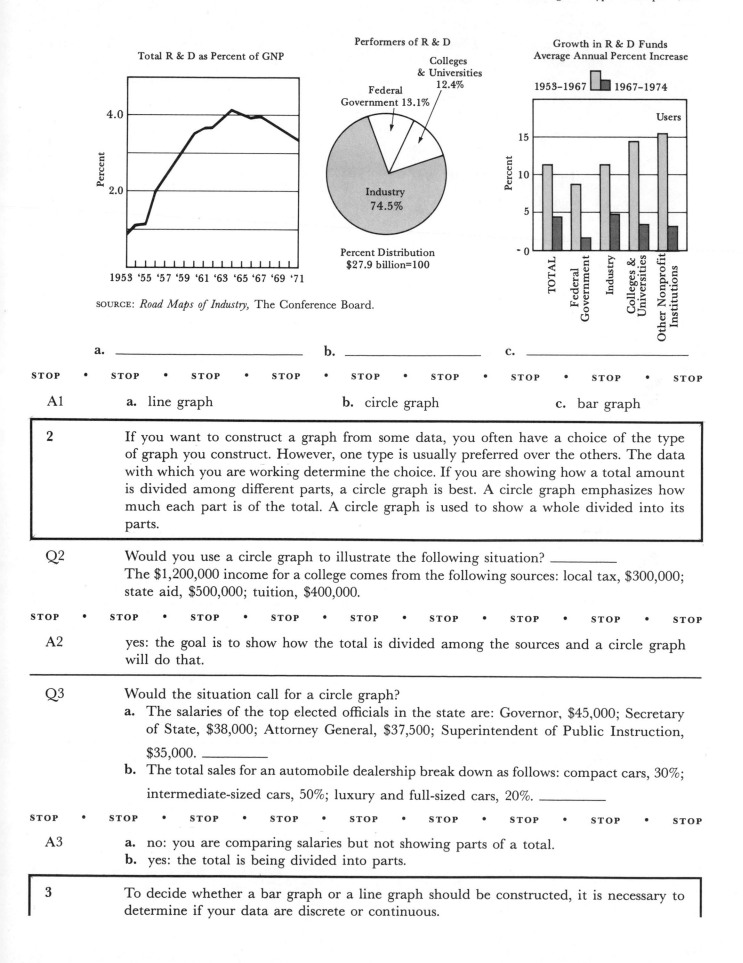

Total R & D as Percent of GNP

Performers of R & D

Growth in R & D Funds
Average Annual Percent Increase

SOURCE: *Road Maps of Industry,* The Conference Board.

a. _____   b. _____   c. _____

STOP • STOP • STOP • STOP • STOP • STOP • STOP • STOP • STOP

A1   **a.** line graph   **b.** circle graph   **c.** bar graph

| 2 | If you want to construct a graph from some data, you often have a choice of the type of graph you construct. However, one type is usually preferred over the others. The data with which you are working determine the choice. If you are showing how a total amount is divided among different parts, a circle graph is best. A circle graph emphasizes how much each part is of the total. A circle graph is used to show a whole divided into its parts. |

Q2   Would you use a circle graph to illustrate the following situation? _____
The $1,200,000 income for a college comes from the following sources: local tax, $300,000; state aid, $500,000; tuition, $400,000.

STOP • STOP • STOP • STOP • STOP • STOP • STOP • STOP • STOP

A2   yes: the goal is to show how the total is divided among the sources and a circle graph will do that.

Q3   Would the situation call for a circle graph?
a. The salaries of the top elected officials in the state are: Governor, $45,000; Secretary of State, $38,000; Attorney General, $37,500; Superintendent of Public Instruction, $35,000. _____
b. The total sales for an automobile dealership break down as follows: compact cars, 30%; intermediate-sized cars, 50%; luxury and full-sized cars, 20%. _____

STOP • STOP • STOP • STOP • STOP • STOP • STOP • STOP • STOP

A3   **a.** no: you are comparing salaries but not showing parts of a total.
b. yes: the total is being divided into parts.

| 3 | To decide whether a bar graph or a line graph should be constructed, it is necessary to determine if your data are discrete or continuous. |

Continuous data consist of measurements that can be made more precise if a person chooses. For example, if the data are the amount of time spent on a project, it might be measured in days. If this is not satisfactory for some reason, time could be measured in hours or, more precisely, in minutes or seconds. Therefore, time is a continuous variable.

Discrete data allow no interpretation between the values of the data. For example, in a football game there may be 2 field goals scored or 3, but it is impossible to score $2\frac{1}{2}$ field goals. The number of field goals is a discrete variable.

Examples of continuous data:

1.   Height: 60 in, 60.25 in, 6.2546 in (the number of decimal places is dependent on the precision of the measurement)
2.   Time measured in hours: 4 hours, $4\frac{1}{2}$ hours, 4.2576 hours

Examples of discrete data:

1.   Makes of automobiles: Ford, Chevrolet, VW (there is no interpretation of halfway between a Chevrolet and a Ford)
2.   Number of people in a classroom: 42, 16, 4 (you would not count people in fractional parts)

Sometimes there is no natural order for discrete data. For example, it makes no difference which make of automobile is listed first. When there is an order to discrete data, the numbers change in incremental steps, usually increasing by one each time.

Q4   Indicate whether the variable is continuous or discrete:

    a.   number of swimmers in a race   _____

    b.   dosage of insulin in a vein   _____

    c.   inches of snowfall in a year   _____

    d.   number of bulls in a bullfight   _____

STOP   •   STOP   •   STOP   •   STOP   •   STOP   •   STOP   •   STOP   •   STOP   •   STOP

A4   **a.** discrete   **b.** continuous   **c.** continuous   **d.** discrete

4   Both bar graphs and line graphs have a vertical axis and a horizontal axis. Both bar and line graphs are used to show a numerical value measured on the scale on the vertical axis for various values located on the horizontal axis. Bar graphs are used when the horizontal axis of the graph represents discrete values. Line graphs are used when the horizontal axis of the graph represents continuous values.

The horizontal axis on the following figure represents new prescriptions (Rx's), refill prescriptions, and total prescriptions. These are discrete values, so a bar graph is used.

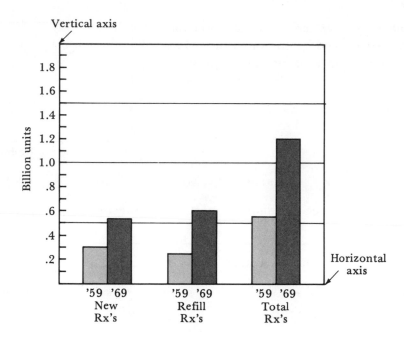

The horizontal axis on the following figure represents time measured in hours. Since time is a continuous variable, the graph should be readable between numbers on the horizontal axis. Therefore, a line graph is constructed.

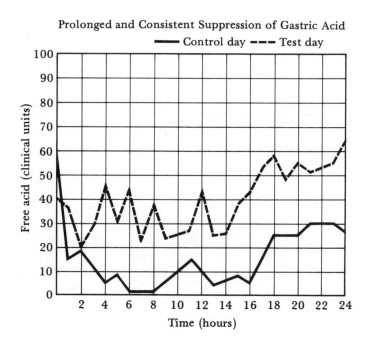

Therefore, in order to tell what type of graph to construct, first determine the type (continuous or discrete) of data on the horizontal axis.

Q5   Suppose that the amount of food consumed per day was being compared for college freshmen of various weights. Are the data on the horizontal axis continuous or discrete? _____

Food consumed

Weight of freshmen

STOP • STOP • STOP • STOP • STOP • STOP • STOP • STOP • STOP

A5   continuous: you can interpret the data no matter how precisely it is reported. 100 lb, 112.5 lb, 149.275 lb.

Q6   What type of graph would you use in Q5? _____

STOP • STOP • STOP • STOP • STOP • STOP • STOP • STOP • STOP

A6   line graph

Q7   Suppose that a graph is to be made that shows the maximum legal weight allowed in a truck with various numbers of axles.
a. Is the number of axles continuous or discrete?

_____

b. What type of graph would be drawn?

_____

Weight

Number of axles

STOP • STOP • STOP • STOP • STOP • STOP • STOP • STOP • STOP

A7   a. discrete: a truck could not have one-half an axle.
b. bar graph

Q8   Assume that the variable listed is on the horizontal axis of a graph. Indicate whether the data are continuous or discrete and what type of graph would be drawn.

	Data	Type data	Type graph
a.	countries: USA, France, USSR	_____	_____
b.	months: 3 mo, $3\frac{1}{2}$ mo, $4\frac{2}{3}$ mo	_____	_____
c.	attendance at ball games	_____	_____
d.	temperature	_____	_____
e.	age	_____	_____

STOP • STOP • STOP • STOP • STOP • STOP • STOP • STOP • STOP

A8   a. discrete, bar graph
b. continuous, line graph
c. discrete, bar graph
d. continuous, line graph
e. continuous, line graph

(*Note:* Age is continuous, because $2\frac{1}{2}$ years old makes sense; however, the general practice of not changing your age until your birthday causes it to be used like a discrete variable.)

This completes the instruction for this section.

13.1 Exercises

1. Indicate whether each of the following variables is continuous or discrete:
   a. number of television sets sold
   b. cubic centimeters of blood in the body
   c. gasoline pumped from a gasoline-station pump
   d. bricks
   e. ice cream cones
   f. soda pop from a tap
   g. votes cast in an election
2. Indicate what type of graph should be used to illustrate each of the following sets of data:
   a. You want to show how the fuel usage increases as you increase speed from 0 to 90 mph.
   b. A drug is made of a mixture of several ingredients. You want to show the percent of each ingredient as a part of the total.
   c. A department store has seven departments. You want to show how each of these departments contributes to the total sales for the store.
   d. The dosage of a medicine changes according to the amount of fever a patient has. You want to illustrate the dosage for body temperature between 98.6 and 105 °F.
   e. A chain store with 27 locations wants to compare the total sales of three of its outlets.

13.1 Exercise Answers

1. a. discrete   b. continuous   c. continuous   d. discrete
   e. discrete   f. continuous   g. discrete
2. a. line graph: because speed is a continuous variable
   b. circle graph: because you are showing parts of a total
   c. circle graph: because you are showing parts of a total
   d. line graph: because temperature is a continuous variable
   e. bar graph: because the outlets are discrete values (You would not use a circle graph, because you are not showing parts of a total, only comparing outlets.)

13.2 Circle Graphs

1   A circle graph has the general appearance of a pie cut into uneven wedges. The geometric name of each wedge is a *sector*. The angle that is located at the center of the circle in each sector is called the *central angle* of the sector.

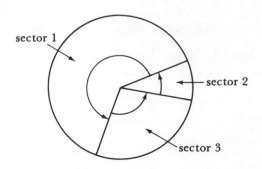

To interpret a circle graph you must know that the complete circle represents the total amount and that each sector represents a part of the total amount in the same ratio as the measure of its central angle is to 360°.

Q1   If a circle graph has a sector with a central angle of 90°, what percent of the total does that sector represent?

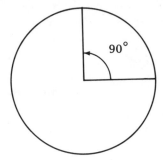

STOP • STOP • STOP • STOP • STOP • STOP • STOP • STOP • STOP

A1   The sector represents 25 percent because $\dfrac{90}{360} = \dfrac{1}{4} = 25\%$

2   Just as the sum of the parts must equal the total amount represented, the sum of the percent of all sectors must be 100 percent and the sum of the central angles must be 360°.

Q2   Consider the circle graph and find the percent for the category "miscellaneous."

Distribution of Expenses for an Average Family

STOP • STOP • STOP • STOP • STOP • STOP • STOP • STOP • STOP

A2   20%: 35% + 25% + 20% = 80%
100% − 80% = 20% for miscellaneous

3

Usually the total amount that the whole circle represents is given in the title of the circle graph. To find the amount that each sector represents, multiply the total amount represented times the percent of the total included in the sector. For example, in the circle graph shown, to find the number of dollars used for housing in a family with income $8,500, we would find $8,500 \times 35\% = 8,500 \times 0.35 = \$2,975$.

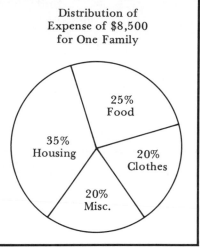

Distribution of
Expense of $8,500
for One Family

Q3

a. Find the amount spent on food for the family whose expenses are shown in the preceding circle graph.

b. Find the amount spent on clothes for the same family.

STOP • STOP • STOP • STOP • STOP • STOP • STOP • STOP • STOP

A3

a. $2,125: $8,500 \times 25\% = 8,500 \times 0.25 = \$2,125$
b. $1,700: $8,500 \times 20\% = 8.500 \times 0.20 = \$1,700$

4

To construct a circle graph you must find the number of degrees to use for the central angle for each sector. This can be done with a proportion.

$$\frac{X}{360} = \frac{\text{amount of part}}{\text{amount of total}}$$

where X = number of degrees in the sector

If you substitute in the proportion and solve for X, you will have the degrees in the central angle of the sector. To actually draw this, you will need a protractor.

For example, to find the number of degrees for the gas and oil sector of a circle graph of the information below you would proceed as shown.

Cost of Used-Car Ownership for Two-Year Period

Initial cost of car	$1,500
Gas and oil	850
Upkeep and repairs	750
Insurance	500
Total	$3,600

To find central angle of gas and oil sector:

$$\frac{X}{360} = \frac{850}{3,600}$$

$$X = \frac{850}{\underset{10}{\cancel{3,600}}} \times \frac{\overset{1}{\cancel{360}}}{1}$$

$$X = 85°$$

Q4   a. Determine the central angle for initial cost of the car, upkeep and repairs, and insurance for the information given in Frame 4.

   b. Construct a circle graph by using the results of part a and the example in Frame 4. Be sure to write a title and label each sector.

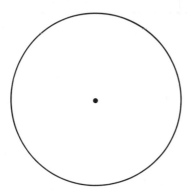

STOP • STOP • STOP • STOP • STOP • STOP • STOP • STOP • STOP

A4   a. $\dfrac{X}{360} = \dfrac{1,500}{3,600}$ $\dfrac{X}{360} = \dfrac{750}{3,600}$ $\dfrac{X}{360} = \dfrac{500}{3,600}$

   $X = \dfrac{1,500}{3,600} \cdot 360$ $X = \dfrac{750}{3,600} \cdot 360$ $X = \dfrac{500}{3,600} \cdot 360$

   $X = 150°$ $X = 75°$ $X = 50°$

   b. Cost of Used-Car Ownership for Two-Year Period

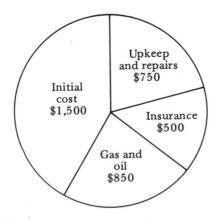

This completes the instruction for this section.

13.2   Exercises

Use the circle graph below to answer problems 1–5.

Distribution of 538 Electoral Votes by Region, 1972

538 Electoral Votes = 100%

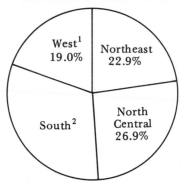

[1] Includes Alaska (3 electoral votes) and Hawaii
(4 electoral votes)
[2] Includes the District of Columbia (3 electoral votes)

SOURCE: *Road Maps of Industry,* The Conference Board.

1. Determine the percent of electoral votes in the South.
2. Find the number of electoral votes in the Northeast. (*Note:* The answer should be rounded off to a whole vote, since it is a discrete variable.)
3. Use your answer to problem 1 to determine the number of electoral votes in the South.
4. Use your answer in problem 2 to find the number of degrees in the central angle of the sector that represent electoral votes in the Northeast.
5. Determine the number of degrees in the central angle of the sector that represent the electoral votes of the South.
6. Use the circle graph below to find the approximate number of black people in the U.S. population.
7. Use the circle graph below to find the number of Spanish-speaking people in the U.S. population.

Distribution of U.S.
Population, 1970

Total 203 million = 100%

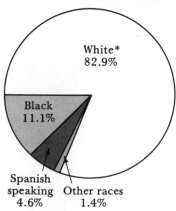

*Excludes Spanish-speaking persons.

SOURCE: *Road Maps of Industry,* The Conference Board.

13.2   **Exercise Answers**

1. 31.2%
2. 123 votes (you would round off 123.202 to 123, because it is a discrete variable):
    $0.229 \times 538 = 123.202$
3. 168 votes: $0.312 \times 538 = 167.856$
4. $$\frac{X}{360} = \frac{123}{538}$$

    $$X = \frac{123}{538} \cdot 360$$

    $$X = 82°$$
5. 112°: can be done by two methods:

    First method   Second method
    $0.312 \times 360 = 112.320$ $$\frac{X}{360} = \frac{168}{538}$$

    $$X = \frac{168}{538} \cdot 360$$

    $$X = 112.4$$

6. 22,533,000
7. 9,338,000

13.3   **Bar Graphs**

1   A bar graph is useful when a quick visual impression of the relative sizes of various quantities is to be communicated. If the actual numbers are desired corresponding to the heights of the bars, they should be obtainable on a scale given as a part of the graph. The values will, by necessity, be approximate. The following bar graph is appropriate for the data being considered because the cities are discrete data.

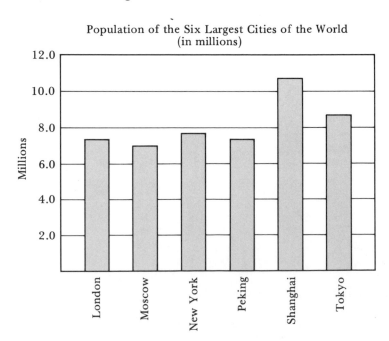

Population of the Six Largest Cities of the World
(in millions)

If a short bar is placed next to a tall bar, it accents the shortness of the bar. To avoid the criticism that the maker of the graph is deliberately trying to distort the relationships among the numbers, two procedures have been developed for arranging discrete data that have no natural order. The first procedure is to arrange the data alphabetically. The second procedure is to arrange the bars from tallest to shortest. Notice that the first procedure was used in the bar graph in this frame.

Q1   Use the graph in Frame 1 to answer the following questions.

    a. What is the most populous city in the world? _____

    b. What city is third largest? _____

    c. What two cities seem almost tied for the
        fourth largest in the world? _____

    d. How many people live in Shanghai? _____

    e. How many people live in Moscow? _____

    f. How many people live in New York? _____

STOP • **STOP** • **STOP** • **STOP** • **STOP** • **STOP** • **STOP** • **STOP** • **STOP**

A1   **a.** Shanghai   **b.** New York   **c.** London, Peking   **d.** 10,800,000
     e. 7,000,000   **f.** 7,700,000

2   A bar graph consists of a series of vertical or horizontal bars drawn according to scale. When constructing a bar graph one must be careful not to introduce a distortion in the impression left with the reader. The visual impression should match the actual relationships between the numbers being represented. An example will show how a bar graph can leave the wrong impression when not constructed correctly. Study the data in the table and the bar graph and see if you can locate the distortion.

Average Weekly Earnings
Ardis $300
Baker $400
Chambers $600

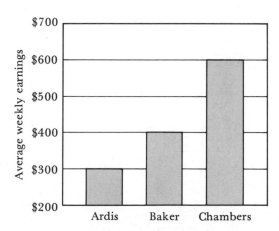

The relationships that exist in the chart (Chambers making twice as much as Ardis, Baker making $\frac{2}{3}$ of Chambers) are not reflected in the height of the bars. In the graph it appears that Chambers makes 4 times the earnings of Ardis and Baker makes $\frac{1}{2}$ the earnings of Chambers. Although the top of the bar corresponds to the correct dollar amounts, the visual impression (some say this is what people usually remember) is inaccurate. The graph shown is actually only the top portion of a graph as it should be drawn. Compare the two graphs.

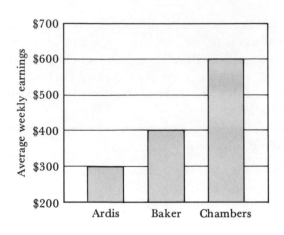

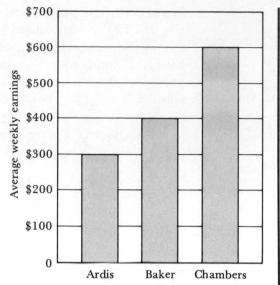

Notice that starting the vertical scale at zero instead of a larger number causes the heights of the bars to be in the same proportion as the numbers they represent. Using any other number as the smallest value on the vertical scale will distort the impression. Therefore, remember—*always start the vertical scale on a bar graph at the number zero.* Removing a section of a bar or removing a section of the vertical scale also distorts the ratios of the bars of a bar graph. The omissions are usually indicated with a jagged line. If you see such a jagged line, be on the lookout for a distorted graph.

Q2   For each of the graphs below (1) indicate the actual ratio of the heights of the bars and the ratio of the numbers represented by the bars, and (2) what in the construction of the graph has caused the distortion?

a.

(1) _____ _____
(2) _____

b.

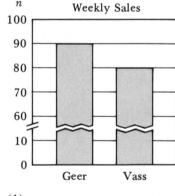

(1) _____ _____
(2) _____

c.

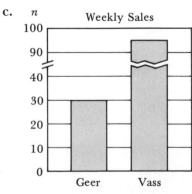

(1) _____ _____
(2) _____

STOP • STOP • STOP • STOP • STOP • STOP • STOP • STOP • STOP

A2 **a.** (1) $\dfrac{5}{3}$, $\dfrac{70}{50} = \dfrac{7}{5}$

(2) vertical scale starts at 20 instead of 0

b. (1) $\dfrac{5}{4}$, $\dfrac{90}{80} = \dfrac{9}{8}$

(2) section removed from bars

c. (1) $\dfrac{3}{5\frac{1}{2}} = \dfrac{6}{11}$, $\dfrac{30}{95} = \dfrac{6}{19}$

(2) section removed from bar

Q3 For each of the bar graphs in Q2, indicate whether the distortion favors Geer or Vass.

a. _____ **b.** _____ **c.** _____

STOP • **STOP** • **STOP** • **STOP** • **STOP** • **STOP** • **STOP** • **STOP** • **STOP**

A3 **a.** Geer: the graph makes it look like he sold $\dfrac{5}{3}$ as much as Vass, but he only sold $\dfrac{7}{5}$ as much.

b. Geer: the graph makes it look like he sold $\dfrac{5}{4}$ as much as Vass, but he only sold $\dfrac{9}{8}$ as much.

c. Geer: the graph makes it look like he sold $\dfrac{6}{11}$ as much as Vass, but he sold only $\dfrac{6}{19}$ as much.

---

3 The most important part of drawing a bar graph is determining the scale. You must be able to represent the largest number by a bar that fits into the available space. Suppose that you were to graph the data from the chart below on the space provided.

Population of the Five Largest Metropolitan
  Areas (in millions of persons)

New York	11.5
Los Angeles	6.7
Chicago	6.5
Philadelphia	4.7
Detroit	4.0

The population of the New York metropolitan area would have to be represented by a bar no longer than 6 units because there are six spaces available. *To find the scale, divide the amount to be represented by the longest bar by the number of available spaces.* (If your paper is unlined, divide by the number of inches available.) Round the resulting value up to a convenient *larger* value. In the example given:

$$11.5 \div 6 = 1.9\frac{1}{6} = 2.0 \text{ million rounded up}$$

Before you are given an opportunity to complete the bar graph of this example, you will be given a chance to practice determining the scale for various graphs.

Q4   Suppose that the largest number to be represented was 426 and you wanted to draw your graph on a location that had 10 vertical spaces. What would you use for a scale?

STOP • **STOP** • **STOP** • **STOP** • **STOP** • **STOP** • **STOP** • **STOP** • **STOP**

A4   Use 50 units per space. 425 ÷ 10 = 42.6 rounded upward would be 50. (45 would not be incorrect.)

Q5   If the longest bar on a bar graph was to represent 45,050 and you had 7 inches in the vertical direction available, what would you use for a scale?

STOP • **STOP** • **STOP** • **STOP** • **STOP** • **STOP** • **STOP** • **STOP** • **STOP**

A5   7,000: 45,050 ÷ 7 = 6,436 rounded up to 7,000. Therefore, use 7,000 per inch. In this case 8,000, or even 10,000, would not be incorrect. Use of 10,000 would cause your bars to be much shorter, and you would not be using as much of your available space.

Q6   The longest bar should represent $44, and there are 6 vertical spaces available. What should be used for the vertical scale?

STOP • **STOP** • **STOP** • **STOP** • **STOP** • **STOP** • **STOP** • **STOP** • **STOP**

A6 $8 per space: $44 ÷ 6 = $7.33 rounded upward would be 8. It would also be correct to use $10 per space.

Q7   Use the scale of 2.0 million people per vertical space and complete the bar graph in Frame 3. Be sure that both scales are labeled and that a title is included.

STOP • **STOP** • **STOP** • **STOP** • **STOP** • **STOP** • **STOP** • **STOP** • **STOP**

A7

Population of the Five Largest Metropolitan Areas
(millions of persons)

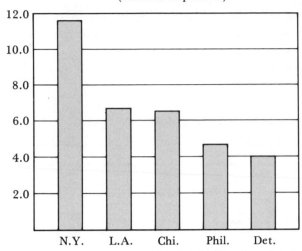

(*Note:* The lengths of each bar must be approximated as nearly as possible.)

Q8   Use the space provided to draw a bar graph of the data given here.

Gold Reserves by Countries (in billions)

Europe and Japan (combined)	$33.3
United Kingdom	7.1
United States	14.3
Less-developed areas	15.0

STOP • **STOP** • **STOP** • **STOP** • **STOP** • **STOP** • **STOP** • **STOP** • **STOP**

A8   To determine the scale:

$33.3 \div 5 = 6.6$
$= 7$ rounded

Gold Reserves by Countries
(in billions of dollars)

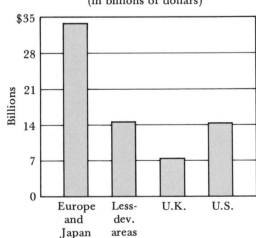

(*Note:* Rounding to 8 or even 10 would not be incorrect. Your graph would look different, however.)

This completes the instruction for this section.

13.3 Exercises

A distortion is sometimes introduced into a bar graph by removing a section of a bar. Since this type of graph appears often, you should read the graph very critically. The questions following the graph will assist your thinking. The graph is more "sophisticated" than the ones that you have been reading, but the same principles apply.

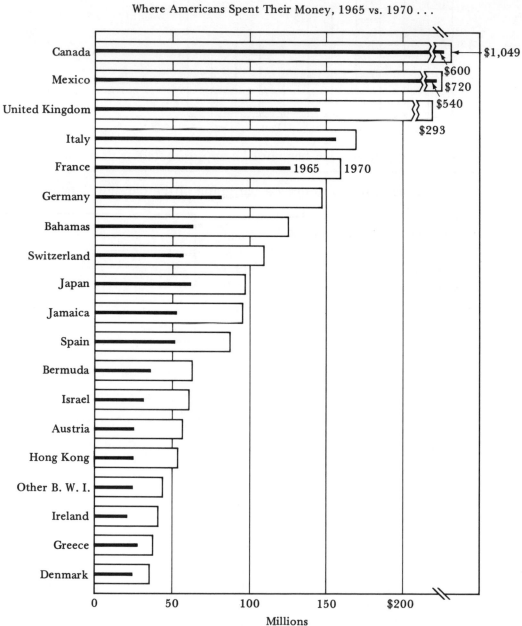

Where Americans Spent Their Money, 1965 vs. 1970 . . .

SOURCE: *Road Maps of Industry,* The Conference Board.

1. How much money was spent in France in 1970?
2. How much money was spent in France in 1965?
3. How much money was spent in Spain in 1965?
4. The amount spent in Italy in 1970 was approximately how many times the amount spent in Austria in 1970?

5. Does the ratio of the apparent length of the bar for 1970—Canada to the bar for 1970—France equal the ratio of the dollar amounts?
6. Was the same length of bar omitted from the first three bars?
7. The 1970 bar for Mexico should be approximately how many times as long as the 1970 bar for Switzerland?
8. Should the length of the 1965 bar for Canada actually reach as close to the top of the separated section of the bar as is shown on the graph?

Proceed as the number items indicate to construct a graph of the data in the chart.

Six Largest Employee Organizations

Teamsters	1,800,000
Automobile workers	1,500,000
Steel workers	1,200,000
National Education Association	1,100,000
Electrical Workers (BEW)	900,000
Machinists	850,000

9. What kind of graph is appropriate, and why?
10. Determine a scale.
11. Label the horizontal and vertical scales. Arrange organizations from largest to smallest.
12. Draw the bars.
13. Write an appropriate title.

13.3   Exercise Answers

1. 160,000,000
2. 125,000,000
3. 50,000,000
4. 3 times
5. no
6. no
7. almost 7 times
8. no
9. bar graph, because organizations are discrete data

10. 1,800,000 ÷ 7 = 257,000, so round to 300,000 per space
11, 12, and **13** are shown below.

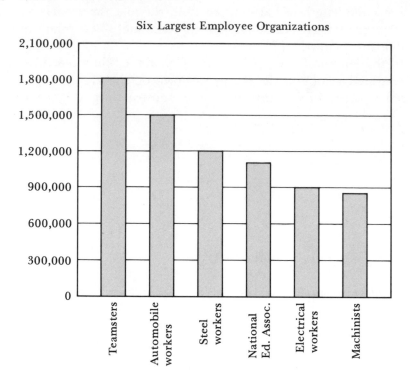

Six Largest Employee Organizations

13.4   Line Graphs

1

Line graphs are used when the horizontal scale indicates *continuous* data. The line graph is particularly useful for showing a continuous trend over a period of time. Time is then located on the horizontal axis. One can interpret a line graph for values in between the numbers shown on the horizontal scale.

The following line graph shows the distances that a car has traveled since the beginning of its trip at 12:00.

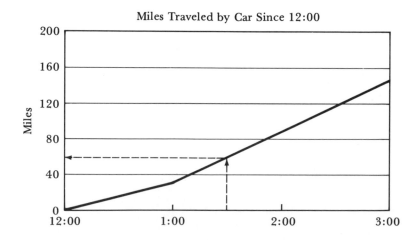

Miles Traveled by Car Since 12:00

The distance traveled at a given time is found by locating the time on the horizontal scale, moving vertically to the line, then moving horizontally to read the height of the intersection point on the vertical scale. The dashed line on the graph shows how to locate the distance the car had traveled at 1:30. Reading the vertical scale corresponding to 1:30 on the horizontal scale, we see that the car had traveled 60 miles.

Q1   Use the line graph in the previous example to approximate the distance traveled at:

     a. 12:30 _____   **b.** 1:15 _____   **c.** 2:45 _____

STOP • STOP • STOP • STOP • STOP • STOP • STOP • STOP • STOP

A1   **a.** 15 miles
     b. 45 miles
     c. 135 miles

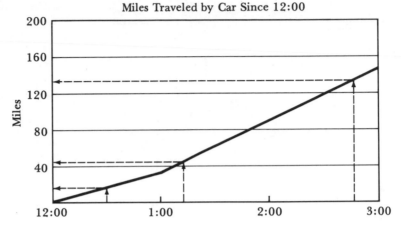

Miles Traveled by Car Since 12:00

(answers may vary by 5 miles and still be correct)

2   Notice the change in the slope (slant) of the line at 1:00. This indicates that the car is traveling faster, such as would happen if the car left the city streets and entered an expressway at 1:00. The car traveled from 0 miles at 12:00 to 30 miles at 1:00 for an average of 30 miles per hour. From 1:00 to 2:00 the car traveled from 30 miles to 90 miles for an average of 60 miles per hour. From this example you can see that the steeper the slope of the line of the graph, the faster the variable on the vertical scale is changing.

     A line graph is useful for reading particular values, showing trends that are increasing or decreasing, and for showing the rate of change of a quantity.

Q2   The line graph shows the total amount of money collected by the end of each day in a United Fund campaign. Find the following:

     a. Money collected by the 4th day. _____

     b. Money collected by the 10th day. _____

     c. Is money coming in faster on the 3rd day or the 7th? _____

     d. Is money coming in faster on the 7th day or the 13th? _____

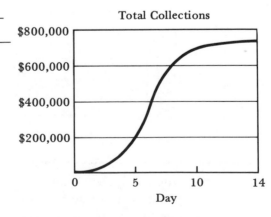

Total Collections

STOP • STOP • STOP • STOP • STOP • STOP • STOP • STOP • STOP

A2   **a.** $100,000   **b.** $700,000   **c.** 7th   **d.** 7th

| 3 | When constructing a line graph, determine the vertical scale in the same manner as the bar graph. (Divide the largest number to be represented by the number of spaces available.) Lay off the horizontal scale corresponding to the appropriate variable. Time is located on the horizontal scale. Locate points corresponding to the given data and then connect the points with straight lines from left to right. |

Q3   Construct a line graph from the given data.

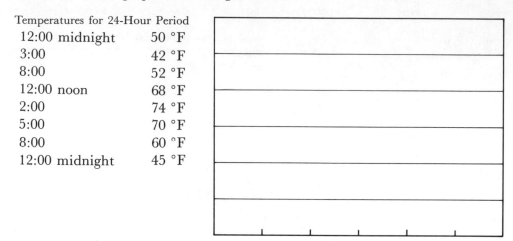

Temperatures for 24-Hour Period

12:00 midnight	50 °F
3:00	42 °F
8:00	52 °F
12:00 noon	68 °F
2:00	74 °F
5:00	70 °F
8:00	60 °F
12:00 midnight	45 °F

(*Note:* The data for the horizontal scale are not evenly spaced. Do not make them evenly spaced on your graph.)

STOP • **STOP** • **STOP** • **STOP** • **STOP** • **STOP** • **STOP** • **STOP** • **STOP**

A3   The scale is determined by $74 \div 6 = 12\frac{1}{3}$, which is rounded to 15.

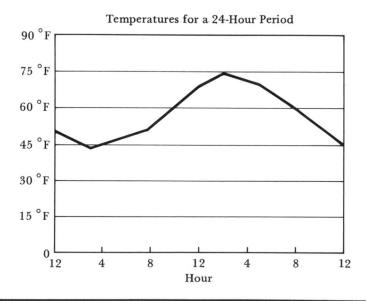

Temperatures for a 24-Hour Period

| 4 | Often it is important to know a value in between values of a chart. For example, one might be interested in the temperature at 10:00 A.M. in the situation presented in Q3. The line graph allows you to approximate this temperature quite accurately. The graph of A3 indicates that the temperature at 10:00 A.M. was 60°; 60° is a much more reasonable answer than giving the temperature of either of the two closest times that are in the chart. |

Approximation between two known values is called *interpolation*. Interpolation is used a great deal in mathematics and everyday life. There are some hazards in interpolation which you should be aware of. The following example will illustrate.

Graph I: Motor Vehicle Production, Weekly for Selected Weeks (in thousands)

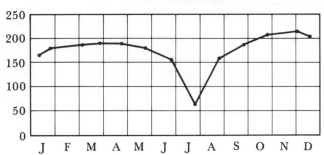

Graph II: Motor Vehicle Production, Weekly (in thousands)

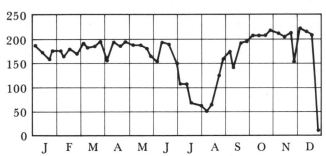

Notice that the first line graph is good for showing the general trend over the year. It shows the significant decrease in production in July and August when car companies change production to the new models. It misses the sudden changes in production that the more sensitive second graph illustrates. Answer the following questions to see the accuracy of interpolation.

Q4 Use the two graphs in the previous frame to answer the questions below:
a. Find the production for the third week in March.

 Graph I—interpolation _____ Graph II _____
b. Find the production for the fourth week in November.

 Graph I—interpolation _____ Graph II _____
c. What was the amount of the lowest weekly production of the year?

 Graph I _____ Graph II _____
d. Which graph would be most useful for predicting the week of the Easter holiday?

 _____

e. What week would you say included the Easter holiday? _____

STOP • STOP • STOP • STOP • STOP • STOP • STOP • STOP • STOP

A4 **a.** Graph I, 190, 000; graph II, 180,000
 b. Graph I, 220,000; graph II, 150,000
 This fluctuation was probably due to Thanksgiving.
 c. Graph I, 70,000; graph II, 10,000
 d. Graph II
 e. first week in April

5   From the preceding questions you can tell that interpolation will be inaccurate when there are rapid changes that are not reflected in the graph. In situations where the data change smoothly, interpolation is more accurate.

Another type of approximation is much less likely to be accurate but is still useful. Approximating beyond the available data is called *extrapolation*. If a person had only graph I in Frame 4 available to him and wanted to know the most likely production for the fourth week of December, at the end of the year, he would probably have predicted about 200,000. (Check this yourself.) Such a prediction would have been disastrous, however, because that was the week of the Christmas holidays and production went to zero. (Check graph II.) A prediction on the other end of the year for the first week in January would have been much more accurate. Extrapolating on the curve of graph I would have given an approximation of about 160,000, whereas from graph II we see that it was actually 180,000. You will work with extrapolation further in the exercises at the end of this section.

This completes the instruction for this section.

13.4   Exercises

Selected Indicators of the Demand for Labor

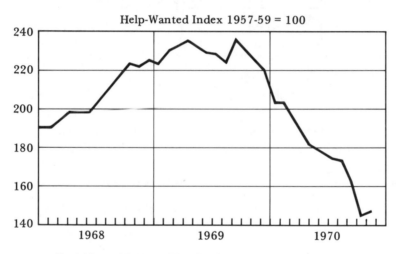

SOURCE: *Road Maps of Industry,* The Conference Board.

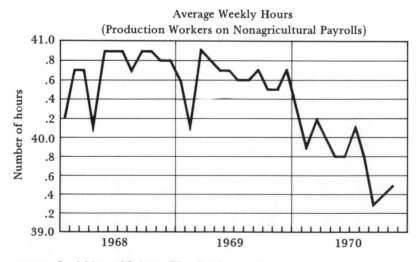

SOURCE: *Road Maps of Industry,* The Conference Board.

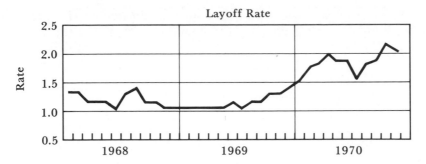

SOURCE: *Road Maps of Industry,* The Conference Board.

1. Which of the years had the highest average help-wanted index?
2. Which year was the layoff rate the highest?
3. Which year did workers have the highest average weekly hours?
4. As the layoff rate increased, what happened to the help-wanted index?
5. What were the average weekly hours for June 1968?
6. What was the layoff rate for July 1970?
7. Extrapolate to find the layoff rate for December 1970.

When the population predictions are made for the future, we are really extrapolating from our existing data. Different assumptions can be made to help improve the extrapolation. The graph shows examples of four different possibilities. The assumed growth rates are given under B, C, D, and E.

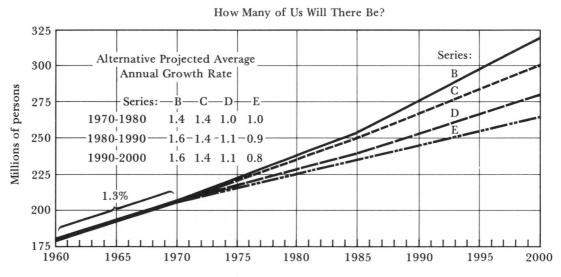

SOURCE: *Road Maps of Industry,* The Conference Board.

8. What would the population be in 1990 under assumption B _____
   C _____ D _____ E _____?

9. What would the population be in 2000 under assumption B _____
   C _____ D _____ E _____?

10. Construct a line graph of the following data showing the demand for engineer/scientists as an index where the 1961 demand is equated to 100. Label the vertical scale and title.

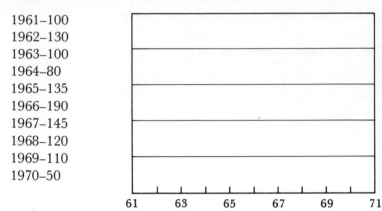

```
1961–100
1962–130
1963–100
1964–80
1965–135
1966–190
1967–145
1968–120
1969–110
1970–50
```

13.4 Exercise Answers

1. 1969
2. 1970
3. 1968
4. decreased
5. 40.9
6. 1.6
7. 1.9
8. B, 277 million; C, 265 million; D, 255 million; E, 245 million
9. B, 320 million; C, 300 million; D, 280 million; E, 265 million
10. Scale: 190 ÷ 5 = 38 rounded to 40

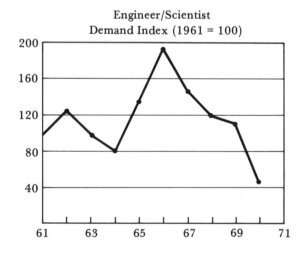

Technical Labor Demand

Engineer/Scientist
Demand Index (1961 = 100)

Chapter 13 Sample Test

At the completion of Chapter 13 it is expected that you will be able to work the following problems.

13.1 Choosing the Type of Graph

1. Identify each of the following variables as continuous or discrete:
   a. measurement in inches of a person's waist
   b. number of players on a hockey team
   c. miles per hour that a motorcycle travels
   d. hourly temperatures throughout a day
2. Choose the type of graph (bar graph or circle graph) that would be used to illustrate the following data:
   a. You want to compare the inventories of a drugstore at the end of each of the last five years.
   b. You want to examine the total inventory of a department store by comparing the inventories of all five departments.
   c. You want to examine the total income from sales of an insurance agency by comparing the total sales of all of its four salesmen.
3. Choose the type of graph (line graph or bar graph) that would be used to illustrate the following data:
   a. You want to show the total sales for Chrysler, Ford, and General Motors last year.
   b. You want to show the changes in the price of a stock over the course of a year.
   c. You want to compare the dollars per capita that each of the fifty states in the United States spent on education last year.

13.2 Circle Graphs

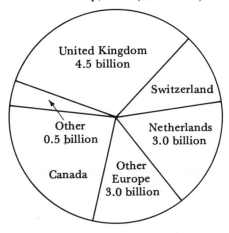

Distribution of Foreign Direct
Investments in the U.S. by Country
of Ownership, 1971 ($15 billion)

4. If Switzerland controls 10 percent of the foreign investments in the United States, what is its investment in the United States?
5. Find the dollars invested by Canada in the United States.
6. What is the sum of all the central angles of the sectors of a circle?
7. Find the number of degrees in the central angle of the sector representing Switzerland.
8. Find the number of degrees in the central angle of the sector representing Netherlands.

13.3   Bar Graphs

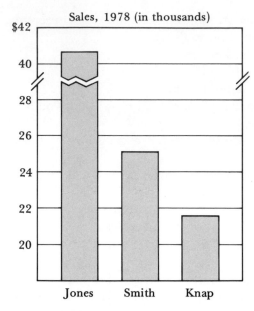

Sales, 1978 (in thousands)

Use the graph to answer questions 9–11.

9. Give two reasons why the bar graph is a distortion of the data on which it is based.
10. What were the total sales of Smith in the graph?
11. What were the total sales of Knap in the graph?
12. If you were constructing a bar graph and had to draw the longest bar representing 108 million on a graph with six vertical spaces, what scale would you use?
13. Graph the following data.

Retirement Income/Month
Atwood	$1,200
Baker	600
Calson	420
Dunlap	2,000
Ehen	800

13.4   Line Graphs

A production manager recorded the total production at several times during an 8-hour shift. The shift started at 8:00 A.M. and ended at 5:00 P.M. with an hour off for lunch.

Total Production Until a Given
Time in a Shift

Time	Production Units
8:00 A.M.	0
9:00 A.M.	220
10:00 A.M.	400
12:00 noon	630
1:00 P.M.	630
2:00 P.M.	730
4:00 P.M.	850

14. Use the data to construct a line graph of the total production at various times during the shift.

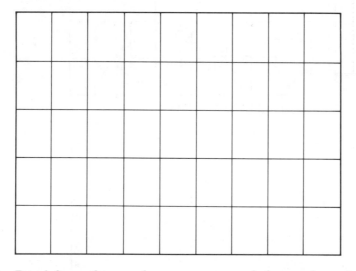

15. Read from the graph you constructed the total production at 11:00 A.M.

16. The process of approximating the production for 11:00 A.M. between the two known values of the production at 10:00 A.M. and noon is called _____ (interpolation or extrapolation).

17. Predict from your graph what the production will be at the end of the shift at 5:00 P.M.

18. The process of approximating the production at 5:00 P.M. beyond the known values is called _____ (interpolation or extrapolation).

19. How could you explain the leveling off of the production from noon to 1:00 P.M.?

20. Is production increasing at a faster or slower rate at 9:00 A.M. than it is at 2:00 P.M.? Explain why this is reasonable.

Chapter 13 Sample Test Answers

1. **a.** continuous   **b.** discrete   **c.** continuous   **d.** continuous
2. **a.** bar graph   **b.** circle graph   **c.** circle graph
3. **a.** bar graph   **b.** line graph   **c.** bar graph
4. 1.5 billion
5. 2.5 billion
6. 360°
7. 36°
8. 72°

9. **1.** The vertical scale has been broken between 28,000 and 40,000.
   2. The vertical scale does not begin at zero.
10. $25,000
11. $21,500
12. 1 space = 20 million
13.

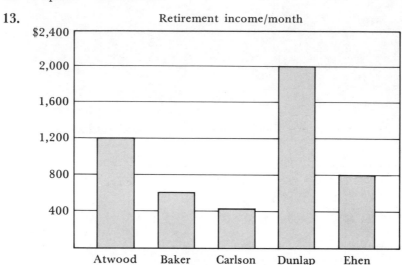

Retirement income/month

14.

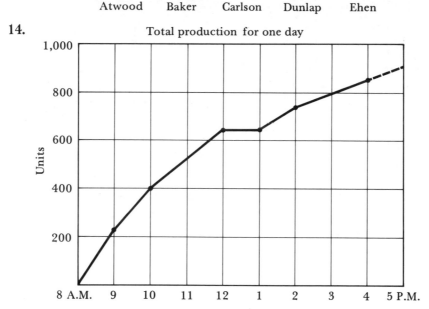

Total production for one day

15. 520
16. interpolation
17. 900
18. extrapolation
19. lunch break
20. Faster rate; people are fresh in the morning and tired later in the day.

Chapter 14

Introduction to Statistics

Statistics is an area of mathematics that has been rather recently developed in the history of mathematics. It is, however, becoming extremely important in our modern world. One needs fundamental knowledge of statistics to read many articles in the daily newspapers. Some fundamental notions of statistics will be presented in this chapter.

14.1 Mean, Median, and Mode

1	The field of statistics was developed to take sets of numbers called *data** and identify the important characteristics of them. One goal is to pick one number that is representative of a set of numbers. This number is called a *measure of central tendency*. Three different ways of deciding on the most representative number have been developed. These produce the mean, the median, and the mode. For some data (sets of numbers) the mean, median, and mode are equal; for others, they are different. The *mode* of a set of numbers is the number that occurs with the greatest frequency. **Example:** Find the mode of the following numbers: 2, 7, 5, 3, 7, 4, 3, 3 *Solution* The number 3 occurs three times, 7 twice, and all other numbers once. Therefore, 3 is the mode. *The word "data" is considered plural. A data set is different from other sets in mathematics, because a number can occur more than once in a data set. The notation { } is not used for data sets, because a number is never listed twice in such a set but may occur several times in a data set.

Q1 Find the mode of the given data: 2, 5, 3, 2, 5, 4, 5. _____

STOP • STOP • STOP • STOP • STOP • STOP • STOP • STOP • STOP

A1 5: 5 occurs three times.

Q2 Find the mode of the given numbers: 4, 3, 7, 4, 8, 6, 5, 7, 4, 2, 3, 4, 4, 8, 3, 2. _____

STOP • STOP • STOP • STOP • STOP • STOP • STOP • STOP • STOP

A2 4: 4 occurs five times.

2	There are some potential problems in finding the mode: There may be no mode or there may be more than one. 1. If each number occurs only once, there is *no mode*.

2. If *two* numbers occur with the same frequency and this frequency is greater than all other frequencies, they are *both* considered to be modes. The data are said to be *bimodal*.

Although it does not occur often, it is possible to have three modes, four modes, and so on. The data are said to be *multimodal*.

Example 1: Find the mode of the following data: 3, 5, 7, 2, 9, 8, 4, 6.

Solution

There is no mode.

Example 2: Find the mode of the distribution* 3, 5, 7, 4, 5, 3, 2, 8, 7, 5, 3.

Solution

The modes are 5 and 3. The distribution is bimodal.

*Data are sometimes referred to as a *distribution,* because any data set forms a uniquely distributed pattern when arranged on a line.

Q3   Find the mode of each distribution of numbers:

    a. 2, 5, 7, 4, 3, 12, 17, 7, 5, 8, 5   _____

    b. 5, 8, 3, 5, 7, 4, 5, 8, 2, 8, 1   _____

    c. 2, 8, 7, 12, 5, 6, 3, 10, 9   _____

STOP • STOP • STOP • STOP • STOP • STOP • STOP • STOP • STOP

A3   **a.** 5   **b.** 5 and 8: bimodal   **c.** no mode

---

3   The second measure of central tendency is the *median*. To obtain the median do the following:

Step 1: Arrange the numbers from smallest to largest.

Step 2: Choose the middle number (for sets with an odd number of values).

Example: Find the median of the data:

5, 8, 3, 5, 7, 4, 5, 8, 2, 8, 1

Solution

Step 1: Arrange from smallest to largest:

1, 2, 3, 4, 5, 5 5, 7, 8, 8, 8

Step 2: The median is 5, because there were an equal number above 5 as were below.

Q4   (1) Arrange the data sets from smallest to largest.
    (2) Determine the median.

    a. 2, 1, 7, 3, 5   (1) _____   (2) ____

    b. 2, 3, 7, 5, 1, 4, 3, 6, 7   (1) _____   (2) ____

STOP • STOP • STOP • STOP • STOP • STOP • STOP • STOP • STOP

A4   **a.** (1) 1, 2, 3, 5, 7
       (2) 3

b. (1) 1, 2, 3, 3, 4, 5, 6, 7, 7
(2) 4

Q5   Given these data, 3, 3, 5, 8, 7, 3, 1, 9, 10, 8, 7, find:

a. The median   _____

b. The mode   _____

STOP • **STOP** • **STOP** • **STOP** • **STOP** • **STOP** • **STOP** • **STOP** • **STOP**

A5   **a.** 7   **b.** 3

---

4   Consider the numbers

1, 5, 6, 9, 11

The median or middle number is 6, because there are two numbers on each side. However, examine what happens when there are eight numbers in a set:

1, 5, 5, 6, 8, 9, 9, 11

There is no number that separates the set with the same number of numbers before it as after it. There are now two middle numbers. The *median* is the number one-half way between the two middle numbers. In the set above the two middle numbers are 6 and 8. The number one-half way between is $\dfrac{6+8}{2}$. The median is therefore 7. The median is found this way whenever there are an even number of numbers in the data.

Q6   Find the median of the following data sets:
a. 4, 9, 7, 10, 11, 1, 4, 9, 6, 9

b. 1, 12, 2, 11, 4, 11, 5, 10, 5, 9, 6, 9

STOP • **STOP** • **STOP** • **STOP** • **STOP** • **STOP** • **STOP** • **STOP** • **STOP**

A6   **a.** 8: $\dfrac{7+9}{2} = 8$ **b.** $7\dfrac{1}{2}$: $\dfrac{6+9}{2} = 7\dfrac{1}{2}$

---

5   The third measure of central tendency is the mean. The *mean* of a set of numbers is the result obtained when the numbers are added and the sum is divided by the number of numbers in the data set. The mean is frequently called "the average." However, "average" is a very imprecise word, because the median and mode are also considered to be averages.

Example: Find the mean of 5, 2, 7, 3, 3, 6.

Solution

$$\frac{5+2+7+3+3+6}{6} = \frac{26}{6} = 4\frac{1}{3}$$

Therefore, the mean of the data is $4\dfrac{1}{3}$.

Q7   Find the mean of 4, 6, 10, 17, 18.

STOP • STOP • STOP • STOP • STOP • STOP • STOP • STOP • STOP

A7   11: $\dfrac{4 + 6 + 10 + 17 + 18}{5} = \dfrac{55}{5} = 11$

Q8   The heights of the 10 members of a basketball team are 67, 70, 72, 75, 77, 72, 79, 72, 73, 75 inches.

    a. Find the mean.   _____

    b. Find the median.   _____

    c. Find the mode.   _____

STOP • STOP • STOP • STOP • STOP • STOP • STOP • STOP • STOP

A8   **a.** 73.2 in   **b.** 72.5 in   **c.** 72 in

Q9   A small business consists of the president making $35,000 per year, a secretary making $6,000 per year, a custodian making $5,000 per year, and five "workers," each making $8,000 per year.

    a. Find the mean salary.   _____

    b. Find the median salary.   _____

    c. Find the mode salary.   _____

STOP • STOP • STOP • STOP • STOP • STOP • STOP • STOP • STOP

A9   **a.** $10,750: $\dfrac{86,000}{8}$ **b.** $8,000   **c.** $8,000

Q10   Answer yes or no. Which of the following would be affected if the president of the business in Q9 gave himself a $5,000 raise and left everyone else at the same salary level?

a. mean _____   **b.** median _____   **c.** mode _____

STOP • **STOP** • **STOP** • **STOP** • **STOP** • **STOP** • **STOP** • **STOP** • **STOP**

A10   **a.** yes   **b.** no   **c.** no

---

6   One might wonder why there are three different ways of calculating an "average." This is because sometimes one method is more appropriate than the others. The mode is preferred when one wants to speak of the most popular choice. The result chosen by more people than any other would be the mode. It does not even have to be a number as the median and mean must be. Consider two examples:

1. The average automobile tire driven on the road today is the 6.70 × 15.
2. The average television set sold in Ann Arbor, Michigan, is a color set.

Q11   One sometimes sees a description of an "average" person in the United States. Such a description follows. Indicate yes if the description would be the result of finding a mode. Answer no, otherwise.

The average man is married, has 2.43 children, owns his own home, drives a Chevrolet, makes $11,234 per year, and is not satisfied with his employment.

a. is married   _____   **b.** has 2.43 children   _____

c. owns his own home   _____   **d.** drives a Chevrolet   _____

e. makes $11,234   _____   **f.** not satisfied with his employment   _____

STOP • **STOP** • **STOP** • **STOP** • **STOP** • **STOP** • **STOP** • **STOP** • **STOP**

A11   **a.** yes   **b.** no   **c.** yes   **d.** yes   **e.** no   **f.** yes

---

7   If one or a few values are so large (or small) that the mean is noticeably increased (or decreased) because of them, the most appropriate measure of central tendency is the median. In the business of Q9, the employees might not appreciate it if their president were to describe their company as paying an average salary of $10,750, when all the employees but himself make considerably less than that. The median of $8,000 seems to represent the salaries paid more accurately in this instance. In cases where there are a few very outstanding values, the median is most appropriate. The median is less affected by wide fluctuations of scores on the extremes of the distribution than the mean. Of course, if both the median and mean are given, one has a better understanding of the data. If the median and mean are equal, the scores are "balanced" about the common value. On the other hand, if the mean is larger than the median, there are a few large values that caused the mean to be greater.

Q12   A local politician had the following eight contributions to his campaign fund: $20, $25, $25, $30, $32, $35, $51, $350.

a. Find the median._____

b. Find the mean. _____

c. Would the mean or the median be the best representative of the "average" contribution to his campaign fund? _____

STOP • STOP • STOP • STOP • STOP • STOP • STOP • STOP • STOP

A12   **a.** $31   **b.** $71   **c.** median

Q13   If the mean height of a group of girls is 5 ft 1 in and the median is 5 ft 4 in, does it indicate that there were a few very short or a few very tall girls? _____

STOP • STOP • STOP • STOP • STOP • STOP • STOP • STOP • STOP

A13   A few very short girls, because the mean is less than the median.

---

8   The mean is the most common measure of central tendency used and most people assume the mean is meant when the word "average" is used. Do not be misled into thinking this way. *All three measures of central tendencies are averages.*

The mean is affected by each value of the data, is easily calculated, and is useful for more advanced statistical procedures. Since other statistical procedures are dependent upon the mean, almost every statistical description of data requires as a part of the analysis the calculation of the mean.

This completes the instruction for this section.

14.1   Exercises

Find the mean, median, and mode for the data sets in problems 1–5:

1. 40, 43, 43, 47, 49
2. 18, 25, 25, 27, 29, 40
3. 240, 280, 230, 200, 360, 310
4. 21, 26, 26, 32, 17, 13, 19, 17, 15
5. 15, 16, 12, 13, 15, 18, 17, 16, 14, 16, 14, 18, 17, 15, 17, 19, 14, 20, 16, 16, 17, 16, 14, 15, 13, 12, 14, 13
6. In a data set with 10 numbers, not all the same, which measure of central tendency could be equal to the largest number in the data set?
7. Sometimes one of the measures of central tendency does not exist. Which one is it and under what circumstances does this happen?
8. When there is no middle number, how is the median found?
9. If a person wants to report an average score that is not unduly affected by one extremely large score, which measure of central tendency should be chosen?
10. What measure of central tendency is used for more advanced statistics?

14.1   Exercise Answers

1. mean, 44.4; median, 43; mode, 43

2. mean, $27\frac{1}{3}$; median, 26 (recall that you must arrange them from smallest to largest); mode, 25

3. mean, 270; median, 260; no mode

4. mean, $20\frac{2}{3}$; median, 19; modes, 17 and 26

5. mean, 15.4; median, 15.5; mode, 16
6. the mode, if the largest number occurred more often than any others
7. the mode, when all numbers occur only once
8. The two middle numbers are added and the sum is divided by 2. (Alternatively, the mean of the two middle numbers is found.)
9. median
10. mean

14.2   Frequency Tables and Frequency Polygons

1   When given a large data set, it is helpful to construct a frequency table in order to more easily make computations. A *frequency table* lists all the possible values of the data with the number of times that value appears in the data beside it.

Example: Construct a frequency table for the following data: 15, 16, 12, 13, 15, 18, 17, 16, 14, 16, 14, 18, 17, 15, 17, 19, 14, 20, 16, 16, 17, 16, 14, 15, 13, 12, 14, 13.

Solution

The largest value in the data is 20 and the smallest is 12. These and all numbers between are listed in a column. A tally mark is made beside each value as you find it occurring in the data. Finally, the frequency of each value is listed beside it. The total frequency is listed below the frequency column.

Value	Tally	Frequency
12	||	2
13	|||	3
14	⦀⦀	5
15	||||	4
16	⦀⦀ |	6
17	||||	4
18	||	2
19	|	1
20	|	1
		28

Q1   Answer the following questions about the frequency table constructed in Frame 1:

a. How many times did the value 17 occur in the data?   _____

b. What value occurred most frequently?   _____

c. What would the value you gave for part b be called?   _____

d. How many times did a value of 15 or less occur?   _____

e. How many times did a value of 16 or larger occur?   _____

STOP • STOP • STOP • STOP • STOP • STOP • STOP • STOP • STOP

A1   a. 4   b. 16   c. mode   d. 14   e. 14

Q2   Construct a frequency table for the values 10, 12, 14, 12, 13, 11, 15, 14, 13, 12, 11, 12, 13, 14, 16, 11, 12, 14, 12, 15, 15.

Value	Frequency
10	

STOP • STOP • STOP • STOP • STOP • STOP • STOP • STOP • STOP

A2

Value	Frequency
10	1
11	3
12	6
13	3
14	4
15	3
16	1
	21

Q3   Determine the mode of the data in Q2. ____

STOP • STOP • STOP • STOP • STOP • STOP • STOP • STOP •STOP

A3   12 is the mode, because it occurred 6 times.

---

2   The three measures of central tendency can be calculated by using the frequency table. The mode can be found as in Q1 and Q3 by finding the value with the greatest frequency. The median can be obtained from a frequency table by finding the middle or the mean of the two middle values. In the frequency table of Q3 there were 21 values. The eleventh score from each end would be the middle score, because there would be 10 values above and 10 below. The values must be arranged from smallest to largest. The frequencies are added until the eleventh value is obtained.

Value	Frequency
10	1
11	3
12	6
median → 13	3
14	4
15	3
16	1
	21

Since there were 10 scores of 12 or below, the eleventh score was one of the three 13s. Therefore, the median is 13.

Q4   Find the median of the data described in the following frequency chart:

Value	Frequency
37	1
38	3
39	7
40	8
41	4
42	2
	25

STOP • **STOP** • **STOP** • **STOP** • **STOP** • **STOP** • **STOP** • **STOP** • **STOP**

A4   40:

Value	Frequency
37	1
38	3
39	7
40	8
41	4
42	2
	25

The 13th score must be one of the scores of 40.

---

3   For large sets of data a system is needed to determine the position of the median in the data when the values are arranged from smallest to largest. The variable n is used to represent the number of values in the data set. If n is odd, there is a middle value. The position of that middle value may be obtained by adding 1 to n and taking half of it.

position of middle value of odd $n = \dfrac{1}{2}\left(n + 1\right)$

If n is even, there are two middle values. Their positions can be found by taking $\dfrac{1}{2}n$ and $\dfrac{1}{2}n + 1$.

Example 1: The median has which position in a data set where $n = 405$?

Solution

middle number $= \dfrac{1}{2}(405 + 1) = 203$

The median is the 203rd number from the smallest.

Example 2: The median is computed from which positions in a data set where $n = 264$?

Solution

middle numbers are $\dfrac{1}{2}(264)$ and $\dfrac{1}{2}(264) + 1$

The median is halfway between the 132nd and 133rd values.

Q5   Find the position of the median in a data set with each of the following values of n:

 a. $n = 27$ **b.** $n = 94$

 c. $n = 482$ **d.** $n = 507$

STOP • STOP • STOP • STOP • STOP • STOP • STOP • STOP • STOP

A5   **a.** 14th from smallest: $\frac{1}{2}(27 + 1) = 14$

 b. halfway between the 47th and 48th value from the smallest: $\frac{1}{2}(94)$ and $\frac{1}{2}(94) + 1$

 c. halfway between the 241st and 242nd from the smallest

 d. 254th from the smallest

---

4   An example with an even number of values will be considered.

Example: Find the median and the mode.

Value	Frequency
24	10
25	15
26	17
27	19
28	24
29	20
30	12
31	5
	122

Solution

Since there are 122 values (be careful *not* to count the values under the heading "value"), the two middle values are $\frac{1}{2}(122)$ and $\frac{1}{2}(122) + 1$. Find the 61st and 62nd values. There are 61 values of 27 or below (found by adding the frequency column). Therefore, the 61st value is 27 and the 62nd value is 28. The median is $\frac{27 + 28}{2} = 27.5$. The mode is 28, because it occurred 24 times, which was more frequent than any other value.

Q6   Find (1) the median and (2) the mode of the following distributions:

a.

Value	Frequency
67	2
68	10
69	12
70	9
71	7
72	5
73	3
	48

(1) Median _____

(2) Mode _____

b.

Value	Frequency
25	4
27	8
28	10
29	6
30	2
	30

(1) Median _____

(2) Mode _____

STOP • STOP • STOP • STOP • STOP • STOP • STOP • STOP • STOP

A6   **a.** (1) 69.5: the 24th value is 69, the 25th is 70.
         (2) 69
       b. (1) 28: the 15th and 16th values are 28.
         (2) 28

5   When finding the mean of a data set with a large *n*, it is easier to first make a frequency chart. Consider the data set below. If the mean was calculated from the original data set, 150 numbers would have to be added. Instead, follow this procedure:

Step 1: Multiply each value by its frequency.

Step 2: Add the products obtained in Step 1.

Step 3: Divide the sum of Step 2 by *n*.

Value	Frequency	Value × Frequency	
17	20	340	
18	30	540	
19	50	950	← Step 1
20	40	800	
21	10	210	
	n = 150	2,840	← Step 2

Step 3: mean $= \dfrac{2,840}{150} = 18.9\frac{1}{3}$

Q7   Find the mean, median, and mode of the following distribution:

Value	Frequency	Value × Frequency
25	2	
26	5	
27	10	
28	8	
29	3	
30	3	
	31	

mean = _____   median = _____   mode = _____

STOP • STOP • STOP • STOP • STOP • STOP • STOP • STOP • STOP

A7   mean = 27.45   median = 27   mode = 27

Value	Frequency	Value × Frequency
25	2 ⎫ 7 ⎫	50
26	5 ⎬ ⎬ 17	130
27	10 ┘ ┘	270
28	8	224
29	3	87
30	3	90
	31	851

$$\text{mean} = \frac{851}{31} = 27.45 \qquad \text{median is the 16th value} = 27$$

Q8   Find the mean, median, and mode of the following distribution:

Value	Frequency
100	10
101	24
102	30
103	32
104	22
105	10
	128

mean _____   median _____   mode _____

STOP • **STOP** • **STOP** • **STOP** • **STOP** • **STOP** • **STOP** • **STOP** • **STOP**

A8   mean = 102.48   median = 102.5   mode = 103

---

6   Statisticians like to be able to look at a distribution of data graphically. Some characteristics of the data are more apparent in graphical form than in tabular form. The graph of a frequency table is called a *frequency polygon*. Shown here is a frequency table and its corresponding frequency polygon.

Value	Frequency
42	2
43	5
44	7
45	4
46	3
	21

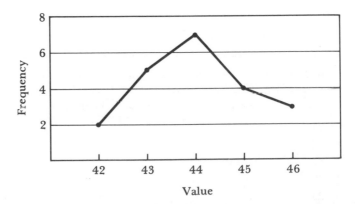

Notice that values are on the horizontal axis and frequencies on the vertical axis. Each point above a value represents the frequency of that value. After the points are plotted they are connected by line segments.

Q9   **a.** Use the frequency table to locate points and draw a frequency polygon.

Value	Frequency
24	10
25	15
26	25
27	30
28	20
29	5
	$n = 105$

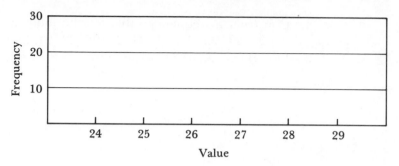

b. Use the frequency polygon to fill in the frequency column in the frequency table.

Value	Frequency
82	
83	
84	
85	
86	
87	
	$n =$

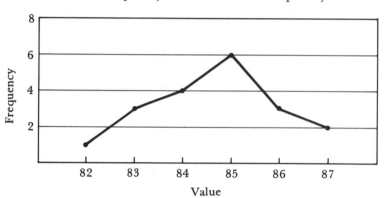

STOP • STOP • STOP • STOP • STOP • STOP • STOP • STOP • STOP

A9   **a.**

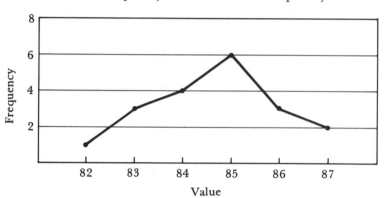

b.

Value	Frequency
82	1
83	3
84	4
85	6
86	3
87	2
	$n = 19$

This completes the instruction for this section.

14.2   Exercises

1. Construct a frequency table for the following values in the data set: 23, 24, 25, 22, 23, 23, 24, 25, 21, 24, 23, 22, 26, 22, 24, 24.

2. Find the mean, median, and mode of the frequency distribution found in the frequency table.

Value	Frequency
101	1
102	3
103	10
104	7
105	4
106	3
	20

3. Plot a frequency polygon corresponding to the frequency table of problem 2.

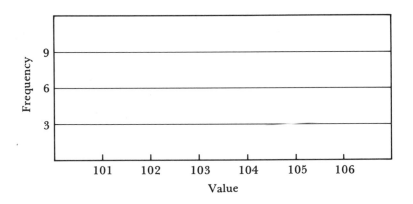

4. Use the following data set to complete each of the requests: 47, 50, 52, 48, 49, 50, 49, 51, 50, 46, 48, 50, 48, 51, 53, 49, 48, 47.
   a. Construct a frequency table.
   b. Find the mode.
   c. Find the median.
   d. Find the mean.
   e. Construct a frequency polygon with the data.

5. Read the frequency polygon and fill in the frequency table.

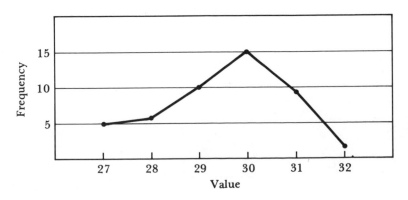

Value	Frequency
27	
28	
29	
30	
31	
32	
$n =$	

14.2   Exercise Answers

1.

Value	Frequency
21	1
22	3
23	4
24	5
25	2
26	1
	$n = 16$

2.

Value	Frequency			Value × Frequency
101	1 ⎫			101
102	3 ⎬ 4 ⎫			306
103	10⎭ ⎬ 14			1,030
104	7			728
105	4			420
106	3			318
	28			2,903

$$\text{mean} = \frac{2,903}{28} = 103.7 \text{ (rounded off to tenths)}$$

median = 103.5 (halfway between 14th and 15th values)
mode = 103

3.

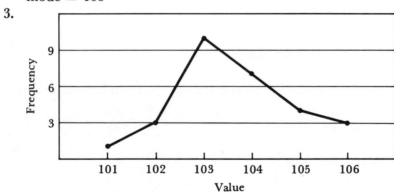

4. a.

Value	Frequency			Value × Frequency
46	1 ⎫			46
47	2 ⎬ 3 ⎫			94
48	4⎭ ⎬ 7 ⎫			192
49	3⎭ ⎬ 10			147
50	4			200
51	2			102
52	1			52
53	1			53
	18			886

b. modes = 48 and 50
c. median = 49 (9th and 10th values are both 49)

d. mean $= \dfrac{886}{18} = 49.2$

e.

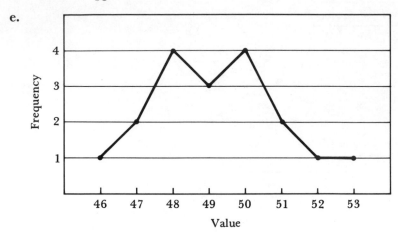

5.

Value	Frequency
27	5
28	6
29	10
30	15
31	9
32	2
	$n = 47$

Chapter 14 Sample Test

At the completion of Chapter 14 it is expected that you will be able to work the following problems.

14.1   Mean, Median, and Mode

Find the mean, median, and mode for each of the following data sets:

1. 8, 9, 11, 12, 13, 13, 17
2. 5, 3, 7, 3, 12, 6, 7, 2
3. 5, 17, 1, 49, 11, 33

Choose the measure of central tendency that has been or should be used in each situation:

4. The average man in America is a city dweller.
5. The average test score of 100 students.
6. The average height of a basketball team.
7. The average salary of a small company with 1 president, 4 salesman, 2 secretaries, and 1 custodian.
8. The middle value.
9. The most popular value.
10. The average used for advanced statistics.

14.2   Frequency Tables and Frequency Polygons

11. Use the following data to complete parts a and b: 20, 22, 26, 24, 21, 23, 25, 23, 20, 24, 21, 22, 20, 23, 23, 22, 21, 20, 21, 24, 22, 23, 25, 23.

     a. Construct a frequency table.

     b. Construct a frequency polygon.

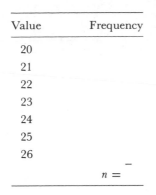

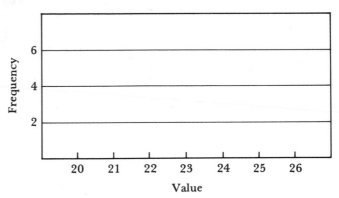

12. Use the frequency table to complete a, b, and c:

Value	Frequency	Frequency × Value
40	5	
41	7	
42	6	
43	9	
44	5	
45	4	
	$n =$	

     a. Find the mean of the data.

     b. Find the median of the data.

     c. Find the mode of the data.

Chapter 14 Sample Test Answers

1. mean, $11\frac{6}{7}$; median, 12; mode, 13

2. mean, $5\frac{5}{8}$; median, 5.5; mode, 3 and 7

3. mean, $19\frac{1}{3}$; median, 14; mode, none

4. mode   **5.** mean or median   **6.** mean or median

7. median   **8.** median   **9.** mode

10. mean

11. a.

Value	Frequency
20	4
21	4
22	4
23	6
24	3
25	2
26	1
	$n = \overline{24}$

b.

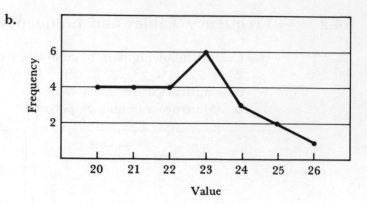

12. a. $42\frac{7}{18}$ **b.** 42.5   **c.** 43

Appendix

Table I Multiplication Facts

1	2	3	4	5	6	7	8	9	10	11	12	13	14	15	16	17	18	19	20	21	22	23	24	25
2	4	6	8	10	12	14	16	18	20	22	24	26	28	30	32	34	36	38	40	42	44	46	48	50
3	6	9	12	15	18	21	24	27	30	33	36	39	42	45	48	51	54	57	60	63	66	69	72	75
4	8	12	**16**	20	24	28	32	36	40	44	48	52	56	60	64	68	72	76	80	84	88	92	96	100
5	10	15	20	**25**	30	35	40	45	50	55	60	65	70	75	80	85	90	95	100	105	110	115	120	125
6	12	18	24	30	**36**	42	48	54	60	66	72	78	84	90	96	102	108	114	120	126	132	138	144	150
7	14	21	28	35	42	**49**	56	63	70	77	84	91	98	105	112	119	126	133	140	147	154	161	168	175
8	16	24	32	40	48	56	**64**	72	80	88	96	104	112	120	128	136	144	152	160	168	176	184	192	200
9	18	27	36	45	54	63	72	**81**	90	99	108	117	126	135	144	153	162	171	180	189	198	207	216	225
10	20	30	40	50	60	70	80	90	**100**	110	120	130	140	150	160	170	180	190	200	210	220	230	240	250
11	22	33	44	55	66	77	88	99	110	**121**	132	143	154	165	176	187	198	209	220	231	242	253	264	275
12	24	36	48	60	72	84	96	108	120	132	**144**	156	168	180	192	204	216	228	240	252	264	276	288	300
13	26	39	52	65	78	91	104	117	130	143	156	**169**	182	195	208	221	234	247	260	273	286	299	312	325
14	28	42	56	70	84	98	112	126	140	154	168	182	**196**	210	224	238	252	266	280	294	308	322	336	350
15	30	45	60	75	90	105	120	135	150	165	180	195	210	**225**	240	255	270	285	300	315	330	345	360	375
16	32	48	64	80	96	112	128	144	160	176	192	208	224	250	**256**	272	288	304	320	336	352	368	384	400
17	34	51	68	85	102	119	136	153	170	187	204	221	238	255	272	**289**	306	323	340	357	374	391	408	425
18	36	54	72	90	108	126	144	162	180	198	216	234	252	270	288	306	**324**	342	360	378	396	414	432	450
19	38	57	76	95	114	133	152	171	190	209	228	247	266	285	304	323	342	**361**	380	399	418	437	456	475
20	40	60	80	100	120	140	160	180	200	220	240	260	280	300	320	340	360	380	**400**	420	440	460	480	500
21	42	63	84	105	126	147	168	189	210	231	252	273	294	315	336	357	378	399	420	**441**	462	483	504	525
22	44	66	88	110	132	154	176	198	220	242	264	286	308	330	352	374	396	418	440	462	**484**	506	528	550
23	46	69	92	115	138	161	184	207	230	253	276	299	322	345	368	391	414	437	460	483	506	**529**	552	575
24	48	72	96	120	144	168	192	216	240	264	288	312	336	360	384	408	432	456	480	504	528	552	**576**	600
25	50	75	100	125	150	175	200	225	250	275	300	325	350	375	400	425	450	475	500	525	550	575	600	**625**

Table II English Weights and Measures

Units of Length

1 foot (ft or ′) = 12 inches (in or ″)
1 yard (yd) = 3 feet
1 rod (rd) = $5\frac{1}{2}$ yards = $16\frac{1}{2}$ feet
1 mile (mi) = 1,760 yards = 5,280 feet

Units of Area

1 square foot = 144 square inches
1 square yard = 9 square feet
1 acre = 43,560 square feet
1 square mile = 640 acres

Units of Weight

1 pound (lb) = 16 ounces (oz)
1 ton (t) = 2,000 pounds

Units of Capacity (Volume)
Liquid

1 pint (pt) = 16 fluid ounces (fl oz)
1 quart (qt) = 2 pints
1 gallon (gal) = 4 quarts = 8 pints

Dry

1 quart (qt) = 2 pints (pt)
1 peck (pk) = 8 quarts
1 bushel (bu) = 4 pecks = 32 quarts

Units of Capacity

1 cubic foot = 1,728 cubic inches
1 cubic yard = 27 cubic feet
1 gallon = 231 cubic inches
1 cubic foot = 7.48 gallons

Table III   Metric Weights and Measures

Units of Length

1 millimeter (mm) = 1000 micrometer (μm)
1 centimeter (cm) = 10 millimeters
1 decimeter (dm) = 10 centimeters
1 meter (m) = 10 decimeters
1 dekameter (dam) = 10 meters
1 hectometer (hm) = 10 dekameters
1 kilometer (km) = 10 hectometers

Units of Weight

1 milligram (mg) = 1000 micrograms (μg)
1 centigram (cg) = 10 milligrams
1 decigram (dg) = 10 centigrams
1 gram (g) = 10 decigrams
1 dekagram (dag) = 10 grams
1 hectogram (hg) = 10 dekagrams
1 kilogram (kg) = 10 hectograms
1 megagram (Mg) = 1000 kilograms
(metric ton)

Units of Area

1 square centimeter = 100 square millimeters
1 square decimeter = 100 square centimeters
1 square meter = 100 square decimeters
1 square kilometer = 1 000 000 square meters
1 hectare (ha) = 10 000 square meters
1 are (a) = 100 square meters

Units of Capacity (Volume)

1 milliliter (ml) = 1 cubic
centimeter (cc or cm^3)
1 centiliter (cl) = 10 milliliters
1 deciliter (dl) = 10 centiliters
1 liter (l) = 10 deciliters
= 1000 cubic centimeters
1 deckaliter (dal) = 10 liters
1 hectoliter (hl) = 10 dekaliters
1 kiloliter (kl) = 10 hectoliters
1 megaliter (Ml) = 1000 kiloliters

Table IV   Metric and English Conversion   **585**

Table IV   Metric and English Conversion

Length

English	Metric
1 inch	= 2.54 cm
1 foot	= 30.5 cm
1 yard	= 91.4 cm
1 mile	= 1610 m
1 mile	= 1.61 km
0.0394 in	= 1 mm
0.394 in	= 1 cm
39.4 in	= 1 m
3.28 ft	= 1 m
1.09 yd	= 1 m
0.621 mi	= 1 km

Capacity

English	Metric
1 gallon	= 3.79 l
1 quart	= 0.946 l
0.264 gal	= 1 l
1.05 qt	= 1 l

Weight

English	Metric
1 ounce	= 28.3 g
1 pound	= 454 g
1 pound	= 0.454 kg
0.0353 oz	= 1 g
0.00220 lb	= 1 g
2.20 lb	= 1 kg

Area

English	Metric
1 square inch	= 6.45 square centimeters
1 square foot	= 929 square centimeters
1 square yard	= 8361 square centimeters
1 square rod	= 25.3 square meters
1 acre	= 4047 square meters
1 square mile	= 2.59 square kilometers
10.76 square feet	= 1 square meter
1,550 square inches	= 1 square meter
0.0395 square rod	= 1 square meter
1.196 square yards	= 1 square meter
0.155 square inch	= 1 square centimeter
247.1 acres	= 1 square kilometer
0.386 square mile	= 1 square kilometer

Volume

English	Metric
1 cubic inch	= 16.39 cubic centimeters
1 cubic foot	= 28.317 cubic centimeters
1 cubic yard	= 0.7646 cubic meter
0.06102 cubic inch	= 1 cubic centimeter
35.3 cubic feet	= 1 cubic meter

Table V   Decimal Fraction and Common Fraction Equivalents

$\frac{1}{64} = 0.015625$	$\frac{9}{32} = 0.28125$	$\frac{17}{32} = 0.53125$	$\frac{25}{32} = 0.78125$
$\frac{1}{32} = 0.03125$	$\frac{19}{64} = 0.296875$	$\frac{35}{64} = 0.546875$	$\frac{51}{64} = 0.796875$
$\frac{3}{64} = 0.046875$	$\frac{5}{16} = 0.3125$	$\frac{9}{16} = 0.5625$	$\frac{4}{5} = 0.8$
$\frac{1}{16} = 0.0625$	$\frac{21}{64} = 0.328125$	$\frac{37}{64} = 0.578125$	$\frac{13}{16} = 0.8125$
$\frac{5}{64} = 0.078125$	$\frac{1}{3} = 0.33\bar{3}$	$\frac{19}{32} = 0.59375$	$\frac{53}{64} = 0.828125$
$\frac{3}{32} = 0.09375$	$\frac{11}{32} = 0.34375$	$\frac{3}{5} = 0.6$	$\frac{27}{32} = 0.84375$
$\frac{7}{64} = 0.109375$	$\frac{23}{64} = 0.359375$	$\frac{39}{64} = 0.609375$	$\frac{55}{64} = 0.859375$
$\frac{1}{8} = 0.125$	$\frac{3}{8} = 0.375$	$\frac{5}{8} = 0.625$	$\frac{7}{8} = 0.875$
$\frac{9}{64} = 0.140625$	$\frac{25}{64} = 0.390625$	$\frac{41}{64} = 0.640625$	$\frac{57}{64} = 0.890625$
$\frac{5}{32} = 0.15625$	$\frac{2}{5} = 0.4$	$\frac{21}{32} = 0.65625$	$\frac{29}{32} = 0.90625$
$\frac{11}{64} = 0.171875$	$\frac{13}{32} = 0.40625$	$\frac{2}{3} = 0.66\bar{6}$	$\frac{59}{64} = 0.921875$
$\frac{1}{5} = 0.2$	$\frac{27}{64} = 0.421875$	$\frac{43}{64} = 0.671875$	$\frac{15}{16} = 0.9375$
$\frac{3}{16} = 0.1875$	$\frac{7}{16} = 0.4375$	$\frac{11}{16} = 0.6875$	$\frac{61}{64} = 0.953125$
$\frac{13}{64} = 0.203125$	$\frac{29}{64} = 0.453125$	$\frac{45}{64} = 0.703125$	$\frac{31}{32} = 0.96875$
$\frac{7}{32} = 0.21875$	$\frac{15}{32} = 0.46875$	$\frac{23}{32} = 0.71875$	$\frac{63}{64} = 0.984375$
$\frac{15}{64} = 0.234375$	$\frac{31}{64} = 0.484375$	$\frac{47}{64} = 0.734375$	
$\frac{1}{4} = 0.25$	$\frac{1}{2} = 0.5$	$\frac{3}{4} = 0.75$	
$\frac{17}{64} = 0.265625$	$\frac{33}{64} = 0.515625$	$\frac{49}{64} = 0.765625$	

Table VI   Formulas from Geometry   **587**

Table VI   Formulas from Geometry

		Perimeter	Area

Rectangle

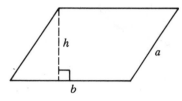

$p = 2l + 2w$ $A = lw$

Square

$p = 4s$ $A = s^2$

Parallelogram

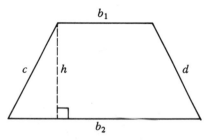

$p = 2a + 2b$ $A = bh$

Trapezoid

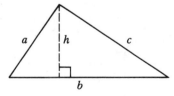

$A = \dfrac{1}{2} h(b_1 + b_2)$

Triangle

$p = a + b + c$ $A = \dfrac{1}{2} bh$

The sum of the measures of the angles of a triangle = 180°. In a right triangle:

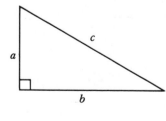

$c^2 = a^2 + b^2$
(Pythagorean theorem)

		Circumference	Area

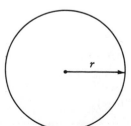

$C = 2\pi r$ $A = \pi r^2 \ (d = 2r)$
$C = \pi d$

Circle

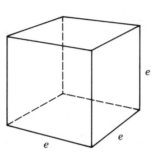

Triangular prism

Volume

$$V = \frac{1}{2}bh_1h_2$$

or

$V = Bh$, where B is the area of the base

Cube

$$V = e^3$$

Rectangular prism

$V = lwh$

or

$V = Bh$, where B is the area of the base

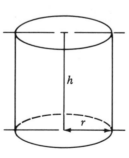

Cylinder

$$V = \pi r^2 h$$

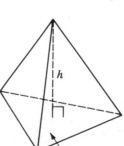

Pyramid B (area of the base)

$$V = \frac{1}{3}Bh$$

Table VI   Formulas from Geometry   **589**

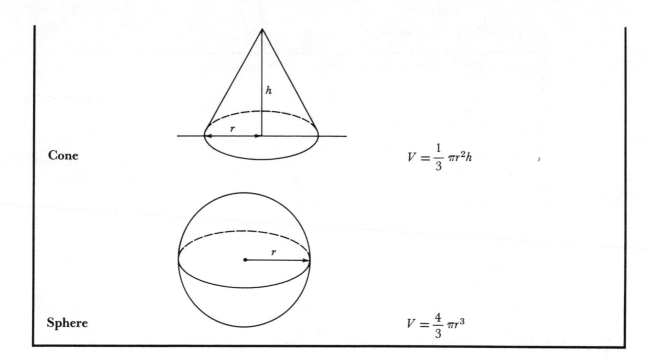

Cone

$$V = \frac{1}{3}\,\pi r^2 h$$

Sphere

$$V = \frac{4}{3}\,\pi r^3$$

Table VII   Symbols

The section number indicates the first use of the symbol.

1.2	$+$	"plus"	Indicates addition (sum) of two numbers
1.2	$=$	"is equal to"	
1.2	$-$	"minus"	Indicates subtraction (difference) of two numbers
1.3	$\times$	"times"	Indicates multiplication (product) of two numbers
1.3	$\div$	"divided by"	Indicates division (quotient) between two numbers
1.3	$a\,R\,b$	"a remainder b"	
1.4	a^b	"a to the bth power"	Exponential notation
2.1	$\dfrac{a}{b}$	"a over b"	Fraction indicating a of b equal parts
2.3	LCD	"least common denominator"	
2.3	$,\ldots$	"and so forth"	Indicates that list continues
3.1	.	"point" or "and"	Decimal point used to separate the whole-number part of a number from the decimal (fraction) part of the number
3.5	$0 \cdot \overline{ab}$	"bar"	Indicates that the group of numbers under the bar repeat endlessly
4.1	$\%$	"percent"	Abbreviation for the word "percent"
5.1	$\{a, b, c\}$	"braces"	Used to denote a set
5.1	$\{\ \}$ or $\varnothing$	"empty set"	Set with no elements
5.1	$\in$	"is an element of"	
5.1	$\notin$	"is not an element of"	
5.2	$(\ )$	"quantity"	Parentheses used as a grouping symbol
5.2	$a(b) = (a)(b) = (a)b$		Product of a and b
5.2	$a \cdot b$	"times"	Raised dot to indicate the product of a and b
5.3	N		Denotes the set of natural numbers, $\{1, 2, 3, \ldots\}$
5.3	W		Denotes the set of whole numbers, $\{0, 1, 2, \ldots\}$
5.3	$\neq$	"is not equal"	
6.1	$^-$	"negative"	Raised minus sign to denote a negative number

Table VII   Symbols   **591**

6.1	I		Denotes the set of integers, $\{\ldots, {}^-2, {}^-1, 0, 1, 2, \ldots\}$
6.1	$+$	"positive"	Raised plus sign to denote a positive number
6.2	${}^-a$	"opposite of a"	Raised minus sign to indicate the opposite of a number
9.1	l_1	"l sub-one"	Denotes lines
9.1			Indicates that l_1 and l_2 are perpendicular
9.1	$\parallel$	"is parallel to"	
9.1	$\perp$	"is perpendicular to"	
9.1	$\overleftrightarrow{AB}$	"line AB"	
9.1	$\overline{AB}$	"line segment AB"	
9.2	$\overrightarrow{AB}$	"ray AB"	
9.2	$\measuredangle$	"angle"	
9.2	$\circ$	"degree"	Unit of measure of an angle
9.2	$m\measuredangle A$	"measure of angle A"	
10.2	$\doteq$	"is approximately equal to"	
10.3	π	"pi"	Constant whose value is generally given as approximately equal to 3.14 or $\frac{22}{7}$
10.4	b_1 and b_2	"b sub-one and b sub-two"	Subscripts used to distinguish between the two bases of a trapezoid
10.5	$\sqrt{}$	"radical" or "square root of"	
12.1	$a:b$	"a to b" or "a is to b"	Colon used in expressing a ratio of two numbers or quantities

Table VIII   Squares and Square Roots

Number	Square	Square Root	Number	Square	Square Root	Number	Square	Square Root
1	1	1.000	51	26 01	7.141	101	1 02 01	10.050
2	4	1.414	52	27 04	7.211	102	1 04 04	10.100
3	9	1.732	53	28 09	7.280	103	1 06 09	10.149
4	16	2.000	54	29 16	7.348	104	1 08 16	10.198
5	25	2.236	55	30 25	7.416	105	1 10 25	10.247
6	36	2.449	56	31 36	7.483	106	1 12 36	10.296
7	49	2.646	57	32 49	7.550	107	1 14 49	10.344
8	64	2.828	58	33 64	7.616	108	1 16 64	10.392
9	81	3.000	59	34 81	7.681	109	1 18 81	10.440
10	1 00	3.162	60	36 00	7.746	110	1 21 00	10.488
11	1 21	3.317	61	37 21	7.810	111	1 23 21	10.536
12	1 44	3.464	62	38 44	7.874	112	1 25 44	10.583
13	1 69	3.606	63	39 69	7.937	113	1 27 69	10.630
14	1 96	3.742	64	40 96	8.000	114	1 29 96	10.677
15	2 25	3.873	65	42 25	8.062	115	1 32 25	10.724
16	2 56	4.000	66	43 56	8.124	116	1 34 56	10.770
17	2 89	4.123	67	44 89	8.185	117	1 36 89	10.817
18	3 24	4.243	68	46 24	8.246	118	1 39 24	10.863
19	3 61	4.359	69	47 61	8.307	119	1 41 61	10.909
20	4 00	4.472	70	49 00	8.367	120	1 44 00	10.954
21	4 41	4.583	71	50 41	8.426	121	1 46 41	11.000
22	4 84	4.690	72	51 84	8.485	122	1 48 84	11.045
23	5 29	4.796	73	53 29	8.544	123	1 51 29	11.091
24	5 76	4.899	74	54 76	8.602	124	1 53 76	11.136
25	6 25	5.000	75	56·25	8.660	125	1 56 25	11.180
26	6 76	5.099	76	57 76	8.718	126	1 58 76	11.225
27	7 29	5.196	77	59 29	8.775	127	1 61 29	11.269
28	7 84	5.292	78	60 84	8.832	128	1 63 84	11.314
29	8 41	5.385	79	62 41	8.888	129	1 66 41	11.358
30	9 00	5.477	80	64 00	8.944	130	1 69 00	11.402
31	9 61	5.568	81	65 61	9.000	131	1 71 61	11.446
32	10 24	5.657	82	67 24	9.055	132	1 74 24	11.489
33	10 89	5.745	83	68 89	9.110	133	1 76 89	11.533
34	11 56	5.831	84	70 56	9.165	134	1 79 56	11.576
35	12 25	5.916	85	72 25	9.220	135	1 82 25	11.619
36	12 96	6.000	86	73 96	9.274	136	1 84 96	11.662
37	13 69	6.083	87	75 69	9.327	137	1 87 69	11.705
38	14 44	6.164	88	77 44	9.381	138	1 90 44	11.747
39	15 21	6.245	89	79 21	9.434	139	1 93 21	11.790
40	16 00	6.325	90	81 00	9.487	140	1 96 00	11.832
41	16 81	6.403	91	82 81	9.539	141	1 98 81	11.874
42	17 64	6.481	92	84 64	9.592	142	2 01 64	11.916
43	18 49	6.557	93	86 49	9.644	143	2 04 49	11.958
44	19 36	6.633	94	88 36	9.695	144	2 07 36	12.000
45	20 25	6.708	95	90 25	9.747	145	2 10 25	12.042
46	21 16	6.782	96	92 16	9.798	146	2 13 16	12.083
47	22 09	6.856	97	94 09	9.849	147	2 16 09	12.124
48	23 04	6.982	98	96 04	9.899	148	2 19 04	12.166
49	24 01	7.000	99	98 01	9.950	149	2 22 01	12.207
50	25 00	7.071	100	1 00 00	10.000	150	2 25 00	12.247

Table IX   Fraction, Decimal, and Percent Equivalents   **593**

Table IX   Fraction, Decimal, and Percent Equivalents

$\dfrac{1}{16} = 0.06\dfrac{1}{4}$ or $0.0625 = 6\dfrac{1}{4}\%$ or 6.25% $\dfrac{3}{8} = 0.37\dfrac{1}{2}$ or $0.375 = 37\dfrac{1}{2}\%$ or 37.5%

$\dfrac{1}{12} = 0.08\dfrac{1}{3} = 8\dfrac{1}{3}\%$ $\dfrac{2}{5} = 0.4 = 40\%$

$\dfrac{1}{10} = 0.1 = 10\%$ $\dfrac{1}{2} = 0.5 = 50\%$

$\dfrac{1}{9} = 0.11\dfrac{1}{9} = 11\dfrac{1}{9}\%$ $\dfrac{3}{5} = 0.6 = 60\%$

$\dfrac{1}{8} = 0.12\dfrac{1}{2}$ or $0.125 = 12\dfrac{1}{2}\%$ or 12.5% $\dfrac{5}{8} = 0.62\dfrac{1}{2}$ or $0.625 = 62\dfrac{1}{2}\%$ or 62.5%

$\dfrac{1}{7} = 0.14\dfrac{2}{7} = 14\dfrac{2}{7}\%$ $\dfrac{2}{3} = 0.66\dfrac{2}{3} = 66\dfrac{2}{3}\%$

$\dfrac{1}{6} = 0.16\dfrac{2}{3} = 16\dfrac{2}{3}\%$ $\dfrac{7}{10} = 0.7 = 70\%$

$\dfrac{1}{5} = 0.2 = 20\%$ $\dfrac{3}{4} = 0.75 = 75\%$

$\dfrac{1}{4} = 0.25 = 25\%$ $\dfrac{4}{5} = 0.8 = 80\%$

$\dfrac{3}{10} = 0.3 = 30\%$ $\dfrac{5}{6} = 0.83\dfrac{1}{3} = 83\dfrac{1}{3}\%$

$\dfrac{1}{3} = 0.33\dfrac{1}{3} = 33\dfrac{1}{3}\%$ $\dfrac{7}{8} = 0.87\dfrac{1}{2}$ or $0.875 = 87\dfrac{1}{2}\%$ or 87.5%

Glossary

1. When a bar is placed over a vowel, the vowel says its own name.
2. A curved mark over a vowel indicates the following sounds:
   a. ă as in at
   b. ĕ as in bed
   c. ĭ as in it
   d. ŏ as in ox
   e. ŭ as in rug

Acute (ŭ kūt′)
(**angle**) Angle whose measure is less than 90°.
(**triangle**) Triangle with three angles less than 90°.

Add (ăd) To combine into a sum.

Addition (ă dĭ′ shŭn) Act of adding.

Adjacent (ŭ jā′ sĕnt) (**sides of a quadrilateral**) Any two sides that intersect at a vertex.

Algebraic (ăl′ jŭ brā′ ĭk) (**open expression**) Expression in which the position of an unknown number is held by a letter (variable).

Altitude (ăl′ tĭ tōod)
(**of a parallelogram**) Perpendicular distance between two parallel sides.
(**of a pyramid**) Line segment from the apex perpendicular to the base.
(**of a right-circular cone**) Length of the axis.
(**of a right-circular cylinder**) Length of the axis.
(**of a trapezoid**) Perpendicular distance between the two parallel sides.
(**of a triangle**) Line segment from the vertex perpendicular to the base.

Angle (ăng′ g'l) Two rays with a common endpoint.

Apex (ā′ pĕx) (**of a pyramid**) Common vertex of the faces which is not part of the base.

Area (air′ ē ŭ) Measure (in square units) of the region within a closed curve (including polygons) in a plane.

Associative (ŭ sō′ shŭ tĭv) Indicates that the grouping of three numbers in an addition or multiplication can be changed without affecting the sum or product.

Axis (ăk′sĭs)
(**of a cone**) Line segment connecting the vertex to the center of the base.
(**of a cylinder**) Line segment connecting the centers of the bases.

Base (bās)
(**of a power**) Number being raised to a power.
(**of a triangle**) Side opposite a vertex.

Bases (bās′ ĭz)
(**of a cylinder**) Two circular regions that are parallel.
(**of a prism**) Parallel faces.
(**of a trapezoid**) Parallel sides.

Celsius (sĕl′ sē ŭs) (**scale**) Temperature scale on which the interval between the freezing point and boiling point of water is divided into 100 parts or degrees, so that 0 °Celsius corresponds to 32 °Fahrenheit and 100 °Celsius to 212 °Fahrenheit.

Centi (sĕn′ tĭ) Latin prefix indicating one-hundredth.

Circle (sur′ kŭl) Set of all points in a plane whose distance from a given point (center) is equal to a positive number, *r* (radius).

Circumference (sur kŭm′ fur ĕns) **(of a circle)** Distance around.

Commutative (kŭ mū′ tŭ tĭv) Indicates that the order of two numbers in an addition or multiplication can be changed without affecting the sum or product.

Compass (kŭm′ pŭs) Instrument used to draw circles and compare distances.

Complementary (kŏm plŭ mĕn′ tŭ rē) **(angles)** Two angles whose measures added together result in a sum of 90°.

Composite (kŏm pŏz′ ĭt) **(number)** Natural number that has more than two natural number factors.

Cone (kōn) **(right-circular)** Circular region (base) and the surface made up of line segments connecting the circle with a point (vertex) located on a line through the center of the circle and perpendicular to the plane of the circle.

Constant (kŏn′ stănt) Numbers without a literal coefficient (numbers that do not vary in value).

Continuous (kŭn tĭn′ ū ŭs) **(data)** Measurements that can be made more precise if a person chooses, such as the length of a line.

Cube (kūb) Regular hexahedron (polyhedron of six equal faces).

Cylinder (sĭl′ ĭn dur) **(right-circular)** Two circular regions with the same radius in parallel planes (bases) connected by line segments perpendicular to the planes of the two circles.

Data (dā′ tŭ) Something, actual or assumed, used as a basis of reckoning, such as sets of numbers.

Decagon (dĕk′ ŭ gŏn) Polygon with 10 sides.

Deci (dĕs′ ĭ) Latin prefix indicating one-tenth (the decimal number system is a system based on the use of 10 digits, where each position in a numeral has a value one-tenth the position to its left).

Decimal (dĕs′ ĭ măl) **(fraction)** Fraction whose denominator is a power (multiple) of 10; such as $0.5 = \dfrac{5}{10}$, $2.67 = \dfrac{267}{100}$, etc.

Degree (dŭ grē′) **(of angular measure)** $\dfrac{1}{360}$ of a complete revolution of a ray around its endpoint.

Deka (dĕk′ ŭ) Greek prefix indicating a multiple of 10.

Denominator (dŭ nŏm′ ĭ nā tur) Bottom number in a fraction, such as 3 in $\dfrac{2}{3}$.

Diameter (dī ăm′ ŭ tur) **(of a circle)** Twice the radius (of the circle).

Difference (dĭf′ ur ĕns) Result of a subtraction of two numbers, such as 5 in $7 - 2 = 5$.

Digit (dĭj′ ĭt) Any one of the symbols 0, 1, 2, 3, 4, 5, 6, 7, 8, or 9.

Discrete (dĭs krēt′) **(data)** Allows no interpretation between the values of the data, such as the number of people in a room.

Divide (dĭ vīd′)

Dividend (dĭv′ ĭ dĕnd) Number being divided, such as 15 in $15 \div 3$.

Divisibility (dĭ vĭz′ ĭ bĭl′ ĭ tē)

Divisible (dĭ vĭz′ ĭ b′l)

Division (dĭ vĭ′ zhŭn) Act of dividing.

Divisor (dĭ vī′ zur) Number by which another number (dividend) is divided, such as 3 in 15 ÷ 3.

Dodecagon (dō dĕk′ ŭ gŏn) Polygon with twelve sides.

Dodecahedron (dō′ dĕk ŭ hē′ drŭn) Polyhedron that has twelve faces.

Edges (ĕ′ jĕz) **(of a polyhedron)** Sides of the faces of the polyhedron.

Elements (ĕl′ ŭ mĕnts) Things that make up a set.

Empty (ĕmp′ tē) **(set)** Set with no elements.

Equilateral (ē′ kwĭ lăt′ ur ŭl) **(triangle)** Triangle with three equal sides.

Equivalent (ŭ kwĭv′ ŭ lĕnt) **(expressions)** Two or more expressions that have the same evaluation for all replacements of the variable.

Exponent (ĕks′ pō nŭnt) Indicates the number of times the base is used as a factor. For example, in 2^3, 3 is the exponent and indicates that 2 (base) is being used three times as a factor. That is, $2^3 = 2 \cdot 2 \cdot 2$.

Extrapolation (ĕks′ tră pŭ lā′ shŭn) Approximating beyond the available data.

Extremes (ĕks trēmz′) **(of a proportion)** First and fourth terms of a proportion, such as 2 and 35 in $2:5 = 14:35$.

Faces (fās′ ĭz) **(of a polyhedron)** Polygonal regions that form the surface of the polyhedron.

Factor (făk′ tur) One of the values in a multiplication expression.

Fahrenheit (făr′ ĕn hīt) **(scale)** Temperature scale on which the interval between the freezing point and boiling point of water is divided into 180 parts or degrees, so that 32 °Fahrenheit corresponds to 0 °Celsius and 212 °Fahrenheit to 100 °Celsius.

Finite (fī′ nīt) **(set)** Set in which the elements can be counted and the count has a last number.

Fraction (frăk′ shŭn) Number which indicates that some whole has been divided into a number of equal parts and that a portion of the equal parts are represented.

Great circle (of a sphere) Intersection of the sphere and a plane containing the center of the sphere.

Hectare (hĕk′ tair) 10 000 square meters.

Hecto (hĕk′ tŭ) Greek prefix indicating a multiple of 100.

Hexagon (hĕk′ sŭ gŏn) Polygon with six sides.

Hexahedron (hĕk′ sŭ hē′ drŭn) Polyhedron that has six faces.

Hypotenuse (hī pŏt′ ŭ no͞os) Side opposite the right angle in a right triangle.

Icosahedron (ī′ kō sŭ hē′ drŭn) Polyhedron that has twenty faces.

Improper (ĭm prŏp′ ur) **(fraction)** Fraction that has a value greater than or equal to 1.

Infinite (ĭn′ fĭ nĭt) **(set)** Set whose count is unending.

Integer (ĭn′ tŭ jur) Any number that is in the following list: . . . , ⁻2, ⁻1, 0, 1, 2, ...

Interpolation (ĭn tur′ pŭ lā′ shŭn) Approximation between two known values.

Intersecting (ĭn′ tur sĕkt′ ĭng) **(lines)** Two (or more) lines that have one point in common.

Isosceles (ī sŏs′ ĕ lēz) **(triangle)** Triangle with at least two equal sides.

Kelvin (kĕl′ vĭn) **(scale)** Scale of absolute temperature, in which the zero is approximately −273.15 °Celsius. The interval between the freezing point and boiling point of water is divided into 100 parts or degrees, so 273.15 Kelvin corresponds to 0 °Celsius and 373.15 Kelvin to 100 °Celsius.

Kilo (kĭl′ ŭ) Greek prefix indicating a multiple of 1000.

Kilogram (kĭl′ ŭ grăm) Base unit of weight in the metric system (1 kilogram = 1000 grams).

Lateral (lăt′ ur ŭl) **(surface of a cylinder)** Curved surface connecting the bases.

Least common denominator Smallest number that is exactly divisible by each of the original denominators of two or more fractions.

Like terms Terms of an algebraic expression which have exactly the same literal coefficients (including exponents).

Line (līn) Set of points represented with a picture such as ⟵——————————⟶ (a line is always straight, has no thickness, and extends forever in both directions).

Linear (lĭn′ ē ur) **(measurement)** Measurement along a (straight) line.

(line) segment (sĕg′ mĕnt) Portion of a line between two points (including its end points).

Liter (lē′ tur) Base unit of volume or capacity in the metric system.

Literal coefficient (lĭt′ ŭr ŭl kō′ ĕ fĭsh′ ŭnt) Letter factor of an indicated product of a number and one or more variables.

Mean (mēn) **(of a set of numbers)** Result obtained when all numbers of the set are added and the sum is divided by the number of numbers.

Means (mēnz) **(of a proportion)** Second and third terms of a proportion, such as 5 and 14 in $2:5 = 14:35$.

Median (mē′ dē ăn) **(of a set of numbers)** Middle number of a set of numbers that have been arranged from smallest to largest; in a set with an even number of numbers, the median is one-half way between the two middle numbers.

Meter (mē′ tŭr) Base unit of length in the metric system.

Micrometer (mī′ krō mē′ tŭr) 0.000 001 meter or $\dfrac{1}{1\,000\,000}$ meter.

Midpoint Point on a line segment that divides its length by 2.

Milli (mĭl′ ĭ) Latin prefix indicating one-thousandth.

Mixed number Understood sum of a whole number and a proper fraction.

Mode (mōd) **(of a set of numbers)** Number that occurs with the greatest frequency.

Multiple (mŭl′ tĭ p'l) Product of a quantity by a whole number.

Multiplication (mŭl′ tĭ plī kā′ shun) Act of multiplying.

Multiply (mŭl′ tĭ plī)

Natural (năt′ yū răl) **(number)** Any number that is in the following list: 1, 2, 3, ...

Negative (nĕg′ ŭ tĭv) **(number)** Any number that is located to the left of zero on a horizontal number line.

Numeral (noō′ mur ŭl) A symbol that represents a number, such as V, ⊥⊦⊦⊤, and 5, which are all numerals naming the number five.

Numerator (noō′ mur ā′ tur) Top number in a fraction, such as 2 in $\dfrac{2}{3}$.

Numerical (noō mēr′ ĭ kŭl) **(coefficient)** Number factor of an indicated product of a number and a variable.

Obtuse (ŏb toōs′)
(angle) Angle whose measure is between 90° and 180°.
(triangle) Triangle that has one angle greater than 90°.

Octagon (ŏk′ tŭ gŏn) Polygon with eight sides.

Octahedron (ŏk′ tŭ hē′ drŭn) Polyhedron that has eight faces.

Open sentence Equation that does not contain enough information to be judged as either true or false, such as $x + 2 = 7$.

Opposite (ŏp′ ŭ zĭt) **(sides of a quadrilateral)** Pairs of sides that do not intersect.

Opposites (ŏp′ ŭ zĭtz) Two numbers on a number line which are the same distance from zero, such as $^-5$ and 5, $^-6\frac{2}{3}$ and $6\frac{2}{3}$, etc.

Parallel (păr′ ŭ lĕl) **(lines)** Two (or more) lines in the same plane that have no points in common.

Parallelogram (păr′ ŭ lĕl′ ŭ gram) Quadrilateral with opposite sides parallel.

Pentagon (pĕn′ tŭ gŏn) Polygon with five sides.

Pentahedron (pĕn′ tŭ hē′ drŭn) Polyhedron that has five faces.

Percent (pur sĕnt′) Fraction with a denominator of 100, such as $\frac{7}{100} = 0.07 = 7\%$.

Perimeter (pĕ rĭm′ ŭ tur) **(of a polygon)** Total length of all the sides of the polygon.

Perpendicular (pur′ pĕn dĭk′ ū lur) **(lines)** Two intersecting lines that form a right angle.

Plane (plān) Flat surface (such as a table top) that extends infinitely in every direction.

Point Location in space that has no thickness.

Polygon (pŏl′ ĭ gŏn) Plane closed figure of three or more angles.

Polygonal (pŭ lĭg′ ŭ nŭl) **(region)** Polygon and its interior.

Polyhedra (pŏl′ ĭ hē′ dră) Plural of polyhedron.

Polyhedron (pŏl′ ĭ hē′ drŭn) Geometric solid where surfaces are polygonal regions.

Positive (pŏz′ ĭ tĭv) **(number)** Any number that is located to the right of zero on a horizontal number line.

Power (pow′ ur) Expression used when referring to numbers with exponents, such as 3^5, "the fifth power of three."

Prime (prīm) **(number)** Natural number other than 1 divisible by exactly two natural-number factors, itself and 1.

Prism (prĭz′m) Polyhedron that has two faces which have the same size and shape and which lie in parallel planes, and the remaining faces are parallelograms.

Product (prŏd′ ŭkt) Result of a multiplication, such as 15 is the product of 3×5.

Proper (prŏp′ ur) **(fraction)** Fraction that has a value less than 1.

Proportion (prŭ por′ shun) Statement of equality between two ratios.

Protractor (prō trăk′ tur) Instrument used to measure angles.

Pyramid (pĭr′ ŭ mĭd) Polygonal region (base), the surface made up of line segments connecting the polygon with a point outside the plane of the base (apex), and all interior points.

Pythagoras (Pŭ thag′ or us) Greek philosopher and mathematician.

Quadrilateral (kwŏd′ rĭ lăt′ ur ŭl) Polygon with four sides.

Quantity (kwŏn′ tĭ tē) Indicates an expression that has been placed within parentheses or that is being considered in its entirety.

Quotient (kwō′ shŭnt) Result of a division, such as 5 is the quotient of $15 + 3$.

Radical (răd′ ĭ kŭl) **(sign)** The symbol $\sqrt{}$ (used to indicate square root).

Radii (rā′ dē ī) **(of a circle)** Plural of radius.

Radius (rā′ dē ŭs) **(of a circle)** Line segment connecting the center of a circle to any point on the circle.

Ratio (rā′ shō) Quotient of two quantities or numbers.

Rational (răsh′ ŭn ŭl) **(number)** Any number that can be written in the form $\frac{p}{q}$, where p and q are integers and $q \neq 0$.

Ray (rā) Part of the line on one side of a point which includes the endpoint.

Rectangle (rĕk′ tăng′ g'l) Parallelogram whose sides meet at right angles.

Reduce (rŭ dōōs′) Process of dividing both numerator and denominator of a fraction by some common factor.

Regular (rĕg′ ū lur)
  (polygon) Polygon that has sides and angles of equal measure.
  (polyhedron) Polyhedron that has faces which are regular polygons.

Remainder (rŭ mān′ dur) Amount left over when it is not possible to divide the dividend into an equal number of groups exactly.

Replacement set Set of permissible values of a variable in an algebraic expression.

Right
  (angle) Any of the angles formed by two intersecting lines, where all four angles are equal (the measure of a right angle is 90°).
  (triangle) Triangle with one angle of 90°.

Scalene (skā lēn′) **(triangle)** Triangle that has no pair of sides equal.

Set (sĕt) Well-defined (membership in the set is clear) collection of things.

Set notation Use of braces, { }, to indicate a set.

(slant) height (hīt) **(of a right circular cone)** Length of a line segment from the vertex to a point on the circle of the base.

Sphere (sfēr) Set of points in space at a fixed distance (radius) from a given point (center).

Square (skwair) Rectangle whose sides are of equal length.

Subset (sŭb′ sĕt) Set in which each element is also an element of another set.

Subtract (sŭb trăkt′)

Subtraction (sŭb trăk′ shŭn) Act of subtracting.

Sum (sŭm) Result of an addition of two or more numbers.

Supplementary (sŭp′ lŭ mĕn′ tŭ rē) **(angles)** Two or more angles whose measures added together result in a sum of 180°.

Terms (tŭrmz) Parts of an expression separated by an addition symbol.

Tetrahedron (tĕt′ rŭ hē′ drŭn) Polyhedron that has four faces.

Trapezoid (trăp′ ŭ zoid) Quadrilateral in which only one pair of opposite sides are parallel.

Triangle (trī′ ăng′ g'l) Polygon with three sides.

Variable (vair′ ĭ ŭ b'l) Letter in an algebraic expression that can be replaced by any one of a set of many numbers.

Vertex (vur′ tĕks) **(of an angle)** Common endpoint of two rays that form an angle.

Vertices (vur′tĭ sēz) Plural of vertex.
  (of a polygon) Vertices of the sides of the polygon.
  (of a polyhedron) Vertices of the faces of the polyhedron.

Volume (vŏl′ yŭm) Measure (in cubic units) of the space within a closed solid figure.

Whole (number) Any number that is in the following list: 0, 1, 2, 3, . . . (the next whole number is formed by adding 1 to the previous whole number).

Index